Water Quality Engineering

Water Quality Engineering

Edited by Raven Spoon

Syrawood Publishing House
New York

Published by Syrawood Publishing House,
750 Third Avenue, 9th Floor,
New York, NY 10017, USA
www.syrawoodpublishinghouse.com

Water Quality Engineering
Edited by Raven Spoon

International Standard Book Number: 978-1-68286-445-6 (Hardback)

Cataloging-in-publication Data

Water quality engineering / edited by Raven Spoon.
p. cm.
Includes bibliographical references and index.
ISBN 978-1-68286-445-6
1. Water quality. 2. Water quality management. 3. Water--Purification. 4. Water-supply engineering.
5. Hydraulic engineering. I. Spoon, Raven.
TD370 .W38 2017
628.161--dc23

Printed in China.

TABLE OF CONTENTS

PREFACE

This book explores all the important aspects of water quality engineering in the present day scenario. It is an emerging field which studies the varied characteristics and processes of water. This book will provide interesting topics for research which interested readers can take up. It strives to provide a fair idea about this discipline and to help develop a better understanding of the latest advances within this field. This book is a compilation of topics that not only deal with water quality engineering but provide a detailed and comprehensive analysis of the multiple branches related to it. Scientists and students actively engaged in this field will find this book full of crucial and unexplored concepts.

This book was inspired by the evolution of our times; to answer the curiosity of inquisitive minds. Many developments have occurred across the globe in the recent past which has transformed the progress in the field.

This book was developed from a mere concept to drafts to chapters and finally compiled together as a complete text to benefit the readers across all nations. To ensure the quality of the content we instilled two significant steps in our procedure. The first was to appoint an editorial team that would verify the data and statistics provided in the book and also select the most appropriate and valuable contributions from the plentiful contributions we received from authors worldwide. The next step was to appoint an expert of the topic as the Editor-in-Chief, who would head the project and finally make the necessary amendments and modifications to make the text reader-friendly. I was then commissioned to examine all the material to present the topics in the most comprehensible and productive format.

I would like to take this opportunity to thank all the contributing authors who were supportive enough to contribute their time and knowledge to this project. I also wish to convey my regards to my family who have been extremely supportive during the entire project.

Editor

The Bactericidal Effect of Dendritic Copper Microparticles, Contained in an Alginate Matrix, on *Escherichia coli*

Simon F. Thomas[1], Paul Rooks[2], Fabian Rudin[1], Sov Atkinson[2], Paul Goddard[3], Rachel Bransgrove[2], Paul T. Mason[1], Michael J. Allen[2]*

1 Plymouth Martine Laboratory Applications Ltd, Plymouth, United Kingdom, **2** Plymouth Marine Laboratory, Plymouth, United Kingdom, **3** Protein Technologies Ltd, Manchester, United Kingdom

Abstract

Although the bactericidal effect of copper has been known for centuries, there is a current resurgence of interest in the use of this element as an antimicrobial agent. During this study the use of dendritic copper microparticles embedded in an alginate matrix as a rapid method for the deactivation of *Escherichia coli* ATCC 11775 was investigated. The copper/alginate produced a decrease in the minimum inhibitory concentration from free copper powder dispersed in the media from 0.25 to 0.065 mg/ml. Beads loaded with 4% Cu deactivated 99.97% of bacteria after 90 minutes, compared to a 44.2% reduction in viability in the equivalent free copper powder treatment. There was no observed loss in the efficacy of this method with increasing bacterial loading up to 10^6 cells/ml, however only 88.2% of *E. coli* were deactivated after 90 minutes at a loading of 10^8 cells/ml. The efficacy of this method was highly dependent on the oxygen content of the media, with a 4.01% increase in viable bacteria observed under anoxic conditions compared to a >99% reduction in bacterial viability in oxygen tensions above 50% of saturation. Scanning electron micrographs (SEM) of the beads indicated that the dendritic copper particles sit as discrete clusters within a layered alginate matrix, and that the external surface of the beads has a scale-like appearance with dendritic copper particles extruding. *E. coli* cells visualised using SEM indicated a loss of cellular integrity upon Cu bead treatment with obvious visible blebbing. This study indicates the use of microscale dendritic particles of Cu embedded in an alginate matrix to effectively deactivate *E. coli* cells and opens the possibility of their application within effective water treatment processes, especially in high particulate waste streams where conventional methods, such as UV treatment or chlorination, are ineffective or inappropriate.

Editor: Chien-Sheng Chen, National Central University, Taiwan

Funding: This work was supported by grants awarded to MJA from the Bill & Melinda Gates Foundation (OPP1044451, OPP1095464). The findings and conclusions contained within are those of the author(s) and do not necessarily reflect positions or policies of the Bill & Melinda Gates Foundation. The funders had no role in study design, data collection and analysis, decision to publish, or preparation of the manuscript.

Competing Interests: The authors expect to develop a technology platform to treat contaminated water based on the results described herein. PG of Protein Technologies Ltd (PTL) will be responsible for the protection and commercialisation of this technology; PML Applications and Plymouth Marine Laboratory (of which SFT/FR/PTM and PR/SA/RB/MJA are employees, respectively) have a royalty agreement with PTL for any revenue generated through such activities. It is envisioned that the technology platform will be made freely available in the developing world for humanitarian purposes. This does not alter the authors' adherence to all the PLOS ONE policies on sharing data and materials.

* E-mail: mija@pml.ac.uk

Introduction

Elemental copper has been used as an anti-spoiling agent and preservative for over two millennia, predating the discovery of the role of microbes in disease during the late nineteenth century [1]. The use of copper as an antimicrobial agent has recently provoked renewed interest, due to its rapid and extensive activity, especially on dry surfaces. For example, *Salmonella enterica* and *Campylobacter jejuni* are destroyed within two hours on a dry copper surface [2] and Santo *et al* showed that *Staphlococcus haemolyticus* was deactivated within minutes, accompanied by extensive membrane damage [3]. This has resulted in the increasing popularity of copper surfaces in hospital and institutional environments as an active antimicrobial surface, on items such as door knobs and bench tops. The anti-microbial mechanism of copper in such applications is poorly characterised, although recent work suggests that toxicity is due to the oxidation of the surface layer and consequent donation of electrons to cell wall components, such as lipids and proteins, resulting in a rapid reduction of cell integrity [3]. Other mechanisms such as internal protein binding and DNA mutation have also been suggested [4], [5] and indeed *in vitro* experimentation has showed that Cu can bind to DNA. However, Macomber *et al* indicated that *in vivo*, Cu^{2+} ions were sequestered by the glutathione naturally produced by *E. coli* thus preventing oxidative damage to DNA through Fenton-like reactions [6].

Unfortunately, the use of free Cu salts as a water sterilisation method is restricted by toxicity to both aquatic organisms and humans, although a combination of Cu with Ag is often used to inhibit *Legionella* sp. in cooling tower water [7]. Accumulation of Cu in the gills of freshwater fish has been shown to inhibit Na^+ influx and reduce Na-K ATPase activity, and just 1.2 µg/l Cu^{2+} has been shown to produce chronic toxicity for the freshwater mussel, *Lamsilis siliquoidea* [8] and 37 µg/l Cu^{2+} is toxic to the

Minnesota Sculpin [9]. The use of elemental copper is therefore preferential, but surface coating of large vessels with copper results in a slower antimicrobial action due to reduced contact time. Micro and nano copper particles offered increased relative surface areas, but are difficult to evenly distribute in water and the entrapment of these particles in a matrix is desirable. Alginate is obtained from a variety of brown macroalgae, such as *Laminaria* sp. and *Macrocystis pyrifera*, and is widely used as an emulsifier, thickener and as a matrix for the binding of proteins, enzymes and metals [10], [11]. It is an anionic polysaccharide consisting of homopolymeric blocks of (1–4)-linked β-D-mannuronate and its C-5 epimer α-L-guluronate residues, covalently linked together in different sequences or blocks [12]. Alginate complexes with metal ions via its carboxylate groups, and alginate binding with Cu salts and subsequent antimicrobial activity has been shown previously [13].

In the following article, the use of dendritic copper microparticles bound within an alginate matrix as an antimicrobial agent is assessed and the effect of a key environmental parameter, dissolved oxygen content, on the antimicrobial activity ascertained.

Methods

Bead Formation

Beads were prepared according to a modification of the method described by Albarghouthi *et al* [14]. Briefly 4% (w:v) sodium alginate (Sigma Aldrich, Poole, UK) was dissolved in deionised water by heating to 60°C whilst stirring on a magnetic stirrer at 100 rpm. Once dissolved, the alginate solution was autoclaved at 15 psi for 15 minutes. Once cooled, dendritic copper microparticles (Specific surface 1600 cm^2/g, <63 μm) were added where appropriate at a concentration between 1–4%. The particles were dispersed by sonication using a VibraCell sonic probe (Sonics and materials, Dansbury, Ct, USA) for 60 minutes. The mixture was stored at 4°C until further processing.

Beads were formed by the drop wise addition of alginate/copper mixture via a 0.8 mm gauge needle into a 4°C solution of $CaCl_2$ (2.5% w:v). Once formed, beads were washed in deionised water three times and stored in deionised water at 4°C until used.

Growth of *E. coli* ATCC 11775

E. coli ATCC 11775 from −80°C stored stocks was inoculated into 50 ml of sterilised Luria-Bertani (LB) broth (Sigma Aldrich, Poole, UK) and incubated overnight at 37°C in a shaking incubator (Bibby scientific, Staffordshire, UK) at 150 rpm; 1 ml of this stock was then added to 250 ml of LB broth, which was incubated overnight under the same conditions. The bacterial cell numbers were then determined using the Gram staining method [15] and subsequent enumeration using an improved Neubauer hemocytometer.

Antimicrobial Assay Procedure

All assays were performed in a media containing 0.5 g/l Bacto peptone and 0.2 g/l Yeast extract (both Sigma Aldrich, Poole, UK). Briefly, *E. coli* was added at the appropriate concentration, to 50 ml of assay media, and then shaken at 37°C in a shaking incubator (Bibby scientific, Staffordshire, UK) at 150 rpm for 30 minutes to allow acclimation. Alginate beads (1 g) were then aseptically added giving a final Cu copper concentration of between 0.125 and 1 mg/ml. The samples returned to the shaking incubator, and 1 ml of sample was then removed at 15 minute intervals for microbial examination. Each sample was serial diluted at 4°C in 1/8 strength Ringers solution (Fisher, Leicester, UK) and plated onto 90 mm Petri dishes containing MacConkey media number 2 (Sigma Aldrich, Poole, UK). The Petri dishes were then incubated overnight at 37°C and the distinctive pink colonies (indicative of *E. coli*) counted. All treatments were prepared in triplicate, and technical triplicate replicates also performed. Minimum inhibitory concentrations (MIC) of Cu to *E. coli* were determined using the methods described by Wiegand *et al* [16]. Free copper was dispersed into the media prior to bacterial inoculation by sonication.

The effects of dissolved oxygen (DO) on copper-induced microbial toxicity were assessed using the same methods as above, but at a 2.5 litre scale in an Infors Minifor fermenter (Reigate, UK) and DO was measured via the manufacturer's proprietary probe and adjusted by air injection and variable stirring rate via the equipment's software.

Scanning Electron Microscopy (SEM)

Samples were prepared for analysis by sputter coating with chrome (for bacteria) and carbon (for beads and copper identification), using an EMTECH K450X sputter coater (Wrexham, UK). Prepared samples were subsequently analysed on a JEOL JSM-700F field emission scanning electron microscope (JEOL, Welwyn Garden City, UK).

Quantification of Cu^{2+} Leaching and Absorption by Alginate Beads

Cu-Alginate beads (5 g) were placed in a 250 cm^3 Schott bottle with 50 cm^3 of MilliQ water. Lids were screwed on to prevent additional air from entering the system. 100 μl aliquots of each sample were taken at 0, 1, 2, 4, 8, 12, 24, 36, 48 hours and placed in Eppendorfs for [Cu^{2+}] analysis. Analysis of [Cu^{2+}] in MilliQ water used in all the samples and MilliQ water with 5 g beads without copper was performed, to account for any Cu^{2+} from beads and water. These readings were subsequently deducted from those of the samples.

Absorption of Cu^{2+} by alginate beads was determined by measuring the free copper concentration over time using a Palintest 7500 photometer (Gateshead, UK) with the proprietary free copper test kit, whilst using the manufacturer's instructions.

Results

Copper/Alginate Bead Complex as an Antimicrobial Agent

The 4% copper bead treatment produced a rapid reduction in viability of $10^8\ ml^{-1}$ *E. coli* cells, with 87.41% and 99.97% of bacteria rendered non-viable after 30 and 90 minutes respectively (Figure 1). The other treatments resulted in a dose related reduction of *E. coli* viability, with between 79 and 97% of bacteria non-viable after 90 minutes (Figure 2). However the 4% Cu bead treatment produced the most rapid viability reduction compared to the other treatments, with 87% of cells non-viable after 30 minutes compared to between 47–48.2% for the other treatments. Indeed, there was an almost instantaneous 49.7% reduction in viability immediately upon exposure to the 4% Cu beads. There was no reduction in *E. coli* viability observed in samples exposed to alginate beads containing no copper, and 4% copper particles not associated with alginate, only reduced *E. coli* viability by 44.2% after 90 minutes treatment (Figure 1).

MICs for Cu in alginate compared to Cu not associated with alginate also differed, with a value of 0.065% w:v Cu in alginate (0.02 mg/ml final concentration), preventing visible growth in an overnight culture of *E. coli*, whilst a dose of 0.5 mg/ml Cu powder, not associated with alginate, was required to similarly prevent visible growth.

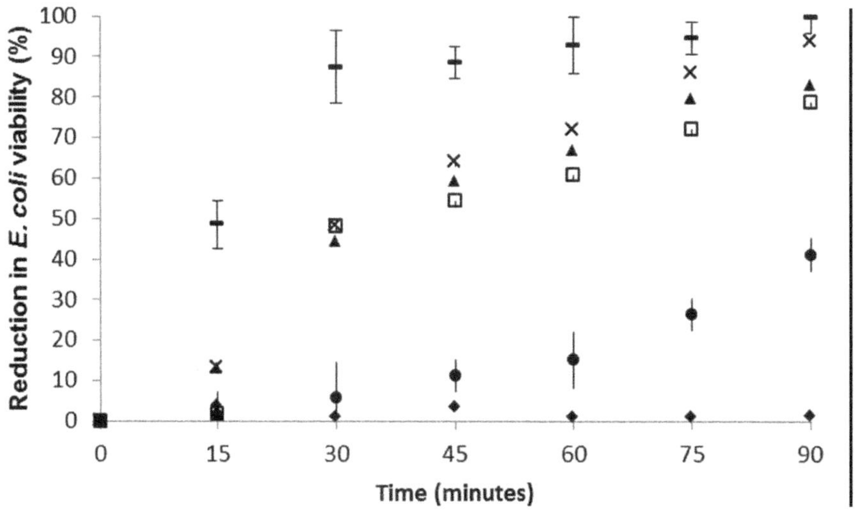

Figure 1. The biocidal effect of differing Cu concentrations entrapped in alginate beads on *E. coli* ATCC 11775 over time. Black diamond = Plain alginate beads. White square = 1% Cu W:W. Black triangle = 2% Cu. Black cross = 3% Cu, Black dash = 4% Cu. Black circle = 4% copper powder not associated with alginate. Bars denote standard errors.

There was a linear dose response of increasing Cu concentrations within the beads to the viability of *E. coli* cells from 0 to 1% Cu (R^2 = 0.993[Figure 2]). Just 21.6% of cells were non-viable after 90 minutes exposure to 0.25% Cu, and this figure increased to 46.1% following exposure to 0.5% Cu. At the 1% Cu treatment, 76.5% of cells were non-viable. Increases in Cu concentrations above 1% led to a slower rate of loss of *E. coli* viability from 81% achieved at 1.5% Cu dose to >99.999% at the 4% Cu dose (Figure 2).

Effect of Bacterial Numbers on Copper Efficacy

At *E. coli* numbers up to 10^6/ml more than 99.9% of cells were rendered non-viable after 90 minutes exposure (Figure 3), with no viable cells observed from an initial inoculum of 10^3 cells/ml and 99.9997% of cells were non-viable at a level of 10^4 cells/ml. However, this figure decreased as bacterial abundance increased above this figure, and at a loading of 10^8 cells/ml, 88.2% of cells did not grow on MacConkey media and at 10^9 cells/ml, this figure decreased further to 79.6% (Figure 3).

The Relationship between Dissolved Oxygen (DO) Concentration in Media and the Reduction in *E. coli* Viability

When 10^6 *E. coli* cells/ml were exposed to 4% Cu (in alginate beads) under anoxic conditions, no loss in viability was observed in the bacteria after 90 minutes exposure (Figure 4). Indeed an average 4.01% increase in viable bacteria was observed indicating that bacterial cell division occurred in the presence of bactericidal levels of copper, under low oxygen tensions. Once DO concentrations increased above 50% of saturation bacterial viability was rapidly reduced, and at 75 and 100% saturation, 99.7 and 99.997% respectively, of bacteria were no longer viable on MacConkey media (Figure 4). However, upon exposure for 180 minutes to Cu, *E. coli* cell viability began to decrease in the

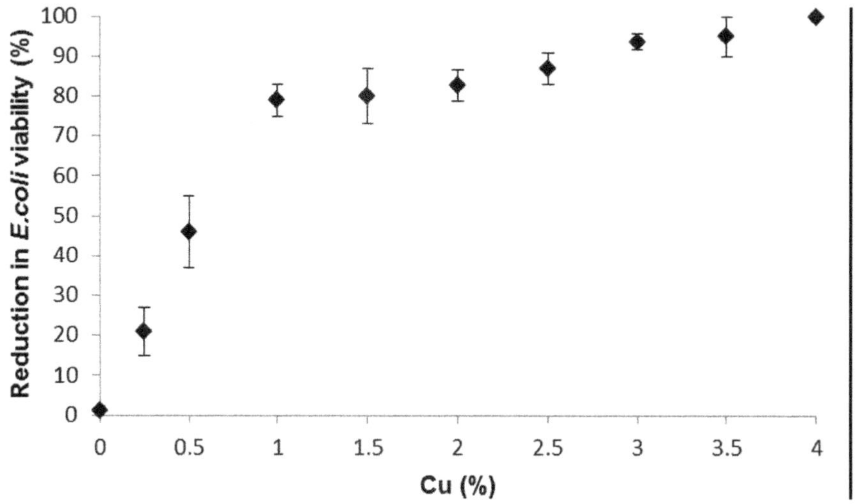

Figure 2. The response of *E. coli* ATCC 11775 to increasing copper concentrations in alginate beads after 90 minute exposure. Bars denote standard errors.

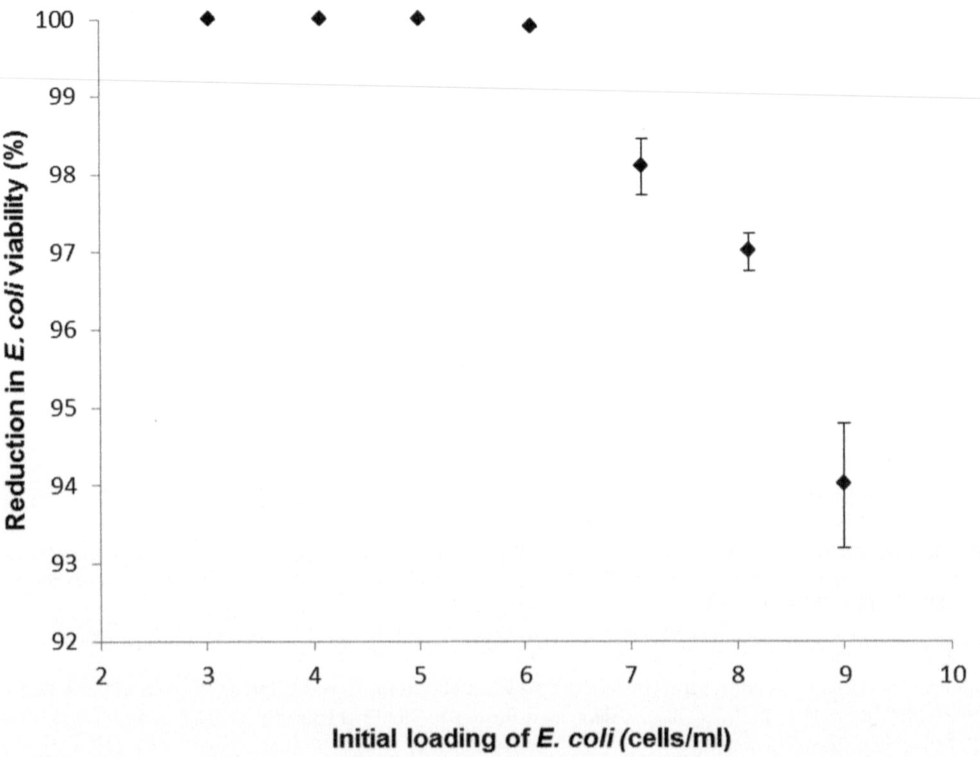

Figure 3. The relationship between increased bacterial loading and reduction in viability upon 90 minute exposure to 4% dendritic copper in alginate beads. Bars denote standard errors.

anoxic conditions and 96% of cells were no longer able to grow on agar plates (Figure 5).

Quantification of Cu^{2+} Leaching and Absorption by Alginate Beads

The leaching of Cu^{2+}, from dendritic Cu encapsulated in the Cu-Alginate beads, was below the limit of detection 48 hours post bead production (data not shown), despite minor leaching occurring in the initial 12 hours. The maximal leaching occurred between 8 and 9 hours post production and was measured at 0.4 mg/l at a rate of 0.2 mg/l/hr, however the Cu^{2+} concentration in the media subsequently decreased to below detection limits from 12 hours onwards, accompanied by a blue discolouration on the bead surface. The Cu^{2+} content of the media then remained unchanged and below the limit of detection until the termination of the experiment (48 hours).

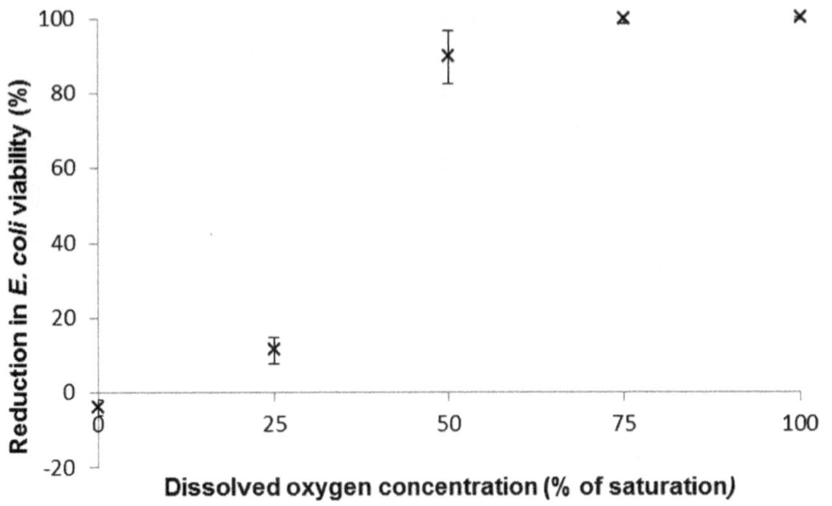

Figure 4. The relationship between dissolved oxygen concentration and *E. coli* ATCC 11775 viability using 4% Cu in alginate beads. The experiment was performed over 90 minutes at 37°C, with a bacterial loading of 10^6 cells/ml. Bars denote standard errors.

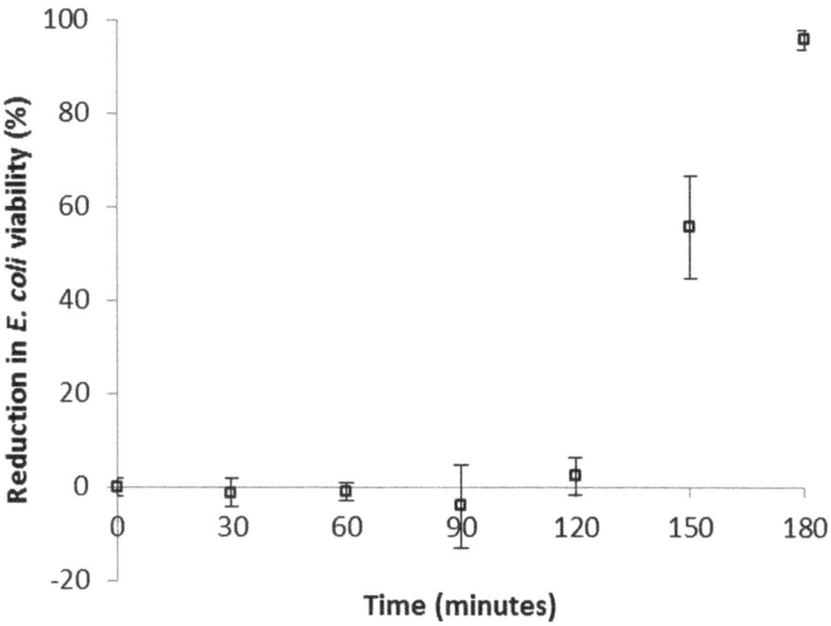

Figure 5. The viability of *E. coli* ATCC 11775 during treatment with 4% Cu (in alginate beads) under anoxic conditions (DO<0.1 mg/l). The experiment was performed at 37°C, with a bacterial loading of 10^6 cells/ml.

Electron Microscopy of *E. coli* and Alginate Beads

Morphological differences were observed between the 4% copper-alginate bead treated and untreated cells (Figure 6), with obvious membrane disruption observed in *E. coli* treated with copper after 90 minutes of exposure (Figure 6*i*). No morphologically intact cells were observed in the treated sample, whilst no membrane disruption was seen in any of the cells of the untreated sample. The copper-treated cells displayed clear blebbing, presumably due to the escape of intracellular material, which was not observed in either the untreated or time zero of the treated samples.

The SEM visualisation of the alginate beads indicated a smooth, plated exterior, exhibiting a honey combed interior structure of the beads (Figure 7c). The dendritic copper particles are embedded in the layered alginate shells (Figure 7a), and the presence of copper was confirmed by x-ray analysis of the particles within the beads. Also, copper was detected evenly distributed throughout the bead interior, and this was thought to represent the Cu^{2+} ion produced from the oxidation of the elemental copper present. The dendritic structure of the copper could clearly be observed within the alginate matrix (Figure 7b).

Discussion

During this study, the efficacy of alginate beads containing copper for the destruction of *E. coli* in an artificial media was established. The presence of dendritic Cu particles alone, did not exhibit the same degree of antimicrobial properties as the Cu bound within the alginate matrix. The synergistic effect between copper and alginate could be due to the stability provided by the alginate matrix in terms of shape and size of Cu. Dispersed or weakly aggregated metallic particles in suspension have a more variable internal structure and different reactivity and rates of surface reactions than strongly aggregated nanoparticles [13]. Subsequent changes in the physicochemical properties may alter the interaction of Cu particles with bacteria, thus affecting the antimicrobial activity. The immobilization of colloidal metal

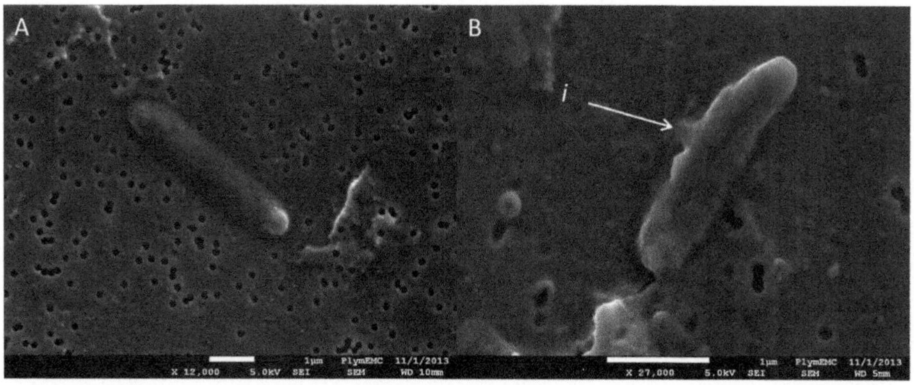

Figure 6. SEM of *E. coli* cells A: before and B: 45 minutes post treatment with Cu-containing alginate beads. *i* indicates disruption to cell membrane with associated loss in cellular integrity.

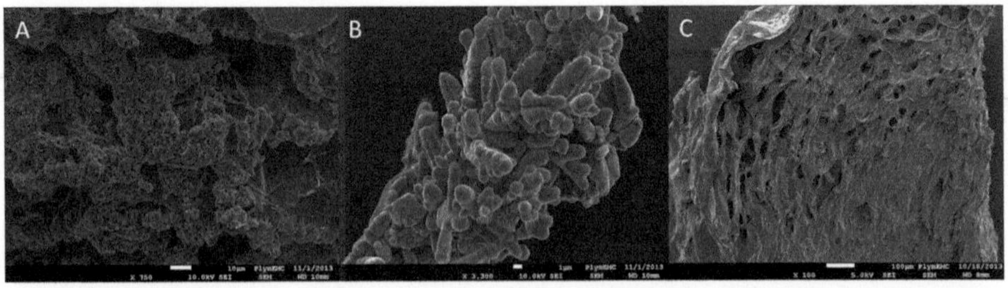

Figure 7. SEM of internal surface of alginate bead structure. A: Internal bead surface at x 750 magnification of dendritic copper particles visible, B: Internal bead surface at x 3300 times magnification of dendritic copper particle protruding from alginate matrix and C: A cross section of the Cu alginate bead.

nanoparticles therefore prevents aggregation and provides positional stability to the nanoparticles on a surface, or in a structure, thus coordinating the interaction of Cu, bacteria and other functional groups present [17], [18].

The rapid kill rate was affected by low oxygen tensions in the media, supporting the observations of Baker *et al* that the initial damage caused to microbial cells is by the Cu^{2+} species and is oxidative in nature [19]. In anaerobic conditions the predominant copper species is Cu^{+} and the Cu(I)-translocating P-type ATPase responsible for regulating Cu homeostasis under such conditions is Cu^{+} specific [20]. However, under aerobic conditions the Cu^{2+} ion predominates and in most aqueous conditions this ion will be strongly bound to organic components. Indeed, Cu^{2+} is used as a catalyst in waste water treatment systems to oxidise organic waste [21], and such reactions are capable of producing reactive oxygen species through multiple pathways such as the Fenton-type reaction, which are highly toxic to many bacterial cells [22], [23]. Thus the 700 mg/l dissolved organic carbon present in the test media could provide an additive toxic effect through the production of ROS, thus providing interesting prospects for the use of this technique in the rapid deactivation of bacteria in recalcitrant waste waters with high organic loading, such as those found in many parts of the world that lack modern integrated sewerage treatment systems.

The electron micrographs of treated *E. coli* cells (Figure 6) indicated that cell membrane integrity is rapidly affected by Cu and the leakage of intracellular material can clearly be observed and this corresponds to the finding of Santo *et al* that the initial copper-induced cellular damage is membrane associated with consequent reduction in cellular osmotic stability [3]. An antimicrobial effect under anaerobic conditions was observed 2 hours post treatment, which suggests that the varying findings of different researchers, as to the exact nature of Cu-induced microbial lysis, may be due to different mechanisms, dependent on the redox status of the environment.

As the primary antimicrobial activity of Cu observed during this study appears oxidative in nature, the delay in observed lysis of *E. coli* under anoxic conditions may be due to a switch in microbial metabolism from respiratory to fermentative. Indeed Fung *et al* found that *E. coli* detoxification efflux mechanisms were more effective at removing Cu^{+} from the cell under anaerobic conditions than Cu^{2+} under aerobic [24]. However, under anaerobic conditions the turnover of Fe-S cluster enzymes appears to be especially sensitive to Cu-induced toxicity in *E. coli.* However, this toxicity was induced by Cu^{+} not Cu^{2+} as for aerobically-cultured organisms [24].

During this study the antimicrobial effect of micro as opposed to nano Cu particles in an alginate matrix is indicated for the first time. This is especially important at a time of increased interest in the microbial properties of Cu in the fields of medicine, public health and waste water treatment. The use of dendritic micro copper has advantages in terms of both cost and ease of manufacture over nanoparticles, for example dendritic copper sells for around $8–11 per kilo, whilst 50 nm Cu particles cost around $1200–1500 per kilo. This study suggests that Cu microparticles embedded within an alginate matrix have a strong potential for use as an antimicrobial in a wide range of clinical and industrial environments.

Author Contributions

Conceived and designed the experiments: SFT SA MJA. Performed the experiments: SFT PR FR SA RB PTM MJA. Analyzed the data: SFT SA PG MJA. Contributed reagents/materials/analysis tools: SFT PG MJA. Wrote the paper: SFT SA PG MJA.

References

1. HHA Dolwet, JRJ Sorenson (1985) Historic uses of copper compounds in Medicine. J Trace Elem Med Biol 2.
2. Faundez G, Troncoso M, Navarette P, Figueroa G (2004) Antimicrobial activity of copper surfaces against suspensions of *Salmonella enterica* and *Campylobacter jejuni.* BMC Microbiol 4: 19.
3. Santo EC, Quaranta D, Grass G (2012) Antimicrobial metallic copper surfaces kill *Staphylococcus haemolyticus* via membrane damage. Microbiologyopen 1: 46–52.
4. Harwood V, Gordon A (1994) Regulation of extracellular copper-binding proteins in copper-resistant and copper-sensitive mutants of *Vibrio alginolyticus.* Appl Environ Microbiol 60: 1749–1753.
5. Gordon A, Howell D (1994) Responses of Diverse Heterotrophic Bacteria to Elevated Copper Concentrations. Can J Microbiol 40: 408–411.
6. Macomber L, Rensing C, Imlay JA (2007) Intracellular copper does not catalyze the formation of oxidative DNA damage in *Escherichia coli.* J Bacteriol 189: 1616–1626.
7. Lin Y, Vidic R, Stout J, Yu V (2002) Negative effect of high pH on biocidal efficacy of copper and silver ions in controlling *Legionella pneumophila.* Appl Environ Microbiol 68: 2711–2715.
8. Jorge M, Loro V, Bianchini A, Wood C, Gillis P (2013) Mortality, bioaccumulation and physiological responses in juvenile freshwater mussels (*Lampsilis siliquoidea*) chronically exposed to copper. Aquat Toxicol 126: 137–147.
9. Besser J, Mebane C, Mount D, Ivey C, Kunz J (2007) Sensitivity of mottled sculpin (*Cottus bairdi*) and rainbow trout (*Onchorhynchus mykiss*) to acute and chronic toxicity of cadmium, copper, and zinc. Environ Toxicol Chem 26: 1657–1665.
10. Castro M, Orive G, Hernández R, Bartkowia A, Brylak W, et al. (2009) Biocompatibility and in vivo evaluation of oligochitosans as cationic modifiers of alginate/Ca microcapsules. J Biomed Mater Res 91: 1119–1130.
11. Anal A, Bhopatkar D, Tokura S, Tamura H, Stevens W (2003) Chitosan-alginate multilayer beads for gastric passage and controlled intestinal release of protein. Drug Dev Ind Pharm 29: 713–724.
12. Sun J, Tan H (2013) Alginate-Based Biomaterials for Regenerative Medicine Applications. Materials 6: 1285–1309.
13. Dîaz-Visurraga J, Daza C, Pozo C, Becerra C, von Plessing C, et al. (2012) Study on antibacterial alginate-stabilized copper nanoparticles by FTIR and 2D-IR correlation spectroscopy. Int J Nanomedicine 7: 3597–3612.

14. Albarghouthi M, Fara D, Saleem M, El-Thaher T, Matalka K, et al. (2000) Immobilization of antibodies on alginate-chitosan beads. Int J Pharm 206: 23–34.
15. Holt JG, Krieg NR, Sneath PHA, Staley JT, Williams ST (2010) Bergey's Manual of Determinative bacteriology, 9th Ed, Lippincott Williams and Wilkins, London.
16. Wiegand I, Hilpert K, Hancock R (2008) Agar and broth dilution methods to determine the minimal inhibitory concentration (MIC) of antibiotic substances. Nat Protoc 3: 163–175.
17. Xu Q, Liu Y, Liu F, Meng W, Cui K, et al (2014) Aluminum plasmonic nanoparticles enhanced dye sensitized solar cells, Opt Express 22: A301–A310.
18. Liu Y, Chen S, Zhong L Wu G (2009) Preparation of high-stable silver nanoparticle dispersion by using sodium alginate as a stabilizer under gamma radiation. Radiat Phys Chem Oxf Engl 1993 78: 251–255.
19. Baker J, Sitthisak S, Sengupta M, Johnson M, Jayaswal RK, et al. (2010) Copper stress induces a global stress response in *Staphylococcus aureus* and represses *sae* and *agr* expression and biofilm formation. Appl Environ Microbiol 76: 150–160.
20. Rensing C, Fan B, Sharma R, Mitra B, Rosen B (2000) CopA: An *Escherichia coli* Cu(I)-translocating P-type ATPase. Proc Natl Acad Sci USA 97: 652–656.
21. Chen Y, Wang D, Zhu X, Zheng X, Feng L (2012) Long-term effects of copper nanoparticles on wastewater biological nutrient removal and N_2O generation in the activated sludge process. Environ Sci Technol 46: 12452–12458.
22. Held D, Sylvester F, Hopcia K, Biaglow J (1996) Role of Fenton Chemistry in Thiol-Induced Toxicity and Apoptosis. Radiat Res 145: 542–553.
23. Karlsson H, Cronholm P, Gustafsson J, Moller L (2008) Copper oxide nanoparticles are highly toxic: a comparison between metal oxide nanoparticles and carbon nanotubes. Chem Res Toxicol 21: 1726–1732.
24. Fung D, Lau W, Chan W, Yan A (2013) Copper efflux is induced during anaerobic amino acid limitation in Escherichia coli to protect iron-sulfur cluster enzymes and biogenesis. J Bacteriol 195: 4556–4568.

2

Patterns in Temporal Variability of Temperature, Oxygen and pH along an Environmental Gradient in a Coral Reef

Òscar Guadayol[1*¤], Nyssa J. Silbiger[1,2], Megan J. Donahue[1], Florence I. M. Thomas[1]

1 Hawai'i Institute of Marine Biology, School of Ocean and Earth Science and Technology, University of Hawai'i at Mānoa, Kāne'ohe, Hawai'i, United States of America,
2 Department of Biology, University of Hawai'i at Mānoa, Honolulu, Hawai'i, United States of America

Abstract

Spatial and temporal environmental variability are important drivers of ecological processes at all scales. As new tools allow the *in situ* exploration of individual responses to fluctuations, ecologically meaningful ways of characterizing environmental variability at organism scales are needed. We investigated the fine-scale spatial heterogeneity of high-frequency temporal variability in temperature, dissolved oxygen concentration, and pH experienced by benthic organisms in a shallow coastal coral reef. We used a spatio-temporal sampling design, consisting of 21 short-term time-series located along a reef flat-to-reef slope transect, coupled to a long-term station monitoring water column changes. Spectral analyses revealed sharp gradients in variance decomposed by frequency, as well as differences between physically-driven and biologically-reactive parameters. These results highlight the importance of environmental variance at organismal scales and present a new sampling scheme for exploring this variability *in situ*.

Editor: Judi Hewitt, University of Waikato (National Institute of Water and Atmospheric Research), New Zealand

Funding: Funding for this project came from National Oceanic and Atmospheric Administration (NOAA) Office of National Marine Sanctuaries-HIMB partnership (MOA #2009-039/7932). This paper is also funded in part by a grant/cooperative agreement from the National Oceanic and Atmospheric Administration, Project R/EL-33, which is sponsored by the University of Hawaii Sea Grant College Program, SOEST, under Institutional Grant No. NA09OAR4170060 from NOAA Office of Sea Grant, Department of Commerce. The views expressed herein are those of the author(s) and do not necessarily reflect the views of NOAA or any of its subagencies. UNIHI-SEAGRANT-JC-12-24. NJS was supported by NOAA Dr. Nancy Foster Scholarship Award No. NA12NOS4290140. This is HIMB contribution #1574 and SOEST contribution #9042. The funders had no role in the study design, data collection and analysis, decision to publish, or preparation of the manuscript.

Competing Interests: The authors have declared that no competing interests exist.

* E-mail: oscar@guadayol.cat

¤ Current address: School of Biological, Biomedical and Environmental Sciences, University of Hull, Kingston-upon-Hull, United Kingdom

Introduction

The spatial and temporal distribution of aquatic organisms is tightly coupled to underlying physical, chemical, and biological changes that are distributed unequally over a wide range of scales (e.g. [1], [2]). Most of this variability is concentrated in large-scale phenomena, such as seasonal cycles or latitudinal gradients. Thus, the effects of low frequency environmental variability, expressed in terms of means, trends and seasonality, have received more attention than variance at high frequencies (e.g. [3]). However, behavioral and physiological responses of individuals occurs mainly at small scales (e.g. [4–9]), and environmental variability at high frequencies may be important to structuring aquatic systems. Therefore, characterizing the whole spectrum of variance experienced by organisms is important if we want to understand how responses of individuals to perturbations scale up into an ecosystem response [10]. Focusing on smaller scale variability is particularly relevant in the present context of environmental change because the scale and frequency of variation is expected to shift along with mean changes in global climate [11].

The development of faster, more accurate and affordable autonomous sensors has triggered studies describing patterns of natural variability in the framework of global change, particularly for pH [12–15]. A number of these studies have focused on determining ranges of biologically relevant variables across contrasting marine environments [13] or in seasonal and/or diurnal variability [14], [16–17], but intermediate and smaller scales have received little attention. Thus, studies exploring the whole spectrum of temporal environmental variability, and how this changes along spatial gradients, are lacking. Such multiscale studies can help us understand the distribution of organisms and can provide a background for testing, *in situ*, hypotheses about how organisms will acclimate and adapt [18] to upcoming changes in environmental variability.

Tropical coral reefs are particularly appropriate for an investigation of high-frequency environmental fluctuations because dramatic changes in temporal variability can occur over very short distances for several reasons. First, coastal systems are boundaries between ocean, land, and atmosphere and are subjected to a wide array of fluctuations from all three sources. Second, reef shorelines can be very abrupt, and steep physical and chemical gradients are established across changes in depth. Concurrent with these physical gradients are gradients in primary productivity, which can drive fluctuations in oxygen and pH over relatively small temporal and spatial scales. Third, tropical systems are characterized by a relatively low seasonality and strong diurnal forcing; therefore, high-frequency variability may be comparatively more important than in temperate environments. In addition, the need to understand patterns of natural variability is particularly acute on coral reefs, because corals show narrow tolerance levels and are sensitive to global change [19–21]. Reef

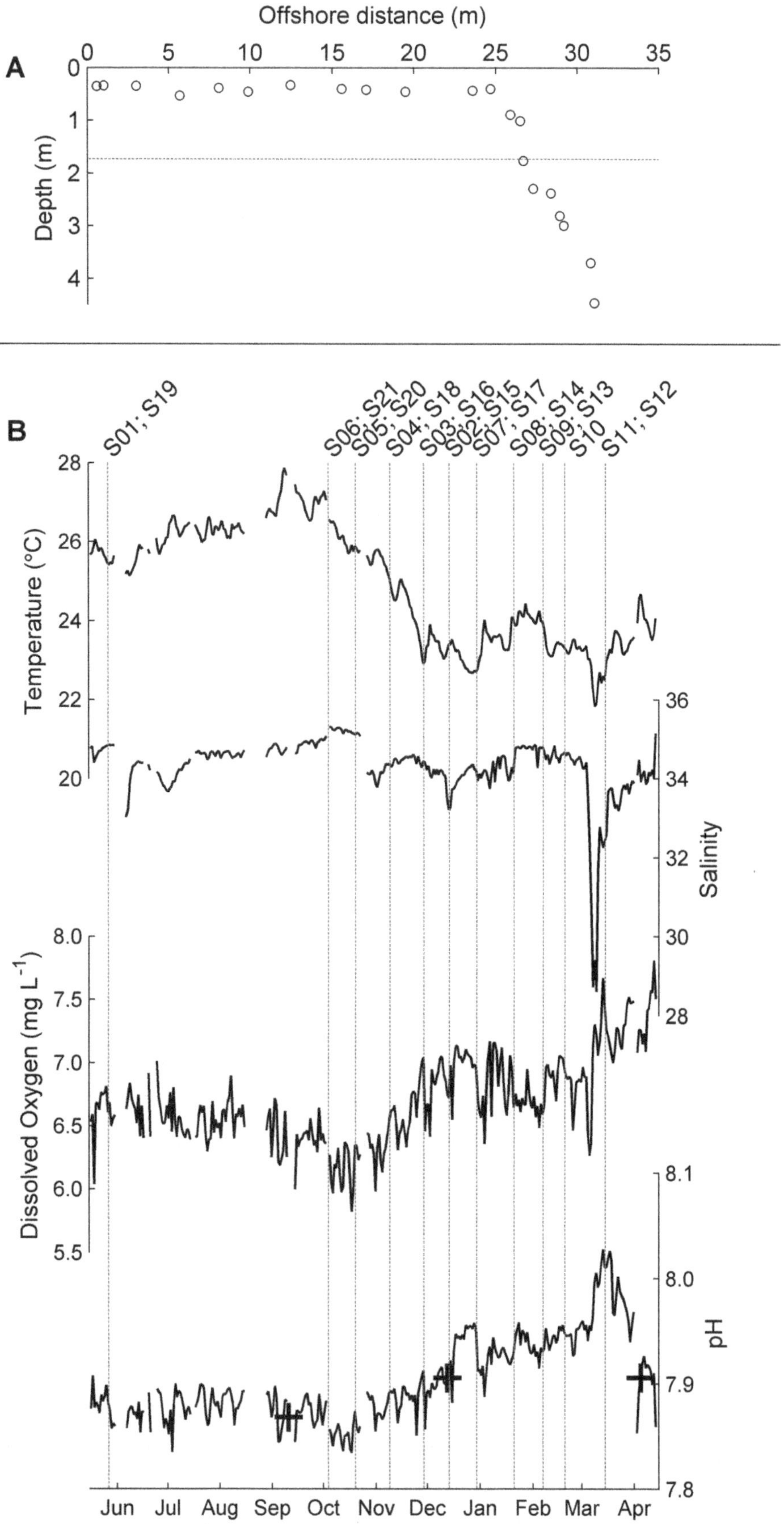

Figure 1. Timing and position of sensor deployments. (A) Position of sites in the reef-flat to reef-slope transect on distance to shore-depth axes. Sites are numbered consecutively (S1–S21) from left to right. Horizontal dotted line marks the depth at which the long-term time series was

located. (B) Time series plots of daily averaged temperature, salinity, dissolved oxygen concentration and pH from the long-term station. Vertical dotted lines mark the starts of the two-weeks deployments along the transect. Crosses on top of the pH time series correspond to in situ water measurements of pH. The names of the sites used in the deployments are shown on top.

organisms may undergo acclimatization to small-scale spatial and temporal variation in the environment [22–23]. Thus, describing the levels of natural variability these organisms are facing at all scales is critical for predicting their resilience to global change [14]. Here we examine the spatial distribution of the whole spectra of environmental fluctuations across a coral reef in Kāne'ohe Bay, Hawai'i, using a novel spatial and temporal sampling scheme that combines short time series from sites distributed across a transect, with a long-term time series from an adjacent permanent station.

Methods

Site Description

Moku o Lo'e (Coconut Island), home to the Hawai'i Institute of Marine Biology, is a small island of about 12 ha located in Kāne'ohe Bay, a tropical semi-enclosed embayment at the eastern shore of the island of O'ahu in the Hawaiian archipelago. It is situated on the southern basin of the bay lagoon, where water residence time is relatively long [24–25] and advection is minimal. The biochemistry of this basin is heavily influenced by land, and subject to episodic releases of nutrients [26–27].

Seasonality is defined by dry (May to October) and rainy (October to April) seasons. NE trade winds are prevailing throughout the year, but occasionally intense south "Kona" winds occur in winter. A weak thermocline is usually established in summer at a depth of about 8.5 m [28].

Diurnal and semidiurnal components are the only significant tidal constituents [29]. Mean and maximum tidal ranges are 0.7 and 1.1 m, respectively, and currents are generally mild and tidally driven [25], [30].

Spatio-temporal Sampling Design

In order to explore distribution in variance across spatial and temporal scales, we established 21 random-stratified sampling locations along a ~32 m reef flat-to-reef slope transect across a fringing reef. At each site along the transect, two-week long times series of environmental parameters were obtained non-simultaneously throughout a period of 11 months. At an adjacent long-term station, we obtained a similar time series for the entire experimental period. This is an approximation to an idealized (albeit inefficient and expensive) setup with a grid of sensors deployed simultaneously.

The transect was oriented eastward off the southeastern shore of Coconut Island (N21°25.975′, W157°47.175′). Starting 0.5 meters from shore, the transect ran along the reef flat for ~24 m and then down the reef slope for another ~7 m to a depth of 4.5 m (Fig. 1A). Sites were stratified between the reef flat and reef slope, with sites S1–S11 on the flat and S12–S21 on the slope. The transect falls mostly within the sublittoral zone, although the shallowest parts of the reef flat are occasionally exposed during the lowest tides. During each sampling period, measurements were taken simultaneously at two randomly selected locations on the transect, one in the reef slope and the other in the flat, using multi-parametric probes. The random order helped ensure that the spatial gradient along the transect was not systematically confounded by long-term temporal trends or seasonality. The instruments were deployed at 5 to 10 cm above the bottom for a minimum of 2 weeks between May 2011 and March 2012 (Fig. 1B). Each probe contained temperature, conductivity, dissolved oxygen concentration ([O_2]) and pH sensors (Sonde 600XLM, YSI Incorporated). Only temperature, pH and [O_2] data were used in this study. Salinity could not be used in the statistical analyses because its distribution was not normal, and the time series were spiky. Therefore, the spectral analyses resulted in flat, anomalous spectral densities.

The long-term station was located a few meters south of the transect, just off the easternmost point of Moku o Lo'e, on a site with a shorter reef flat than the transect. It consisted of a multi-parametric instrument (Sonde 6600V2-4, YSI) with sensors for temperature, conductivity, pH and [O_2]. As the aim was characterizing water column background conditions, rather than bottom conditions, the instrument was placed in midwater, attached facing down to a pole at a depth of ~1.7 m over a ~3 m deep bottom.

In addition, an acoustic Doppler current profiler (ADCP, 2.0 MHz Aquadopp Profiler, Nortek A.S.) was placed at about 14.0 meters in depth on average, a few meters offshore of the long-term station. It measured at 20 bins of 0.5 m. No specific permits were required for sensor deployments at our study site.

Sensor Performance and Configuration

Sampling rates were limited by power and memory to ~0.1 min^{-1}, resulting in a Nyquist frequency of 3 h^{-1}. Temperature sensors (YSI 6560) were thermistors with an accuracy of ±0.15°C and a resolution of 0.01°C according to manufacturer [31]. Oxygen sensors (YSI 6150 ROX) were optical probes based on the luminescence lifetime method and had ±0.1 $mg{\cdot}L^{-1}$ accuracy and 0.01 $mg{\cdot}L^{-1}$ resolution for the ranges encountered in this study [31]. Sensors for pH (pH/ORP YSI 6565) were bulb-shaped glass electrodes with ±0.2 accuracy and 0.01 resolution [31]. All multi-parametric probes were calibrated periodically in the laboratory using standard procedures and calibration solutions. The long-term station was recovered, cleaned, calibrated, and re-deployed 3 times during the study, and the probes used for the transect were calibrated 7 times. Pre-calibration measurements of commercial standard solutions were conducted to detect any relevant drifts in the sensors. In all cases the difference between the post and pre-calibration values was equal or better than the accuracy of the sensors, and thus it was concluded that no detectable drift was observed. pH sensors were calibrated in the laboratory against NBS standards, and then converted into total pH scale using CO2SYS program [32]. To further account for any undetected drift and to increase the accuracy of pH measures, we calibrated sensor values against water samples collected every six hours over 24 h periods in December and April. pH samples were analyzed using standard spectrophotometric techniques and calibrated against a Tris buffer following [33].

Spectral Analyses

The distribution of variance across frequencies was explored using spectral analyses of the time series. We then compared these distributions among the sites along the spatial transect. Spectral analyses are a set of mathematical tools that transform a data series from the temporal to the frequency domain (e.g. Fig. 2A & C). Results are usually plotted as periodograms graphing the density of variance (or, depending on context, power or energy) against frequency (Fig. 2C). The integral of a density spectrum over all frequencies gives an estimate of the total variance of the signal.

Similarly, cross-spectral analyses of two simultaneous time series can be used to produce periodograms of density of covariance vs. frequency, which integrate to total covariance. We examined the spectra in two ways: 1) by decomposing the variance (or covariance) over different frequency ranges, and 2) by comparing the scaling exponents of the power-law decay of variance against frequency.

The first way, decomposing the variance over different frequency ranges, provides information about the contribution of given frequencies to total variance or covariance. This approach is particularly useful to ascertain the importance of specific periodic processes, such as daily and tidal patterns, that show in a periodogram as peaks in spectral density. The decomposition of variance or covariance is accomplished by integrating the spectra over different frequency ranges. Just as the integration of the whole spectrum gives an estimation of the total variance or covariance of the signal, integrating specific ranges of frequencies within the spectra yields estimates of the contribution to total variance or covariance accounted for by each range (e.g. [34–35]). Based on the most prominent peaks observed in the periodograms (see Fig. 2C), the ranges used were: [0, 1/27), [1/27, 1/21), [1/21, 1/14), [1/14, 1/10), and [1/10, 3) hours^{-1}. These frequency ranges are named throughout the paper as weekly (1d-2w period), daily (21–27 h), daily to half-daily (14–21 h), half-daily (10–14 h), and hourly (1/3–10 h) components respectively. Decomposition was done on raw periodograms obtained from fast Fourier transforms of the detrended time series. Raw spectra were divided by total variance or covariance to obtain the percentages reported here.

Power-law exponents of spectral density functions were used to identify patterns of variability at the highest end of frequencies resolved by our sampling, namely periods from 20 minutes to 8 hours. In this range, no significant peaks and no changes in the slope were observed. Spectral slopes are commonly used in the oceanographic literature to test for underlying physical models. Under isotropic turbulent conditions for example, and over certain frequency ranges, passive tracers such as temperature are expected to show the same $-5/3$ slope as kinetic energy [36]. Spectral densities were estimated using Welch's method on the detrended signals [37], with 4-day segments and 50% overlap. To remove the effect of irradiation and diurnal and semidiurnal tidal constituents, which are overwhelmingly dominant in this system [29], a high-pass Butterworth filter with $1/10\ h^{-1}$ cutoff frequency was applied to each series before performing the spectral analyses. Then, linear least-squares models were fitted to the log-log transformed spectral densities in the range between $1/8\ h^{-1}$ and $3\ h^{-1}$ after resampling the log-log to keep it evenly spaced in the frequency axis and avoid a disproportionate weight of the highest frequencies.

Fast Fourier transformations require uninterrupted time series, which was not the case in some of the deployments. To maximize the length of the usable time series, gaps $\leq$2 hours (i.e. 12 data points) were linearly interpolated. The minimum uninterrupted time series length to perform spectral analyses was set to 7 days.

Statistical Analyses

The focus of this study is a spatial comparison of time series' spectra across 21 locations. However, since only two locations were measured in any two-week period, sampling period could influence comparisons between locations. To assess the relative contribution of sampling period and spatial location to variation between time series' spectra, we used variance components analyses (VCA) on the mean, total variance, decomposed variances, and spectral slopes of the measured environmental variables (pH, dissolved oxygen, and temperature). We fitted a random effects model with deployment date (10 levels) as random factor using REML estimation in the R package *nlme* (nlme version 3.1–105, [38]) in R (version 2.14.1, [39]). We excluded from this analysis the one sampling period when only a single block was measured (S10, Fig. 1B). This model generates two variance components: variance attributed to deployment date and residual variance attributed to location. In some cases, variance components approached zero. When fitted using traditional ANOVA estimation [40], these variance components were negative, indicating a true estimate of zero.

Moving averages and variances were obtained from the long-term time series using a two-week window, which makes them directly comparable to statistics computed from the two-week

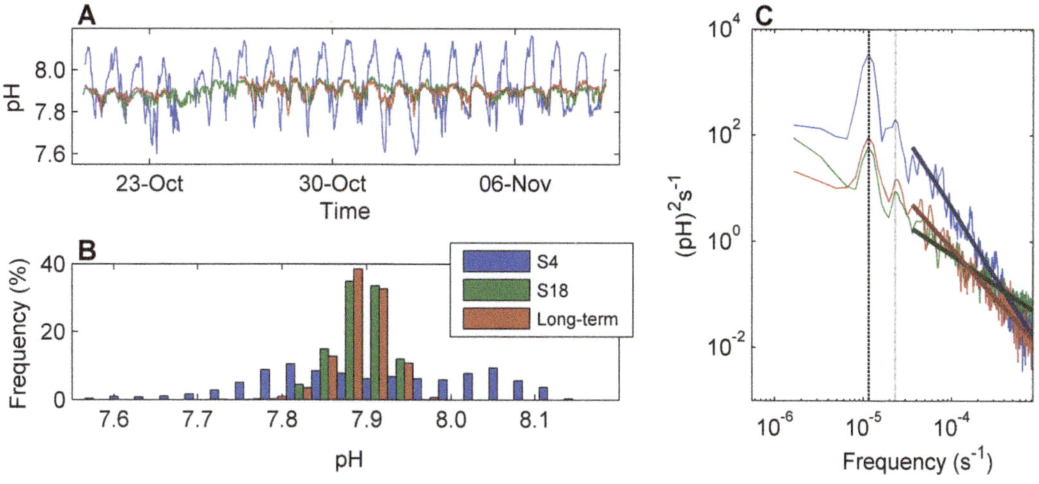

Figure 2. Example of 3 simultaneous time-series of pH. Simultaneous time-series of pH, taken from October 20th to November 9th 2011. The sites shown are: S4 (in blue), located 8.2 m offshore at a depth of 0.4 m, S18 (in green), located 31.6 m offshore at a depth of 3.7 m, and the long-term station (in red), located at 1.7 m depth over a bottom 3 meters deep. (A) Time series plot. (B) Histograms of frequencies of pH values during this period. (C) Power spectral densities. Darker straight lines in (C) depict the best fit power-law models between frequencies $1/8\ h^{-1}$ and $3\ h^{-1}$ obtained using linear least squares method. The vertical dashed and dotted lines in the spectra mark the diurnal and semi-diurnal frequencies respectively.

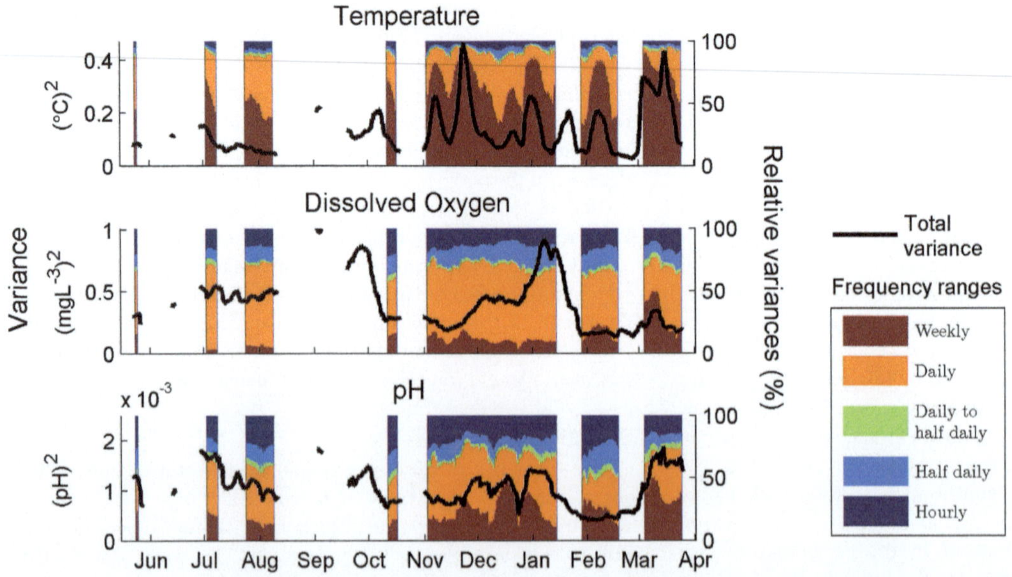

Figure 3. Time series of decomposed variances. Variance decomposition of temperature, dissolved O_2 concentration and pH in the long-term time series station, obtained using a running window of two weeks. Left axes scale absolute variances (black lines), and right axes the relative contribution (in %) of each frequency range to total variance. To perform the spectral analyses the time-series need to be uninterrupted; thus short gaps in the signals resulted in the long gaps observed in the figure.

deployments. Segments with gaps accounting for up to 10% of the total values were accepted to compute moving statistics in figure 3.

Results

Variance vs. Covariance

In order to discriminate spatial gradients in the statistics under study from seasonal and other long-term temporal trends, two approaches are possible. The first is using the long-term station to standardize each one of the two-week long time series along the spatial gradient. In the spectral domain, this may be accomplished using cross-spectral analyses, which partition the covariance between two series (in our case, the long-term times series vs. each one of the sites) across different frequencies. Using covariance unveils spatial patterns in high-frequency temporal fluctuations, but does not provide an easy physical interpretation. An alternative approach is using auto-spectral analyses of each short-term time series. This will yield variances, which are more easily interpreted and compared than covariances. In the present study, the patterns observed using both approaches were similar (Fig. 4), and therefore, for ease of interpretation, the results reported in the present paper correspond exclusively to auto-spectral analyses. However, the auto-spectral approach is valid only if the decomposed variances are stationary in time, or more practically, if variability in time is much lower than in space, as in the present study. Otherwise, cross-spectral analysis between each location and the long-term station must be used despite the more difficult interpretation.

Before using the auto-spectral analyses for our interpretation we tested for stationarity of the parameters derived from such analyses, using the long-term time series. As expected, none of the spectral parameters were constant throughout the period of study (Fig. 3). However, the seasonal variability in the permanent station was considerably smaller than the spatial variability along the transect. For temperature, variance at the station ranged from 0.03 to 0.46°C^2, while variance across the transect ranged from 0.05 to 1.11°C^2. For oxygen concentration, variance ranged from

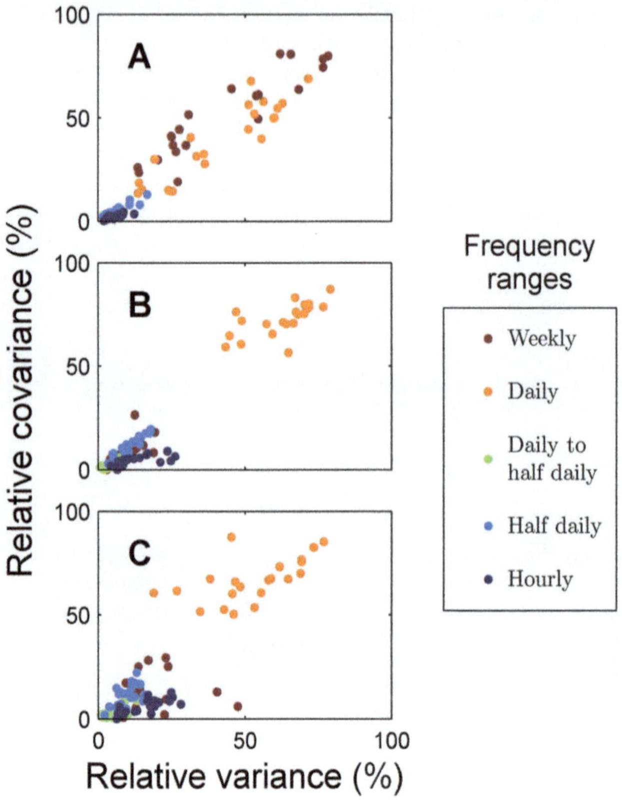

Figure 4. Relative decomposed covariances vs. variances. Relative decomposed covariances derived from cross-spectral analyses between transect sites and long-term station vs. relative decomposed variances derived from auto-spectral analyses of transect sites. (A) Temperature, (B) Dissolved oxygen concentration, (C) pH.

0.13 to 1.00 $mg^2 \cdot L^{-2}$ at the station and from 0.19 to 6.28 $mg^2 \cdot L^{-2}$ along the transect. For pH, variance ranged from 4×10^{-4} to 19×10^{-4} at the station and from 7×10^{-4} to 190×10^{-4} along the transect. Overall, the range of measured variances along the transect were 2.5, 7, and 12 times greater than the range of measured variances from the permanent station for temperature, [O_2], and pH, respectively. In relative terms, the contribution of the different frequencies to total variance was also remarkably constant throughout the year, with one exception: lowest frequency contribution on temperature variance changed considerably (Fig. 3). Spectral slopes also remained comparatively invariant though time in the long-term station, with means ($\pm$standard deviation) indistinguishable from $-5/3$: -1.75 ± 0.22 for temperature, -1.64 ± 0.15 for [O_2], and -1.82 ± 0.24 for pH.

The VCA comparing the relative contribution of spatial location and sampling period to the variance between time series, also showed that seasonal variability was negligible compared to spatial variability (Table 1). In general, the estimates for seasonal variability approached zero for the spectral parameters used in this study; that is, nearly all between-time-series variance was attributed to spatial location. By contrast, time explained most of the variability in mean values of temperature and [O_2]. This is a consequence not only of the seasonal trends, but also of the lack of spatial differences in means.

Spatial Patterns in Time Series Spectra

Overall, daily frequencies contained the most variance (Fig. 5), accounting for 42% of total variance in temperature, 63% in [O_2] and 53% in pH on average. Thus, the relative importance of daily frequencies was higher for biologically-driven parameters (pH and [O_2]), while the relative importance of weekly frequencies was higher for temperature.

Several spatial trends in both total and decomposed variances are observed (Fig. 5). First, total variance of all parameters decreases offshore. Second, the relative contribution of the different frequency ranges changed with the decrease in total variance along the transect. In all 3 parameters, the relative contribution of daily frequencies decreased offshore. By contrast, the relative contribution of weekly frequencies increased with distance to shore, particularly for temperature. Relative contributions of the highest frequencies ($>1/10\ h^{-1}$) also increased offshore, but only in the case of the biologically driven parameters (pH and [O_2]). Finally, distance to shore appeared to be a better linear predictor of changes in all variance components than depth (Figs. 6 and 7).

Spectral analyses provide information about recurrence and persistence of environmental fluctuations (i.e., frequency) vs. their intensity expressed in terms of variance. However, a more natural way to evaluate the contribution of each frequency to overall variability may be achieved by looking at the amplitude of fluctuations. Assuming that values are normally distributed (Fig. 2B), amplitude at a given frequency can be estimated by applying the empirical rule: 95.45% of values lie within 2 standard deviations. Using this approach we estimated, for example, daily oscillations in temperature, [O_2] and pH to be respectively >2.5°C, >8 mgL^{-1} and >0.45 (compare to Fig. 2A and [14], Fig. 2) in the reef flat, and <0.5°C, ~1 mgL^{-1}, and 0.06 in the reef slope (Fig. 6). Spatial gradients at higher frequencies are less steep, but amplitudes may still be biologically significant.

Finally, spectral slopes also increased linearly with distance to shore (Fig. 7). Close to shore spectral slopes approached -3, whereas offshore the slopes fell between $-5/3$ and -1. Spectral slopes were negatively related to the daily component of variance, so that the steepest slopes occurred in those sites with highest

Table 1. Variance components analyses (VCA).

		% of variability explained	
Parameter		**Temporal**	**Spatial (Residual)**
Temperature	Mean	92.1	7.9
	σ^2_{Total}	0.0	100.0
	σ^2_{Weekly}	0.1	99.9
	σ^2_{Daily}	0.0	100.0
	$\sigma^2_{Daily\text{-}to\text{-}half\text{-}daily}$	1.8	98.2
	$\sigma^2_{Half\text{-}daily}$	0.0	100.0
	σ^2_{Hourly}	2.1	97.9
	Slope	0.0	100.0
Oxygen	Mean	64.8	35.2
	σ^2_{Total}	0.0	100.0
	σ^2_{Weekly}	0.0	100.0
	σ^2_{Daily}	0.0	100.0
	$\sigma^2_{Daily\text{-}to\text{-}half\text{-}daily}$	2.9	97.1
	$\sigma^2_{Half\text{-}daily}$	2.3	97.7
	σ^2_{Hourly}	0.0	100.0
	Slope	0.0	100.0
pH	Mean	0.0	100.0
	σ^2_{Total}	0.0	100.0
	σ^2_{Weekly}	0.1	99.9
	σ^2_{Daily}	0.0	100.0
	$\sigma^2_{Daily\text{-}to\text{-}half\text{-}daily}$	2.7	97.3
	$\sigma^2_{Half\text{-}daily}$	3.2	96.7
	σ^2_{Hourly}	0.0	100.0
	Slope	0.0	100.0

Results of the variance components analyses (VCA) performed on all the statistics derived from the auto-spectral analyses of the transect stations time-series. Deployment date was fit as a random effect (temporal) and the residual variance is attributed to location (spatial). The VCs were fitted using REML, and many of the VCs approached zero. When fitted using traditional ANOVA estimation, these VCs were negative, indicating a true estimate of zero.

diurnal variance. This could result from the peaks in daily and half-daily frequencies leaking down to highest frequencies. Note, however, that slopes were estimated after applying a high-pass filter that effectively removed these peaks.

Discussion

We used two complementary spectral analyses to assess the distribution of temporal variability along a cross-reef transect: variance decomposition at different frequency ranges, and fitting of spectral slopes at the high-frequency end of the spectra. The first assesses the contribution of the most important frequencies to variability; the second is a common way to characterize patchiness in the distribution of a scalar (e.g. [41]) and is better suited to explore the highest frequencies where no individual peaks stand out.

Results show dramatic changes in the regime of temporal fluctuations over spatial scales of meters (Figs. 5, 6 and 7). Even when average values are similar along the transect, as is the case with temperature and [O_2], individuals on the reef flat experience a highly variable environment, and those on the reef slope a relatively constant one. Here, we first discuss the suitability of the

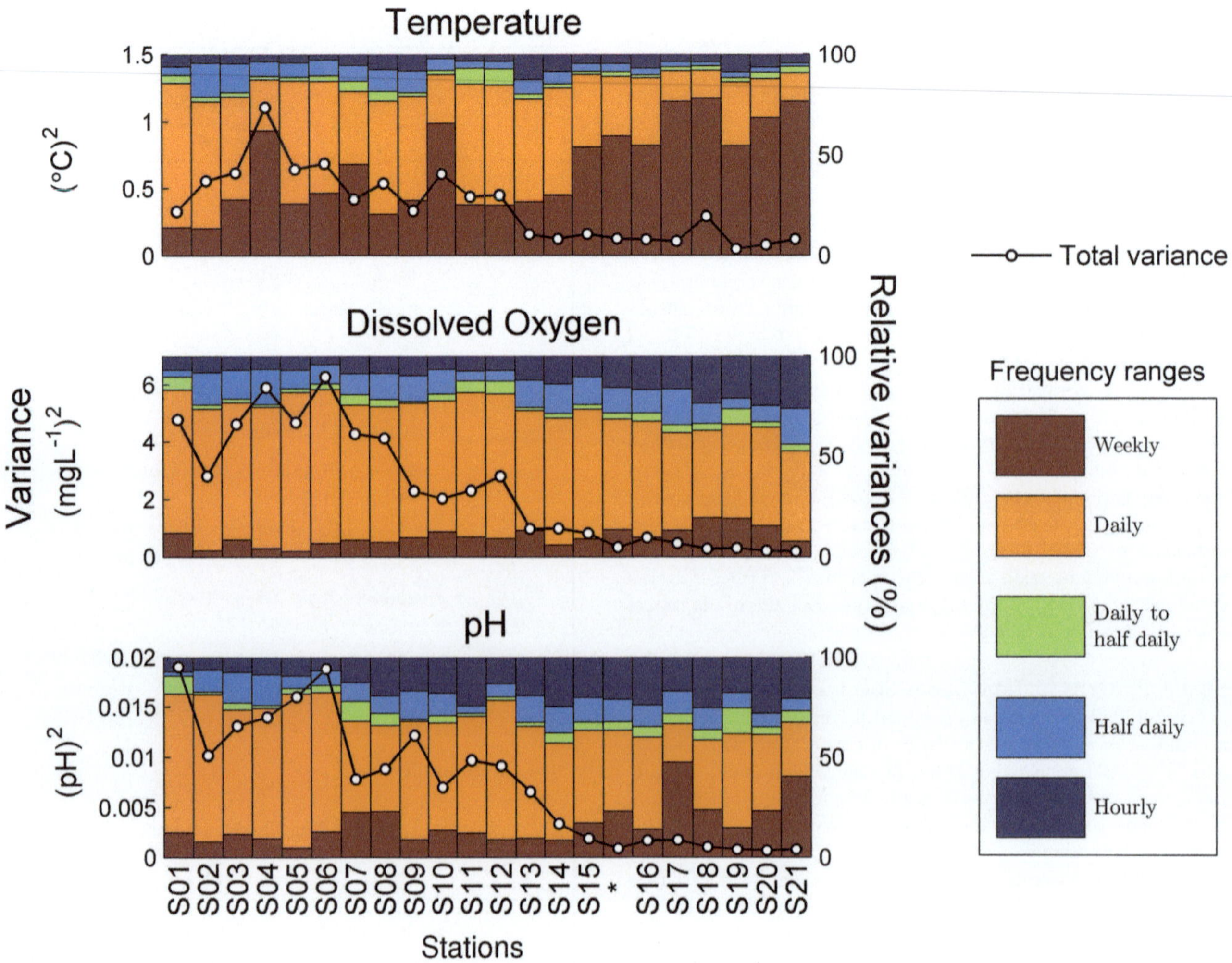

Figure 5. Variance decomposition along the transect. Variance decomposition of temperature, dissolved O_2 concentration and pH for each one of the deployments along the inshore/offshore transect. Left axes scale absolute variances (black lines), and right axes the relative contribution (in %) of each frequency range to total variance. Sites are ordered from shallowest and closest to shore (S1) to deepest and most offshore (S21). The asterisk marks the average values from the long-term water column station, positioned within the transect according to its depth.

spatial and temporal sampling design and its potential application in other systems; second, we address what physical and biological processes may be driving the observed patterns; and finally, we discuss the ecological relevance of these results in the light of the ongoing discussions about the responses of organisms and ecosystems to natural variability and climate change.

Sampling Design

Our observational design, combining mobile short-term deployments of sensors and a long-term stationary suite of sensors, made use of limited resources to provide information on the spectral distribution of environmental variance across a spatial transect. Auto-spectral analyses were adequate tools for characterizing variance distribution in this design because variability along the transect was much higher than the long-term temporal dynamics. However, auto-spectra would not be appropriate in systems where differences between sampling periods are pronounced (e.g., at mid and high latitudes where there is strong seasonality). In such cases, an alternative approach would be to normalize each of the two-week time series on the transect to its corresponding period in the long-term station, using cross-spectral techniques. This normalization allows direct comparison of the spectra of fluctuations across transect locations. In our dataset, the spatial patterns obtained using the cross-spectral approach were very similar to the results from the auto-spectral approach we present (see, for example, Fig. 4). This similarity is additional evidence that temporal changes in the regime of fluctuations are small when compared to spatial changes. We chose to present results of auto-spectra instead of cross-spectra, because they are more naturally interpreted, allowing the calculation of the range of values experienced at each location (Fig. 6). However, the use of cross-spectral analyses between a permanent station and a grid of temporary stations may prove valuable in systems where differences between sampling periods are more pronounced than differences across space.

The analyses conducted here assume that the observed spatial patterns in variance are stationary, that is, that ratios between variances at the long-term station and those at each site along the transect remain constant with time. It is a reasonable assumption, given that measurements periods were randomly spread over half a year, and observed spatial patterns remained clear. Again, caution should be exerted when applying this kind of sampling design in systems with strong seasonal and long-term variability or

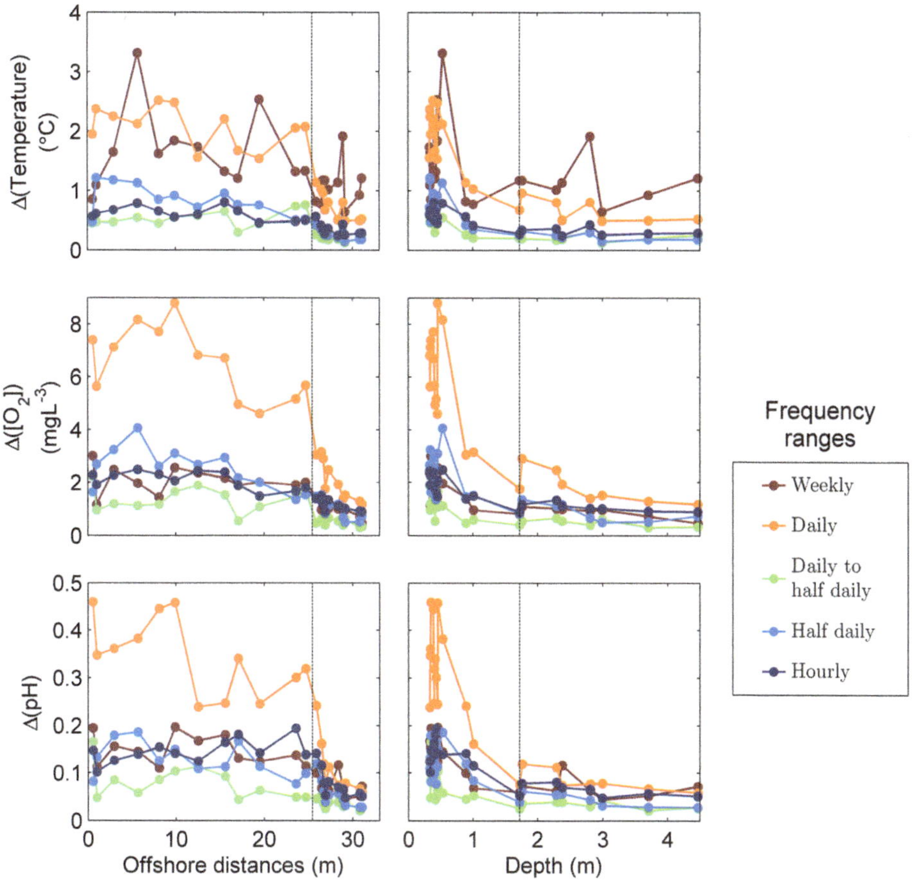

Figure 6. Amplitudes of oscillations by frequency ranges. Amplitudes A are calculated from decomposed variances, assuming that all parameters have a normal distribution, as $A = 4\cdot\sqrt{\sigma 2}$. Vertical dash lines mark the position of the long-term time series station.

across large spatial scales. In such cases, the use of cross-spectral analyses may prove a better choice.

Spatial Patterns and their Causes

Spatial and inter-parameter differences occur at all frequencies (Fig. 5). However, a clear pattern is common: all variance components decrease offshore, and spectral slopes get less steep. All statistics are related to both distance to shore and depth (Figs. 6 and 7). Although depth and distance to shore are obviously correlated, the mechanisms by which they may affect the distribution of temporal variability are different.

Depth plays and important role because energy and matter are much more quickly transferred through the shallow water column (<1 m) over the reef flat than through the deeper water column (2–5 m) of the reef slope. This leads to rapid and intense physical, chemical, and biological responses to high-frequency atmospheric forcing on the reef flat. Most physical and biological processes dominant in coastal systems, such as light, tides, waves, or turbulence, are strongly influenced by water column height. Not surprisingly, the relative importance of daily variance decreases offshore whereas that of weekly variance increases (Fig. 5).

Distance to shore reflects horizontal mixing between the reef flat and the bay waters. Limited horizontal mixing may result in long water residence times in the reef flat, depending mostly on tidal oscillations and wind [42]. The fact that the distribution statistics are linearly related to distance to shore, and exponentially related to depth (Figs. 6 and 7) suggests that limited mixing between the reef flat and the reef slope may best explain the observed patterns in variability.

In any case, the combined effects of depth and distance to shore result in clear differences in fluctuation regime between reef flat and reef slope: 1) processes generating variability are most intense and/or amplified in the reef flat, and 2) there is limited mixing between reef flat and slope. The pattern and scale of temporal variability reflect the underlying processes.

At the weekly scale, the longest temporal scale in our sampling, variability is most likely related to weather (e.g. [26]). This is supported by several observations. First, temperature, the most physically-driven and seasonal parameter, shows the largest relative contributions at this scale (Figs. 3 and 5). Second, the time series of weekly variance (Fig. 3) is related to the meteorological record. For example, the highest peak in weekly variance occurs for all parameters in March 2012, during the strongest rainstorm of the study period; this pattern is signaled by a drop in salinity (Fig. 1B). And finally the weekly component does not show clear spatial patterns along the transect, especially for temperature (Fig. 6). For example, a local peak in weekly variance observed in S10 coincides with the rainstorm in March 2012. This lack of clear patterns is to be expected whenever the low-frequency dynamics are confounding the spatial pattern. Kāneʻohe Bay dynamics are well known to be largely driven by rainstorms [26], [43].

Daily frequencies contributed most to total variance for all parameters (Fig. 5), and decreased offshore, as has been previously

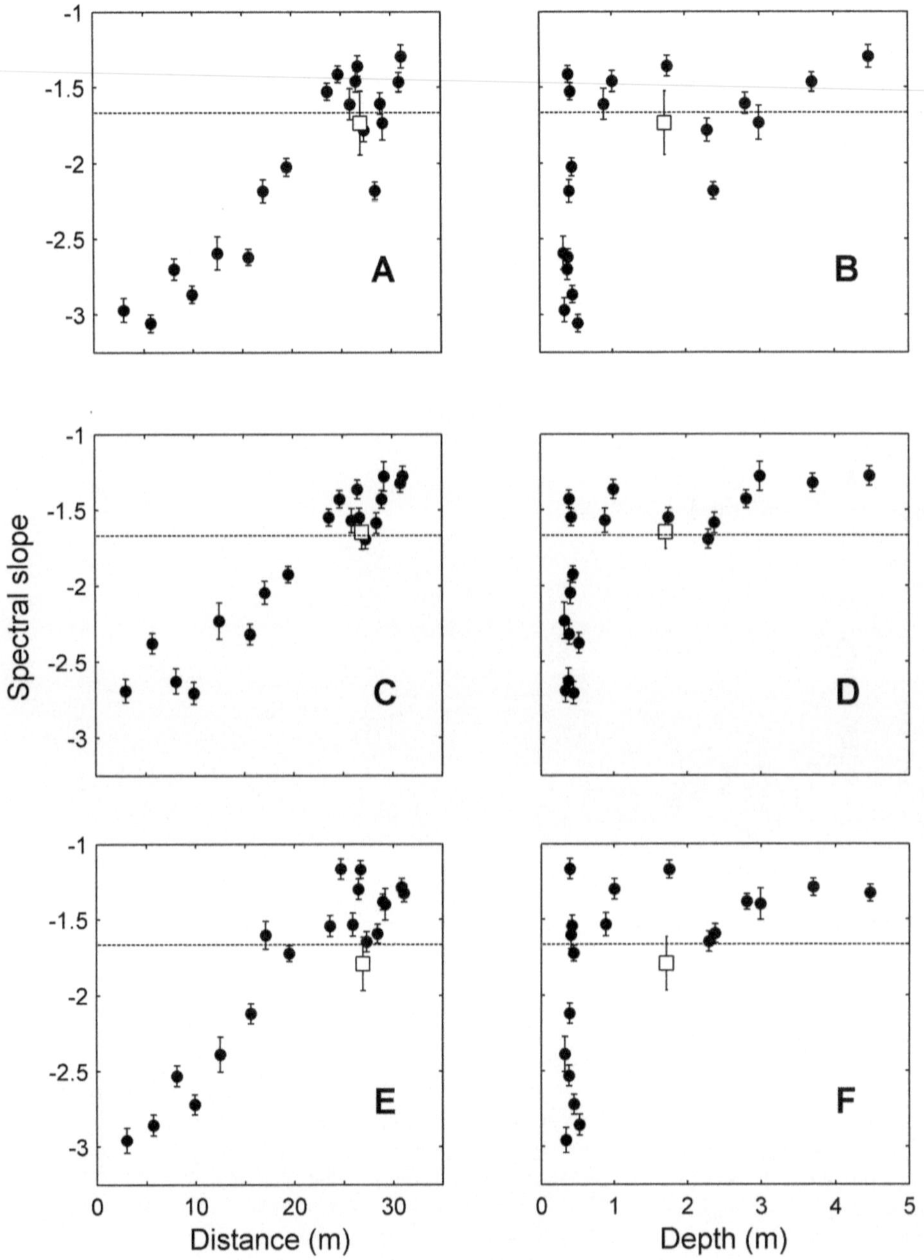

Figure 7. Spectral slopes vs. distance from shore and depth. Spectral slopes of temperature (A & B), dissolved O_2 (C & D) and pH (E & F) vs. distance from shore (left column) and depth (right column) of each station along the transect. Dots with error bars represent the fitted slopes with 95% confidence intervals. Empty squares with error bars show the mean ± standard deviation of all the spectral slopes in the long-term station estimated for each period of deployment. Dashed horizontal lines mark the −5/3 typical of 3D isotropic turbulence spectra.

described [14]. There are several interdependent physical phenomena with intense diurnal periodicity, including solar radiation, tides, waves and winds. The interplay of these phenomena has been shown to drive daily temperature oscillations both in the water column and inside corals [42], [44] potentially leading to local coral bleaching conditions. On the other hand, daily cycles of net photosynthesis/respiration, combined with the depth and limited horizontal mixing outlined above, are likely to drive the variance of daily frequencies in [O_2] and pH (e.g. [12]). Although the biomass of corals and algae is higher on the reef slope than the reef flat, the limited mixing on the reef flat may amplify the effects of the photosynthesis/respiration cycle observed in the water column, contributing to the stronger diurnal cycle on the reef flat.

Finally, at the highest frequencies of our sampling, turbulence is likely to be the most important factor contributing to variability because it is the main mechanism transferring variance from large to small scales and will cause mixing of scalars between reef flat and slope. A scalar under isotropic turbulence is expected to show a spectral slope of $-5/3$ (e.g. [5], [41], [45–46]). Thus, the average $-5/3$ spectral slope at the long-term station (Fig. 5) suggests that turbulence may be important at these scales. However, it is unclear whether the sampling rate (0.1 min^{-1}) was fast enough to resolve any part of the turbulent inertial subrange [41]. It was certainly

Table 2. Amplitude of fluctuations at different frequency ranges.

	Climate change	Seasonal signal	Weekly	Daily	Daily to half-daily	Half-daily	Hourly
ΔTemperature (°C)	0.74±0.18	~6	0.6–3.3	0.5–2.5	0.1–0.7	0.1–1.2	0.2–0.8
$\Delta[O_2]$ (mgL^{-1})	-	~1.5	0.5–3.0	1.2–8.8	0.3–2.2	0.5–4.1	0.9–2.5
ΔpH (units)	0.1	~0.19	0.04–0.20	0.06–0.46	0.02–0.17	0.03–0.19	0.05–0.19

Amplitude of oscillations at particular frequency ranges derived from the partition of variances performed with spectral analyses, assuming normal distributions and applying the empirical rule. The amplitude of climate change for temperature corresponds to the estimated changes between 1906 and 2005 [51]. The climate change value for pH corresponds to the decrease in surface pH between 1750 and 1994 derived from estimated uptake of anthropogenic CO_2 by the ocean [52]. Seasonal signals are derived from the long-term time series.

too slow to resolve the smallest scales, usually ranging between milliseconds and minutes [47]. But it could still be faster than the overturning rates of the largest eddies, and thus spectral analyses could overlap with the inertial subrange. To elucidate this, we assumed that the size of the largest turbulent eddies was limited by the height of the water column, and applied Taylor's frozen hypothesis on the horizontal velocities from the ADCP [48]. This heuristic approach yields scales between a few minutes to about 5 hours in the reef slope, and between tens of seconds to less than 5 minutes in the reef flat. Thus, some overlap with the inertial subrange may occur in the reef slope, but not in the flat.

Spectral slopes offshore lie between $-5/3$ and -1. These values are indicative of turbulent diffusivity driving changes at these scales. However, slopes gradually decreased onshore, to reach values of -3 at the sites nearest to shore. This indicates that physical or biological processes other than water motion are driving variability [5]. Spectral slopes are not by themselves good diagnostics of the mechanics of a system [41], especially in complex, highly energetic environments like the reef flat, where many phenomena, including breaking waves, tides and currents, are interacting. Nonetheless, they provide an easy way to parameterize environmental variability as experienced by organisms. Water motion at these scales, whether turbulent or not, is critical for organisms because it controls the transport and exchange of nutrients and particles not only horizontally but also between water column and bottom [49]. The acute patterns in the spectral slopes between reef flat and reef slope are thus symptomatic of fundamental differences in the environment organisms are sensing, and deserve to be studied in connection to physiological responses.

Ecological Relevance

We have shown robust spatial patterns in the regime of environmental fluctuations across a reef flat at relatively high frequencies. But, to what extent do changes occurring at these frequencies impact the life of reef organisms, and what role do they have in defining their spatial distribution? Such questions are obviously out of the scope of the present study, which solely deals with environmental variability. However, our approach may be useful in defining the limits of natural variability that individuals are experiencing, thus allowing for hypothesis-driven field studies, and providing a basis for better predictions of biological responses to change.

The amplitudes of oscillations in temperature, pH and oxygen observed at short timescales on this small transect are comparable to long-term trends (Table 2). At the long-term station (Fig. 1B), seasonal oscillations in temperature (~6°C) and pH (~0.19) are comparable to daily oscillations on the reef flat. The increase in temperature (0.74°C±0.18°C) and the decrease in pH (0.1) attributed to climate change over the past centuries [50–51], are comparable to changes we observe at hourly frequencies, for example. These hourly variations of pH and temperature are within the range known to impact coral bleaching and growth [52–53]. Of course, perturbations that are too short and/or infrequent may not trigger a response in organisms [54] or have a lasting effect. But natural variability provides the context for ongoing discussion about acclimatization and response of organisms and ecosystems to climate change [15].

For example, corals can acclimatize to changes in fluctuation frequency [21], [55–56], and exposure to high-frequency low-intensity fluctuations may increase tolerance, through acclimatization, to large environmental events [21]. On the other hand, acclimatization of organisms to long-term changes can be diminished by exposure to high-frequency intense fluctuations [57]. Further, our own data shows that these spatial gradients in pH variance are correlated with bioerosion/accretion patterns (Silbiger et al., unpublished data).

Spatial patterns in high-frequency environmental variability could also translate into different distributions of biological organisms and processes. Temporal and spatial variances are correlated across scales (e.g. [58]). Frequent, short, and weak perturbations tend to be local, whereas intense perturbations usually impact large areas for long periods. Thus, temporal variance at high frequencies can potentially be very different in two sites that are spatially very close. Immediate responses to high-frequency changes in the environment may be small, hardly detectable shifts in organism function. However, in the long run, these fluctuations may affect species distributions and shape biological spatial heterogeneity and resilience to large environmental change. Thus, it is likely that the differences in exposure to environmental fluctuation across the reef are driving patterns of genetic and community structure. New molecular tools that allow measuring individual responses as changes in gene and protein expression or enzyme activity (e.g. [8], [55]) offer a possibility to explore the question of how these responses can translate into long-term spatial changes in community structure.

These are important issues to address as global climate change and ocean acidification alter the pattern of environmental fluctuations. Experimental works have mostly addressed how changes in baselines may affect individuals and populations, but there is an increased awareness of the importance of variance at organism scales (e.g. [13], [21], [56], [59–63]). Under climate change scenarios we may expect dispersion statistics (variances, ranges or extremes) to be altered by: 1) an increase in the frequency of intense meteorological events [11], and 2) a multiplicative effect on variances (i.e. variance correlates with mean, as opposed to independent variance typical of additive models). Multiplicative changes in mean temperature and pH are more likely to lead to ranges of variability outside tolerance limits for many of these coral reef organisms. Biologically reactive parameters with nonlinear responses to environmental changes can be hypothesized to show multiplicative variances [64].

Conclusions

The sampling scheme that we used allowed us to explore the spatial distribution of temporal variability making use of limited resources. Spectral analyses of time series reveal patterns in the regime of environmental fluctuations at all frequencies resolved in the study that occur over very short distances making it a good model to explore the acclimatization and adaptation of organisms to high-frequency variability.

Furthermore, the spectral analyses allow a quantification of the amplitude of fluctuations occurring at specific scales. The results show that the fluctuations at relatively high frequencies and also at weekly scales are comparable in magnitude to diurnal, seasonal and longer trends (Table 2). This pattern is likely particular to low latitudes with mild seasonality, but it stresses the importance of looking at the whole spectrum of natural variability rather than focusing on specific frequencies.

In general, spectral analyses cannot by themselves distinguish the underlying mechanisms that generate environmental variance. They can, however, reveal and quantify patterns in variance that organisms in different places will experience. Whether and how organisms may sense and respond to these different regimes of fluctuations, and whether this can translate into long-term spatial changes in community structure, can now be addressed *in situ* with new molecular tools (e.g. [9], [18], [56]). The analyses presented here show that coastal systems, and coral reefs in particular, may be good systems to investigate *in situ* responses to high frequency fluctuations, and provide clues about what scales are worth exploring.

Acknowledgments

We are very grateful to the Point lab at HIMB for providing space and infrastructure to cable the instruments.

Author Contributions

Conceived and designed the experiments: OG NJS MJD FIMT. Performed the experiments: OG NJS. Analyzed the data: OG NJS MJD. Wrote the paper: OG NJS MJD FIMT.

References

1. Steele JH (1985) A comparison of terrestrial and marine ecological systems. Nature 313: 355–358. doi:10.1038/313355a0.
2. Prairie JC, Sutherland KR, Nickols KJ, Kaltenberg AM (2012) Biophysical interactions in the plankton: A cross-scale review. Limnol Oceanogr Fluids Environ 2: 121–145. doi:10.1215/21573689-1964713.
3. Benedetti-Cecchi L (2003) The importance of the variance around the mean effect size of ecological processes. Ecology 84: 2335–2346. doi:10.1890/02-8011.
4. Wing SR, Patterson MR (1993) Effects of wave-induced lightflecks in the intertidal zone on photosynthesis in the macroalgae *Postelsia palmaeformis* and *Hedophyllum sessile* (Phaeophyceae). Mar Biol 116: 519–525. doi:10.1007/BF00350069.
5. Powell TM, Okubo A (1994) Turbulence, Diffusion and Patchiness in the Sea. Philos Trans R Soc Lond B Biol Sci 343: 11–18. doi:10.1098/rstb.1994.0002.
6. Falciatore A, Ribera d' Alcalà M, Croot P, Bowler C (2000) Perception of Environmental Signals by a Marine Diatom. Science 288: 2363–2366. doi:10.1126/science.288.5475.2363.
7. Litchman E (2000) Growth rates of phytoplankton under fluctuating light. Freshw Biol 44: 223–235. doi:10.1046/j.1365-2427.2000.00559.x.
8. Song B, Ward BB (2007) Molecular Cloning and Characterization of High-Affinity Nitrate Transporters in Marine Phytoplankton. J Phycol 43: 542–552. doi:10.1111/j.1529-8817.2007.00352.x.
9. Koehl MAR, Hadfield MG (2010) Hydrodynamics of Larval Settlement from a Larva's Point of View. Integr Comp Biol 50: 539–551. doi:10.1093/icb/icq101.
10. Helmuth B, Broitman BR, Yamane L, Gilman SE, Mach K, et al. (2010) Organismal climatology: analyzing environmental variability at scales relevant to physiological stress. J Exp Biol 213: 995–1003. doi:10.1242/jeb.038463.
11. IPCC (2007) Climate change 2007: the physical science basis. Contribution of working group I to the fourth assessment report of the intergovernmental panel on climate change. Cambridge University Press, Cambridge, United Kingdom and New York, NY, USA. 996 p.
12. Wootton JT, Pfister CA, Forester JD (2008) Dynamic patterns and ecological impacts of declining ocean pH in a high-resolution multi-year dataset. Proc Natl Acad Sci 105: 18848–18853. doi:10.1073/pnas.0810079105.
13. Hofmann GE, Smith JE, Johnson KS, Send U, Levin LA, et al. (2011) High-Frequency Dynamics of Ocean pH: A Multi-Ecosystem Comparison. PLoS ONE 6: e28983. doi:10.1371/journal.pone.0028983.
14. Price NN, Martz TR, Brainard RE, Smith JE (2012) Diel Variability in Seawater pH Relates to Calcification and Benthic Community Structure on Coral Reefs. PLoS ONE 7: e43843. doi:10.1371/journal.pone.0043843.
15. Duarte CM, Hendriks IE, Moore TS, Olsen YS, Steckbauer A, et al. (2013) Is Ocean Acidification an Open-Ocean Syndrome? Understanding Anthropogenic Impacts on Seawater pH. Estuaries Coasts 36: 221–236. doi:10.1007/s12237-013-9594-3.
16. Middelboe AL, Hansen PJ (2007) High pH in shallow-water macroalgal habitats. Mar Ecol Prog Ser 338: 107–117. doi:10.3354/meps338107.
17. Provoost P, van Heuven S, Soetaert K, Laane RWPM, Middelburg JJ (2010) Seasonal and long-term changes in pH in the Dutch coastal zone. Biogeosciences 7: 3869–3878. doi:10.5194/bg-7-3869-2010.
18. Kelly MW, Hofmann GE (2013) Adaptation and the physiology of ocean acidification. Funct Ecol 27: 980–990. doi:10.1111/j.1365-2435.2012.02061.x.
19. Hoegh-Guldberg O, Mumby PJ, Hooten AJ, Steneck RS, Greenfield P, et al. (2007) Coral Reefs Under Rapid Climate Change and Ocean Acidification. Science 318: 1737–1742. doi:10.1126/science.1152509.
20. Jokiel PL, Rodgers KS, Kuffner IB, Andersson AJ, Cox EF, et al. (2008) Ocean acidification and calcifying reef organisms: a mesocosm investigation. Coral Reefs 27: 473–483. doi:10.1007/s00338-008-0380-9.
21. Oliver TA, Palumbi SR (2011) Do fluctuating temperature environments elevate coral thermal tolerance? Coral Reefs 30: 429–440. doi:10.1007/s00338-011-0721-y.
22. Edmunds PJ, Gates RD (2008) Acclimatization in tropical reef corals. Mar Ecol Prog Ser 361: 307–310. doi:10.3354/meps07556.
23. Sanford E, Kelly MW (2011) Local Adaptation in Marine Invertebrates. Annu Rev Mar Sci 3: 509–535. doi:10.1146/annurev-marine-120709-142756.
24. Smith SV, Kimmerer WJ, Laws EA, Brock RE, Walsh TW (1981) Kaneohe Bay Sewage Diversion Experiment: Perspectives on Ecosystem Responses to Nutritional Perturbation. Pac Sci 35: 279–395.
25. Lowe RJ, Falter JL, Monismith SG, Atkinson MJ (2009) A numerical study of circulation in a coastal reef-lagoon system. J Geophys Res Oceans 114: C06022. doi:10.1029/2008JC005081.
26. De Carlo EH, Hoover DJ, Young CW, Hoover RS, Mackenzie FT (2007) Impact of storm runoff from tropical watersheds on coastal water quality and productivity. Appl Geochem 22: 1777–1797. doi:10.1016/j.apgeochem.2007.03.034.
27. Drupp P, Carlo EHD, Mackenzie FT, Bienfang P, Sabine CL (2011) Nutrient Inputs, Phytoplankton Response, and CO_2 Variations in a Semi-Enclosed Subtropical Embayment, Kaneohe Bay, Hawaii. Aquat Geochem 17: 473–498. doi:10.1007/s10498-010-9115-y.
28. Bathen KH (1968) A Descriptive Study of the Physical Oceanography of Kaneohe Bay, Oahu, Hawaii. Hawai'i Institute of Marine Biology Tech Rept 14. Available: http://scholarspace.manoa.hawaii.edu/handle/10125/15276. Accessed 2013 Dec 5.
29. Ostrander CE, McManus MA, DeCarlo EH, Mackenzie FT (2008) Temporal and Spatial Variability of Freshwater Plumes in a Semienclosed Estuarine-Bay System. Estuaries Coasts 31: 192–203. doi:10.1007/s12237-007-9001-z.
30. Lowe RJ, Falter JL, Monismith SG, Atkinson MJ (2009) Wave-Driven Circulation of a Coastal Reef–Lagoon System. J Phys Oceanogr 39: 873–893. doi:10.1175/2008JPO3958.1.
31. YSI Incorporated (2011) 6-Series Multiparameter Water Quality Sondes User Manual. Available: http://www.ysi.com/media/pdfs/069300-YSI-6-Series-Manual-RevH.pdf. Accessed 2013 Dec 5.
32. Lewis E, Wallace D (1998) Program developed for CO_2 system calculations. Carbon Dioxide Information Analysis Center, managed by Lockheed Martin Energy Research Corporation for the US Department of Energy. Available: http://cdiac.esd.ornl.gov/oceans/co2rprtnbk. Accessed 2013 Dec 6.
33. Dickson AG, Sabine CL, Christian JR (2007) Guide to best practices for ocean CO_2 measurements. PICES Special Publication 3, 191 pp. Available: http://cdiac.ornl.gov/oceans/Handbook_2007.html. Accessed 2013 Dec 6.
34. Chatfield C (2003) The Analysis of Time Series: An Introduction, Sixth Edition. CRC Press. 345 p.
35. Denny MW, Helmuth B, Leonard GH, Harley CDG, Hunt LJH, et al. (2004) Quantifying scale in ecology: Lessons from a wave-swept shore. Ecol Monogr 74: 513–532. doi:10.1890/03-4043.

36. Batchelor GK (1959) Small-scale variation of convected quantities like temperature in turbulent fluid Part 1. General discussion and the case of small conductivity. J Fluid Mech 5: 113–133. doi:10.1017/S002211205900009X.
37. Welch PD (1967) The use of fast Fourier transform for the estimation of power spectra: A method based on time averaging over short, modified periodograms. IEEE Trans Audio Electroacoustics 15: 70–73. doi:10.1109/TAU.1967.1161901.
38. Pinheiro J, Bates D, DebRoy S, Sarkar D, the R Development Core Team (2013) nlme: Linear and Nonlinear Mixed Effects Models. R package version 3.1–108.
39. Team RDC (2005) R: A language and environment for statistical computing. ISBN 3-900051-07-0. R Foundation for Statistical Computing. Vienna, Austria, 2013. Available: http://www. R-project. Accessed 2013 Dec 6.
40. Searle SR, Casella G, McCulloch CE (1992) Variance Components. John Wiley & Sons. 501 p.
41. Franks PJS (2005) Plankton patchiness, turbulent transport and spatial spectra. Mar Ecol Prog Ser 294: 295–309. doi:10.3354/meps294295.
42. McCabe RM, Estrade P, Middleton JH, Melville WK, Roughan M, et al. (2010) Temperature variability in a shallow, tidally isolated coral reef lagoon. J Geophys Res Oceans 115: C12011. doi:10.1029/2009JC006023.
43. Ringuet S, Mackenzie FT (2005) Controls on nutrient and phytoplankton dynamics during normal flow and storm runoff conditions, southern Kaneohe Bay, Hawaii. Estuaries 28: 327–337. doi:10.1007/BF02693916.
44. Jimenez IM, Larkum AWD, Ralph PJ, Kühl M (2012) In situ thermal dynamics of shallow water corals is affected by tidal patterns and irradiance. Mar Biol 159: 1773–1782. doi:10.1007/s00227-012-1968-8.
45. Platt T (1972) Local phytoplankton abundance and turbulence. Deep Sea Res Oceanogr Abstr 19: 183–187. doi:10.1016/0011-7471(72)90029-0.
46. Yamazaki H, Mitchell JG, Seuront L, Wolk F, Li H (2006) Phytoplankton microstructure in fully developed oceanic turbulence. Geophys Res Lett 33: L01603. doi:10.1029/2005GL024103.
47. Peters F, Marrasé C (2000) Effects of turbulence on plankton: an overview of experimental evidence and some theoretical considerations. MEPS 205: 291–306. doi:10.3354/meps205291.
48. Lumley JL, Terray EA (1983) Kinematics of Turbulence Convected by a Random Wave Field. J Phys Oceanogr 13: 2000–2007. doi:10.1175/1520-0485(1983)013<2000:KOTCBA>2.0.CO;2.
49. Boudreau BP, Jørgensen BB (2001) The benthic boundary layer: transport processes and biogeochemistry. Oxford University Press. 404 p.
50. Trenberth KE, Jones PD, Ambenje P, Bojariu R, Easterling D, et al. (2007) Climate change 2007: the physical science basis. In: Solomon S, Qin D, Manning M, Chen Z, Marquis M, et al., editors. Climate Change 2007: The Physical Science Basis. Contribution of Working Group I to the Fourth Assessment Report of the Intergovernmental Panel on Climate Change. Cambridge University Press, Cambridge, United Kingdom and New York, NY, USA. 235–336.
51. Bindoff NL, Willebrand J, Artale V, Cazenave A, Gregory J, et al. (2007) Observations: Oceanic Climate Change and Sea Level. In: Solomon S, Qin D, Manning M, Chen Z, Marquis M, et al., editors. Climate Change 2007: The Physical Science Basis. Contribution of Working Group I to the Fourth Assessment Report of the Intergovernmental Panel on Climate Change. Cambridge University Press, Cambridge, United Kingdom and New York, NY, USA. 235–336.
52. Dove S, Ortiz JC, Enríquez S, Fine M, Fisher P, et al. (2006) Response of holosymbiont pigments from the scleractinian coral *Montipora monasteriata* to short-term heat stress. Limnol Oceanogr 51: 1149–1158. doi:10.4319/lo.2006.51.2.1149.
53. Kroeker KJ, Kordas RL, Crim RN, Singh GG (2010) Meta-analysis reveals negative yet variable effects of ocean acidification on marine organisms. Ecol Lett 13: 1419–1434. doi:10.1111/j.1461-0248.2010.01518.x.
54. Shaw EC, Munday PL, McNeil BI (2013) The role of CO_2 variability and exposure time for biological impacts of ocean acidification. Geophys Res Lett 40: 4685–4688. doi:10.1002/grl.50883.
55. Barshis DJ, Stillman JH, Gates RD, Toonen RJ, Smith LW, et al. (2010) Protein expression and genetic structure of the coral *Porites lobata* in an environmentally extreme Samoan back reef: does host genotype limit phenotypic plasticity? Mol Ecol 19: 1705–1720. doi:10.1111/j.1365-294X.2010.04574.x.
56. Mayfield AB, Chan P-H, Putnam HM, Chen C-S, Fan T-Y (2012) The effects of a variable temperature regime on the physiology of the reef-building coral *Seriatopora hystrix*: results from a laboratory-based reciprocal transplant. J Exp Biol 215: 4183–4195. doi:10.1242/jeb.071688.
57. Niehaus AC, Angilletta MJ, Sears MW, Franklin CE, Wilson RS (2012) Predicting the physiological performance of ectotherms in fluctuating thermal environments. J Exp Biol 215: 694–701. doi:10.1242/jeb.058032.
58. Dickey TD (2003) Emerging ocean observations for interdisciplinary data assimilation systems. J Mar Syst 40–41: 5–48. doi:10.1016/S0924-7963(03)00011-3.
59. Miner BG, Vonesh JR (2004) Effects of fine grain environmental variability on morphological plasticity. Ecol Lett 7: 794–801. doi:10.1111/j.1461-0248.2004.00637.x.
60. Gagliano M, McCormick MI, Moore JA, Depczynski M (2010) The basics of acidification: baseline variability of pH on Australian coral reefs. Mar Biol 157: 1849–1856. doi:10.1007/s00227-010-1456-y.
61. Gorospe KD, Karl SA (2010) Small-Scale Spatial Analysis of In Situ Sea Temperature throughout a Single Coral Patch Reef. J Mar Biol 2011: 719580. doi:10.1155/2011/719580.
62. Joint I, Doney SC, Karl DM (2011) Will ocean acidification affect marine microbes? ISME J 5: 1–7. doi:10.1038/ismej.2010.79.
63. Pincebourde S, Sanford E, Casas J, Helmuth B (2012) Temporal coincidence of environmental stress events modulates predation rates. Ecol Lett 15: 680–688. doi:10.1111/j.1461-0248.2012.01785.x.
64. Shaw EC, McNeil BI, Tilbrook B, Matear R, Bates ML (2013) Anthropogenic changes to seawater buffer capacity combined with natural reef metabolism induce extreme future coral reef CO_2 conditions. Glob Change Biol 19: 1632–1641. doi:10.1111/gcb.12154.

The Effect of Diel Temperature and Light Cycles on the Growth of *Nannochloropsis oculata* in a Photobioreactor Matrix

Bojan Tamburic[1]*[⁹], Supriya Guruprasad[1⁹], Dale T. Radford[1], Milán Szabó[1], Ross McC Lilley[1], Anthony W. D. Larkum[1], Jim B. Franklin[1], David M. Kramer[2], Susan I. Blackburn[3], John A. Raven[1,4], Martin Schliep[1], Peter J. Ralph[1]

1 Plant Functional Biology and Climate Change Cluster, University of Technology Sydney, Sydney, New South Wales, Australia, **2** Department of Biochemistry and Molecular Biology, Michigan State University, East Lansing, Michigan, United States of America, **3** Marine and Atmospheric Research, Commonwealth Scientific and Industrial Research Organisation, Hobart, Tasmania, Australia, **4** Division of Plant Sciences, University of Dundee at the James Hutton Institute, Invergowrie, Scotland, United Kingdom

Abstract

A matrix of photobioreactors integrated with metabolic sensors was used to examine the combined impact of light and temperature variations on the growth and physiology of the biofuel candidate microalgal species *Nannochloropsis oculata*. The experiments were performed with algal cultures maintained at a constant 20°C versus a 15°C to 25°C diel temperature cycle, where light intensity also followed a diel cycle with a maximum irradiance of 1920 µmol photons m^{-2} s^{-1}. No differences in algal growth (Chlorophyll *a*) were found between the two environmental regimes; however, the metabolic processes responded differently throughout the day to the change in environmental conditions. The variable temperature treatment resulted in greater damage to photosystem II due to the combined effect of strong light and high temperature. Cellular functions responded differently to conditions before midday as opposed to the afternoon, leading to strong hysteresis in dissolved oxygen concentration, quantum yield of photosystem II and net photosynthesis. Overnight metabolism performed differently, probably as a result of the temperature impact on respiration. Our photobioreactor matrix has produced novel insights into the physiological response of *Nannochloropsis oculata* to simulated environmental conditions. This information can be used to predict the effectiveness of deploying *Nannochloropsis oculata* in similar field conditions for commercial biofuel production.

Editor: Shin Yamazaki, University of Texas Southwestern Medical Center, United States of America

Funding: Funds were provided by a C3 start-up grant, UTS Faculty of Science and UTS Strategic research support; The University of Dundee is a registered Scottish charity, No. SC015096. The funders had no role in study design, data collection and analysis, decision to publish, or preparation of the manuscript.

Competing Interests: The authors have read the journal's policy and have the following conflicts: John A Raven has share options in Scottish Bioenergy; David M Kramer was a co-founder of Phenometrics but he is no longer involved with this company. This does not alter their adherence to all the PLOS ONE policies on sharing data and materials.

* E-mail: bojan.tamburic@uts.edu.au

⁹ These authors contributed equally to this work.

Introduction

Algal Biofuel Production with *Nannochloropsis*

Microalgae have the potential to produce sustainable and affordable transport fuels for the future. These unicellular photosynthetic organisms capture light energy and use it to fix atmospheric carbon dioxide into proteins, carbohydrates and lipids. The use of algal biofuels leads to considerably lower carbon dioxide emissions per unit energy than for petrochemical-based fuel [1], making algal biofuels a 'green' solution to the global energy challenge. Lipid production is the primary focus of the algal biofuels industry [2,3], but other biofuel precursors such as terpenes are also under development [4]. These secondary metabolites store chemical energy, and they can be chemically processed into biodiesel and aviation fuel [5,6].

Microalgae of the genus *Nannochloropsis* from the class Eustigmatophyceae are leading candidates for algal biofuel production. The unusual photosynthetic apparatus of these algae contains only chlorophyll *a*. The genome sequence of *Nannochloropsis* species is known, and a genetic transformation method utilising homologous recombination in *Nannochloropsis* has been demonstrated [7], opening up new possibilities in biofuels research using *Nannochloropsis*.

Nannochloropsis sp. normally accumulates lipids to 28.7% of the dry weight, or 65 to 70% of the dry weight under nutrient deprivation, resulting in lipid productivities ranging from 25.8 to 60.9 mg L^{-1} d^{-1} [8–10]. This productivity compares favourably with other microalgal species that have faster growth rates but lower lipid content, such as *Spirulina maxima* (8.6 mg lipid L^{-1} d^{-1}), *Chlorella vulgaris* (9.2 mg lipid L^{-1} d^{-1}) and *Dunaliella tertiolecta* (20.0 mg lipid L^{-1} d^{-1}) [11]. In general, oil yields from microalgae are several orders of magnitude better than those of higher plants, including traditional biofuel crops such as corn, rape, palm and sugarcane [12].

Many algae operate an active carbon-concentrating mechanism when growing in suboptimal carbon dioxide concentrations. This mechanism transports inorganic carbon from outside the cell to the chloroplast stroma, thereby increasing the rate of photosynthetic carbon fixation. Central to the carbon-concentrating mechanism in *Nannochloropsis* is the uptake of bicarbonate (HCO_3^-) across the cell membrane, which results in the accumulation of high intracellular carbon dioxide concentrations, and little benefit was achieved in adding carbon dioxide when attempting to optimise the growth of *N. gaditana* [13].

The alga *Nannochloropsis oculata* was selected for this study based on its ability to grow in ponds of saline, brackish and hypersaline water [8]. This ensures that *N. oculata* biofuel production facilities will never compete with food crops for freshwater or arable land. Australia is perfectly suited for the development of algal biofuels based on *N. oculata* due to its extensive coastline with adjacent marginal or non-arable land, low population density and high sunlight availability [14]. In order to quantify the biofuel production potential of *N. oculata*, it is imperative to understand the physiological processes that govern its growth under naturally variable environmental conditions.

Photobioreactor (PBR) Matrix

Algae have been grown commercially in outdoor ponds and photobioreactors (PBRs) [15,16]. Conducting field trials in these large-scale facilities is expensive and time-consuming. A more practical approach is to use laboratory-scale PBRs to simulate natural environmental conditions at the anticipated commercial location; however, most indoor laboratory-scale devices do not fully simulate the outdoor daily cycles of light intensity and temperature [17–19]. A series of custom-built PBRs (ePBR, Phenometrics, Lansing, MI, USA) have been integrated with metabolic sensors to simulate outdoor algal growth conditions in the laboratory. Using this innovative technology, it is now possible to mimic diel irradiance and temperature cycles, as well as seasonal variations. The PBR capabilities have been further enhanced with the latest *in vivo* measurement technologies, including dissolved oxygen optodes (bare-fibre oxygen minisensor, Firesting O_2, PyroScience, Aachen, Germany) and fluorometers (Pocket PAM, Walz, Effeltrich, Germany). It is important to analyse the combined effect of several interacting environmental factors in order to understand how they affect algal growth [20]; therefore, a number of PBRs are operated as a matrix, so that different growth conditions can be explored simultaneously using the same parent culture.

Light and Temperature Variations

Incident irradiance is the most important environmental factor governing algal biomass production rates [21,22]. *Nannochloropsis* has been found to have a flexible photosynthetic apparatus, which can acclimate to a wide range of constant and sinusoidally varying light intensities, as well as higher frequency fluctuations in irradiance [23]. The rate of photosynthesis increases linearly with increasing light intensity, and then asymptotically until the photosynthetic electron transport chain, or the enzyme ribulose bisphosphate carboxylase-oxygenase (RuBisCO), becomes light saturated [24,25]. Irradiances higher than the light saturation level can damage algal cells through photoinhibition [26]. As a result, algal biomass production rates under diel light cycles in outdoor ponds are typically lower than those maintained under constant illumination in the laboratory [27]. On the other hand, *Nannochloropsis* can utilise very intense light provided that subsequent dark periods occur to allow for re-oxidation of the key electron transport components of the photosynthetic apparatus [23]. This light flickering effect can be introduced by controlling the mixing rate in systems grown under natural illumination [23].

Temperature is another important environmental factor because it affects the rate of all enzymatic, electron transport and solute movement reactions within algal cells, and influences the properties of cellular components such as lipids, proteins and carbohydrates [28,29]. Photosynthesis is one of the most thermally sensitive processes in plants and algae, and it incurs greater damage from occasional extreme temperature events [28,30]. Aquatic microalgae are frequently forced to adapt to large variations in temperature owing to diel and seasonal cycles, although the temperature differences are not as large as those experienced by desert crust algae or even leaves on vascular plants. These thermal adaptations can determine the variability in biomass and lipid productivity [31]. The optimum temperature for *Nannochloropsis* sp. has been reported as 24–26°C [32,33]. The effect of temperature variations on the growth rate and photosynthetic performance of *Nannochloropsis* sp. have also been reported [34,35]. A comprehensive comparison between *N. oculata* growth at constant temperature in the laboratory and at variable temperatures in nature has not been performed. There also remains a distinct knowledge gap in understanding the combined effect of diel light and temperature variations on algal physiology and growth.

Aims and Objectives

The aim of this study is to compare the growth of *N. oculata* in PBRs (i) at constant temperature (ii) with variable temperature following a diel pattern, where both treatments were held under varying light intensity that also follows the diel cycle. The objective is to understand the combined effects of light and temperature on algal physiology, and to determine the importance of temperature in controlling large-scale biofuel production. Chlorophyll *a* content is measured to determine algal growth rates, dissolved oxygen concentrations and pH are monitored continuously *in situ*, and variable chlorophyll fluorescence and net photosynthesis measurements are used to understand algal physiology.

Materials and Methods

Microalgal Culture and Medium

Nannochloropsis oculata (Droop) Green (CS-179) was obtained from the Australian National Algae Culture Collection (ANACC; CSIRO, Hobart, Australia) and grown in f/2 medium (0.2 µm filtered artificial seawater enriched with: 8.82×10^{-4} M $NaNO_3$, 3.62×10^{-5} M $NaH_2PO_4.7H_2O$, trace metal solution and vitamin solution) [36]. Stock cultures were maintained at 20°C under a 12 h:12 h light:dark cycle at 40 µmol photons $m^{-2}\,s^{-1}$ of photosynthetically active radiation (PAR). Sparging with small bubbles of ambient air ensured that the exchange of atmospheric carbon dioxide and oxygen in the culture medium is maximised. This prevented the steady-state concentrations of these gasses from shifting too far from air equilibrium levels due to photosynthetic and respiratory activities, allowing examination of how irradiance and temperature impact growth. *Nannochloropsis* cells are approximately 3 µm in diameter.

Photobioreactor Setup

N. oculata was cultured in cylindrical photobioreactors (ePBR, Phenometrics, Lansing, MI, USA) with a 450 mL working volume (Figure 1). Experiments were performed over a period of 12 days following an additional 2 day acclimation period of *N. oculata* cultures in the PBRs. Seven PBRs were inoculated with 10% v/v of stock culture. All PBRs were illuminated by LEDs with white light of spectral composition similar to sunlight under a 12 h:12 h light:dark cycle. A comparison between the LED spectral output and the solar spectrum is shown in Figure s1; while the LED provides steady white light across the PAR region, it is particularly rich in blue wavelengths. Vieler *et al.* [37] have recently found evidence for blue light receptors and components of a circadian rhythm in the genome sequences of *N. oceanica* and *N. gaditana*. This makes it likely that *N. oculata* will respond physiologically in some way, when growing under higher intensities of blue light. Whether most or all of that response will be complete within two days, or take much longer, will require further experimental investigation. Sinusoidal irradiance levels during the light phase ranged from 0 to1920 μmol photons m^{-2} s^{-1} PAR, with maximum light intensity occurring at midday. This maximum light intensity corresponds to the approximate solar PAR maximum under direct sunlight in the tropics [38]. Lights were turned on at 06:00 h and turned off at 18:00 h. The PBRs were operated in batch mode. Two different temperature regimes were used: constant temperature maintained at 20±1°C and sinusoidal temperature ranging between 15±0.5°C and 25±0.5°C; the peak temperature occurred around dusk (18:00±30 mins). This 6 h lag between peak ambient light intensity and seawater temperature is common in coastal regions [39]. Four PBRs were operated under constant temperature ($n = 4$) and three PBRs were operated under sinusoidal temperature ($n = 3$). All PBRs were aerated with filtered (oil-free) ambient air that was bubbled into the culture via gas dispersion tubes (product no. 9435–39, porosity D, ACE Glass Inc., Vineland, NJ, USA). However, individual PBRs behave differently due to small variations in sparging rate, mixing and algal culture composition. Measurements of temperature (Phenometrics), pH (Van London pH electrode, Houston, TX, USA) and dissolved oxygen (OXB430-OI bare-fibre oxygen minisensor, Firesting O_2, PyroScience, Aachen, Germany) were performed continuously in all PBRs. Data were collected every 5 min, so that effects on algal physiology could be closely monitored. Dissolved oxygen measurements were corrected for temperature changes in the PBRs in accordance with theoretical oxygen solubilities in seawater (Figure S2) [40]; pH measurements were additionally corrected for electrode baseline drift.

Growth Measurements

A direct microscopic cell count was carried out using a haemocytometer (Neubauer, Germany). Cell counts were performed on undiluted 10 μL samples and 5 out of 25 haemocytometer squares were counted twice; each cell count was performed in triplicate on alternate days.

Chlorophyll *a* (Chl *a*) determinations are vital in biofuel research since they provide a fast estimation of algal cell density and physiology [41,42]. Chl *a* was measured using a modified extraction method [43]. Briefly, a sample of algal culture was filtered (0.2 μm) and the filter extracted in the dark for 24 h in 100% ethanol saturated with $MgCO_3$. Samples were then centrifuged and the supernatant analysed spectrophotometrically. Extinction values at 632, 649, 665 and 750 nm were recorded using a UV-Vis spectrophotometer (Varian Cary 50 Bio, Palo Alto, CA, USA). The Chl *a* concentration was calculated using the equation below [43].

$$Chla = 1.3382\,Abs_{632} - 4.7709\,Abs_{649} + 13.1047\,Abs_{665} - 1.8402\,Abs_{750}$$

An additional chlorophyll assay was performed, based on Chl *a* fluorescence, with a Turner fluorometer (using the *in vivo* module), where raw fluorescence was measured (Trilogy, Turner Design, Sunnyvale, CA, USA). The excitation wavelength was 485 nm and the fluorescence was detected at >685 nm.

Fluorescence Measurements

The quantum yield of photosystem II (Y(II)) was determined using a pulse-amplitude modulated fluorometer (Pocket PAM, Walz, Effeltrich, Germany). The Pocket PAM was placed against the outside of the chamber at approximately 1/3 height. Measurements were recorded at six time points throughout the diel cycle: 05:30 (pre-dawn), 08:30 (mid-morning), 12:00 (midday), 16:15 (mid-afternoon), 17:30 (pre-dusk) and 18:15 (early-night). Pocket PAM settings were as follows: blue light, measuring light intensity of ~0.2 μmol photons m^{-2} s^{-1} PAR, saturation pulse intensity of 2600 μmol photons m^{-2} s^{-1} PAR, and saturation pulse width of 0.8 s.

Net Photosynthesis Measurements

Net photosynthesis was estimated by measuring the oxygen evolution rates of *N. oculata* at different times of day; analysis was performed at the same time as fluorescence measurements. Aeration of the PBR was temporarily switched off (~15 mins) whilst minisensors (PyroScience) measured the initial rate of change in dissolved oxygen concentrations. Oxygen evolution was determined by fitting a linear curve to the rate of dissolved oxygen increase. The oxygen evolution rate was normalised against the corresponding Chl *a* content of the *N. oculata* culture.

Results and Discussion

N. oculata Growth

Three different techniques, microscopic cell counts, *in vitro* Chl *a* extraction and *in vivo* Chl *a* fluorometry were used to measure the growth of *N. oculata* cultures under constant and sinusoidal temperature regimes (Figure 2). Both temperature treatments showed growth curves with a distinct lag phase (day 0 to day 2), exponential phase (day 2 to day 8) and stationary phase (day 8 to day 12).

Microscopic cell counts proved to be unreliable (Figure 2A and 2D). Cell counting with a haemocytometer is susceptible to various sources of systematic user error; preparation of non-uniform cell suspensions, improper filling of measuring chambers and failure to adopt a convention for counting cells in contact with boundary lines and each other [35]. Notable outliers were obtained on day 8 of the experiment. The percentage error was also large; as high as 15% for constant temperature measurements and 34% for sinusoidal temperature measurements.

In vitro Chl *a* extraction measurements (Figure 2B and 2E) are the universally accepted standard for microalgal chlorophyll determination [16]. Chl *a* content is strongly correlated to the density of green microalgal cells [43]. *In vitro* measurements are time-intensive and potentially disruptive to the experiment. *In vivo* Chl *a* fluorescence measurements (Figure 2C and 2F) are a more practical alternative to manual cell counts [44]. The Pearson product-moment correlation coefficient between all growth rate measurements obtained using these two techniques was 0.85. This correlation was higher under constant temperature (Figure 2B and 2C); logistic curve fits [45] give a specific growth rate of 0.60 d^{-1} for Chl *a* extraction, and a comparable 0.62 d^{-1} in the case of *in vivo* Chl *a* fluorescence.

However, this relationship breaks down in the case of sinusoidal temperature (Figure 2E and 2F). Chl *a* extraction measurements

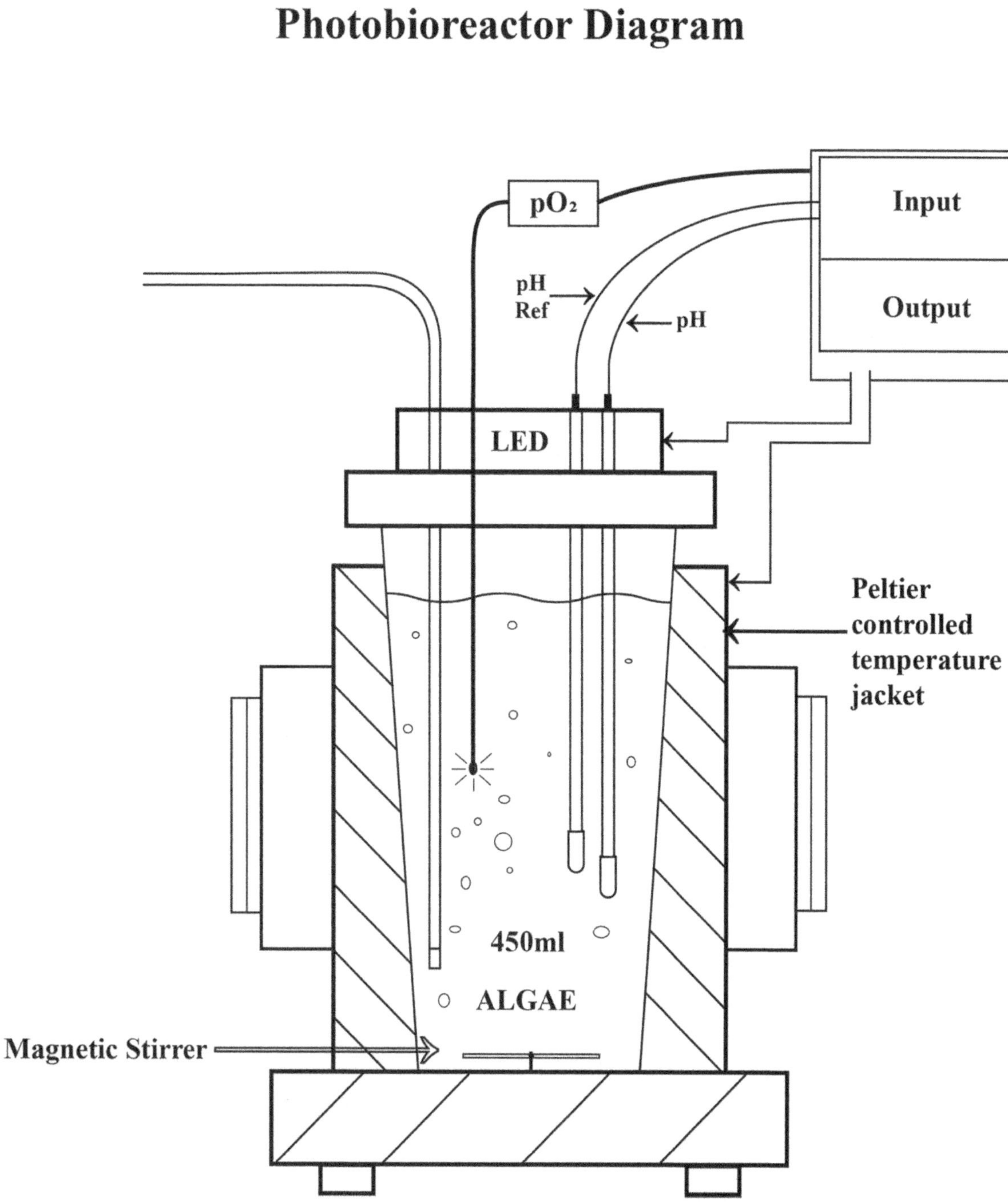

Figure 1. Phenometrics ePBR schematic diagram. Light intensity is controlled using a white light LED array and culture temperature is controlled with a Peltier controlled temperature jacket. The PBR is comprised of a magnetically-stirred chamber, filled with a 10% v/v *N. oculata* stock culture inoculum in f/2 medium. Aeration with ambient air is supplied through a gas dispersion tube. Dissolved oxygen (pO_2) is measured optically (PyroScience) and pH is measured electrochemically (Phenometrics).

show a faster growth rate during the first 6 days of experiment ($0.77\ d^{-1}$), followed by a linear depletion of the Chl *a* content. The reliability of this growth rate estimate is questionable, since the logistic curve is based on only five data points (seven in all other cases). The corresponding *in vivo* Chl *a* fluorescence measurements showed no sign of this Chl *a* depletion, and the logistic fit gives a reduced growth rate of $0.44\ d^{-1}$. The most likely reason for this discrepancy is that the *in vivo* Chl *a* fluorescence module cannot distinguish between Chl *a* and Chl *a* breakdown products, such as chlorophyllide *a* [44]. It is therefore generally appropriate to use *in vivo* fluorescence to estimate Chl *a* content during the exponential growth phase, but not during the subsequent stationary phase. Chl *a* extraction will therefore be used as the default measure of *N. oculata* growth in this study.

In general, *N. oculata* growth rates under constant and sinusoidal temperature regimes are very similar. There is some indication that growth may be slightly faster under sinusoidal temperature, but also that this condition may result in faster degradation of Chl

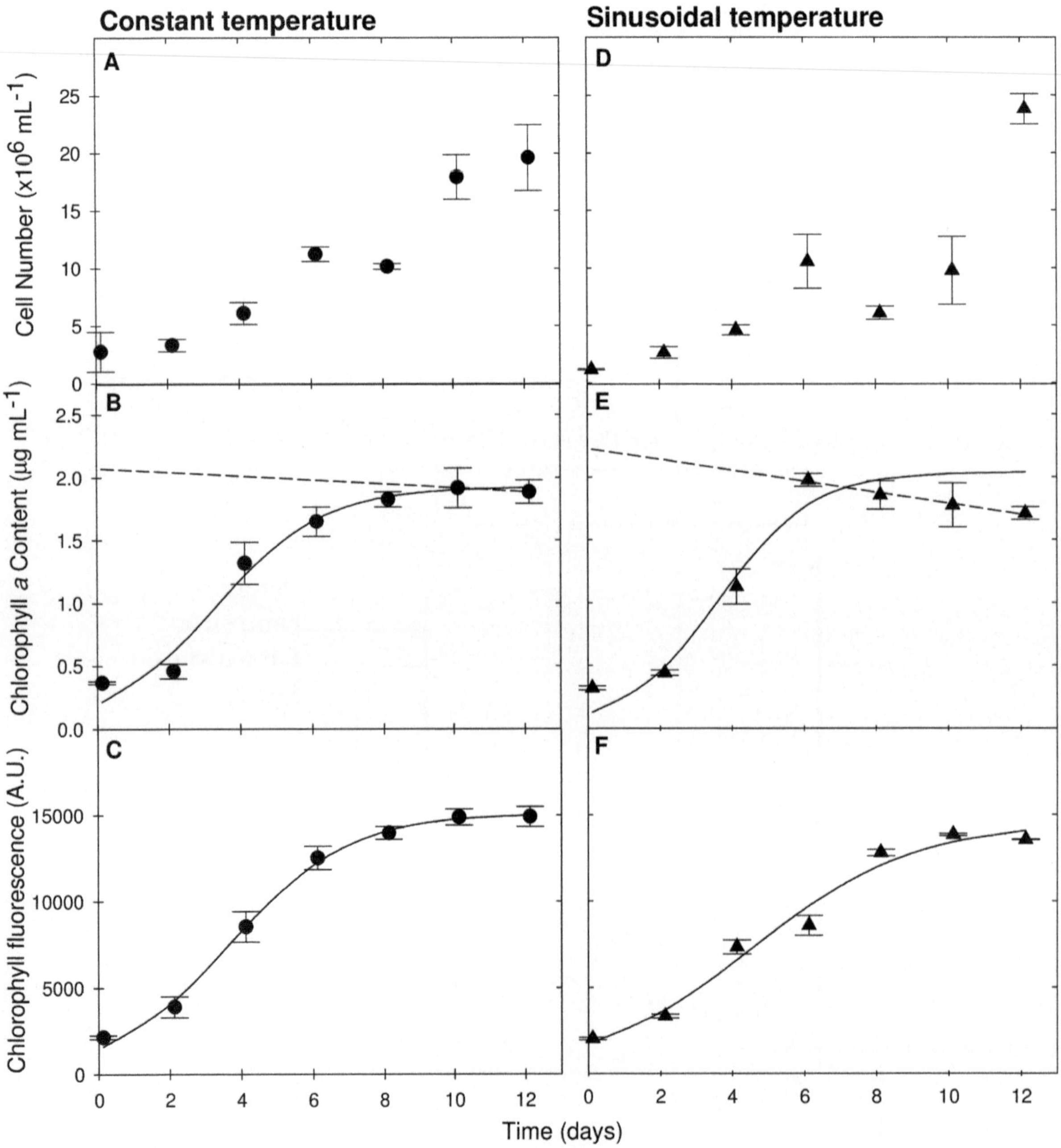

Figure 2. Measurements of *N. oculata* growth. *N. oculata* grown in Phenometrics ePBRs under two different temperature regimes: constant temperature (circles) in A-C and sinusoidal temperature (triangles) in D-F. Growth was measured by microscopic cell counts (A, D), Chl *a* extraction (B, E) and *in vivo* fluorometry (C, F). For constant temperatures n = 4, for sinusoidal temperatures n = 3; error bars represent ±1 standard error. A logistic fit based on the method of least squares [45] was used to model algal growth (B, C, E, F; solid line). Linear regression was used to model subsequent Chl *a* depletion, which occurred after the exponential growth phase (B, C; dotted line).

a during the stationary phase (Figures 2B and 2E). There is minimal difference in cell densities following the exponential growth phase, indicating that the algal culture is either light or nutrient limited. Previous experiments have shown that *Nannochloropsis* can be grown effectively over a wide temperature range: between 15 and 30°C [4,32,33]. Our study shows that *N. oculata* also copes well with natural diel variations in temperature between 15 and 25°C. It would appear that fine temperature control is not necessary in industrial *N. oculata* cultivation, apart from the need to control the intense temperature gradients observed in shallow ponds [3]. This result may be different in the case of low carbon dioxide availability or other nutrient limitations. However, it would be premature to draw conclusions at this stage before first examining the effect of diel temperature variations on algal physiology and long-term health.

Dissolved Oxygen (pO_2) Profiles

In situ pO_2 profiles of *N. oculata* were measured optically under constant temperature (Figure 3) and sinusoidal temperature regimes (Figure 4); all data was post-processed to correct for changes in oxygen solubility at different temperatures. Diel trends are consistent between replicates of identical temperature regimes; however, there is a noticeable difference in the shape of oxygen evolution and cycle duration between the different treatments. The pO_2 profiles under constant temperature show a 12-hour daylight cycle, compared to the 24-hour cycle evident under the sinusoidal temperature regime.

Under the constant temperature regime, pO_2 responds strongly to the presence of light, reaching a maximum concentration between 09:30 and 10:30 (Figure 3). pO_2 decreases sharply until the early afternoon (approximately 13:00), where the rate of decrease slows for a couple of hours, before once again decreasing rapidly until sunset (Figure 3B, D, F). Since the culture is consistently sparged with air there are no obvious signs of oxygen consumption by respiration during the dark period.

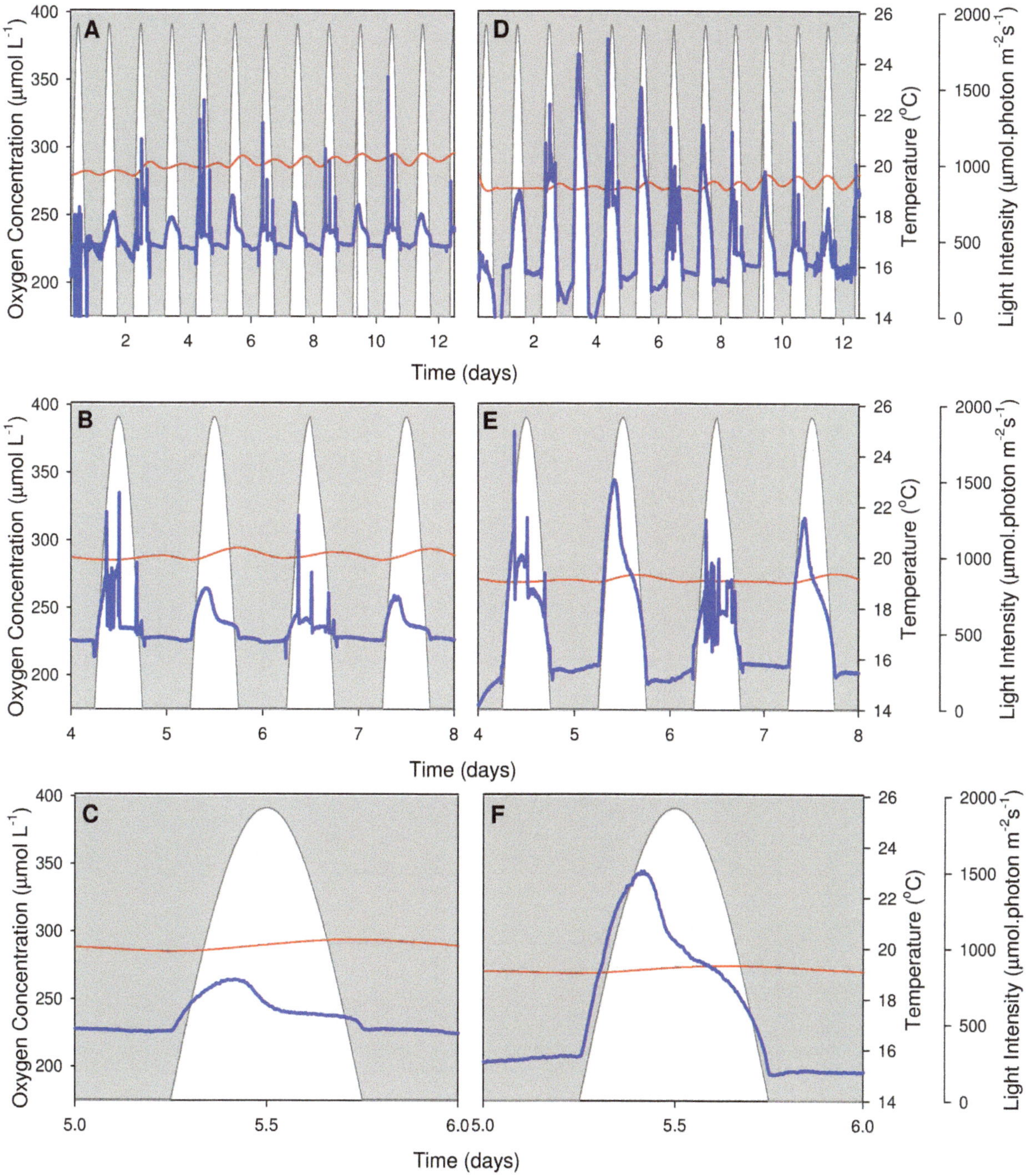

Figure 3. Constant temperature pO_2 profiles. Dissolved oxygen profiles of *N. oculata* when exposed to constant temperature. Two representative PBRs are shown (A, B, C and D, E, F). Top panels (A, D) display the entire experiment, middle panels (B, E) present in more detail the pO_2 profile during the exponential growth phase (day 4 to day 8), and bottom panels (C, F) highlight a single day (day 5). White regions represent the sinusoidal light regime, thin red lines give the smoothed culture temperature and thick blue lines show the pO_2 profile. Large spikes on even-numbered days are caused by short interruptions to the air supply during periodic oxygen evolution measurements.

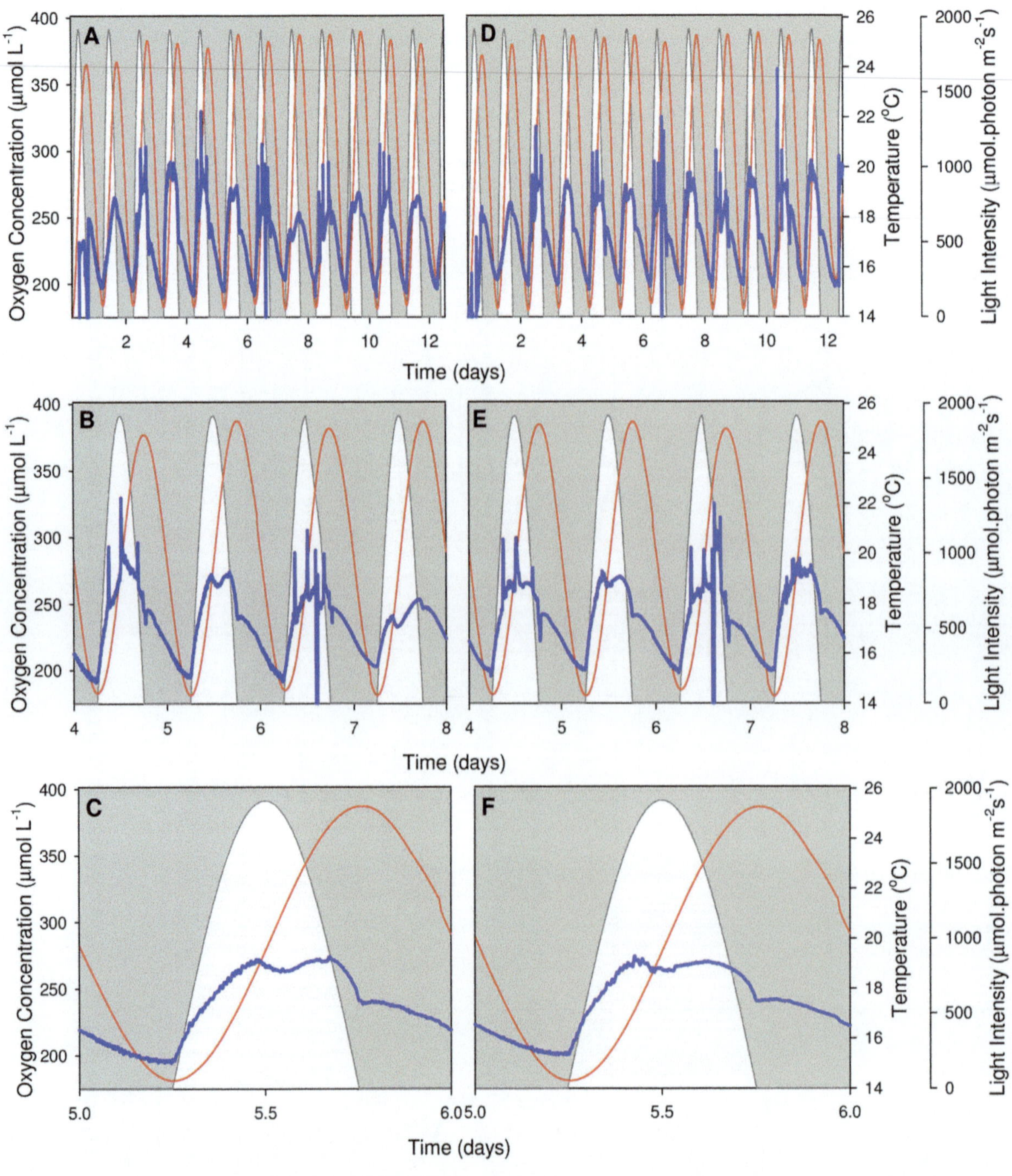

Figure 4. Sinusoidal temperature pO_2 profiles. Dissolved oxygen profiles of *N. oculata* when exposed to sinusoidal temperature. Two representative PBRs are shown (A, B, C and D, E, F). Top panels (A, D) display the entire experiment, middle panels (B, E) present in more detail the pO_2 profile during the exponential growth phase (day 4 to day 8), and bottom panels (C, F) highlight a single day (day 5). White regions represent the sinusoidal light regime, thin red lines give the smoothed culture temperature and thick blue lines show the pO_2 profile. Large spikes on even-numbered days are caused by short interruptions to the air supply during periodic oxygen evolution measurements.

Under the sinusoidal temperature regime, pO_2 values are less dictated by light than those under constant temperature. The increase of pO_2 at the onset of irradiance at dawn is slower than under constant temperature, despite the fact that the PAR fluctuation stays the same under both treatments. This is because temperature is lower at this early stage of the diel cycle, leading to reduced net photosynthesis [30]. The pO_2 reaches a primary maximum just before midday (Figure 4B, D, F). It then decreases slightly in accordance with the pattern observed under constant temperature. It is at this stage that a decoupling between pO_2 and the light intensity occurs. From 14:00 until the simulated sunset, pO_2 correlates strongly with the sinusoidal temperature cycle, reaching a secondary maximum shortly prior to the temperature maximum at 18:00. This maximum can be explained by the increased net photosynthesis at higher temperatures [30]. It is nevertheless surprising that pO_2 increases while PAR is decreas-

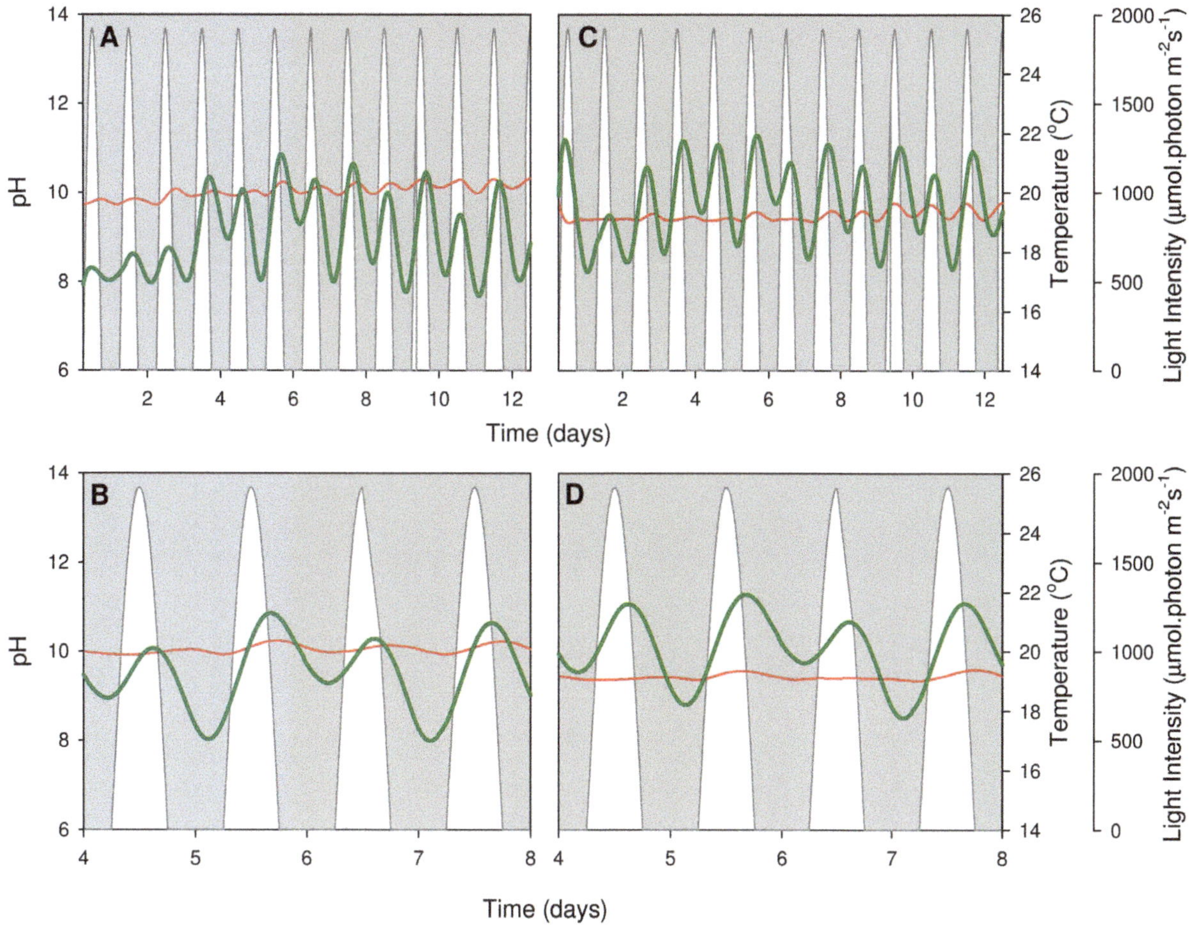

Figure 5. Constant temperature pH profiles. Smoothed (moving average) pH profiles of *N. oculata* when exposed to constant temperature. Two representative PBRs are shown (A, B and C, D). Top panels (A, C) display the entire experiment, while bottom panels (B, D) present in more detail the pH profile during the exponential growth phase (day 4 to day 8). White regions represent the sinusoidal light regime, thin red lines give the smoothed culture temperature and thick green lines show the pH profile.

ing, since PAR provides the energy for the water-splitting reaction. Following the secondary maximum, pO_2 levels decrease rapidly until night, but there is not enough time for them to return to the baseline values of dawn that day. During the subsequent dark period, the rate of pO_2 decrease slows and pO_2 levels eventually return to initial (pre-sunrise) levels due to the combined influence of respiration and air sparging. More detailed studies are needed to determine if there is evidence for an initial rapid O_2 uptake after cessation of illumination followed by a lower rate of respiration [46].

Absolute pO_2 values are difficult to interpret because pO_2 is a function of physical (sparging) and biological (photosynthesis and respiration) processes. The physical sparging process is constant throughout the day and generally acts to remove excess oxygen produced by net photosynthesis. Dissolved oxygen goes down during the second half of the light period because the rate of oxygen removal by sparging is faster than the rate of oxygen production by net photosynthesis. This effect is exemplified by the significant variations in pO_2 amplitudes between Figures 3B and 3D. Although all PBRs use identical air dispersion tubes and air pressures, variations are introduced due to the propensity of *N. oculata* to colonise small apertures such as those in glass sinters (sparging bars), leading to time and PBR dependent changes in the air dispersion rate; this limitation will be addressed in future experiments. It is, however, possible to compare trends in pO_2 amplitude in selected PBRs (Figure S3: constant temperature regime; Figure S4: sinusoidal temperature regime). Under both temperature regimes, pO_2 concentration is highest during the exponential growth phase (day 3 and day 7). This effect will be explored in more detail using net photosynthesis measurements. There is also a temporal shift in the pO_2 maxima under constant temperature; in general, the maximum occurs earlier, and it is more pronounced, when the algal growth rate is faster (Figure S3). This result indicates that *N. oculata* cells in the exponential growth phase require less light to produce the same amount of oxygen by photosynthesis. The corresponding sinusoidal temperature results are influenced by both light intensity and temperature variations (Figure S4). The hysteresis effect is clear in Figure S3, where the degree of photosynthetic down-regulation is greater in the afternoon than in the morning. This is caused when oxygen evolution is measured under increasing steps in irradiance, rather than decreasing steps in irradiance, as occurs in the afternoon. The shape of the hysteresis is dependent upon irradiance; however as temperature alters the rate of D1 protein repair [47], this could also alter the hysteresis.

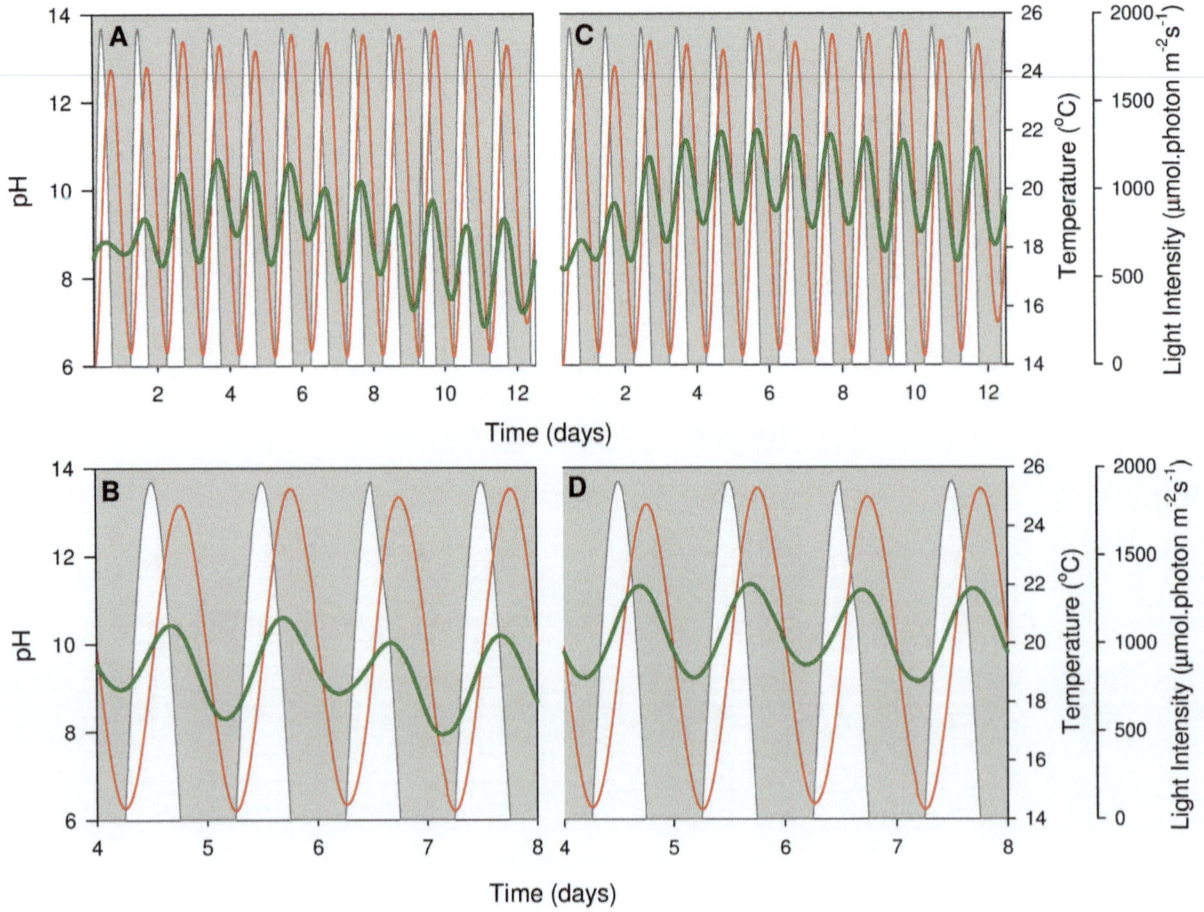

Figure 6. Sinusoidal temperature pH profiles. Smoothed (moving average) pH profiles of *N. oculata* when exposed to sinusoidal temperature. Two representative PBRs are shown (A, B and C, D). Top panels (A, C) display the entire experiment, while bottom panels (B, D) present in more detail the pH profile during the exponential growth phase (day 4 to day 8). White regions represent the sinusoidal light regime, thin red lines give the smoothed culture temperature and thick green lines show the pH profile.

pH Profiles

pH profiles were determined electrochemically under constant temperature (Figure 5) and sinusoidal temperature regimes (Figure 6); all data were smoothed and post-processed to account for probe drift and temperature variations. Under both temperature regimes, pH fluctuations occur in 24-hour cycles, with higher pH levels measured during the light phase. At night, a decrease in pH is seen due to a decrease in the pO_2:pCO_2 ratio as a result of the absence of photosynthesis and presence of respiration [48]. The maximum rate of pH increase occurs at approximately 10:00 under constant temperature and just before midday under sinusoidal temperature. This strong correlation between pH increase and maximum pO_2 concentration (c.f. Figures 3, 4, 5 and 6) indicates that both parameters provide a reasonable estimate for the diel variations in photosynthesis rate.

Further to this, a 48 h pattern occurs where the pH shows a more gradual nocturnal decrease on every even-numbered day. This trend is more apparent under the constant temperature regime (Figure 5B, D), where diel temperature variations do not affect the pH measurements. The 48 h trend coincides with the oxygen evolution measurements that took place on alternate days and caused interruptions to the air dispersion.

N. oculata can tolerate alkaline pH fluctuations, since growth occurs between pH 8 and pH 13 [4,35]. The experimental variations in this study are comfortably within this pH tolerance range, indicating that pH variations did not limit *N. oculata* growth.

Figures 5A, C and 6A, C show curved responses to time in the overall pH amplitudes of cultures; this pattern is more prevalent in the sinusoidal regime. This result was also observed in the *in vitro* Chl *a* measurements (Figure 2B, E): the curvature of the pH response to time appears to follow the Chl *a* density. A strong pH (and Chl *a* density) decrease was observed in the last four days of the experiment under sinusoidal temperature conditions, whereas the constant temperature curve appears more linear. As discussed previously (Results and Discussion; *N. oculata* growth), the sinusoidal temperature regime likely leads to faster rates of Chl *a* breakdown in the stationary growth phase (Figure 2E). This increases the respiration-to-photosynthesis ratio, leading to the observed reductions in pH levels (Figure 6A, C) [35]. The pH data show no evidence of an increase in culture medium pH (at a given CO_2 partial pressure) with algal growth, which could result from the production of alkali in the assimilation of external nitrate into organic nitrogen [49,50].

Quantum Yield of Photosystem II and Net Photosynthesis

Algal photosynthetic performance was determined using Pocket PAM fluorescence and net photosynthesis measurements (Figure 7 for constant temperature and Figure 8 for sinusoidal temperature).

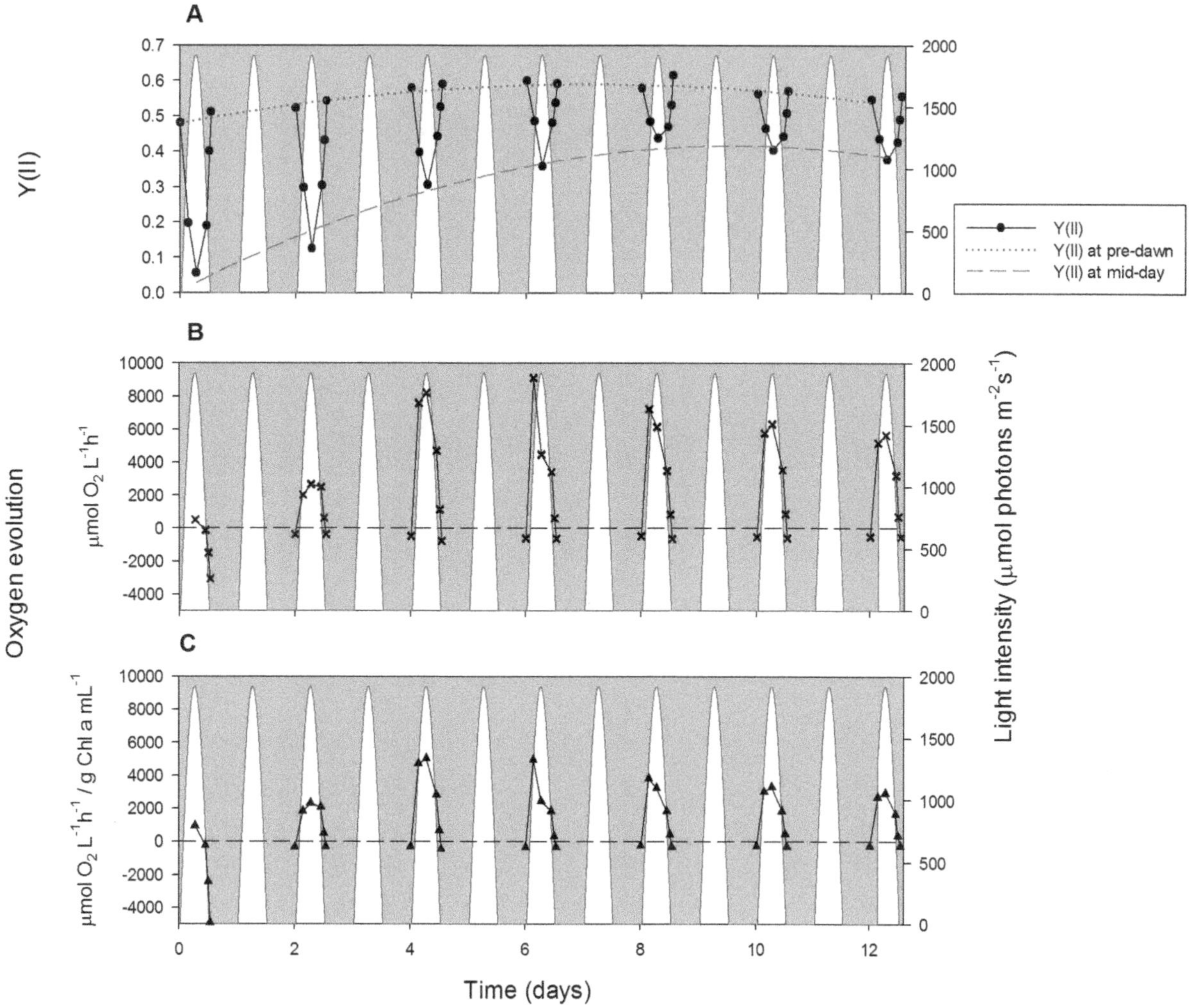

Figure 7. Quantum yield of photosystem II and net photosynthesis under constant temperature. Algal physiology parameters, including: the quantum yield of photosystem II (YII) measured using PAM fluorometry (A, circles), the rate of oxygen concentration change when positive, representing net photosynthesis (B, crosses), and net photosynthesis normalised against Chl *a* content (C, triangles). Data were collected on alternate days, at 6 time points throughout the diel cycle. White areas represent the sinusoidal light regime.

These measurements were very consistent between different PBRs (see Figure S5 and Figure S6 for additional replicates).

The PAM fluorescence parameter chosen was Y(II), the light-adapted effective quantum yield of photosystem II (PSII), i.e. the efficiency at which light absorbed by PSII is used for reduction of the primary quinone electron acceptor, Q_A [51]. Sforza *et al.* [23] found overnight estimates of maximum quantum yield of PSII to average 0.6 in *Nannochloropsis salina*, which is similar to our results obtained in *N. oculata*. The V-shaped pattern of Y(II) in Figures 7A and 8A represents the decreasing effective quantum yield of PSII as light intensities increase during the day, followed by increased effective quantum yield at lower light intensities in the evening. This hysteresis is linked with the widely observed afternoon depression of Y(II) [34]. It has been shown that *Nannochloropsis* sp. responds to variable light intensity by regulating its pigment composition and the ratio between enzyme concentration and irradiation, in an attempt to minimise photodamage [48,52,53]. Figures 7 and 8 illustrate the shade-adapted response that needs to be identified when Y(II) is measured in PBR vessels with a heterogeneous light climate (also shown in Figure S5 and Figure S6). The lowest effective quantum yield is represented by the Y(II) minima at midday.

The pre-dawn Y(II) measurements are similar between the constant and sinusoidal temperature regimes, although the sinusoidal temperature quantum yields are slightly higher, with values ranging from 0.52 to 0.65 (Figure 8A), compared with 0.49 to 0.60 under constant temperature (Figure 7A). In general, pre-dawn Y(II) values increase slowly over the first eight days of experiment before stabilising. Under constant temperature, pre-dawn Y(II) is 0.49 ± 0.01 ($n = 4$) on day 0 and 0.58 ± 0.01 ($n = 4$) on day 8; under sinusoidal temperature, the corresponding Y(II) increase is between 0.49 ± 0.04 ($n = 3$) on day 0 and 0.63 ± 0.02 ($n = 3$) on day 8. Mid-day Y(II) values also increase over time with increasing algal cell density and remain very similar between the two temperature regimes. Under constant temperature, mid-day Y(II) is a very low 0.06 ± 0.02 on day 0, but increases to 0.34 ± 0.06 on day 8; similarly, mid-day Y(II) under sinusoidal temperature starts at 0.08 ± 0.06 on day 0 and recovers to 0.33 ± 0.06 by day 8.

It is highly likely that the *N. oculata* photosystem II in the early afternoon under the sinusoidal temperature regime becomes photodamaged by both the exceedingly high light intensity and

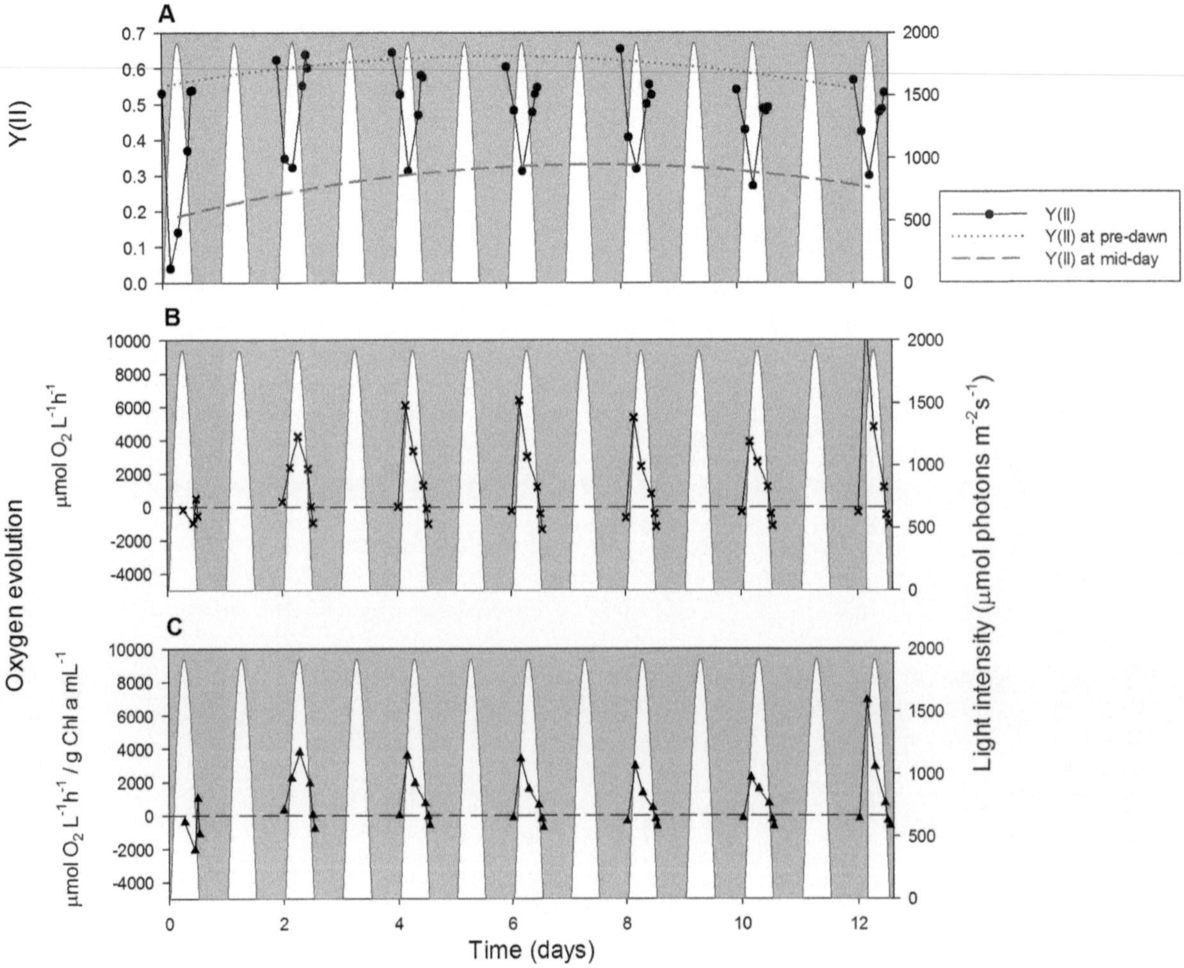

Figure 8. Quantum yield of photosystem II and net photosynthesis under sinusoidal temperature. Algal physiology parameters, including: the quantum yield of photosystem II (YII) measured using PAM fluorometry (A, circles), the rate of oxygen concentration change when positive, representing net photosynthesis (B, crosses), and net photosynthesis normalised against Chl *a* content (C, triangles). Data were collected on alternate days, at 6 time points throughout the diel cycle. White areas represent the sinusoidal light regime.

the elevated temperature. This view is supported by the apparent hysteresis observed in the Y(II) V-shaped curves; excess temperature after midday has a negative influence on Y(II), relative to the lower temperatures in the morning. Nevertheless, sinusoidal Y(II) values consistently recover by pre-dawn, indicating that sinusoidal temperature (Figure 8A) may stimulate faster photosystem recovery in low light and darkness compared with constant temperature (Figure 7A). During the dark phase, the sinusoidal temperature regime begins with a relatively high temperature and finishes at the temperature minimum of 15°C by pre-dawn. Future experiments will investigate whether these temperature changes really do assist photosystem recovery.

The light climate within the PBR is heterogeneous; both with time as the culture ages and vertically with increased absorption in the upper regions of the PBR and more shading towards the bottom of the chamber. The incremental increase of Y(II) values at midday is potentially caused by greater levels of self-shading due to the increased cell density (through increased cell count numbers and Chl *a* content; Figure 2) during *N. oculata* growth. As cultures increase in density, mixing becomes more important whereby cells are moved between areas of high light to more shaded regions of the PBR; this can provide an overall increase in growth. Net photosynthesis rates, based on oxygen evolution measurements (Figures 7B and 8B), provide an accurate representation of the growth rate [35], particularly when normalised against the corresponding Chl *a* content of the cells (Figures 7C and 8C) in agreement with Flameling & Kromkamp [54]. Under constant temperature, maximum net photosynthesis typically occurs during the highest light intensity at midday; the exceptions are days 6 and 8, where the maxima are at mid-morning. Under sinusoidal temperature, maximum net photosynthesis occurs consistently at mid-morning from day 4 of the experiment. This is the result of the reversible down-regulation of photosynthesis by photoinhibition, which occurs after midday and is caused by the combination of high light intensity and temperature. From day 8, the net photosynthesis and Y(II) no longer recover to their initial mid-morning peaks, suggesting that photodamage has occurred due to the combined long-term light and temperature stress (Figure 7A and 8A). Afternoon net photosynthesis values also show an additional hysteresis under the sinusoidal temperature regime, with the oxygen evolution rate remaining high later on in the day under the influence of higher temperatures (Figure 8B). The respiration that occurs after sunset is greater than during pre-dawn under the sinusoidal temperature regime, but not at constant temperature (Figure 7B). This is likely to reflect the relationship between mitochondrial respiration and temperature given the much higher temperature maintained at the end of the light period in the sinusoidal regime [46].

The trends in maximum normalised net photosynthesis (Figure 7C and 8C), pre-dawn Y(II) (Figure 7A and 8A) and algal growth rates measured as a function of *in vitro* Chl *a* content (Figure 2B and 2E) are all consistent across both temperature regimes. All of these measurements provide a reasonable estimate of *N. oculata* growth rates, but they also combine to deliver a more detailed understanding of the physiology of algal cells exposed to different environmental conditions. For example, it becomes clear that the reduction in Chl *a* content under sinusoidal temperature stems from the combined photoinhibitive effects of high light intensity and temperature in the early-to-mid afternoon. Similarly, high afternoon and evening temperatures in the sinusoidal regime lead to a positive hysteresis in the oxygen evolution rate, resulting in higher relative photosynthetic activity towards the end of the day, which may increase the possibility of full pre-dawn Y(II) recovery.

Conclusions

The growth and physiology of the biofuel candidate microalgal species *N. oculata* was examined under two different temperature regimes: constant temperature at 20°C and sinusoidal temperature varying between 15°C and 25°C. The experiments were performed in a matrix of seven PBRs (ePBR, Phenometrics, Lansing, MI, USA; Figure 1), under sinusoidal light intensity varying between 0 and 1920 μmol photons m^{-2} s^{-1} PAR.

The *N. oculata* growth rates, measured as a function of Chl *a* accumulation, were similar under the two regimes (Figure 2). This result implies that no temperature control is necessary to compensate for ±5°C sinusoidal temperature variations in industrial *N. oculata* growth facilities. However, a decline in Chl *a* content under the sinusoidal temperature regime (Figure 2E) raised questions about the long-term health and physiology of *N. oculata* cultures. Dissolved oxygen (Figures 3 and 4) and pH (Figures 5 and 6) profiles showed that algal cells were more active beyond midday under elevated afternoon and early-evening temperatures of the sinusoidal temperature regime.

Furthermore, measurements of the midday Y(II) and the corresponding net photosynthesis rates (Figures 7 and 8) showed that photosynthesis in *N. oculata* cells is down-regulated to a greater extent under the combined influence of high light intensity and temperature in the early afternoon of the sinusoidal temperature regimes. However, the sinusoidal temperature regime also facilitated enhanced photosystem recovery at night, resulting in slightly higher pre-dawn Y(II) values. This effect is likely to be related to increased photosynthetic activity after midday under sinusoidal temperature. The combined effects of increased photosynthetic down-regulation during the day and improved Y(II) recovery under low light and at night appear to cancel each other out, resulting in similar growth rates under sinusoidal and constant temperature regimes. Improved understanding of the photobiological mechanisms that govern these processes is required. This study demonstrates that our PBRs, with their environmental control features and high-resolution monitoring of algal growth and physiology, have the potential to answer many of the unresolved questions in algal biofuel production.

Supporting Information

Figure S1 Comparison of spectral irradiance by Phenometrics ePBR LED (red) with solar spectrum (blue). Total PAR was normalised to 500 μmol photons m^{-2} s^{-1}; the ePBR LED provides a steady white output across the PAR range, but it is particularly rich in blue wavelengths, with a strong peak at 450 nm.

Figure S2 Effect of temperature correction on baseline pO_2 measurements. Raw experimental baseline measurements (black lines), taken in the f/2 medium in the absence of *N. oculata*, were corrected for sinusoidal variations in oxygen solubility at different temperatures to give the temperature-corrected baseline measurement (blue lines); two representative PBRs shown.

Figure S3 pO_2 profile evolution at constant temperature. Data were recorded on alternate days between 05:00 and 24:00 for one representative PBR.

Figure S4 pO_2 profile evolution at sinusoidal temperature. Data were recorded on alternate days between 05:00 and 24:00 for one representative PBR. Strong hysteresis in pO_2 during second part of the day results from increasing temperature.

Figure S5 Quantum yield of photosystem II and net photosynthesis under constant temperature (second replicate). Algal physiology parameters, including: the quantum yield of photosystem II (YII) measured using PAM fluorometry (A, circles), the rate of oxygen concentration change when positive, representing net photosynthesis (B, crosses), and net photosynthesis normalised against Chl *a* content (C, triangles). Data were collected on alternate days, at 6 time points throughout the diel cycle. White areas represent the sinusoidal light regime.

Figure S6 Quantum yield of photosystem II and net photosynthesis under sinusoidal temperature (second replicate): Algal physiology parameters, including: the quantum yield of photosystem II (YII) measured using PAM fluorometry (A, circles), the rate of oxygen concentration change when positive, representing net photosynthesis (B, crosses), and net photosynthesis normalised against Chl *a* content (C, triangles). Data were collected on alternate days, at 6 time points throughout the diel cycle. White areas represent the sinusoidal light regime.

Acknowledgments

The authors would like to thank University of Technology Sydney student Kieran Parker for optimising the Chl *a* extraction method for *N. oculata*, which was a part of his Honours project and which has also been used in this study.

Author Contributions

Conceived and designed the experiments: BT M. Szabo RMCL AWDL JBF M. Schliep PJR. Performed the experiments: SG DTR. Analyzed the data: BT SG DTR M. Szabo RMCL AWDL JBF DMK SIB JAR M. Schliep PR. Wrote the paper: BT SG DTR.

References

1. Greenwell HC, Laurens LML, Shields RJ, Lovitt RW, Flynn KJ (2010) Placing microalgae on the biofuels priority list: a review of the technological challenges. J R Soc Interface 7: 703–726.
2. Stephens E, Ross IL, Mussgnug JH, Wagner LD, Borowitzka MA, et al. (2010) Future prospects of microalgal biofuel production systems. Trends Plant Sci 15: 554–564.
3. Wijffels RH, Barbosa MJ (2010) An outlook on microalgal biofuels. Science 329: 796–799.
4. Georgianna DR, Mayfield SP (2012) Exploiting diversity and synthetic biology for the production of algal biofuels, Nature 488: 329–335.
5. Chisti Y (2007) Biodiesel from microalgae. Biotechnol Adv 25: 294–306.
6. Patel B, Tamburic B, Zemichael FW, Dechatiwongse P, Hellgardt K (2012) Algal Biofuels: A Credible Prospective? ISRN Renewable Energy 10.5402/2012/631574.
7. Kilian O, Benemann CSE, Niyogi KK, Vick B (2011) High-efficiency homologous recombination in the oil-producing alga *Nannochloropsis sp.* Proc Natl Acad Sci U S A 108: 21265–21269.
8. Boussiba S, Vonshak A, Cohen Z, Avissar Y, Richmond A (1987) Lipid and biomass production by the halotolerant microalga *Nannochloropsis salina*. Biomass 12: 37–47.
9. Hodgson PA, Henderson RJ, Sargent JR, Leftley JW (1991) Patterns of variation in the lipid class and fatty-acid composition of *Nannochloropsis oculata* (Eustigmatophyceae) during batch culture.1. The Growth Cycle. J Appl Phycol 3: 169–181.
10. Rodolfi L, Chini ZG, Bassi N, Padovani G, Biondi N, at al. (2009). Microalgae for oil: strain selection, induction of lipid synthesis and outdoor mass cultivation in a low-cost photobioreactor. Biotechnol Bioeng 102: 100–112.
11. Gouveia L, Oliveira AC (2009) Microalgae as a raw material for biofuels production. J Ind Microbiol Biot 36: 269–274.
12. Chisti Y (2008) Biodiesel from microalgae beats bioethanol. Trends Biotechnol 26: 126–131.
13. Huertas E, Montero O, Lubian LM (2000) Effects of dissolved inorganic carbon availability on growth, nutrient uptake and chlorophyll fluorescence of two species of marine microalgae. Aquacult Eng 22: 181–197.
14. Puri M, Abraham RE, Barrow CJ (2012) Biofuel production: Prospects, challenges and feedstock in Australia. Renew Sust Energ Rev 16: 6022–6031.
15. Borowitzka MA (1999) Commercial production of microalgae: ponds, tanks, tubes and fermenters. J Biotechnol 70: 313–321.
16. Larkum AWD (2010) Limitations and prospects of natural photosynthesis for bioenergy production. Curr Opin Biotech 21: 271–276.
17. Lehr F, Posten C (2009) Closed photo-bioreactors as tools for biofuel production. Curr Opin Biotech 20: 280–285.
18. Tamburic B, Zemichael FW, Crudge P, Maitland GC, Hellgardt K (2011) Design of a novel flat-plate photobioreactor system for green algal hydrogen production. Int J Hydrogen Energ 36: 6578–6591.
19. Tamburic B, Dechatiwongse P, Zemichael FW, Maitland GC, Hellgardt K (2013) Process and reactor design for biophotolytic hydrogen production. Phys Chem Chem Phys 15: 10783–10794.
20. Sandnes JM, Kallqvist T, Wenner D, Gislerod HR (2005) Combined influence of light and temperature on growth rates of *Nannochloropsis oceanica*: linking cellular responses to large-scale biomass production. J Appl Phycol 17: 515–525.
21. Janssen M, Tramper J, Mur LR, Wijffels RH (2003) Enclosed outdoor photobioreactors: light regime, photosynthetic efficiency, scale-up, and future prospects. Biotechnol Bioeng 81: 193–210.
22. Tamburic B, Zemichael FW, Maitland GC, Hellgardt K (2012a) Effect of the Light Regime and Phototrophic Conditions on Growth of the H_2-producing Green Alga *Chlamydomonas reinhardtii*. Energy Procedia 29: 710–719.
23. Sforza E, Simionato D, Giacometti GM, Bertucco A, Morosinotto T (2012) Adjusted light and dark cycles can optimize photosynthetic efficiency in algae growing in photobioreactors. PLOS ONE, 7: e38975.
24. Falkowski PG, Raven JA (2007) Aquatic Photosynthesis. Princeton University Press.
25. Losh JL, Young JN, Morel FMM (2013) Rubisco is a small fraction of total protein in marine phytoplankton. New Phytol 198: 52–58.
26. Raven JA (2011) The cost of photoinhibition. Physiol Plantarum 142: 87–104.
27. Zijffers JWF, Janssen M, Tramper J, Wijffels RH (2008) Design process of an area-efficient photobioreactor. Mar Biotechnol 10: 404–415.
28. Richmond A (1986) Handbook of Microalgal Mass culture: Cell response to environmental factors. Boca Raton Florida: CRC Press. 9969 p.
29. Raven JA, Geider RJ (1998) Temperature and algal growth. New Phytol 110: 441–461.
30. Davison I (1991) Environmental effects on algal photosynthesis: temperature. J Phycol 8: 82–88.
31. Guschina IA, Harwood JL (2006) Lipids and lipid metabolism in eukaryotic algae. Prog Lipid Res 45: 160–186.
32. Yamasaki S, Hirata H (1995) CO_2 concentration change in *Nannochloropsis* sp. Culture medium. Aquacult Eng 14: 357–365.
33. Tawfiq S, Abu-Rezq Al-Musallam L, Al-Shimmari J, Dias P (1999) Optimum production conditions for different high-quality marine algae. Hydrobiologia 403: 97–107.
34. Kromkamp JC, Beardall J, Sukenik A, Kopecky J, Masojidek J, et al. (2009). Short-term variations in photosynthetic parameters of *Nannochloropsis* cultures grown in two types of outdoor mass cultivation systems. Aquat Microb Ecol 56: 309–322.
35. Sukenik A, Beardall J, Kromkamp J, Kopecký J, Masojídek J, et al. (2009) Photosynthetic performance of outdoor *Nannochloropsis* mass cultures under a wide range of environmental conditions. Aquat Microb Ecol 56: 297–308.
36. Guillard RRL, Ryther JH (1962) Studies of marine planktonc diatoms: I. Cyclotella nana hustedt, and Detonula confervacea (cleve) gran. Can J Microbiol 8: 229: 239.
37. Vieler A, Wu G, Tsai CH, Bullard B, Cornish AJ, et al. (2012) Genome, functional gene annotation, and nuclear transformation of the heterokont oleaginous alga *Nannochloropsis oceanica* CCMP1779. PLOS Genetics 8: e1003064.
38. ANACC (2013) Australian National Algae Culture Collection, Commonwealth Scientific and Industrial Research Organisation (CSIRO). Available: http://www.marine.csiro.au/microalgae/methods/Light%20Physical%20Units.htm.
39. Pawlowicz R. (2013) Key Physical Variables in the Ocean: Temperature, Salinity, and Density. Nature Education Knowledge 4: 13.
40. Ramsing N, Gundersen J (2013) Seawater and gasses: tabulated physical parameters of interest to people working with microsensors in marin systems. Unisense. Available: http://www.unisense.com/files/PDF/Diverse/Seawater%20&%20Gases%20table.pdf.
41. Genty B, Briantais JM, Baker NR (1989) The relationship between the quantum yield of photosynthetic electron transport and quenching of chlorophyll fluorescence. Biochim Biophys Acta 990: 87–92.
42. Kolber Z, Falkowski PG (1993) Use of active fluorescence to estimate phytoplankton photosynthesis in situ. Limnol Oceanogr 38: 1646–1665.
43. Ritchie RJ (2008) Universal chlorophyll equations for estimating chlorophylls a, b, c, and d and total chlorophylls in natural assemblages of photosynthetic organisms using acetone, methanol, or ethanol solvents. Photosynthetica 46: 115–126.
44. Gregor J, Maršálek B (2004) Freshwater phytoplankton quantification by chlorophyll a: a comparative study of in vitro, in vivo and in situ methods. Water Res 38: 517–522.
45. Tamburic B, Zemichael FW, Maitland GC, Hellgardt K (2012b) A novel nutrient control method to deprive green algae of sulphur and initiate spontaneous hydrogen production. Int J Hydrogen Energ 37: 8988–9001.
46. Beardall J, Burger-Wiersma T, Rijkboer M, Sukenik A, Lemoalle J, et al. (1994) Studies on enhanced post-illumination respiration in microalgae. J Plankton Res 16: 1401–1410.
47. Hill R, Brown CM, DeZeeuw K, Campbell DA, Ralph PJ (2011). Increased rate of D1 repair in coral symbionts during bleaching is insufficient to counter accelerated photo-inactivation. Limnol Oceanogr 56: 139–146.
48. Simionato D, Sforza E, Corteggiani CE, Bertucco A, Giacometti GM, et al. (2011). Acclimation of *Nannochloropsis gaditana* to different illumination regimes: Effects on lipids accumulation. Bioresource Technol 102: 6026–6032.
49. Brewer PG, Goldman JC (1976) Alkalinity changes generated by phytoplankton growth. Limnol Oceanogr 21: 108–117.
50. Goldman JC, Brewer PG (1980) Effect of nitrogen source and growth rate on phytoplankton-mediated changes in alkalinity. Limnol Oceanogr 25: 552–557.
51. Baker NR (2008). Chlorophyll fluorescence: a probe of photosynthesis in vivo. Annu Rev Plant Biol 59: 89–113.
52. Falkowski PG, LaRoche J (1991) Acclimation to spectral irradiance in algae. J Phycol 27: 8–14.
53. Pal D, Khozin-Goldberg I, Cohen Z, Boussiba S (2011) The effect of light, salinity, and nitrogen availability on lipid production by *Nannochloropsis* sp. Appl Microbiol Biot 90: 1429–1441.
54. Flameling IA, Kromkamp J (1998) Light dependence of quantum yields for PSII charge separation and oxygen evolution in eucaryotic algae. Limnol Oceanogr 43: 25.

4

The Effects of Spatial Scale on Breakdown of Leaves in a Tropical Watershed

Renan S. Rezende[1,2]*, Mauricio M. Petrucio[1], José F. Gonçalves, Jr.[2]

1 Graduate Program in Ecology of Department of Ecology and Zoology, Institute of Biological Sciences, Federal University of Santa Catarina, Florianópolis, Santa Catarina, Brazil, **2** Department of Ecology, Institute of Biology, University of Brasília, Brasília, Federal District, Brazil

Abstract

The objective was to assess the effects of natural variation in the physical structure of the environment on biological communities and on the processing of *Eucalyptus cloeziana* and *Inga laurina* and to identify the controlling factors at different scales along stream order gradients. The study area consisted of 14 sampling sites distributed within a tropical watershed (1st, 2nd, 3rd and 4th order streams replicated in 4 sub-basins). Our samples consisted of 3 g of leaves of *E. cloeziana* (high-quality) and *I. laurina* (low-quality) placed in 252 bags with 10mm mesh (measured by the chemical composition of the detritus). Four samples of each leaf type were collected periodically (three times) over a period of 75–125 days and washed on a sieve to separate the invertebrates. A series of leaf disks were cut to determine ash-free dry mass, polyphenol, lignin, cellulose, total microbial biomass and fungal biomass, and the remaining material was oven-dried to determine the dry weight. We performed analyses within and between spatial scales (regional and local) to assess which watershed scale was the more import determinant of the leaf breakdown rate (*k*). The microbial and shredder were most influenced at the local scale (stream order). Shredders were influenced by microorganisms, with stronger interactions between them than were found to drive the *k* at the local scale. Moreover, differences in the overall *k* and abiotic variables were more strongly influenced at the regional scale (sub-basin), showing that the study scale alters the response of the studied variables. We found higher *k* values at higher values of water velocity, dissolved oxygen and temperature, all of which accelerate biological metabolism in response to variations on the regional scale. Watersheds with warmer microclimates and streams with higher nutrient levels and oxygen could be accelerating the ecosystem metabolism, independent of the detritus quality.

Editor: Kerrie Swadling, University of Tasmania, Australia

Funding: The sources of funding that have supported the work were PROCAD-NF/CAPES (No. 173/2010), PNADB/Edital 17/2009/CAPES (No. 1098/2010), PELD/CNPq (No. 558233/2009-0), FAPEMIG (No. APQ-00274-12), Federal University of Santa Catarina (UFSC) and funds of University of Brasília (UnB - DPP; No. 121366/2011). The funders had no role in study design, data collection and analysis, decision to publish, or preparation of the manuscript.

Competing Interests: The authors have declared that no competing interests exist.

* E-mail: renanrezende30@gmail.com

Introduction

The characteristics of lotic ecosystems show natural patterns along an upstream-downstream gradient due to variations in geomorphology and topography in the watershed [1,2]. Natural changes along this gradient (upstream to downstream) include an increase in the dimensions of the stream (width), changes (increases or decreases) in the velocity of the water, and openings in the riparian canopy that allow greater light penetration [3,4]. Greater luminosity increases the temperature and accelerates photosynthetic production and autotrophic metabolism [3]. Along this gradient, therefore, the relative abundance of micro-organisms increases but that of invertebrates decreases (primarily shredders of organic matter), decreases from the headwaters to the downstream sections [1]. These natural changes also modify the energy input and cycling of organic matter in space and time [1,5]. Senescent leaves are an important source of nutrients and food resources in heterotrophic metabolic environments, primarily in headwaters and small streams [6,7]. However, this material is mineralized and available for use by primary producers and other trophic levels after its decomposition. As a result, leaf breakdown is a key process in lotic ecosystems [6,8]. Leaf breakdown can be influenced by many factors, such as physical and chemical variables (associated with water and detritus) and the activities of communities of decomposers (micro-organisms and aquatic invertebrates) [6–13].

The study of leaf breakdown at the scale of a watershed allows us to observe emerging patterns and identify certain factors that structure the ecosystems at different scales [14,15]. It is evident that a series of successively smaller and nested geomorphologic units can have various patterns and structures depending on the scale that is being analyzed [14,16]. These patterns and structures can be observed in riffles and pools within continuous stretches, which are nested within large rivers that make up a watershed [15,17,18]. Studies that address only one scale are subject to problems because certain variables are measured directly in small areas or across short time intervals, whereas few can be measured at fine resolution over large areas [17,18]. In addition, changes at smaller scales are not maintained at larger scales [14]. Therefore, the issue is that unless patterns are consistent at all scales, the findings at one scale cannot be extrapolated to yield accurate predictions at other scales. Accordingly, tests at multiple scales are needed for confident extrapolation. From this perspective, the evaluation of leaf breakdown at different scales enables the development of an integrated vision of the landscape during this important ecological process [19,20].

The streams in a watershed can be considered within a hierarchical framework that presents organized view of spatial and temporal variations among and within stream systems along the "riverscapes" [21–23]. Therefore, several studies have examined leaf processing at large spatial scales across biomes [24,25], latitudinal gradients [24], and altitudinal gradients [2] and influences of land use [26,27]. Moreover, several recurrent topics emerge from considerations of several spatial scales. These topics include the relative importance of fungi and invertebrates [28], the use of bioindicators [20] and the hierarchical nature of lotic ecosystems [15]. Studies assessing allochthonous leaf breakdown at a watershed scale are rare worldwide, but they have been performed in temperate systems [15,20,23,29].

In tropical streams, individual riffles or short stream reaches continue to be the most frequent sites for studies of leaf processing based on the traditional conceptual model [30,31]. Several factors are known to cause variation in the rates of processing within and among tropical stream reaches [13,32]. These factors include the effects of species mixing [33,34], litter quality [35,36], micro-organism communities [37,38], invertebrate communities [39–41], detritivores and shredders [10,42] and seasonal effects [8,43,44]. However, systematic assessments of variability in allochthonous leaf breakdown rates across multiple spatial scales using the watershed as the sampling unit have not been performed in tropical stream systems. This study could help to answer important questions, such as "how does spatial structure influence ecosystem function and how do we integrate within and between spatial scales to assess function", suggested by Sutherland et al. [45] as one of 100 fundamental ecological questions.

Based on the premise that leaf breakdown is the result of the activity of decomposer organisms and the physical and chemical processes occurring in the stream water, which vary along the scale investigated in the study scale [6,12,13], the following hypotheses were tested in this study: (i) natural differences in the physical nature of the stream (increasing canopy opening, water velocity, temperature and nutrient concentrations) accelerate biological metabolism and leaf breakdown from upstream to downstream; (ii) shredders decrease and micro-organisms increase in importance from upstream to downstream; and (iii) differences in the overall *k* values will be more clearly understandable (strongly explained) at an increased spatial scale. The objective of the study was to assess the natural effects of variation in the physical environment on biological communities and the leaf breakdown rates of *Eucalyptus cloeziana F. Muell* and *Inga laurina Sw. Willd* and to identify the controlling factors at different scales along the stream order gradient.

Methods

The Study System

The study area consisted of 14 sampling sites distributed along the Gama-Cabeça do Veado watershed, a part of the Federal District in west central Brazil, comprising 1st, 2nd, 3rd and 4th order streams replicated in 4 sub-basins (Figure 1). The area includes important waterways that form the Paraná basin, a part of the Cerrado biome (Brazilian Savannah). The climate is tropical and has distinct rainy-hot (October to April) and cold-dry seasons (May to September). The mean annual temperature is 20°C, and the altitude varies between 1025 and 1150 m above sea level. The study area included three conservation units protecting the entire watershed and all sampling sites studied (Ecological Station of the University of Brasília, Ecological Reserve of IBGE and Ecological Station of the Botanical Gardens of Brasília). The study was approved by Ministry of Environment of Brazil through the System of Information and Authorization on Biodiversity (SISBIO) for activities with scientific purpose (code: 39629-1), and also was approved by the Scientific and Technical Council of the Ecological Station of the University of Brasília (code: 05–12), IBGE Ecological Reserve (code: 54 PC - PAD 1) and Botanical Gardens of Brasília (code: 13/2011).

Procedures

The experiment was conducted from June through October 2011 (the winter season), a period of zero rainfall, resulting in a high level of homogeneity in the physical and chemical features of the environment. Based on their chemical composition, we selected leaves from two species for use as detritus. The leaves of an exotic species (*Eucalyptus cloeziana F. Muell*) were used to represent high-quality detritus. The planting of *Eucalyptus* mono-cultures in place of the native vegetation has potential repercussions for stream basins. The area in which the study was conducted does not contain *Eucalyptus* plantations. However, the substitution of this monoculture for native vegetation has occurred in neighboring basins, where this monoculture has expanded, as it has throughout Brazil, e.g., to supply charcoal for steelmaking and pulp for papermaking [35]. The low-quality detritus from a native species (abundant in riparian vegetation; *Inga laurina Sw. Willd*) was also used in the study, as this species best represents the plants of the Cerrado.

The leave The two types of leaves were chemically characterized by the mean values of total polyphenols (22.80 ± 2.5; 18.29 ± 1.8 mg/g^{-1}), total tannic acids (0.003 ± 0.0002; 0.002 ± 0.0004 mg/g^{-1}), lignin (42.61 ± 0.7; 45.94 ± 0.5%), cellulose (24.69 ± 1.5; 37.39 ± 1.2%), hardness (0.17 ± 0.1; 0.6 ± 0.3 cm/g^{-1}), nitrogen (13.16 ± 1.3; 16.41 ± 1.0 g/kg^{-1}) and phosphorus (0.46 ± 0.05; 0.53 ± 0.07 g/kg^{-1}) in *E. cloeziana* and *I. laurina*, respectively. The breakdown rates for these two leaf types (collected in nets 1 m^2 in area placed 1.5 m from the ground) were measured individually by the loss of weight of 3 g (± 0.1 g dry weight) of leaves, correcting for the initial humidity and transport loss [46], incubated in litter bags (15 × 15 cm, 10 mm mesh size).

In total, 252 litter bags were placed at a depth of 0.3 m in pool areas at the 14 sampling sites in 1st, 2nd, 3rd and 4th order streams [47] in the 4 sub-basins (Gama – SB1, Capetinga – SB2, Taquaras – SB3 and Cabeça de Veado – SB4 sub-basins). The removal of the *E. cloeziana* and *I. laurina* leaves occurred initially after 10 days of incubation, which corresponds to the expected time required for leaching and initiating microbial colonization [46]. After this sampling, the principal leaf breakdown rate (*k*) was used to estimate the next sampling time for each detritus type (at approximately 75% of the remaining mass), which was determined to be 40 days for *E. cloeziana* and 85 days for *I. laurina*. This stage of decomposition occurs when microbial activity is high and the invertebrate community is already established. Subsequently, the *k* value was used a second time to perform corrections and acquire new values to obtain approximately 50% of the remaining mass for each detritus type. This sampling time was determined to be 75 days for *E. cloeziana* and 125 days for *I. laurina*. After this period, the community established during the ecological process of degradative succession is affected by the reduction in the available detritus (additional information about k, see also Chapter 6 of [46]).

The sampling times were calculated by dividing the initial weight (W0) by the estimated value of *k*. This calculation yields the time for the total course of leaf processing (TLP, days). From the

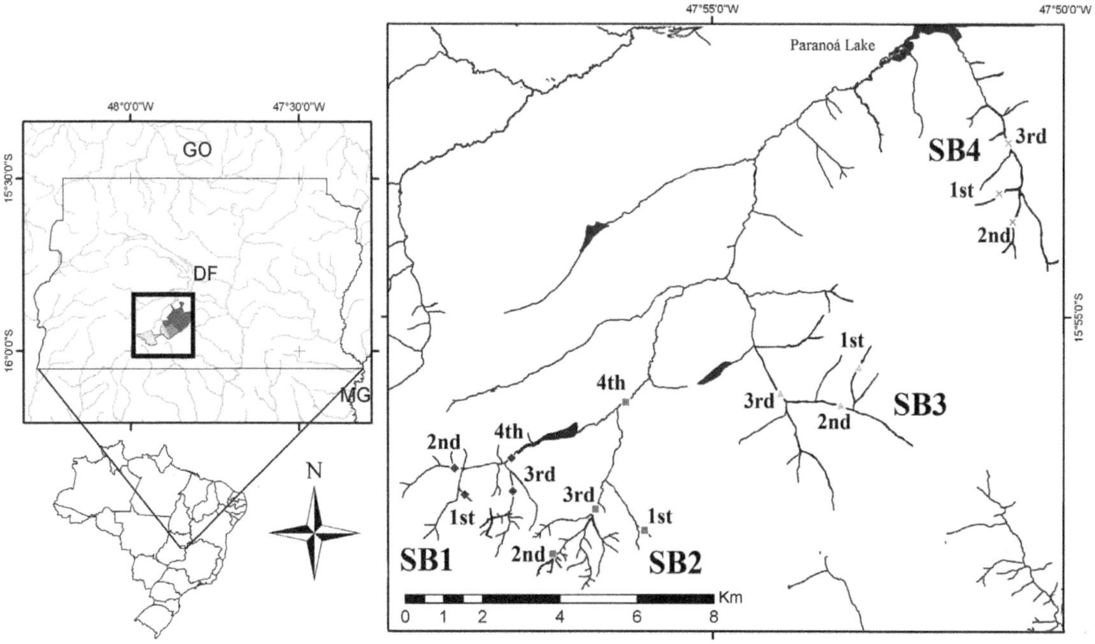

Figure 1. Sampling sites. Geographic location of the sampling sites consisting of streams of the 1st, 2nd, 3rd and 4th orders replicated in 4 sub-basins in the Gama-Cabeça de Veado watershed, Federal District (Brazil).

equation $W0/k = TLP$, we can calculate how many days will be required to reach a desired percentage of the initial weight (Wt). The first sample was collected after 10 days of incubation for both species, so that TLP for 10 days/0.25 = day on which Wt = 75%. The next sample was collected after 40 days for *E. cloeziana* and after 75 days for *I. laurina*, so that TLP 40/75 days/0.5 = day by which Wt = 50%. The above procedure was performed for each sample site (based on the mean value) and type of detritus. However, it was not possible to determine the final value for *I. laurina* because the dry season ended after 120 days, before 50% of the mass had been lost. Measurements after the end of the dry season would not have been meaningful because variations in rainfall and associated variations in other physical and chemical conditions would have influenced the results.

On removal from the streams, the litter bags were placed individually into insulated plastic bags and transported in thermal containers (± 4°C) to the laboratory. Temperature, electrical conductivity, pH, dissolved oxygen and water turbidity were obtained *in situ* with a multi-analyzer measured each time leaf bags were removed. The depth and average speed of the right, left and central portions of the watercourse were measured with a flow-meter, and the instantaneous discharge of water was then calculated. We collected 1 L of water to determine the nitrate [48], ammonia [49] and orthophosphate [50] concentrations. The canopy openings were quantified using hemispherical photographs taken with a digital camera equipped with a fish-eye lens. These photographs were later analyzed using Gap Light Analyzer software (2.0). The leaves were washed with tap water in a 120 µm mesh sieve. The invertebrates retained on the sieve were preserved in 70% alcohol for later identification and counting [51,52]. The numbers of *taxa* and individuals were calculated for the aquatic invertebrate community, and biomass was obtained by desiccation at 60°C for 72 h. The invertebrates were classified into five feeding categories [51–53]: gathering-collectors (G–C), filtering-collectors (F–C), shredders (Sh), scrapers (Sc) and predators (P).

Five leaves from each sample were randomly collected, and three disks (1.2 cm diameter) were extracted from each leaf, resulting in three five-disk sets. One set was used to determine the remaining ash-free dry mass (AFDM; calculated after incineration in a muffle furnace at 550°C for 4 h), and the other sets were used to assess the ergosterol and ATP concentrations. The remaining material was oven-dried at 60°C for 72 h to determine its dry weight. The leaf breakdown rates (*k*) were calculated using the negative exponential model of percent mass lost over time ($W_t = W_{0}e^{-kt}$; Wt = remaining weight; W0 = initial weight; −k = decay rate; t = time). After the leaves had been dried and weighed, they were pulverized for further analysis of the total polyphenol and tannic acid concentration [54], lignin and cellulose contents [55] and the resistance of leaves to rupture (hardness of intact leaves [46]). Values for total nitrogen were obtained using a CHN basic analyzer (Carlo Erba 1500 for WI; Thermo Electron Corp. Milan, Italy), and values for total phosphorus were obtained using the ascorbic acid method after acid digestion. The total micro-organism biomass was measured by quantifying ATP [56]. The biomass of aquatic Hyphomycetes was evaluated by quantifying ergosterol, a lipid exclusive to fungal membranes in this community [57].

Data Analysis

An analysis of variance (function lm, package stats for R version 2.12.1; [58]) was used to analyze the physical and chemical parameters of the water (temperature, electrical conductivity, pH, dissolved oxygen, turbidity, nitrites, nitrates, orthophosphates and mean velocity) and the structure of stream stretches (instantaneous discharge of the stream and canopy openings in riparian vegetation) as dependent variables, using two categorical factors, namely, sub-basins and stream order. Stream order was also used as a co-variate (continuous variable). We also used the leaf mass remaining, invertebrate communities (number of *taxa*, density and biomass), the relative abundance of functional trophic groups of invertebrates (gathering-collectors, filtering-collectors, shredders, scrapers and predators) and microbial biomass (ATP and ergosterol) as dependent variables against the same two categorical factors. Stream order was also used as a co-variate (continuous variable). This procedure was performed similarly for both types of detritus. All models used a Gaussian distribution (link = log; test = F). We used an analysis of contrasts to discriminate among categorical variables. The normality of the data was tested using a Kolmogorov-Smirnov test, the homogeneity of variance was determined with a Levene test, and the data were transformed whenever necessary with the Naperian logarithm (ln) to obtain the best fit [58].

Results

Abiotic Variables

The values of instantaneous discharge, electrical conductivity and nitrates were the highest in the 3rd and 4th order streams. In contrast, the 1st order streams had the highest values for temperature, canopy opening and nitrite concentrations in the water, and the water velocity was the lowest. Dissolved oxygen, pH, turbidity and orthophosphates did not differ among the stream orders (analysis of contrasts, $p < 0.05$; Table S1, Table 1, Figure 2). In sub-basin 2 (SB2), we observed high electrical conductivity, high nitrite and nitrate concentrations and low water temperatures. The values for canopy opening and water velocity were highest in sub-basins 1 (SB1) and 4 (SB4), respectively. Dissolved oxygen and orthophosphates were highest in sub-basin 3 (SB3). Instantaneous discharge, pH and turbidity did not differ among the sub-basins (analysis of contrasts, $p < 0.05$; Table S1, Figure 2). We observed that the higher percentages of sums of squares and variance in instantaneous discharge and water velocity could be explained by differences in the stream order. However, dissolved oxygen, electrical conductivity, temperature, pH, turbidity, canopy opening, nitrates, nitrites and orthophosphates exhibited a high level of variance among the sub-basins (Table 1).

Leaf Breakdown Rates

The leaf breakdown rates (k) were the highest in the 2nd order streams, with values of −0.0083 and −0.0022 for *E. cloeziana* and *I. laurina*, respectively, followed by the 3rd order stream sections, with values of −0.0071 and −0.0022. We also observed higher k values in 1st order streams (−0.0053 and −0.0015) than in 4th order streams (−0.0051 and −0.0018) for *E. cloeziana* and *I. laurina*, respectively. However, the remaining mass did not differ among stream orders for either of the detritus types (Figure S1, Table 2, Figure 3A and 4A). The highest k values were observed in SB4 (−0.0105 and −0.0030), followed by SB3 (−0.0088 and −0.0022 for *E. cloeziana* and *I. laurina*, respectively). For *E. cloeziana*, the values were −0.0062 and −0.0049, whereas *I. laurina* exhibited values of −0.0018 and −0.0016 (for SB1 and SB2, respectively). The remaining mass showed the lowest values in SB4 and SB3 among the sub-basins studied. The variance in the remaining mass was higher (by sums of squares) and also explained the variations in the sub-basins (Figure S1, Table 2, Figure 3A and 4A).

Biotic Community

The density of invertebrates was higher in 4th order streams for both detritus types (means of 27 and 21 ind/g for *E. cloeziana* and *I. laurina*, respectively). Differences for *I. laurina* were observed among sub-basins, with high values in SB3 (mean 24 ind/g) and SB4 (mean 26 ind/g). The number of *taxa* was significantly different among the stream orders and sub-basins, with the highest values in SB3 (mean of 7 and 6 *taxa* for *E. cloeziana* and *I. laurina*, respectively) and SB4 (mean of 6 *taxa* for *E. cloeziana* and *I. laurina*), primarily in 3rd order streams for both detritus types (mean of 7 and 6 *taxa* for *E. cloeziana* and *I. laurina*, respectively). However, the biomass (total mean 0.003 and 0.002 ind/g for *E. cloeziana* and *I. laurina*, respectively) did not differ among the stream orders or sub-basins for either of the detritus types. The high variances in density, richness and biomass (in terms of the percentage of the sums of squares) were explained by differences in the sub-basins for both detritus types, except for the density of invertebrates in *I. laurina* (Figure S2 and S3, Table 2).

The functional trophic groups differed significantly among stream orders only for the filtering-collectors, with the highest values in the 3rd order streams (mean 20% for *E. cloeziana* and *I. laurina*) and the lowest in the 1st order streams (mean of 10% for *E. cloeziana* and *I. laurina*) for both detritus types. The relative abundance of predators was higher in SB3 (mean 24 and 25% for *E. cloeziana* and *I. laurina*, respectively) and SB4 (mean of 32 and 34% for *E. cloeziana* and *I. laurina*, respectively), whereas the values for shredders were higher in SB3 (mean of 27 and 25% for *E. cloeziana* and *I. laurina*, respectively) for both detritus types. However, SB1 exhibited high abundances of gathering-collectors (mean of 41 and 50% for *E. cloeziana* and *I. laurina*, respectively) and scrapers (mean of 18 and 16% for *E. cloeziana* and *I. laurina*, respectively) but a low abundance of filtering-collectors (mean of 6 and 8% for *E. cloeziana* and *I. laurina*, respectively) for both detritus types. The high variance in the relative abundance for all functional trophic groups (by the sums of squares) was also explained by changes in sub-basins for both detritus types (Figure S2 and S3, Table 2, Figure 3B and 4B).

The ATP values differed only among sub-basins for both detritus types, with the highest values in SB3 (mean of 2155.8 nmoles/g AFDM) for *E. cloeziana* (total mean 991.8 nmoles/g AFDM) and in SB3 (mean 633.3 nmoles/g AFDM) and SB4 (mean 2023.9 nmoles/g AFDM) for *I. laurina* (total mean 847.8 nmoles/g AFDM) (Table 2; Fig. 2C and 3C). There were no differences in the ergosterol concentrations among the hydrological stream orders and sub-basins for *E. cloeziana* (total mean 541.2 µg/g). However, we found higher ergosterol concentrations for *I. laurina* (total mean 382.9 µg/g) in SB3 (mean 392.6 µg/g) and SB4 (mean 464.1 µg/g), although they did not differ among the hydrological stream orders. The variances in ATP and ergosterol concentrations were also explained by changes in sub-basins for both detritus types, except for ergosterol in *E. cloeziana*, which showed a high level of variation with stream order (Figure S2 and S3, Table 2, Figure 3D and 4D).

Discussion

Scale Analysis

The instantaneous discharge, water velocity, turbidity and nitrogen series (nitrate and nitrite) were more influenced by changes in stream order (high heterogeneity), with higher values downstream (increasing from 1st to 4th order), as expected according to Vannote et al. [1], except that nitrites exhibited the inverse pattern. The finding of relatively few influences at local scales can be explained by the large discontinuities inherent in smaller geomorphological units (habitat patches create discontinuities in space) that increase the potential influence from the local characteristics of the environment [4,5,14,18]. Flow changes, for example, create hydrological discontinuities along stream corridors and isolate habitats. However, the other abiotic variables were influenced by changes in sub-basins (high homogeneity) that correspond to regional scales [19,21]. These factors worked at the watershed level and may increase its fragility in the face of intense

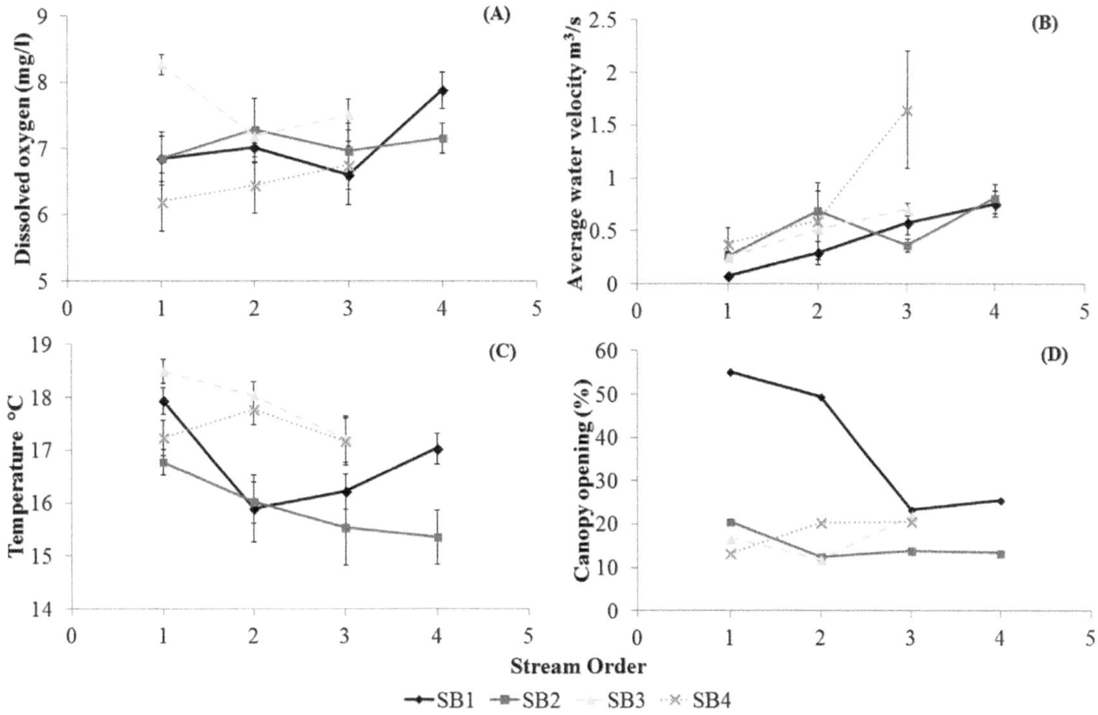

Figure 2. Abiotic variables. Mean values and standard errors for dissolved oxygen (A), water velocity (B), water temperature (C) and canopy opening (D) for the stream orders and among sub-basins.

climatic changes because the climate is the primary controlling factor at large scales [22]. The basins are influenced by environmental factors that systematically change across longitudinal (upstream/downstream), vertical (sediment/water) and lateral (terrestrial/aquatic) gradients, forming different spatial and temporal patterns at regional and local scales [5,17,18].

As expected, the high-quality detritus (*E. cloeziana*) showed more rapid leaf breakdown rates (*k*) than the low-quality (*I. laurina*) detritus, indicating that the rate could be driven by micro-scale processes [14]. These findings represent important evidence that the riparian vegetation could be responsible for determining the ecosystem characteristics [30,31,59], as also proposed by Gonçalves et al. [35] for tropical systems, highlighting the need to study this vegetation. For example, if riparian vegetation is composed of plant species that have a higher stoichiometric ratio (higher quality), we expect more rapid response for organic matter cycling (higher decomposition). This direct relationship between terrestrial and aquatic ecosystems demonstrates that any modification in a riparian ecosystem would affect the function, primarily in areas composed of palatable plants (high quality and decomposition), corroborating the work of Frauendorf et al. [60]. Thus, lower quality vegetation (slower decomposition) will be less sensitive to other factors, and this may explain the resistance of the Brazilian savanna in comparison with other Brazilian tropical systems [61]. Moreover, the leaf breakdown at the macro-scale was more influenced by the regional scale (sub-basin) than by the local scale (stream order) [14], confirming our hypothesis. This result indicates that the patterns observed in studies covering a given time period (timely studies), common in tropical literature (for more see also [13,32]), cannot be generalized from local to regional scales [17,29,62] or to whole watersheds [14–16,18]. In addition, this finding may indicate fragility in the upstream basins due to the slower leaf breakdown rates. However, the upstream area is a source of nutrients and organisms for the downstream basins [1]. The upstream basins can give support productivity and may be responsible for extending the depuration capability of the system (as represented by the microbial pools) and maintaining functionality downstream [60]. Therefore, we believe that the association between detritus quality (important at the micro-scale) and the environmental features of the watershed (important at the macro-scale) is responsible for shaping organic matter cycling in the watershed and should be further investigated in future studies.

Previous studies indicating that the microbial community is the principal decomposer [6,30,35], are confirmed by our results, as we found that the high-quality leaf species was also more susceptible to leaching and microbial action, whereas the low-quality leaf species was influenced primarily by fungal colonization. Both leaf samples were consumed by shredders, but a higher abundance of shredders was observed in the *E. cloeziana* detritus. We also found a higher loss of mass due to the high water velocity (mechanical fragmentation and leaching), dissolved oxygen and temperatures, which accelerate biological metabolism [37,63]. The detritus quality is important only for defining the local rates and their pathways for leaf breakdown [6,35,64]. However, the detritus quality has little influence on the general pattern along the "riverscape" and at any specific scale [15]. Therefore, based on an analysis of the samples after a certain percentage of mass has been lost (25, 75 and 50%) and not simply at predefined time points (e.g., 7, 15, 30 days), it is possible to show a clear colonization effect independent of quality. To be sure, detritus quality is a highly important determinant of the abundance of shredders and explains the importance of shredders for both detritus types [39]. We cannot study the variations associated with spatial scale in terms only of the local context because there are many factors in the ecological levels (community and ecosystem) that are responsible for variability found in the large scale [18,29].

Table 1. Degrees of freedom (DF), residuals, sums of squares (%), F tests and analyses of variance and contrasts (AC) for dissolved oxygen, electrical conductivity, water temperature, pH, turbidity, water velocity, canopy openness and nitrites, nitrates and orthophosphates in sub-basins and along the stream orders in the Gama-Cabeça de Veado Basin.

		Df	Sum Sq (%)	F value	Pr(>F)	AC
Instantaneous discharge	Sub-Basin	3	5.71	1.97	0.127	
	Order	1	31.56	32.71	**<0.001**	$1^{st} = 2^{nd} < 3^{rd} = 4^{th}$
	Residuals	65	62.72			
Dissolved oxygen	Sub-Basin	3	11.34	2.80	**0.047**	SB4 = SB2 = SB1 < SB3
	Order	1	0.87	0.65	0.425	
	Residuals	65	87.78			
Electrical conductivity	Sub-Basin	3	42.63	27.22	**<0.001**	SB4 = SB1 = SB3 < SB2
	Order	1	23.42	44.86	**<0.001**	$1^{st} = 2^{nd} < 3^{rd} = 4^{th}$
	Residuals	65	33.93			
Temperature	Sub-Basin	3	22.29	6.66	**<0.001**	SB2 < SB1 = SB4 = SB3
	Order	1	5.15	4.62	**0.035**	$4^{th} = 3^{rd} = 2^{nd} < 1^{st}$
	Residuals	65	72.55			
pH	Sub-Basin	3	5.44	1.25	0.298	
	Order	1	0.47	0.33	0.567	
	Residuals	65	94.07			
Turbidity	Sub-Basin	3	2.66	0.62	0.605	
	Order	1	4.23	2.96	0.090	
	Residuals	65	93.10			
Average water velocity	Sub-Basin	3	10.44	3.79	**0.014**	SB1 = SB3 = SB2 < SB4
	Order	1	29.77	32.37	**<0.001**	$1^{st} < 2^{nd} = 4^{th} = 3^{rd}$
	Residuals	65	59.77			
Canopy opening	Sub-Basin	3	59.83	41.13	**<0.001**	SB2 = SB3 = SB4 < SB1
	Order	1	8.64	17.83	**<0.001**	$4^{th} = 3^{rd} = 2^{nd} < 1^{st}$
	Residuals	65	31.52			
Nitrate	Sub-Basin	3	42.56	32.68	**<0.001**	SB3 = SB4 = SB1 < SB2
	Order	1	29.22	67.32	**<0.001**	$1^{st} = 2^{nd} < 3^{rd} < 4^{th}$
	Residuals	65	28.21			
Nitrite	Sub-Basin	3	35.50	23.86	**<0.001**	SB4 = SB3 < SB1 < SB2
	Order	1	32.25	65.02	**<0.001**	$4^{th} < 2^{nd} = 3^{rd} < 1^{st}$
	Residuals	65	32.24			
Orthophosphate	Sub-Basin	3	11.62	2.87	**0.043**	SB3 < SB1 = SB4 = SB2
	Order	1	0.54	0.40	0.527	
	Residuals	65	87.83			

However, the local approach has been used in all previous tropical studies of leaf breakdown [12,13].

Leaf Breakdown Rates

In agreement with the proposal of Gonçalves et al. [13] for tropical systems, the k values of *E. cloeziana* were classified as intermediate ($-0.0173 > k > -0.0041$), and those of *I. laurina* were classified as slow ($k < -0.0041$) for all sampling sites, indicating a strong influence of detritus quality. The high leaching due to the high solubility of polyphenols and tannins (secondary compounds) in *E. cloeziana* can accelerate the decay rate [64,65]. Therefore, the rapid leaching of these secondary compounds, which has an inhibitory effect on detritivores, as well as the lower hardness of *E. cloeziana*, did not limit biotic colonization due to the low residence time in this type of detritus [35,64]. Additionally, *Eucalyptus* sp. (an exotic species) is rapidly colonized and decomposed in the Brazilian savannah. It is possible that this pattern is due to the high quality of *Eucalyptus* relative to native species [35,39,66].

In contrast, we found lower breakdown rates for *I. laurina* which were most likely a consequence of a high content of structural compounds (lignin and cellulose) and relative hardness (cuticle thickness), hindering the release of other chemical compounds (e.g., polyphenols, nitrogen and phosphorus [35,65,66]). Therefore, the chemical characteristics of detritus determine the speed of processing (primarily at local scales), showing that leaf breakdown rates increase with quality and palatability [6,7,35,39]. Detritus quality is of lower importance when if it is observed at different

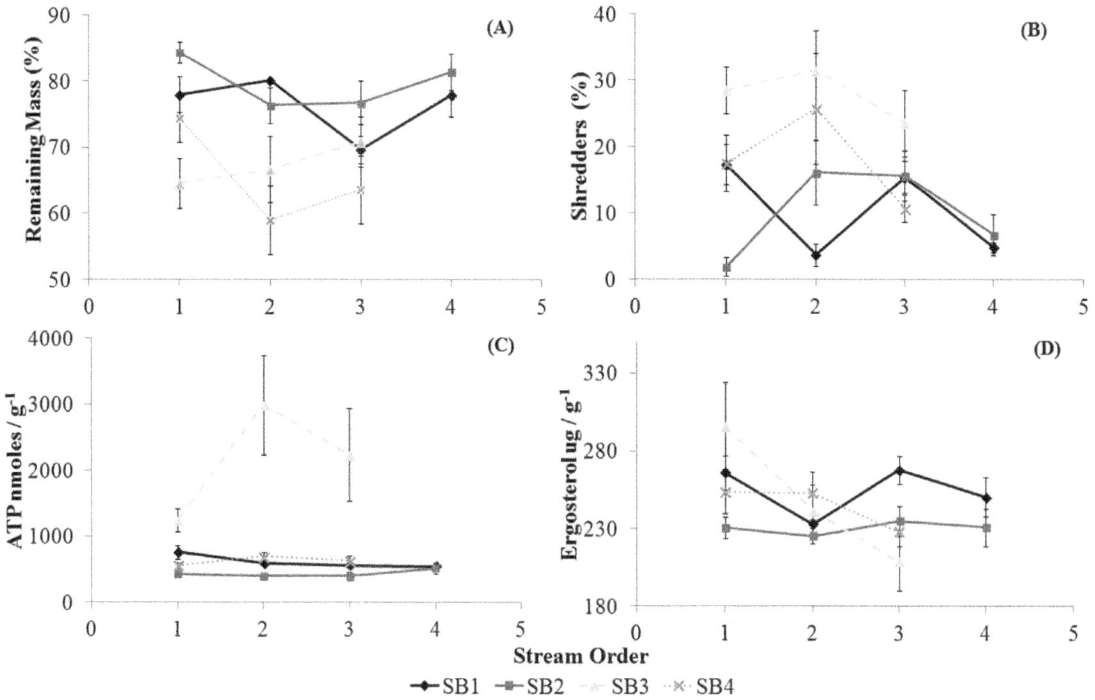

Figure 3. Leaf breakdown process in *E. cloeziana*. Mean values and standard errors for the remaining mass (A), shredder abundance (B), total microbial biomass (ATP; C) and fungal hyphomycetal biomass (Ergosterol; D) for the stream orders and sub-basins for *E. cloeziana*.

scales. In the study area, we observed that the regional scale is decisive for driving the general pattern of this important ecological process along the "riverscapes" [21,22].

Abiotic and Biotic Variables in Leaf Breakdown

The natural environmental changes that occur across stream orders [4] were not sufficient to modify the remaining mass, and

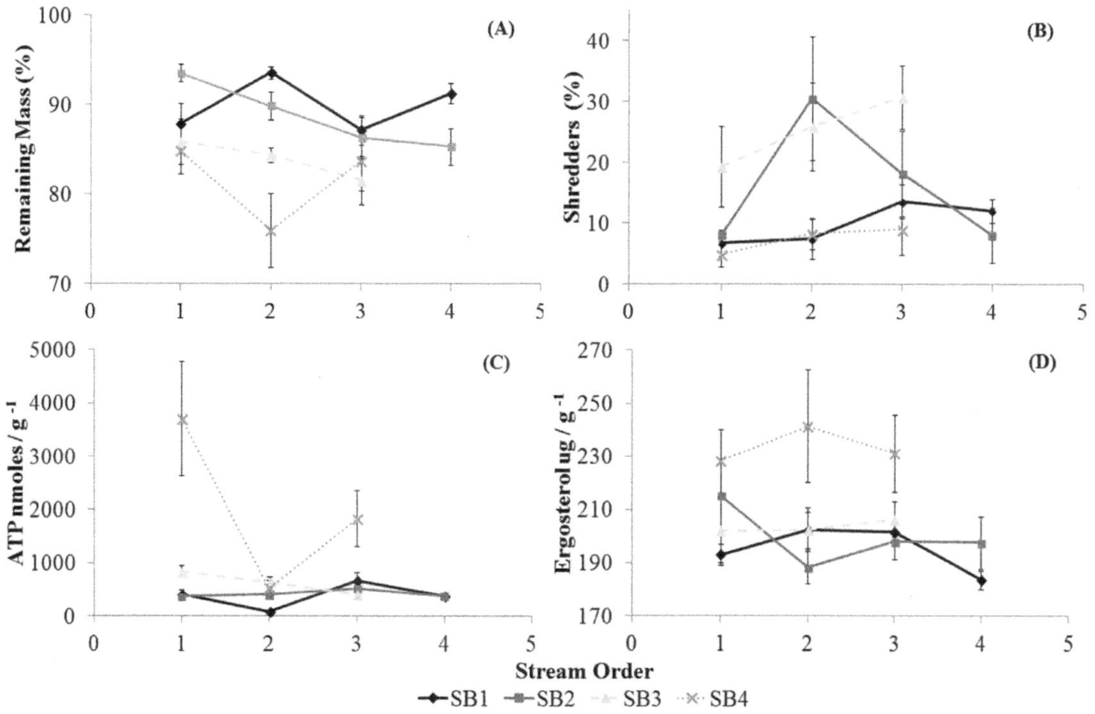

Figure 4. Leaf breakdown process in *I. laurina*. Mean values and standard errors for the remaining mass (A), shredder abundance (B), total microbial biomass (ATP; C) and fungal hyphomycetal biomass (Ergosterol; D) for the stream orders and sub-basins for *I. laurina*.

Table 2. Degrees of freedom (DF), residuals, sums of squares (%), F test and analyses of variance and contrasts (AC) for the remaining mass, density, richness and biomass of invertebrates, functional trophic groups of invertebrates (predators, shredders, gatherer-collectors, filtering-collectors and scrapers), ATP and ergosterol in sub-basins and along the stream orders in the Gama-Cabeça de Veado Basin for *E. cloeziana* and *I. laurina*.

		E. cloeziana					*I. laurina*				
		Df	**Sum Sq (%)**	**F value**	**Pr(>F)**	**AC**	**Df**	**Sum Sq (%)**	**F value**	**Pr(>F)**	**AC**
Remaining mass	Sub-Basin	3	15.05	6.51	**<0.001**	SB4 = SB3 < SB1 = SB2	3	11.01	4.72	**0.004**	SB4 = SB3 < SB2 = SB1
	Order	1	0.18	0.24	0.628		1	1.95	2.51	0.116	
	Residuals	110	84.76				112	87.04			
Invertebrate Density	Sub-Basin	3	0.63	0.25	0.864		3	9.29	4.14	**0.008**	SB2 = SB1 < SB3 = SB4
	Order	1	5.49	6.43	**0.013**	$2^{nd} = 1^{st} = 3^{rd} < 4^{th}$	1	6.88	9.19	**0.003**	$2^{nd} = 1^{st} = 3^{rd} < 4^{th}$
	Residuals	110	93.88				112	83.83			
Invertebrate Richness	Sub-Basin	3	12.42	5.65	**0.001**	SB2 = SB1 < SB4 = SB3	3	25.53	15.24	**<0.001**	SB2 < SB1 = SB3 = SB4
	Order	1	7.00	9.55	**0.003**	$1^{st} = 4^{th} = 2^{nd} < 3^{rd}$	1	11.92	21.33	**<0.001**	$1^{st} = 2^{nd} < 4^{th} = 3^{rd}$
	Residuals	110	80.58				112	62.56			
Invertebrate Biomass	Sub-Basin	3	1.29	0.48	0.697		3	4.10	1.62	0.189	
	Order	1	0.18	0.20	0.652		1	1.47	1.74	0.190	
	Residuals	110	98.53				112	94.43			
Predators	Sub-Basin	3	9.44	3.84	**0.012**	SB1 = SB2 = SB3 < SB4	3	11.87	5.09	**0.002**	SB1 = SB2 < SB3 = SB4
	Order	1	0.45	0.55	0.461		1	1.12	1.45	0.232	
	Residuals	110	90.11				112	87.01			
Shredders	Sub-Basin	3	13.89	5.99	**0.001**	SB2 = SB1 = SB4 < SB3	3	9.76	4.09	**0.008**	SB4 = SB1 = SB2 < SB3
	Order	1	1.12	1.45	0.232		1	1.26	1.58	0.211	
	Residuals	110	84.99				112	88.99			
Gatherer-Collectors	Sub-Basin	3	22.51	10.68	**<0.001**	SB4 = SB3 < SB2 = SB1	3	27.15	14.29	**<0.001**	SB3 = SB2 = SB4 < SB1
	Order	1	0.23	0.33	0.565		1	1.91	3.02	0.085	
	Residuals	110	77.26				112	70.93			
Filtering-Collectors	Sub-Basin	3	11.14	4.89	**0.003**	SB1 < SB3 = SB4 = SB2	3	10.61	4.75	**0.004**	SB1 = SB2 = SB3 < SB4
	Order	1	5.23	6.88	**0.010**	$1^{st} = 2^{nd} = 4^{th} < 3^{rd}$	1	5.98	8.03	**0.005**	$1^{st} < 4^{th} = 2^{nd} = 3^{rd}$
	Residuals	110	83.63				112	83.41			
Scrapers	Sub-Basin	3	21.70	10.21	**<0.001**	SB3 = SB4 = SB2 < SB1	3	11.79	5.07	**0.002**	SB3 = SB4 = SB2 < SB1
	Order	1	0.34	0.47	0.493		1	1.50	1.94	0.166	
	Residuals	110	77.96				112	86.71			
ATP	Sub-Basin	3	26.40	13.19	**<0.001**	SB2 = SB1 = SB4 < SB3	3	17.59	8.07	**<0.001**	SB2 = SB1 = SB3 < SB4
	Order	1	0.20	0.31	0.581		1	1.06	1.46	0.230	
	Residuals	110	73.40				112	81.35			
Ergosterol	Sub-Basin	3	2.25	0.87	0.457		3	9.82	4.07	**0.009**	SB1 = SB2 = SB3 < SB4
	Order	1	3.27	3.81	0.053		1	0.10	0.13	0.720	
	Residuals	110	94.47				112	90.08			

the local scale could not affect the decomposer communities (shredders and microorganisms) for either of the detritus types. Decomposer communities are the driving factors for leaf breakdown, and their absence leads to similar breakdown patterns along the stream order gradient [6]. This finding might indicate that ecological functioning in headwater streams (1st to 3rd order) was similar within the same sub-basin [1]. However, this process might change over a large spatial gradient, as represented by the sub-basin scale [22]. Nevertheless, increases in richness and the density of invertebrates and a decrease in the abundance of filtering-collectors for both detritus types were observed across this large spatial gradient. These results demonstrate that these variables had no effect on leaf breakdown [52,53].

The values of the remaining mass for both detritus types were lower in SB4 and SB3 (high decomposition) than in the other sub-basins. It is probable that the reason for this difference was the higher temperatures, dissolved oxygen concentrations and water velocities resulting from the microclimate of the geographic location (within a valley). Therefore, the higher temperatures [63] and oxygen concentrations [37] observed in SB4 and SB3 may elevate the metabolic activity of the decomposer community [9], especially microorganisms. The higher metabolic activity of the decomposer community, associated with high water velocity (mechanical fragmentation and leaching), which increased the degree of physical abrasion [67], accelerated the leaf breakdown rates. In SB4 and SB3, higher density and richness of invertebrates and higher shredder abundance, with the greatest densities in *E. cloeziana*, were also observed. The shredders directly utilize leaf tissues for feeding, and increasing biological fragmentation [7,42] can also accelerate the leaf breakdown rates [26]. Certain shredders in these locations (genus *Phylloicus*) can build their capsules from leaf tissue, and this use of leaf material also contributes to fragmentation [52,53].

The relative abundance of shredders was influenced by variation, primarily among the sub-basins. A greater relative abundance of shredders in comparison with other tropical systems was observed in SB3 for both detritus types [10,39,42]. A low relative abundance of shredders was found in the other sub-basins. This result is consistent with the findings of previous studies in the Cerrado [13,36,39]. The importance of shredders for leaf breakdown is unclear in the tropics due to their low abundance or absence in these streams [10,39,42], but studies have shown little effect in tropical streams [11,33]. From a global perspective, the strong effects observed in the current study were most likely due to the preference of shredders (primarily Trichoptera and Plecoptera) for high altitudes (due to the lower temperatures) in tropical regions [42]. The preference of this group for *E. cloeziana* indicates that detritus quality is also important [10] and that the composition of the vegetation influences the functioning of aquatic systems. Therefore, higher altitudes (low temperature [42]), high dissolved oxygen, the composition of the flora (ideally including *E. cloeziana* [10]) and moderate values of nutrient concentrations in the water [62] favor a high abundance of shredders. The predominance in the Cerrado of leaves that are low in nutrients [66], is associated with hydric and thermic stress and could be responsible for the absence or low abundance of shredders found in most tropical streams [10,42].

The high-quality detritus (*E. cloeziana*) was shown to be most influenced by the total microbial community and the low-quality detritus (*I. laurina*) by the fungal community in SB4 and SB3 (high decomposition). *E. cloeziana* has elevated amounts of labile compounds, facilitating the activity of bacteria (rapid life cycles) that use compounds derived from the leaching of the leaves of labile detritus as their preferred resource [68]. These bacteria could be important during leaf breakdown and not only, as observed by several authors, at the early stages of the process [6,13,35]. However, this pattern is rarely observed in tropical streams. In lower-quality detritus (*I. laurina*), we observed an interaction between the biomass of the two microbial communities, highlighting fungi as the principal component [63]. Due to its high capacity to metabolize refractory molecules (e.g., cellulose and lignin) and to decompose them, the fungal community is the primary decomposer in tropical streams, and this principle explains the great significance of fungi in *I. laurina* decomposition [6,35,68]. Fungal action can increase the palatability of detritus, as well as its nutritional quality, for other decomposers, and the high biomass of fungi might be another factor responsible for the higher abundance of shredders in these sub-basins [7,35].

In general, we conclude that variations in scale contribute to the variation in the leaf breakdown rate, highlighting the importance of similar studies of this type that determine effects at different scales. The variability of the physical structure of streams (primarily temperature, dissolved oxygen and nutrients) accelerates leaf breakdown from upstream to downstream, but this process was only demonstrated at the sub-basin scale in the location studied, partially corroborating the initial hypothesis. The replacement of shredder invertebrates by microorganisms was observed but was contrary to the prediction of our hypothesis. Shredders were favored by microorganisms (primarily in *E. cloeziana*), with stronger interactions between them than those previously found to drive leaf breakdown rates. Based on our interest in the influence of spatial structure on ecosystem functions, we observed that watersheds with warmer microclimates and streams with higher nutrient levels and oxygen in the water could be accelerating the metabolism of the ecosystem in the watershed, with increased negative effects downstream. For the management of tropical watersheds, we noted that the upstream areas are more fragile and sensitive to environmental impacts but show greater importance in the cycling of nutrients. We performed analyses within and between spatial scales to assess the relative importance of various watershed scales in determining the local breakdown rate for leaves. Local characteristics are responsible for the diversification of this process across the "riverscape", and high heterogeneity underscores the difficulty of making predictions based on local studies.

Supporting Information

Table S1 Abiotic variables in sampling sites. Average values and the standard deviation of outflow, dissolved oxygen in the water (mg l^{-1}), electrical conductivity (μS-cm^2), water temperature (Temp. °C), pH, turbidity (NTU), water velocity (m-s), nitrite, nitrate, orthophosphate (mg l^{-1}) and the percentage of canopy openness (%) in sub-basin and stream order along the Gama-Cabeça de Veado Basin.

Figure S1 Remaining mass over time in sampling sites. Percentages of remaining mass along of the day in *E. cloeziana* (A and C) and *I. laurina* (B and D), between stream order (A e B) and sub-basin (C and D).

Figure S2 Biotic Community over time in *E. cloeziana* detritus. Average values and standard error of density (A and B), richness (C and D), biomass (E and F) of aquatic invertebrates, total microbial biomass (ATP; G and H) and fungal hyphomicetos biomass (I and J) along of the days in *E. cloeziana*, among stream order (A, C, E, G and I) and sub-basin (B, D, F, H and J).

Figure S3 Biotic Community over time in *I. laurina* detritus. Average values and standard error of density (A and B), richness (C and D), biomass (E and F) of aquatic invertebrates, total microbial biomass (ATP; G and H) and fungal hyphomicetos biomass (I and J) along of the days in *I. laurina*, among stream order (A, C, E, G and I) and sub-basin (B, D, F, H and J).

Acknowledgments

The first author also received a scholarship from CAPES during his PhD research in the Graduate Program in Ecology of UFSC. We are also grateful to Cássia Alves Lima for valuable logistical support for the field and laboratory work.

Author Contributions

Conceived and designed the experiments: RSR MMP JFG. Performed the experiments: RSR MMP JFG. Analyzed the data: RSR MMP JFG. Contributed reagents/materials/analysis tools: RSR MMP JFG. Wrote the paper: RSR MMP JFG.

References

1. Vannote RL, Minshall GW, Cummins KW, Sedell JR, Cushing CE (1980) River Continuuum Concept. Canadian Journal of Fisheries and Aquatic Sciences 37: 130–137.
2. Fabre E, Chauvet E (1998) Leaf breakdown along an altitudinal stream gradient. Archiv für Hydrobiologie 141: 167–179.
3. Statzner B, Higler B (1985) Questions and comments on the river continuum concept. Canadian Journal of Fisheries and Aquatic Sciences 42: 1038–1044.
4. Benda L, Poff NL, Miller D, Dunne T, Reeves G, et al. (2004) The network dynamics hypothesis: How channel networks structure riverine habitats. BioScience 54: 413–427.
5. Larned ST, Datry T, Arscott DB, Tockner K (2010) Emerging concepts in temporary-river ecology. Freshwater Biology 55: 717–738.
6. Gessner MO, Chauvet E, Dobson M (1999) A perspective on leaf litter reakdown in streams. Oikos 85: 377–384.
7. Graça MAS, Cressa C, Gessner MO, Feio MJ, Callies KA, et al. (2001) Food quality, feeding preferences, survival and growth of shredders from temperate and tropical streams. Freshwater Biology 46: 947–957.
8. Mathuriau C, Chauvet E (2002) Breakdown of leaf litter in a neotropical stream. Journal of the North American Benthological Society 21: 384–396.
9. Suberkropp K, Chauvet E (1995) Regulation of Leaf Breakdown by Fungi in Streams: Influences of Water Chemistry. Ecology 76: 1433–1445.
10. Wantzen KM, Wagner R (2006) Detritus processing by invertebrate shredders: a neotropical-temperate comparison. Journal of the North American Benthological Society 25: 216–232.
11. Moulton TP, Magalhaes-Fraga SAP, Brito EF, Barbosa FA (2010) Macro-consumers are more important than specialist macroinvertebrate shredders in leaf processing in urban forest streams of Rio de Janeiro, Brazil. Hydrobiologia 638: 55–66.
12. Abelho M (2001) From litterfall to breakdown in streams: a review. Scientific World Journal 1: 656–680.
13. Gonçalves JFJ, Martins RT, Ottoni BMP and Couceiro SRM (2013) Uma visão sobre a decomposição foliar em sistemas aquáticos brasileiros. In: N Hamada, J. L Nessimian and R. B Querino, editors. Insetos aquáticos: biologia, ecologia e taxonomia.
14. Wiens JA (1989) Spatial Scaling in Ecology. Functional Ecology 3: 385–397.
15. Tiegs SD, Akinwole PO, Gessner MO (2009) Litter decomposition across multiple spatial scales in stream networks. Oecologia 161: 343–351.
16. Lowe WH, Likens GE, Power ME (2006) Linking scales in stream ecology. Bioscience 56: 591–597.
17. Schneider DC (2001) The rise of the concept of scale in ecology. BioScience 51: 545–554.
18. Wiens JA (2002) Riverine landscapes: taking landscape ecology into the water. Freshwater Biology 47: 501–515.
19. Allen TFH, Starr TB (1982) Hierarchy: Perspectives for Ecological Complexity. Chicago: University of Chicago Press.
20. Del Arco AI, Ferreira V, Graça MAS (2012) The performance of biological indicators in assessing the ecological state of streams with varying catchment urbanisation levels in Coimbra, Portugal. Limnetica 31: 141–154.
21. Frissell CA, Liss WJ, Warren CE, Hurley MD (1986) A Hierarchical Framework for Stream Habitat Classification: Viewing Streams in a Watershed Context. Environmental Management 10: 199–214.
22. Allan JD (2004) Landscapes and Riverscapes: The Influence of Land Use on Stream Ecosystems. Annual Reviews of Ecology, Evolution, and Systematics 35: 257–284.
23. Royer TV, Minshall GW (2003) Controls on leaf processing in streams from spatial-scaling and hierarchical perspectives. Journal of North American Benthological Society 22: 352–358.
24. Irons JG, Oswood MW, Stout RJ, Pringle CM (1994) Latitudinal patterns in leaf litter breakdown: is temperature really important? Freshwater Biology 32: 401–411.
25. Minshall GW, Petersen RC, Cummins KW, Bott TL, Sedell JR, et al. (1983) Interbiome Comparison of Stream Ecosystem Dynamics. Ecological Monographs 53: 2–25.
26. Sponseller RA, Benfield EF (2001) Influences of Land Use on Leaf Breakdown in Southern Appalachian Headwater Streams: A Multiple-Scale Analysis. Journal of the North American Benthological Society 20: 44–59.
27. Hagen EM, Webster JR, Benfield EF (2006) Are leaf breakdown rates a useful measure of stream integrity along an agricultural landuse gradient? Journal of the North American Benthological Society 25: 330–343.
28. Hieber M, Gessner MO (2002) Contribution of Stream Detrivores, Fungi, and Bacteria to Leaf Breakdown Based on Biomass Estimates. Ecology 83: 1026–1038.
29. Young RG, Collier KJ (2009) Contrasting responses to catchment modification among a range of functional and structural indicators of river ecosystem health. Freshwater Biology 54: 2155–2170.
30. Cummins KW (1974) The importance of different energy sources in freshwater ecosystems. In: K. W Cummins, editor editors. Productivity of World Ecosystems. Washington: National Academy of Sciences. pp. 166.
31. Petersen RC, Cummins KW (1974) Leaf processing in a woodland stream. Freshwater Biology 4: 343–368.
32. Bianchini Jr I (1999) Aspectos do processo de decomposição nos ecossistemas aquáticos continentais. In: M. L. M Pompêo, editor editors. Perspectivas da Limnologia no Brasil. São Luis: Gráfica e Editora União. pp. 21–43.
33. Moretti M, Gonçalves JJF and Callisto M (2007) Leaf breakdown in two tropical streams: Differences between single and mixed species packs. Limnologica 37: 250–258.
34. Bruder A, Schindler MH, Moretti MS, Gessner MO (2013) Litter decomposition in a temperate and a tropical stream: the effects of species mixing, litter quality and shredders. Freshwater Biology: n/a-n/a.
35. Gonçalves JF Jr., Rezende RS, Martins NM, Gregorio RS (2012) Leaf breakdown in an Atlantic Rain Forest stream. Austral Ecology 37: 807–815.
36. Gonçalves JF Jr., Graça MAS, Callisto M (2007) Litter decomposition in a Cerrado savannah stream is retarded by leaf toughness, low dissolved nutrients and a low density of shredders. Freshwater Biology 52: 1440–1451.
37. Medeiros AO, Pascoal C, Graça MAS (2009) Diversity and activity of aquatic fungi under low oxygen conditions. Freshwater Biology 54: 142–149.
38. Wright MS, Covich AP (2005) Relative importance of bacteria and fungi in a tropical headwater stream: Leaf decomposition and invertebrate feeding preference. Microbial Ecology 49: 536–546.
39. Gonçalves JF, Jr., Rezende RS, França J, Callisto M (2012) Invertebrate colonisation during leaf processing of native, exotic and artificial detritus in a tropical stream. Marine and Freshwater Research 63: 428–439.
40. Li AOY, Ng LCY, Dudgeon D (2009) Effects of leaf toughness and nitrogen content on litter breakdown and macroinvertebrates in a tropical stream. Aquatic Sciences 71: 80–93.
41. Ligeiro R, Moretti MS, Goncalves JF Jr, Callisto M (2010) What is more important for invertebrate colonization in a stream with low-quality litter inputs: exposure time or leaf species? Hydrobiologia 654: 125–136.
42. Boyero L, Pearson RG, Dudgeon D, Ferreira V, Graça MAS, et al. (2012) Global patterns of stream detritivore distribution: implications for biodiversity loss in changing climates. Global Ecology and Biogeography 21: 134–141.
43. Larned ST (2000) Dynamics of coarse riparian detritus in a Hawaiian stream ecosystem: a comparison of drought and post-drought conditions. Journal of the North American Benthological Society 19: 215–234.
44. Rueda-Delgado G, Wantzen KM, Tolosa MB (2006) Leaf-Litter Decomposition in an Amazonian Floodplain Stream: Effects of Seasonal Hydrological Changes. Journal of the North American Benthological Society 25: 233–249.
45. Sutherland WJ, Freckleton RP, Godfray HCJ, Beissinger SR, Benton T, et al. (2013) Identification of 100 fundamental ecological questions. Journal of Ecology 101: 58–67.
46. Graça MAS, Barlocher F, Gessner MO (2005) Methods to Study Litter Decomposition. Dordrecht: Springer.
47. Strahler AN (1952) Hypsometric (Area Altitude) Analysis of Erosional Topology. Geological Society of America Bulletin 63: 1117–1142.
48. Golterman HL, Clymo RS, Ohnsta DMAM (1978) Methods for chemical analysis of freshwater. Oxford: Blackwell Scientific Publications.
49. Koroleff F (1976) Determination of nutrients. In: K Grasshoff, editor editors. Methods of sea water analysis. Verlag Chemie Weinhein. pp. 117–181.
50. Strickland JDH, Parsons TR (1960) A manual of seawater analysis. Ottawa: Fisheries Research Board of Canada. 188 p.
51. Pérez GP (1988) Guía para el studio de los macroinvertebrados acuáticos del departamento de Antioquia. Bogotá: Editorial Presencia Ltda.

52. Merrit RW, Cummins KW (1996) An introduction to the aquatic insects of North America. Dubuque: Kendall/Hunt Publishing Company.
53. Cummins K, Merritt R, Andrade P (2005) The use of invertebrate functional groups to characterize ecosystem attributes in selected streams and rivers in south Brazil. Studies on Neotropical Fauna and Environment 40: 69–89.
54. Bärlocher F, Graça MAS (2005) Total Phenolics. In: M. A. S Graça, F Barlocher and M. O Gessner, editors. Methods to Study Litter Decomposition. Dordrecht: Springer. pp. 97–100.
55. Gessner MO (2005) Proximate Lignin and Cellulose. In: M. A. S Graça, F Barlocher and M. O Gessner, editors. Methods to Study Litter Decomposition. Dordrecht: Springer. pp. 115–120.
56. Abelho M (2005) Extraction and quantification of ATP as a measure of microbial biomass. In: M. A. S Graça, F Bärlocher and M. O Gessner, editors. Methods to Study Litter Decomposition. Springer. pp. 223–229.
57. Gessner MO (2005) Ergosterol as a measure of fungal biomass. In: M. A. S Graça, F Barlocher and M. O Gessner, editors. Methods to Study Litter Decomposition. Dordrecht: Springer. pp. 189–195.
58. Crawley MJ (2007) The R Book. England: John Wiley & Sons Ltd.
59. Naiman RJ and Décamps H (1997) The ecology of interfaces: riparian zones. Annu Rev Ecol Syst 28: 621–658.
60. Frauendorf TC, Colón-Gaud C, Whiles MR, Barnum TR, Lips KR, et al. (2013) Energy flow and the trophic basis of macroinvertebrate and amphibian production in a neotropical stream food web. Freshwater Biology 58: 1340–1352.
61. Carvalho FMV, De Marco P, Ferreira LG (2009) The Cerrado into-pieces: Habitat fragmentation as a function of landscape use in the savannas of central Brazil. Biological Conservation 142: 1392–1403.
62. Bae M-J, Kwon Y, Hwang S-J, Chon T-S, Yang H-J, et al. (2011) Relationships between three major stream assemblages and their environmental factors in multiple spatial scales. Annales de Limnologie - International Journal of Limnology 47: S91–S105.
63. Gulis V, Suberkropp K (2003) Interactions between stream fungi and bacteria associated with decomposing leaf litter at different levels of nutrient availability. Aquatic Microbial Ecology 30: 149–157.
64. Ardon M, Pringle CM (2008) Do secondary compounds inhibit microbial- and insect-mediated leaf breakdown in a tropical rainforest stream, Costa Rica? Oecologia 155: 311–323.
65. Ardon M, Stallcup LA, Pringle CM (2006) Does leaf quality mediate the stimulation of leaf breakdown by phosphorus in Neotropical streams? Freshwater Biology 51: 618–633.
66. Oliveira-Filho AT, Ratter JA (1994) A study of the origin of central Brazilian forests by the analysis of plant species distribution patterns. Journal of Botany 52: 141–194.
67. Santos Fonseca AL, Bianchini I, Pimenta CMM, Soares CBP, Mangiavacchi N (2012) The flow velocity as driving force for decomposition of leaves and twigs. Hydrobiologia 703: 59–67.
68. Weyers HS, Suberkropp K (1996) Fungal and Bacterial Production during the Breakdown of Yellow Poplar Leaves in 2 Streams. Journal of the North American Benthological Society 15: 408–420.

Perceptional and Socio-Demographic Factors Associated with Household Drinking Water Management Strategies in Rural Puerto Rico

Meha Jain[1]*, Yili Lim[1], Javier A. Arce-Nazario[2], María Uriarte[1]

1 Department of Ecology, Evolution and Environmental Biology, Columbia University, New York, New York, United States of America, 2 Department of Biology, University of Puerto Rico in Cayey, Cayey, Puerto Rico, United States of America

Abstract

Identifying which factors influence household water management can help policy makers target interventions to improve drinking water quality for communities that may not receive adequate water quality at the tap. We assessed which perceptional and socio-demographic factors are associated with household drinking water management strategies in rural Puerto Rico. Specifically, we examined which factors were associated with household decisions to boil or filter tap water before drinking, or to obtain drinking water from multiple sources. We find that households differ in their management strategies depending on the institution that distributes water (i.e. government PRASA vs community-managed non-PRASA), perceptions of institutional efficacy, and perceptions of water quality. Specifically, households in PRASA communities are more likely to boil and filter their tap water due to perceptions of low water quality. Households in non-PRASA communities are more likely to procure water from multiple sources due to perceptions of institutional inefficacy. Based on informal discussions with community members, we suggest that water quality may be improved if PRASA systems improve the taste and odor of tap water, possibly by allowing for dechlorination prior to distribution, and if non-PRASA systems reduce the turbidity of water at the tap, possibly by increasing the degree of chlorination and filtering prior to distribution. Future studies should examine objective water quality standards to identify whether current management strategies are effective at improving water quality prior to consumption.

Editor: Joan Muela Ribera, Universitat Rovira i Virgili, Spain

Funding: Funding was provided by awards to MU from National Science Foundation DEB 0620910 and the Crosscutting Initiatives program, Earth Institute, Columbia University. This study was partially supported by Award Number P20MD006144 from the National Center on Minority Health and Health Disparities. The content is solely the responsibility of the authors and does not necessarily represent the official views of the National Center on Minority Health and Health Disparities or the National Institutes of Health. This work was also supported by the National Science Foundation Graduate Research Fellowship under Grant No. 11-44155 awarded to MJ. The funders had no role in study design, data collection and analysis, decision to publish, or preparation of the manuscript.

Competing Interests: The authors have declared that no competing interests exist.

* E-mail: mj2415@columbia.edu

Introduction

Over 700 million people across the globe do not have access to clean drinking water, leading to high levels of chronic waterborne illnesses [1–3]. This is particularly problematic in rural communities that do not receive adequately treated water from government facilities and may not have access to appropriate technologies to treat water locally [4,5]. Scientists and policymakers have long considered the best ways to improve access to potable water, yet identifying the most effective ways to manage drinking water is difficult given that it is typically managed by multiple public and private agencies [6–8]. Drinking water is often extracted and treated at different spatial scales (e.g. regional, watershed, and household level), resulting in management by various stakeholders that act at each of these scales (e.g. governmental, private, and household sectors; [9,10]. Given the complexity of drinking water management, policy makers and agencies (e.g. World Health Organization) over the past decade have increasingly recognized the importance of household water management, particularly in regions where government and community water treatment facilities are ineffective [11,12]. Households play an important role in determining the water quality experienced by individuals, as households are the last point of management prior to consumption [4,12].

To target the most successful interventions, it is important to understand the socio-cultural context of current household water management decisions [13]; by understanding how households manage their drinking water and why, policymakers can more effectively target intervention strategies to improve water quality prior to consumption. Though most households in a given community face the same water quality at the tap, some may treat their water prior to consumption while others may not [14,15]. This variation in household water management is influenced by a variety of factors, including knowledge of water treatment practices prior to distribution, perceptions of water quality at the tap, and socio-demographic characteristics of the decision-maker [14,16,17]. For example, previous studies have found that households are more likely to treat their tap water when they believe that government or community treatment facilities are ineffective [18,19], or when they believe that water quality is low at the tap [15]. While previous studies have examined the importance of these factors individually, few studies have considered these multiple drivers within the same analysis. Doing so is important because it identifies which factors are the most influential for household decision-making. This knowledge can then be used to identify and target interventions that are in line with current household perceptions, which has been shown to result in a greater rate of intervention uptake and success [20].

Our study assesses which factors most strongly influence household water management decisions, specifically whether households filter or boil their tap water prior to consumption or whether they obtain drinking water from multiple sources, in rural Puerto Rico. It is important to understand household water management in this region because previous studies have suggested that broader water management institutions do not always provide adequate water quality at the tap, particularly in rural, mountainous regions that are far from government treatment facilities [21]. There are two broad categories of institutions that manage drinking water for the island's four million people: government-managed Puerto Rico Aqueduct and Sewer Authority (PRASA) systems (which serve approximately 3.8 million people), and private and community non-PRASA systems (which serve approximately 400 communities, or up to 250,000 people), which are found primarily in mountainous regions that are too far to be connected to PRASA treatment facilities [21,22]. While the non-PRASA category encompasses a range of management strategies, given decentralized management where each community typically develops their own management plan, it is widely believed that non-PRASA communities in general are exposed to low water quality at the tap due to ineffective management of water prior to distribution. The Puerto Rico Department of Health (PRDOH) considers non-PRASA systems to be a health threat since they typically do not comply with federal water quality standards [23]. This is because about fifty percent of non-PRASA systems obtain water from surface sources and there is little or no monitoring of water quality in these communities [24]. Previous studies estimate that 30% of non-PRASA systems lack any water treatment infrastructure [22], and water is not treated consistently even when water infrastructure exists [22,25]. PRASA systems on the other hand typically filter and chlorinate water at treatment facilities before distribution and provide water quality assessments required by the U.S. Federal Potable Water Standards. Despite centralized management, PRASA systems are often plagued by water shortages and high rates of sediment loading and turbidity, which can result in non-compliances with the US Environmental Protection Agency (EPA) water quality standards [26]. This is because many filtration plants, particularly in mountainous regions, are not equipped to handle water filtration during periods of heavy rainfall [23], which is especially problematic given Puerto Rico's high frequency of tropical storms [27].

Given the possibility of inadequate water treatment by non-PRASA and PRASA facilities, some households have developed management strategies that are thought to improve drinking water quality prior to consumption. These strategies include filtering or boiling tap water or obtaining water from alternate sources like private wells and local markets. In this study, we assessed which perceptional factors that have been postulated to be important in previous literature are most associated with households that undertake water management strategies in rural, mountainous Puerto Rico [15,18,19]. Specifically, we predict the following in order of importance:

(1) households will have different management techniques depending on whether water is provided by government (PRASA) or community (non-PRASA) institutions likely due to differences in water quality at the tap;
(2) households that have problems with institutional water management prior to distribution are more likely to treat water;
(3) households are more likely to treat water if they perceive that water from the tap is of low quality;
(4) households that have less knowledge about how their water is treated prior to distribution are more likely to treat their water.

We quantify the relative importance of these various factors for household decision-making to better guide future water quality assessments and interventions in rural Puerto Rico. While our results are specific to Puerto Rico, we argue that our methodology can also be implemented in other regions to better understand the drivers of household water management and more effectively target interventions to those households vulnerable to low water quality.

Methods

Study site

Data were collected in eight different community sectors within the Cayey Mountain range in Puerto Rico from June to August of 2009. Our study focused on communities in this region because they are thought to be at high risk for low water quality given that they are rural and found in mountainous terrain, which makes them difficult to connect to PRASA treatment facilities. We specifically focused on villages found in Cayey and Patillas municipalities (Figure 1), which contain a large number of non-PRASA communities. Both municipalities are similar in socio-economic and development status. The median household income was $10,923 in Cayey and $9,375 in Patillas in 2000, which were lower than the island average of $13,189 [28]. We selected PRASA and non-PRASA communities that were adjacent to one another in each of the two municipalities. This was possible when we interviewed communities at the boundary where PRASA systems stopped serving communities with piped government water. This paired sampling design reduced possible confounding effects from socio-economic and geographic factors and allowed us to better assess whether households make different decisions based on if PRASA or non-PRASA institutions manage their water. Initial communities (n = 2) were selected based on where our field team had previous experience and knew PRASA and non-PRASA communities were adjacent to one other. We then used a snowball technique and visited additional communities (n = 6) that were suggested to us by the initial community contact [29]. While the communities that we selected for sampling were not entirely selected at random given this snowball technique, we believe that they are representative of the broader region given that each of our four pairs of PRASA and non-PRASA communities were spread across a wide geographic area in the Cayey mountain range (up to 15 km between our four sites).

Data collection

We surveyed 218 respondents across the eight community sectors considered in our study. Each community sector ranged in size from 50 to 200 households, but to ensure comparability we selected adjacent PRASA and non-PRASA communities that were approximately the same size. We aimed to interview 20 to 30 households in each community, and selected survey households at random distributed equally throughout each community. A summary of the number of survey respondents in PRASA verus

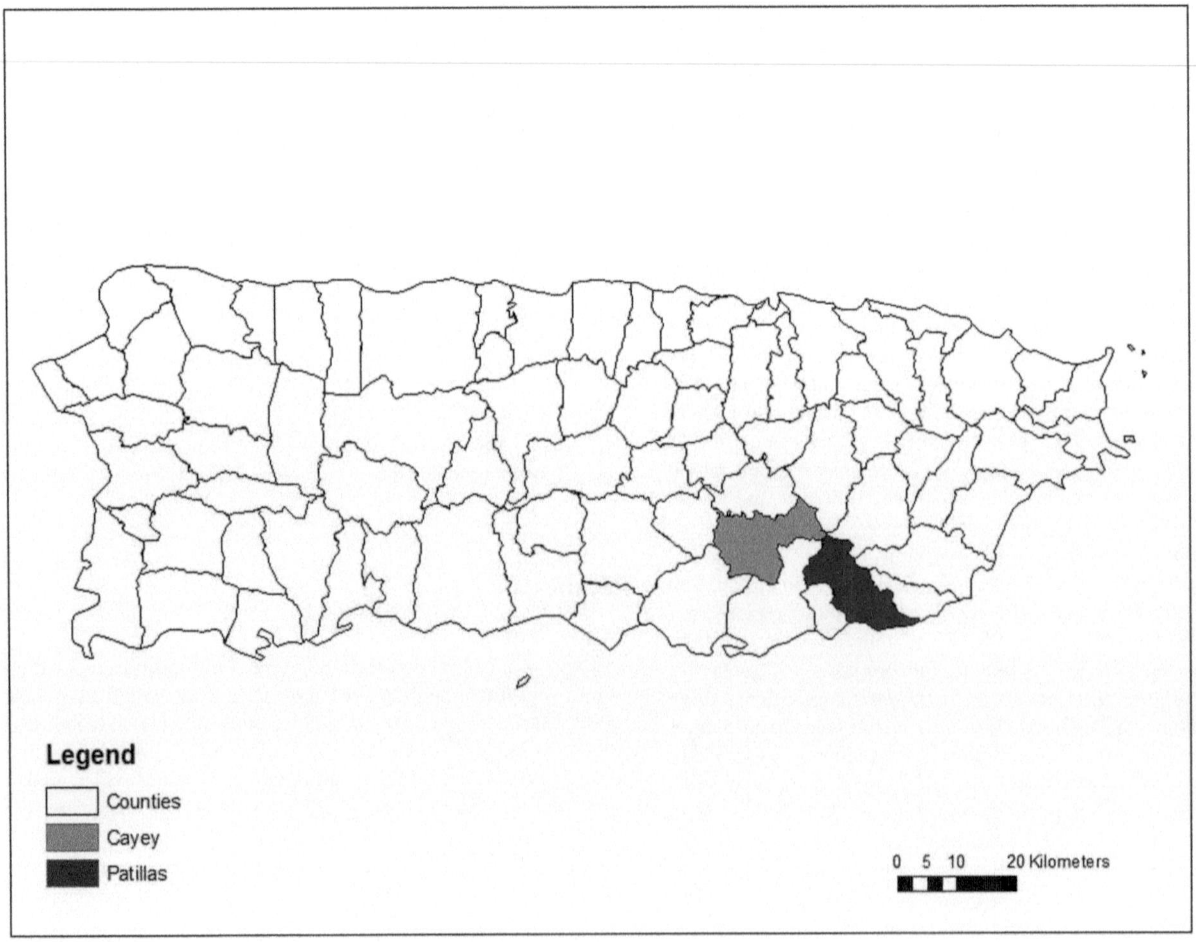

Figure 1. Map of Study Region in Puerto Rico. Municipalities where surveys were conducted are highlighted in gray. We did not list specific communities that we visited to keep the communities we surveyed anonymous.

non-PRASA communities is given in the supplementary information (Table S1). We then spoke to the household member who answered the door and identified which member of the household was in charge of household water management decisions. If that family member was home, we then conducted the oral structured survey with that family member. If the family member in charge of water management decisions was not at home we skipped that household and did not include it in our survey sample.

Ethics statement. Surveys were approved by the Columbia University Institutional Review Board under protocol number IRB-AAAE0079 and informed consent was written. Surveys were conducted in Spanish by local research assistants. We asked all respondents if we could audio record their interviews in order to keep a record of responses and to assist in confirming written responses and only did so if the interviewee gave permission. Our survey instrument contained questions related to whether households undertake any drinking water management prior to consumption, the respondent's perceptions of institutional water management and water quality at the tap, and socio-demographic information for the respondent. Details about each question are listed below, and all data collected were self-reported.

We asked respondents how they managed their drinking water sources prior to consumption, which serves as the dependent variable in our analyses. We grouped responses into two different types of strategies that households may undertake to cope with inadequate water quality. One coping strategy is to increase *the number of drinking water sources* used in the household. Households may diversify sources of drinking water by purchasing bottled water or obtaining drinking water from a personal well. The second coping strategy considered in this study is if households *treat tap water before drinking.* If households believe that their tap water is of inadequate quality, they may filter or boil it before drinking.

We also collected data on the following variables that have been suggested to be important for household water management decisions in previous studies. These variables serve as covariates in our statistical models and we discuss specific data that were collected for each variable of interest. As outlined in the introduction, we believe that management institution type, problems with institutional water management, perceptions of water quality, and knowledge of water treatment will influence household decisions to manage drinking water.

Management institution. We considered *the type of institution that manages water* (i.e. *PRASA or non-PRASA*) as a fixed effect because the way that specific institutions manage water may influence household decision-making. This may occur if institutions influence the behavior of households via uniform rules and norms [30]. Institutions may also affect household decision-making if they expose all households in a given community to the same quality of resource. Previous studies have shown that mismanagement of water treatment by institutions may negatively impact water quality experienced by all households within the distribution system [14].

Problems with institutional management. As a broad measure of whether households believe that institutions effectively manage water, which has been shown to be important in the previous literature [18,19], we asked households whether they *have problems with how their water is managed* by PRASA or non-PRASA operators. We predict that respondents who have more problems with institutional management are more likely to treat tap water since they may believe that their water was inadequately treated before distribution.

Perceptions of water quality at the tap. Even though all households in a given community are exposed to the same water quality at the tap, varying perceptions may lead to heterogeneous behavior among decision-makers. Previous studies have shown that perceptions of water quality are strong drivers of household water management decisions [15]. To assess water quality perceptions, we asked respondents to rank the *quality of their tap water* on a scale of 1 to 4, where 1 equals poor water quality and 4 equals excellent water quality. We predict that households that believe they have poor water quality are more likely to develop coping strategies.

Knowledge of institutional management. Given that previous studies have suggested that increased knowledge of institutional management practices influences individual decision-making [31], we asked respondents whether they *knew how their water was treated* before it is piped to their homes. We predict that households that have less knowledge of how their water was treated by management institutions are more likely to treat water given that they may not trust that their water was treated prior to distribution. Previous studies have suggested a link between increased knowledge, transparency, and trust [32,33].

Socio-economic and demographic variables. Various socio-economic and demographic factors, such as income, age, and gender of the decision-maker, can influence household decisions [34,35]. We considered the age and gender of the respondent as controls in our analysis, but did not include income in our final models because only half of our interviewees responded to this question. Income data were collected as self-reported annual income for the household in $10,000 US increments (e.g. $10,000–$20,000, $20,000–$30,000, etc.). However, to test whether income may be important for water management decisions in our region, we ran our statistical models on the subset of data with income. We found that the income variable was never significant ($p>0.05$), suggesting that it is not a significant driver of water management decisions in this region. Furthermore, since we are interested in quantifying the relative importance of various perceptional and socio-demographic factors for decision-making, excluding income from the analysis should not impact our results; instead, it would at most reduce the amount of variance explained by our models.

Statistical analyses

We conducted three sets of analyses to identify how water management and the drivers of water treatment decisions varied across households in our study. First, we used ANOVA to compare institution types for our two dependent variables of interest: the number of water sources and water treatment. We also compared the distribution of our covariates between institution type using ANOVA analyses. These simple comparisons illustrate whether there were significant differences in coping strategies, perceptions, and socio-demographic factors between households in PRASA and non-PRASA communities.

In a second set of analyses, we used separate logistic regressions to assess the effects of all covariates (Table 1) on the two response variables of interest. To assess whether these covariates have different effects on household decision-making in PRASA and non-PRASA communities, we included interactions between management institution (i.e. PRASA, non-PRASA) and the other covariates. To avoid parameter tradeoffs and clarify interpretation of the results, we dropped covariates that had a correlation >0.4. Based on this criterion, we dropped gender from our analysis. We then conducted stepwise variable selection using AIC_c to select the best model [36]. To facilitate the interpretation of effect magnitudes among covariates, all continuous predictors were standardized by subtracting their mean and dividing by twice their standard deviation [37]. Goodness of fit was calculated using the universal goodness of fit le Cessie and Houwelingen test [38] in the Design package (Version 2.3-0) in R Project Software (R Statistical computing 2012, Version 2.14.1 was used for all analyses).

Finally, to assess the relative importance of each variable, we dropped each variable one at a time from the best logistic regression model and compared the AIC_c from the resulting model with the AIC_c from the best model. Variables that contributed most to model fit, and therefore were the most important in our analysis, had the largest change in AIC_c between the best model and the model with the variable in question dropped [39].

Results

ANOVA results

Several variables differed between PRASA and non-PRASA households (Table 2). Considering water management strategies, non-PRASA households were significantly more likely to obtain water from multiple sources, whereas PRASA households were significantly more likely to treat their tap water before drinking. This simple analysis suggests that households in PRASA and non-PRASA communities mitigate perceived low water quality in different ways. Considering perceptional variables, Non-PRASA households were significantly more likely to know how their institutions managed drinking water prior to distribution and non-PRASA households were also more likely to report higher water quality than PRASA households (Table 2).

Logistic regression models

The most important predictor of household decisions to obtain water from multiple sources was the institution that manages water (e.g. PRASA vs non-PRASA; Table 3, Figure 2A). Respondents in non-PRASA communities were more likely to obtain water from multiple sources than those from PRASA communities. Using the le Cessie and Houwelingen goodness of fit test, there is not a significant difference between observed and predicted values from the model suggesting good model fit ($z = 0.78$, $sd = 0.19$, $p = 0.44$).

The best predictors of household decisions to treat tap water before drinking were the institution that manages water, perceptions of water quality, and the interaction between the institution that manages water and problems with institutional management (Table 3, Figure 2B). PRASA households were significantly more likely to treat their water before drinking than non-PRASA households. Households that reported lower water quality were also more likely to treat their tap water, regardless of water management institution. Finally, the significant interaction between the institution that manages water and whether a household reported problems with institutional management suggests that non-PRASA households that had problems with institutional management were more likely to treat tap water before drinking than PRASA households. Le Cessie and

Table 1. Description and hypothesized relationship for each of the variables considered in our statistical models.

Variable	Variable Code	Description	Hypothesis
Number of Water Sources	Num Source	Number of drinking water sources (0 = one source, 1 = multiple sources)	Dependent Variable
Treat Water	Treat Water	Whether a household filtered or boiled tap water before drinking (0 = No, 1 = Yes)	Dependent Variable
Institution Type	Water System	Which water system the household receives water from (i.e. PRASA = 0, Non-PRASA = 1)	+
Knowledge of Treatment	Treatment Knowledge	Identified if individual had knowledge of how institution (PRASA or Non-PRASA) treated water before it arrives at the tap (i.e. No = 0, Yes = 1)	-
Reported Problems with Institutional Management	Problems	Whether the respondent reported problems with the way institutions manage water (i.e. No = 0, Yes = 1)	+
Perceptions of Water Quality	Water Quality	Self-reported quality of drinking water from the tap (i.e. poor = 1, fair = 2, good = 3, excellent = 4)	-
Demographic Data	Age	Age	Control
Gender	Gender	Gender (0 = Male, 1 = Female)	Control

Variable, coding method, description, and the hypothesized relationship with the likelihood of adopting coping strategies for all covariates considered in both statistical models. A positive relationship indicates that the variable would lead to increased coping, as defined by a higher likelihood of treating water and obtaining water from multiple sources.

Houwelingen goodness of fit test indicated a good fit between predicted and observed data ($z = -1.31$, $sd = 0.14$, $p = 0.19$).

Variable importance

To understand the relative importance of each covariate considered in our logistic models (Table 1), we conducted a full model logistic regression and assessed the importance of each factor based on its contribution to model fit as measured by the change in AIC_c when that variable was dropped from the full model. In the model that predicted which households were more likely to obtain water from multiple sources, we found that the institution that manages water contributed most to model fit (Figure 3A). This suggests that whether households were from PRASA or non-PRASA communities was the most important variable for predicting whether households obtain water from multiple sources. The remainder of the variables in the model contributed little to model fit.

For the model that identified whether households treat or do not treat water, the institution that manages water was also the best predictor (Figure 3B). This suggests that whether households are from PRASA or non-PRASA communities was the most important variable to explain whether households treat or do not treat their water. Perceptions of water quality also contributed significantly to model fit (Figure 3B) suggesting that this variable is also important.

Table 2. Comparison of each variable considered in our statistical models by institution type (PRASA vs non-PRASA).

	Mean value by Institution		ANOVA results		
Variable	PRASA	Non-PRASA	d, f	F	P
Number of Water Sources	0.06	0.26	1, 187	14.28	<0.001*
Treat Water	0.71	0.42	1, 187	16.26	<0.001*
Treatment Knowledge	0.49	0.73	1, 187	11.74	<0.001*
Problems	0.38	0.49	1, 187	2.47	0.12
Water Quality	2.41	2.98	1, 187	21.69	<0.001*
Age	53.20	50.25	1, 187	1.35	0.25

Mean value by institution (i.e. PRASA, Non-PRASA) and ANOVA results (degrees of freedom, F-statistic, p-value) are reported for each variable. * indicates $p<0.05$.

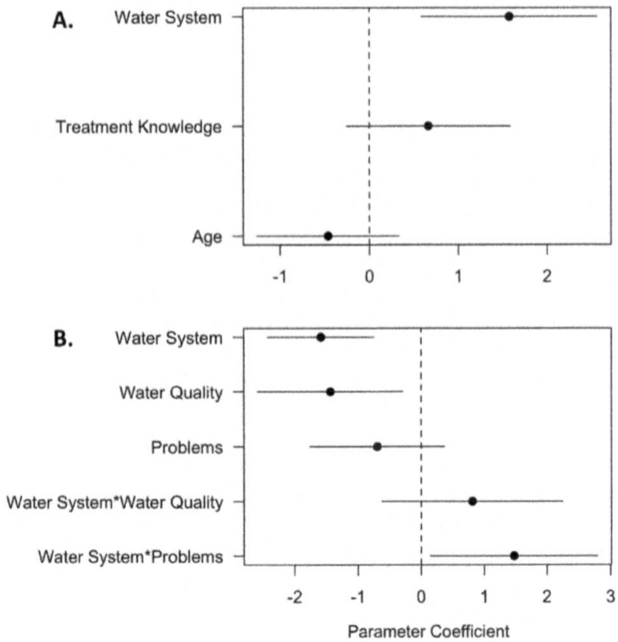

Figure 2. Parameter Estimate Plot of All Variables Considered in the Two Models that Predict Household Water Management Strategies. Standard errors are plotted as black lines. The variable is significant if standard error bars do not cross the zero axis. For the number of water sources (A), institution type is significant ($p<0.005$). For whether households treat water (B), institution type ($p<0.001$), perceptions of water quality ($p<0.05$), and the interaction between institution type and if households have a problem with institutional management ($p<0.05$) are significant.

Table 3. Results for each statistical model predicting which factors are associated with household water management strategies.

Response Variable	Covariates considered in logit model	Parameter Coefficient (Standard Error)	p value	N	GOF (p value)
Number of Water Sources	Water System	1.57 (0.52)	<0.005*	189	0.44
Number of Water Sources	Treatment Knowledge	0.66 (0.48)	0.17	189	0.44
Number of Water Sources	Age	−0.47 (0.42)	0.27	189	0.44
Treat Water	Water System	−1.58 (0.44)	<0.001*	189	0.19
Treat Water	Water Quality	−1.43 (0.60)	0.02*	189	0.19
Treat Water	Problems	−0.69 (0.56)	0.23	189	0.19
Treat Water	Water System*Water Quality	0.81 (0.75)	0.28	189	0.19
Treat Water	Water System*Problems	1.47 (0.70)	0.04*	189	0.19

Variables considered, parameter coefficients with standard error, p values, sample size, and goodness of fit for both of the full models including interaction terms. The first model predicts whether households obtain water from one or more sources, and the second model predicts whether households treat or do not treat their water. Significance of at least 5% is highlighted with a *.

Discussion

Policy-makers and agencies have increasingly recognized the importance of household water management for potable water provisioning given that households are the last point of management prior to consumption [13]. By understanding which factors most influence household water management, policy makers can better identify and target intervention strategies that improve access to clean drinking water. In this study, we examined household water management in rural Puerto Rico. It is important to understand household level management in these communities given that both government PRASA and community non-PRASA water treatment may be ineffective at providing clean water at the tap. Specifically, we analyzed (1) whether households obtained water from multiple sources or filtered or boiled tap water before drinking, and (2) which perceptional and socio-demographic factors were most associated with these management decisions. Our analysis suggests that three of our four initial predictions are correct: households manage water differently based on whether they are in PRASA or non-PRASA communities, households are more likely to treat water if they have problems with institutional management, and households are more likely to treat water if they believe that their tap water is of low quality (Figure 2). The fourth factor we predicted to be important in our analysis, whether households had knowledge of how water was treated prior to distribution, was not significant in our analyses.

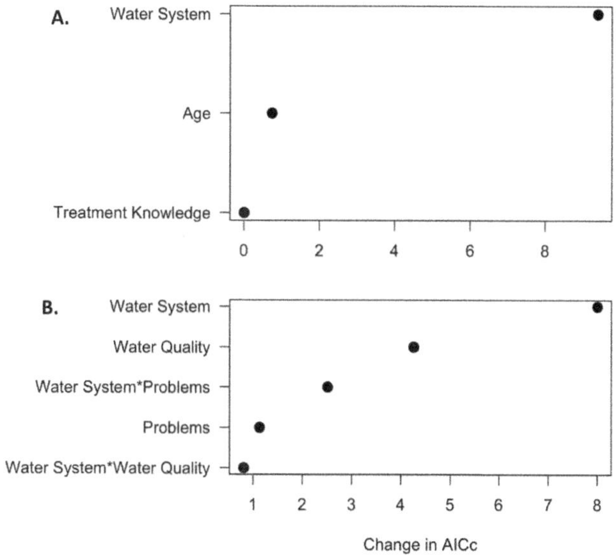

Figure 3. Importance of Each Covariate for Model Fit in the Two Models that Predict Household Water Management Strategies. Change in AIC_c for each of the covariates considered in the full logit model for the number of drinking water sources (A) and whether households treat or do not treat water (B). Larger changes in AIC_c values suggest that the variable contributed more to overall model fit. In both analyses (A and B), the institutional variable Water System (i.e. PRASA, non-PRASA) is the variable that contributes most to overall model fit. In the analysis of whether households treat water (B), water quality perceptions were also an important variable.

The institution that manages water (i.e. PRASA vs non-PRASA) was the strongest driver of household drinking water management (Figure 3). PRASA households were more likely to filter or boil their water before drinking, whereas non-PRASA households were more likely to obtain water from multiple sources (Figure 2A). Differences in management strategies between PRASA and non-PRASA communities may be due to differences in perceptions of low water quality, possibly because of differences in water quality at the tap [14]. In PRASA communities, our informal discussions with community members indicate that perceptions of low water quality are due to the bad taste and odor of tap water, which community members attribute to over-chlorination. PRASA treatment facilities typically add chlorine to water prior to distribution, which has been associated with a reduction in bacteria such as *Escherichia Coli* (E. Coli) [40,41]. However, based on our informal interviews with community members across our survey area, it is possible that PRASA systems are over-chlorinating water in this region; these anecdotal claims are bolstered by objective water quality measures collected by the government for the *barrios* (sub-districts) considered in our study, which show periods when chlorine levels are higher than those recommended by the EPA (> 4.0 ppm, Fig. S1) [24,42,43]. Thus, in PRASA communities, families may filter or boil their tap water in order to improve the smell and taste of water prior to consumption. In non-PRASA communities, discussions with community members suggest that perceptions of low water quality are due to turbidity, which community members attribute to the lack of treatment by non-PRASA institutions. Based on discussions with community members and the operators of non-PRASA systems, it appears as if water was not regularly treated (e.g. via chlorine addition or filters) in storage tanks prior to distribution, which resulted in increased water turbidity at the tap. Households mitigated this perceived low water quality by obtaining water from other sources, like store-bought bottled water or filtered water from friends and relatives in PRASA communities.

Second, households are more likely to treat water if they believe that their water was ineffectively managed by treatment facilities prior to distribution. This corroborates previous studies that show that households and communities increase water management efforts if they believe that government or private agencies ineffectively manage water prior to distribution [18,19]. This result is only significant for non-PRASA communities (Figure 2B), suggesting that perceptions of institutional effectiveness drive decisions to treat water only in non-PRASA households. Institutional perceptions may play a stronger role in non-PRASA relative to PRASA communities because institutional management of drinking water is decentralized; given decentralized management, households in non-PRASA communities often play a stronger role in community-level water management than do households in PRASA communities, where water management is centralized within government agencies. Informal discussions with non-PRASA community members support this interpretation: non-PRASA households state that they feel a strong connection to water management institutions due to increased knowledge of treatment practices (Table 2) and the ability to participate in water management by speaking with local water operators or attending community meetings.

Finally, we found that perceptions of water quality were significant predictors of whether households were more likely to treat their water via filtering and boiling (Figure 2B). These results corroborate previous studies that find that households are more likely to manage their water if they perceive that their tap water is of low quality [15]. It is important to note that we only examined water quality perceptions and not objective water quality metrics at the household level, and it is unclear how well these two measures correlate with one another. If these two measures are not related, this could lead to water management decisions that result in low drinking water quality. For example, households may perceive that their water is of good quality, resulting in no treatment at the tap, when in reality objective water quality measures show that water treatment is required prior to consumption. Future studies should measure objective water quality standards in this region both before and after household treatment of drinking water to determine whether households are accurately perceiving low water quality and treating water effectively.

Based on the three main findings outlined above, we have several recommendations to improve water quality management in this region. First, we argue that both PRASA and non-PRASA institutions would likely improve water quality if they took household perceptions into account and understood how households manage water after it is distributed to the tap. Specifically, PRASA systems may improve water quality if they take steps to improve the taste and odor of tap water. If this low water quality is caused by over-chlorination as many people in PRASA communities believe, these systems should reduce the amount of chlorine used or let chlorinated water sit in storage tanks to allow for dechlorination prior to distribution while controlling for environmental variables that may increase chlorination byproducts [44]. Non-PRASA systems, on the other hand, may benefit by reducing the amount of turbidity at the tap, possibly by filtering water prior to distribution; this, and chlorination, may reduce perceived low water quality at the household scale. Second, objective water quality assessments should be coupled with these household level survey results to focus intervention strategies on the most vulnerable populations, particularly those households that have low water quality but do not treat their water or that treat their water ineffectively. For example, PRASA households perceive low water quality due to bad taste and odor possibly caused by over-chlorination, however, one of the main strategies to mitigate this problem is filtering tap water. Yet to dechlorinate water, expensive active carbon filters are required [45] and these filters were typically not used in this region, suggesting that household strategies to filter water may be ineffective at reducing chlorine content. Finally, given that perceptions of institutional effectiveness appear to influence household management decisions, particularly in non-PRASA communities, we argue that these agencies should strengthen perceptions of institutional effectiveness by increasing the involvement of local community members in water management decisions. If community members have an increased say in how water is managed prior to distribution, it is likely that there will be improved water management given that household-level concerns about water quality are more likely to be addressed [46,47].

It is important to note that this study examined household perceptions of water quality and management, and it is possible that these perceptions are inaccurate when compared to objective measures. For example, most PRASA households believed that the bad taste and odor of tap water were caused by over-chlorination at treatment plants prior to distribution, but it is possible that the bad taste and odor were caused by other factors, like the addition of air or exposure to old pipes during the distribution process [48,49]. Future work should quantify objective water quality and assess whether current management strategies are effective at improving water quality prior to consumption. Second, we conducted our analyses based on survey data collected for over 200 people who live in the Cayey Mountain range. It is possible that our results would differ if we increased the scope of this study, particularly to other regions in Puerto Rico that may have different management strategies in PRASA and non-PRASA systems. Future studies should conduct similar perceptional studies across the island to better identify how universal the findings of this study are. Finally, it is important to note that we used the broad category of non-PRASA to encompass a wide range of institutions. Given that non-PRASA management is decentralized and individual communities are making water management decisions, it is possible that each non-PRASA system managed water slightly differently prior to distribution. We argue, however, that the coarse institutional categorization of non-PRASA is important particularly for policy given that the government uses this coarse categorization in water quality and compliance monitoring [23]. Future work should examine the heterogeneity in water management across non-PRASA systems to identify whether certain management strategies result in different outcomes for water quality and management at the household scale.

In conclusion, this study highlights the importance of social surveys and decision-making analyses to better identify how households currently manage drinking water and which factors influence household management decisions. Our results suggest that both community-level properties, like the type of institution that manages water prior to distribution, and household-level factors, like water quality perceptions, are important for predicting household-level water management behavior. By understanding household perceptions of both water quality and treatment of water prior to distribution, policy-makers can better identify and target intervention strategies that are tailored to current household decision-making. This is important given that previous studies have suggested that policies have a higher chance of uptake and success if they are created considering the local context [20].

Supporting Information

Figure S1 Free chlorine levels in ppm in PRASA and non-PRASA communities across our survey area. Data for PRASA communities were obtained from government databases collected at the barrio level, and data for non-PRASA communities were collected by our field team across several of our study communities of interest. These data suggest that free chlorine levels are typically lower in non-PRASA communities than PRASA communities, and several PRASA measurements have free chlorine levels higher than those recommended by the EPA (4.0 ppm, dotted horizontal line). This suggests that there may be over-chlorination in some PRASA communities.

Table S1 Number of interviewees in Non-PRASA and PRASA communities in our two study municipalities. We do not provide specific names of the communities or sectors surveyed in order to keep anonymity of our participants.

Acknowledgments

We would like to thank our field assistants Natalia Rodriguez and Yazmin Rivera, who conducted the household interviews and helped enter data during the summer of 2009, and Derek Berezdivin, who also helped enter data in 2010.

Author Contributions

Conceived and designed the experiments: MJ YL MU. Performed the experiments: MJ YL. Analyzed the data: MJ. Wrote the paper: MJ MU JA YL.

References

1. Teunis P, Medema GJ, Kruidenier L, Havelaar AH (1997) Assessment of the risk of infection by Cryptosporidium or Giardia in drinking water from a surface water source. Water Research.
2. Hellard ME, Sinclair MI, Forbes AB, Fairley CK (2001) A randomized, blinded, controlled trial investigating the gastrointestinal health effects of drinking water quality. Environ Health Perspect 109: 773–778.
3. Prüss A, Kay D, Fewtrell L, Bartram J (2002) Estimating the burden of disease from water, sanitation, and hygiene at a global level. Environ Health Perspect 110: 537–542.
4. Trevett AF, Carter RC, Tyrrel SF (2004) Water quality deterioration: A study of household drinking water quality in rural Honduras. International Journal of Environmental Health Research 14: 273–283.
5. Hunter PR, Toro GIR, Minnigh HA (2010) Impact on diarrhoeal illness of a community educational intervention to improve drinking water quality in rural communities in Puerto Rico. Bmc Public Health 10,
6. Cash DW, Adger WN, Berkes F, Garden P, Lebel L, et al. (2006) Scale and cross-scale dynamics: Governance and information in a multilevel world. Ecology and Society 11.
7. Berkes F (2006) From community-based resource management to complex systems: The scale issue and marine commons. Ecology and Society 11.
8. Sarker A, Ross H, Shrestha KK (2008) A common-pool resource approach for water quality management: An Australian case study. Ecological Economics 68: 461–471.
9. Lebel L, Garden P, Imamura M (2005) The politics of scale, position, and place in the governance of water resources in the Mekong region. Ecology and Society 10.
10. Saravanan VS (2008) A systems approach to unravel complex water management institutions. Ecological Complexity 5: 202–215.
11. Mintz E, Bartram J, Lochery P, Wegelin M (2001) Not Just a Drop in the Bucket: Expanding Access to Point-of-Use Water Treatment Systems. American Journal of Public Health 91: 1565–1570.
12. Clasen TF, Cairncross S (2004) Editorial: Household water management: refining the dominant paradigm. Tropical Medicine & International Health 9: 187–191.
13. Sobsey MD (2002) Managing water in the home: accelerated health gains from improved water supply. World Health Organization.
14. Gartin M, Crona B, Wutich A, Westerhoff P (2010) Urban Ethnohydrology: Cultural Knowledge of Water Quality and Water Management in a Desert City. Ecology and Society 15: 36.
15. Hu Z, Morton LW, Mahler RL (2011) Bottled Water: United States Consumers and Their Perceptions of Water Quality. Int J Env Res Pub He 8: 565–578.
16. Fielding KS, Russell S, Spinks A, Mankad A (2012) Determinants of household water conservation: The role of demographic, infrastructure, behavior, and psychosocial variables. Water Resour Res 48.
17. Sabau G, Haghiri M (2008) Household willingness-to-engage in water quality projects in western Newfoundland and Labrador: a demand-side management approach. Water and Environment Journal 22: 168–176.
18. Zérah M (2000) Water, unreliable supply in Delhi. Manohar Publishers.
19. Katuwal H, Bohara AK (2011) Coping with poor water supplies: empirical evidence from Kathmandu, Nepal. J Water Health 9: 143–158.
20. Jehu-Appiah C, Aryeetey G, Agyepong I, Spaan E, Baltussen R (2012) Household perceptions and their implications for enrolment in the National Health Insurance Scheme in Ghana. Health Policy and Planning 27: 222–233.
21. Molina-Rivera W (1998) Estimated water use in Puerto Rico, 1995. Washington D.C.: US. Geological Survey Open-File Report.
22. Guerrero-Preston R, Norat J, Rodriguez M, Santiago L, Suarez E (2008) Determinants of compliance with drinking water standards in rural Puerto Rico between 1996 and 2000: a multilevel approach. P R Health Sci J 27: 229–235.
23. Quinones F (2005) PRASA has ample water supplies. Water Industry News.
24. Environmental Protection Agency (2010) Puerto Rico Aqueduct and Sewer Authority (PRASA) Pollutant Discharge Settlement. Environmental Protection Agency.
25. Toro GR, Minnigh HA (2004) Regulation and Financing of Potable Water Systems in Puerto Rico: A Study in Failure in Governance. AWRA Dunde: Scotland.
26. de Cardenas SC (2011) Does private management lead to improvement of water services? Lessons learned form the experience of Bolivia and Puerto Rico. University of Iowa.
27. Boose E, Serrano M, Foster D (2004) Landscape and regional impacts of hurricanes in Puerto Rico. Ecol Monogr 74: 335–352.
28. US Bureau of the Census (2000) Census of population: social and economic characteristics. Washington D.C. USA: Department of Commerce, Economics, and Statistics Administration.
29. Biernacki P, Waldorf D (1981) Snowball Sampling: Problems and Techniques of Chain Referral Sampling. Sociological Methods & Research 10: 141–163.
30. North DC (1991) Institutions. The Journal of Economic Perspectives 5: 97–112.
31. Makutsa P, Nzaku K, Ogutu P, Barasa P (2001) Challenges in implementing a point-of-use water quality intervention in rural Kenya. American Journal of Public Health 91: 1571–1573.
32. Peters RG, Covello VT, McCallum DB (1997) The determinants of trust and credibility in environmental risk communication: an empirical study. Risk Analysis 17: 43–54.
33. Palanski ME, Kahai SS, Yammarino FJ (2011) Team Virtues and Performance: An Examination of Transparency, Behavioral Integrity, and Trust. J Bus Ethics 99: 201–216.
34. Jorgensen B, Graymore M, O'Toole K (2009) Household water use behavior: An integrated model. Journal of Environmental Management 91: 227–236.
35. Below TB, Mutabazi KD, Kirschke D, Franke C, Sieber S, et al. (2012) Can farmers' adaptation to climate change be explained by socio-economic household-level variables? Global Environmental Change 22: 223–235.
36. Hurvich C, Tsai C (1989) Regression and Time-Series Model Selection in Small Samples. Biometrika 76: 297–307.
37. Gelman A, Hill J (2007) Data Analysis Using Regression And Multilevel/Hierarchical Models. Cambridge Univ Pr.
38. le Cessie S, van Houwelingen J (1991) A Goodness-of-Fit Test for Binary Regression-Models, Based on Smoothing Methods. Biometrics 47: 1267–1282.
39. Burnham KP, Anderson DR (2002) Model Selection and Multimodel Inference. 2nd ed. New York: Springer.
40. Payment P, Trudel M, Plante R (1985) Elimination of viruses and indicator bacteria at each step of treatment during preparation of drinking water at seven water treatment plants. Appl Environ Microbiol 49: 1418–1428.
41. Arnold BF, Colford JM (2007) Treating water with chlorine at point-of-use to improve water quality and reduce child diarrhea in developing countries: a systematic review and meta-analysis. Am J Trop Med Hyg 76: 354–364.
42. Autoridad de Acueductos y Alcantarillados (2001-2010) Water Quality Reports.
43. Puerto Rico Department of Public Health (1996-2009) Water Quality Reports.
44. Chowdhury S, Champagne P, McLellan PJ (2009) Science of the Total Environment. Science of the Total Environment, The 407: 4189–4206.
45. Worley JL (2000) Evaluation of Dechlorinating Agents and Disposable Containers for Odor Testing of Drinking Water. Virginia Polytechnic Institute and State University.
46. Olsson P, Folke C, Berkes F (2004) Adaptive Comanagement for Building Resilience in Social-Ecological Systems. Environmental Management 34: 75–90.
47. Larson AM, Soto F (2008) Decentralization of Natural Resource Governance Regimes. Annu Rev Env Resour 33: 213–239.
48. Sangodoyin AY (1993) Water quality in pipe distribution systems. Environmental Management and Health.
49. Young WF, Horth H, Crane R, Odgen T, Arnott M (1996) Taste and Odour Threshold Concetrations of Potential Potable Water Contaminants. Water Research 30: 331–340.

6

Distribution and Dynamic Habitat Use of Young Bull Sharks *Carcharhinus leucas* in a Highly Stratified Northern Gulf of Mexico Estuary

J. Marcus Drymon[1,2]*, Matthew J. Ajemian[1,2¤], Sean P. Powers[1,2]

1 Department of Marine Sciences, University of South Alabama, Mobile, Alabama, United States of America, **2** Center for Ecosystem Based Fishery Management, Dauphin Island Sea Lab, Dauphin Island, Alabama, United States of America

Abstract

Understanding how animals alter habitat use in response to changing abiotic conditions is important for effective conservation management. For bull sharks (*Carcharhinus leucas*), habitat use has been widely examined in the eastern and western Gulf of Mexico; however, knowledge of their movements and the factors influencing them is lacking for populations in the more temperate north-central Gulf of Mexico. To examine how changes in hydrographic conditions affected the presence of young bull sharks in Mobile Bay, Alabama, thirty-five sharks were fitted with internal acoustic transmitters and monitored with an acoustic monitoring array consisting of thirty-three receivers between June 2009 and December 2010. Tagged sharks ranged in size from 60 to 114 cm fork length and were detected between the upper and lower portions of Mobile Bay. Despite a variety of freshwater sources associated with this highly productive estuary, sharks were most consistently detected at the largest input to the system – the Mobile and Tensaw Rivers. Our findings suggest a combination of hydrographic factors interact to influence the distribution of juvenile bull sharks in Mobile Bay. The factors affecting the probability of detecting at least one bull shark varied both temporally (2009 vs 2010) and spatially (upper vs lower bay). Electivity analysis demonstrated that bull sharks showed highest affinity for warm water (29–32°C), moderate salinities (10–11 psu) and normoxic waters (5–7 mg/l), although these patterns were not consistent between regions or across years. We suggest future studies coupling telemetry and hydrographic variables should, when possible, consider the interactions of multiple environmental parameters when defining the dynamic factors explaining the spatial distribution of coastal sharks.

Editor: Athanassios C. Tsikliras, Aristotle University of Thessaloniki, Greece

Funding: This work was funded through a grant from the Shelby Center for Ecosystem Based Fishery Management. The funders had no role in study design, data collection and analysis, decision to publish, or preparation of the manuscript.

Competing Interests: The authors have declared that no competing interests exist.

* E-mail: mdrymon@disl.org

¤ Current address: Harte Research Institute for Gulf of Mexico Studies, Texas A&M University-Corpus Christi, Corpus Christi, Texas, United States of America

Introduction

Coastal ecosystems are composed of a suite of important habitats for many shark species. The habitats within these ecosystems can serve as nursery areas for young-of-the-year and juvenile sharks [1,2] and foraging habitats for all life stages [3]. The use of coastal habitat is influenced by a range of biotic and abiotic factors [4,5]. Biotic factors include prey abundance [6], as well as reduced predation risk leading to decreased mortality [7]. Abiotic factors influencing the use of coastal habitat by sharks include temperature [8], salinity [9,10], and dissolved oxygen [11]. Given the increased urbanization of coastal areas, advancing our understanding of the mechanisms influencing habitat use by sharks can provide predictive capability in the face of changing environmental conditions.

Bull sharks (*Carcharhinus leucas*) are known to inhabit coastal areas circumglobally along warm temperate to subtropical clines. As one of only a few euryhaline sharks [12], bull sharks are able to tolerate significant abiotic fluctuations associated with dynamic coastal habitats and have thus been the focus of several studies within these areas. Using survey data, previous studies from the US have identified the importance of Matagorda Bay, Texas [2] and the Indian River Lagoon, Florida [13] as central nursery areas for bull sharks. Catch data has also illustrated the importance of temperature and salinity in determining habitat use for juvenile bull sharks [14,15]. These relationships have been further supported through acoustic telemetry, including active tracking [16,17] and passive monitoring. Acoustic monitoring arrays in southwest Florida have been used to document bull shark habitat use in Pine Island Sound [18] and the Caloosahatchee River estuary [19,20]. While juvenile bull shark habitat use has been well studied in the western and eastern Gulf of Mexico, knowledge of their movements and the factors influencing them is less abundant for populations in the more temperate north-central Gulf of Mexico. A synthesis of available datasets from coastal Louisiana demonstrates neonate and juvenile bull shark occurrence throughout saline, brackish and freshwater environments [21]. Similarly, the presence of neonate and juvenile bull sharks has been documented through fishery-independent gillnet sampling in the coastal waters of Mississippi and Alabama [22]. The north-central Gulf of Mexico, inclusive of coastal Louisiana and the Mississippi

Bight, is best classified as river dominated. In contrast to the eastern and western extremes of the Gulf of Mexico, dramatic intra and inter-annual variation related to freshwater input dominates the abiotic regime and may be related to climatic oscillations on longer time scales [23]. Understanding habitat use and population dynamics under these conditions is critical to developing a more complete understanding of the ecology of coastal sharks in this region.

As anthropogenic influences continue to alter coastal marine environments, particularly freshwater inputs, habitats that were once suitable may become uninhabitable for estuarine sharks. Our current understanding of habitat use by immature bull sharks is based largely on acoustic tagging studies from south Florida estuaries. Estuaries in the north-central Gulf of Mexico are more dynamic systems, subject to colder temperatures and wider ranges in salinity. Thus, investigating how bull sharks use such estuarine systems is critical if we are to better predict how anthropogenic alterations may affect the spatial distribution of sharks in these different regions. The objectives of this study were to examine spatial patterns in habitat use and describe the relationship between abiotic variables and the distribution of juvenile bull sharks in Mobile Bay, Alabama, monitored with passive acoustic telemetry.

Methods

Ethics Statement

This study was conducted in accordance with the laws of the state of Alabama and under the IACUC protocols (IACUC Board Reference Number 11014) approved by the University of South Alabama. All sampling occurred in state waters under permits granted by the State of Alabama Department of Conservation and Natural Resources Marine Resource Division to the authors. Sharks were collected as part of ongoing, standardized surveys (see *Acoustic Tagging* section below). All efforts were made to reduce animal suffering during handling and tagging procedures.

Study Site

Field sampling occurred in Mobile Bay, Alabama, a dynamic, drowned valley estuary in the north-central Gulf of Mexico. The bay is large and shallow, extending ~50 km north to south and 14–34 km east to west with a mean depth of approximately 3 m. Mobile Bay receives the second largest river discharge of all

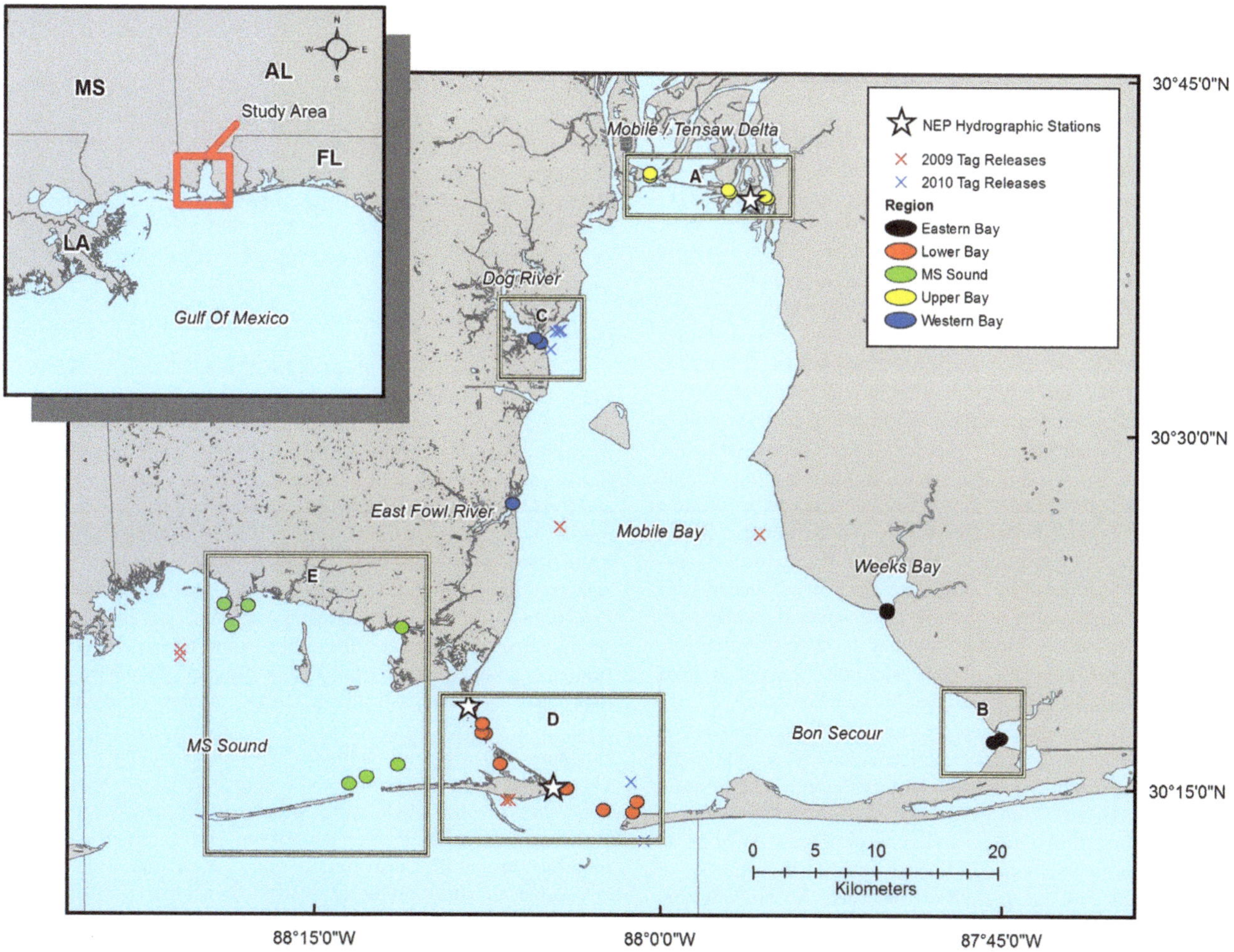

Figure 1. Map of the Coastal Alabama Acoustic Monitoring Program (CAAMP). The circles indicating receiver locations are scaled to the approximate mean detection range, and are coded by region: Upper Bay in yellow (A, n = 6), Eastern Bay in black (B, n = 4), Western Bay in blue (C, n = 5), Lower Bay in red (D, n = 11), and MS Sound in green (E, n = 7). The white stars indicate the location of the National Estuary Program (NEP) mooring stations from which hydrographic data were obtained. X's indicate release locations for the telemetered bull sharks in 2009 (red) and 2010 (blue) subsequently detected in the array.

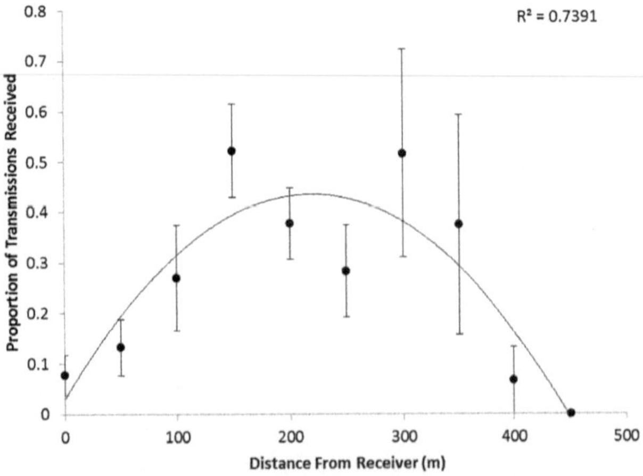

Figure 2. Results of boat-based range testing the WHS 2000 acoustic receivers. Mean proportion of transmissions detected are shown as a function of distance from the receiver, in meters. Error bars are SE.

estuaries in the Gulf of Mexico, primarily from the Alabama and Tombigbee rivers, resulting in large seasonal fluctuations in abiotic parameters [24].

Acoustic Array and Range Testing

Thirty-three Lotek Wireless Hydrophone System (WHS) omni-directional acoustic receivers were deployed in Mobile Bay as part of the Coastal Alabama Acoustic Monitoring Program (CAAMP, Figure 1). The CAAMP array employed two receiver models: Lotek WHS2000s (n = 27) and WHS3050s (n = 6). These receivers detected two separate types of acoustic coding technology; the WHS3050 receivers detected MAP code transmissions, and the WHS2000 receivers detected Rcode transmissions (see transmitter specifications below). Receiver attachment method varied according to location. The majority of receivers were attached to bottom moorings (n = 21), although some were attached to dock pilings (n = 6) or channel markers (n = 6) in regions of high vessel traffic and strong currents. To monitor ingress and egress from Mobile Bay, we placed hydrophones in a "curtain" array at freshwater inputs (Mobile-Tensaw Delta, Dog River, Weeks Bay, etc,) along the periphery of the study area. This array design was chosen based on the objectives of our study, and the cost associated with deploying and maintaining a gridded array across the entire bay [25]. All receivers were present throughout the course of the study area and recorded time, date and the identity of tagged animals that passed within the detection range of the receiver.

Range testing was conducted six times between March and October 2010 to monitor receiver performance over a wide range of hydrographic conditions. Range test tags (same model as those applied to animals; see below) were lowered over the side of a boat (i.e. boat-based method) [26] at varying intervals from the acoustic receiver 1 m from the bottom. At each interval the number of detections per five minute period was recorded. All range testing was conducted with shallow water receivers moored between 2 and 3 m deep, free from surface noise and turbulence. The vessel outboard engine was turned off during testing and the boat was anchored in place. In addition to range testing, acoustic receivers were serviced every 2 months, during which time data were downloaded, biofouling removed, and batteries replaced (as needed).

Specimen Capture and Acoustic Tagging

Sharks were captured with gillnets and longlines as part of ongoing standardized abundance surveys of elasmobranch assemblages in the Mobile Bay estuary [27,28]. Once captured, juvenile bull sharks were removed, measured (pre caudal length, fork length and stretched total length in mm), weighed (to nearest 0.1 kg) and tagged. Sharks were fitted with two tag types; an external, plastic swivel tag (Dalton ID, Henely-on-Thames, UK) and an internal ID only acoustic tag (Lotek model MM-MR-16-50, 16×80 mm, 35 g in air). The MM-MR-16-50 is a multi-mode transmitter, set to pulse every 5 seconds on a MAP code frequency (76.8 kHz, detected by the WHS3050) and every 60 seconds on a an Rcode frequency (69 kHz, detected by the WHS2000). For internal tags, a small incision was made above midline of the ventral surface for implantation. Tags were surgically inserted into the peritoneal cavity, and the incision was closed with surgical sutures (3.0 Ethicon Prolene monofilament). Once closed, an antiseptic wipe was applied to the sutured area. The time needed to apply both the external swivel and the internal acoustic tags ranged from 90–360 seconds.

Hydrographic Datasets

To relate the presence of telemetered bull sharks to abiotic parameters, hydrographic data were obtained from mooring stations maintained by the Dauphin Island Sea Lab (DISL) and the Mobile Bay National Estuary Program (MBNEP) (Figure 1). At each station temperature, salinity and dissolved oxygen data were measured with a YSI 6600 (Yellow Springs Instruments, Inc.) every 30 minutes. These data are displayed in near real-time, and were available for download at http://www.mymobilebay.com. Data were available for every day of the year; for our analysis, we downloaded data from 2009 and 2010, and reduced those data to the range of dates hydrophones were actually deployed and tags were actively transmitting within the array.

Data Analysis

All data were analyzed in XLstat version 13.0 (Addinsoft, Inc.). Detection data from the acoustic receivers were used in conjunction with environmental data from the hydrographic mooring stations to examine potential factors influencing bull shark presence within the acoustic array. For the acoustic receivers, bull sharks were considered present near a hydrophone if >1 detection occurred within a thirty minute period. From these data, presence plots were constructed. To link the acoustic detection data to the hydrographic data, bull shark detections were collated into thirty minute bins and linked to the hydrographic data using the time stamp. Bull shark presence/absence, temperature, salinity and dissolved oxygen data were synchronized every thirty minutes throughout the time bull sharks were detected within the acoustic array. Boxplots and dynamic habitat use plots [29] were used to visualize the relationship between the presence of telemetered bull sharks and single and combinations of hydrographic variables, respectively. A generalized linear model with a binomial probability distribution and a logit link was used to determine if the predictor variables temperature, salinity, dissolved oxygen or their combinations could predict the probability of detecting at least one bull shark. Firth's penalized maximum likelihood method [30] was used to reduce the bias associated with the small (relative) proportion of positive detections in comparison to large periods of absence. The significance of the regression coefficients for individual predictors was evaluated using the Wald statistic and overall performance of the model was assessed by examining the

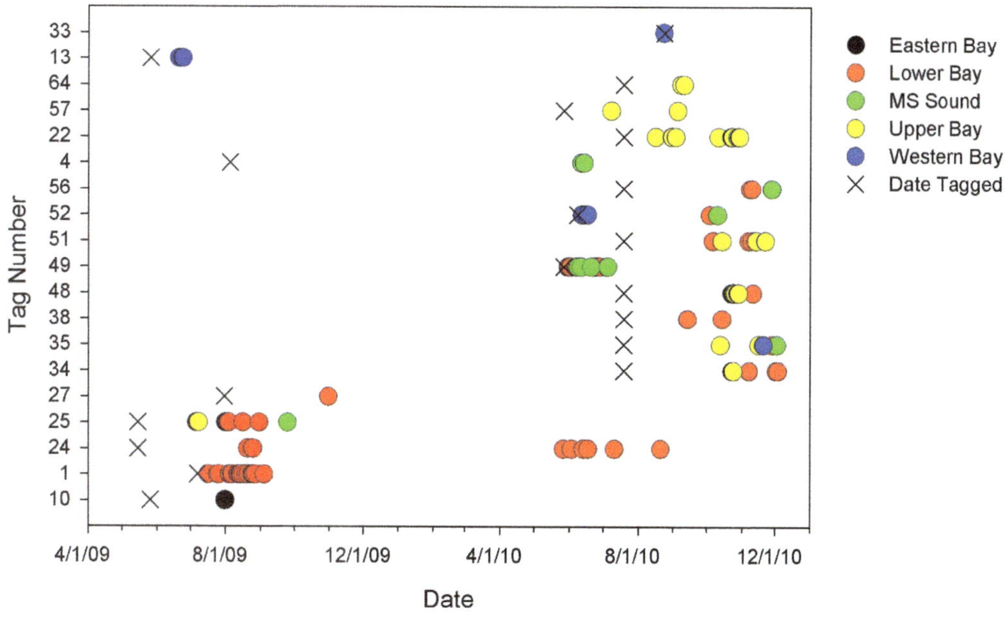

Figure 3. Presence history for acoustically tagged bull sharks. Data are shown for sharks detected in 2009 and 2010, coded by region. The gray area indicates the period when acoustic receivers were not deployed. X's mark the dates tags were deployed on sharks.

receiver operating curves (ROC). Values of ROC range from 0.5 to 1, where 1 suggests perfect discrimination between presence and absence probabilities, and 0.5 suggests the model performance is no better than random. Values greater than 0.8 are considered very good, and values above 0.9 excellent [15]. Given both the wide spatial variation in abiotic data, and the

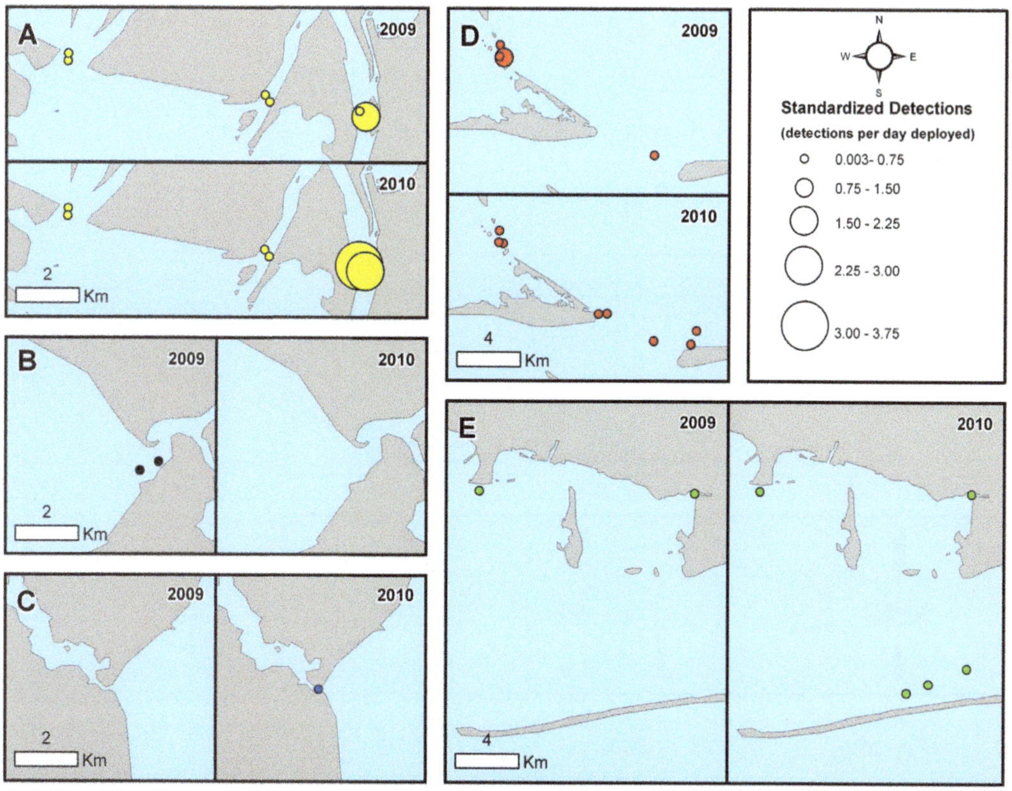

Figure 4. Distribution of bull shark detections. Number of bull shark acoustic detections (detections per day deployed) throughout the CAAMP array in 2009 and 2010. Detections were standardized both to the receiver type (WHS 3050 vs WHS 2000, which had different detection intervals) and the number of days receivers were deployed. Colors (yellow, red, blue, black, and green) and letters (A–E) are consistent with Figure 1.

difference in detection probability resulting from unequal acoustic receiver coverage throughout Mobile Bay, we separated the bay into upper and lower components. In addition, since the hydrophone array was deployed for less time in 2009 than 2010, we separated the two years of acoustic monitoring data. A total of 4 models were constructed: lower bay 2009, lower bay 2010, upper bay 2009, and upper bay 2010. Electivity values were calculated for both regions (lower and upper bay) and for both years (2009 and 2010) following Chesson [31] to determine if telemetered bull sharks displayed dynamic habitat use ranging from avoidance (0) to affinity (1) for various abiotic regimes. Since hydrographic conditions varied between years, neutral values for Chesson's alpha were standardized by subtracting the value 1/(number of categories) following Heupel and Simpfendorfer [19].

Results

Range tests

Analysis of the range test data showed the two receiver types we used had different optimal detection ranges. The vast majority (>80%) of the CAAMP array was composed of WHS2000 receivers. On these receivers, acoustic tags were detected at distances ranging from 0 to 400 meters from the receiver. Optimal detection (the highest proportion of detections) occurred at 150 and 300 m from the receivers, although the overall detection pattern was well described by a second order polynomial fit ($R^2 = 0.74$) (Figure 2). In contrast, range testing of the few WHS3050 receivers used in the array showed a markedly different detection pattern. Acoustic tags were detected between 0 and 300 meters, yet the proportion of tags detected was strongest at the receiver, and decreased with increasing distance from the receiver, a trend best described by an exponential decay ($R^2 = 0.92$).

Acoustic tagging

Between May 2009 and August 2010, thirty-five bull sharks were fitted with internal acoustic transmitters. Nineteen of these fish were subsequently detected in our acoustic array between June 2009 and December 2010 (Figure 3). Sharks acoustically tagged in this study were relatively small, ranging in size from 60 to 114 cm fork length (FL; Table 1), with a median size of 68 cm FL and a mean size of 77 cm FL (±4.07 cm SE). Collectively, this size range is indicative of individuals that are primarily neonate and young-of-the-year [32]. Interestingly, one of the larger sharks (ID number 4) tagged in the summer of 2009 was not detected in the array until the following summer (Table 1). Spatial patterns in acoustic detections were mapped to examine the primary regions of activity within the CAAMP array (Figure 4), and showed that the acoustic receivers in the upper and lower portions of Mobile Bay recorded the most detections (91.4%). In particular, the northeast portion of Mobile Bay appeared to be a "hotspot" for tagged bull sharks, with more relative detections in that area than any other region in the array (Figure 4A). Given the low number of detections throughout Mississippi Sound and the western and eastern portions of Mobile Bay, the remaining analyses focused on the upper and lower regions of Mobile Bay.

Hydrographic data

Temperature, salinity and dissolved oxygen values were recorded during the study periods in 2009 (July – November) and 2010 (March – December) in the upper and lower portions of Mobile Bay. Between these two regions, mean temperature varied little, with values of 26.15°C (±0.04 SE) and 26.98°C (±0.05 SE)

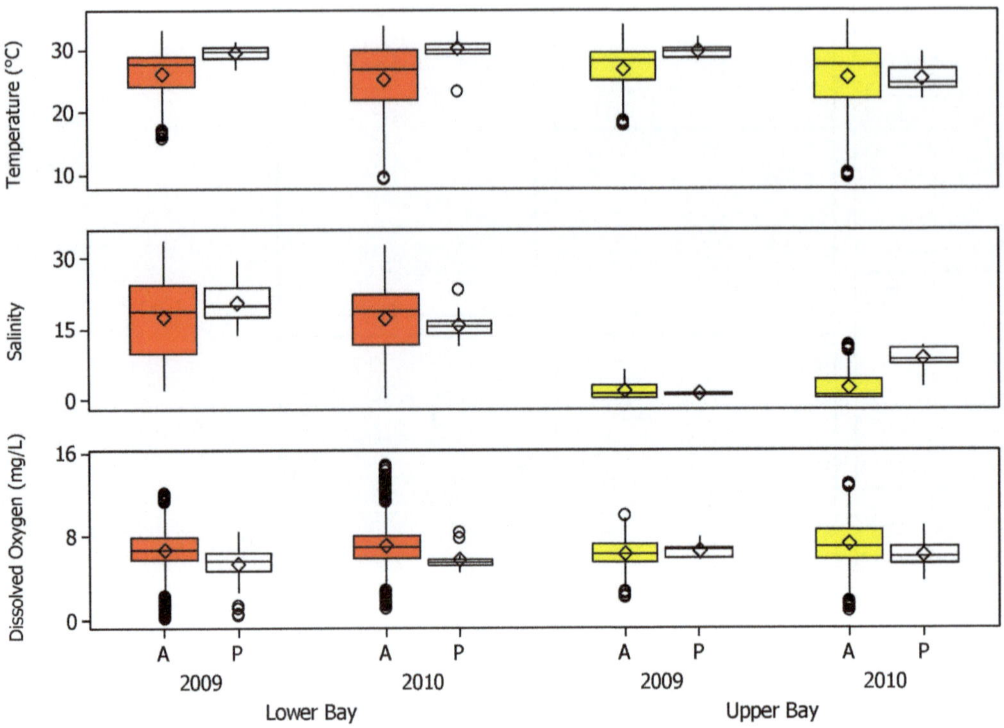

Figure 5. Bull shark presence as a function of hydrographic data. Box and whisker plots (median, interquartile range) of temperature, salinity and dissolved oxygen are shown during periods of telemetered bull shark absence (A) and presence (P) in the lower (red) and upper (yellow) portions of Mobile Bay. Mean values are shown with diamonds, and circles indicate outliers.

Table 1. Biological data for tagged bull sharks.

ID	FL (cm)	Sex	Date tagged
24	74	F	5/15/09
25	110	M	5/15/09
13	60	M	5/26/09
10	64	M	5/26/09
1	96	M	7/8/09
27	68	F	7/31/09
4	95	F	8/5/09
49	106	M	5/27/10
57	76	F	5/27/10
52	80	F	6/8/10
22	67	F	7/19/10
34	61	M	7/19/10
35	60	M	7/19/10
38	67	F	7/19/10
48	65	F	7/19/10
51	69	F	7/19/10
56	68	F	7/19/10
64	67	F	7/19/10
33	690	M	8/23/10

Biological data (length, cm FL and weight in kg) for bull sharks acoustically tagged in 2009 and 2010.

in the upper and lower bay in 2009. A similar trend, yet with slightly lower values was seen in 2010 with mean temperatures of 25.38°C (±0.04 SE) and 25.41°C (±0.05 SE) in the lower and upper bay, respectively. Similarly, dissolved oxygen values showed little difference between region, where mean DO was 6.78 mg/l (±0.02 SE) and 6.34 mg/l (±0.02 SE) in the lower and upper bay, respectively. As was the case with temperature, dissolved oxygen values in both regions were slightly higher in 2010, where mean DO was 7.09 mg/l (±0.01 SE) and 7.20 mg/l (±0.02 SE) in the upper and lower bay, respectively. As expected, salinity values showed the most marked differences between regions, as well as between years. In 2009, mean salinity values in the lower and upper regions were 17.58 psu (±0.08 SE) and 1.52 psu (±0.02 SE), respectively. In 2009, the range of salinity values in the lower bay was 31.87 and 6.02 in the upper bay. In 2010, the salinity regime in the lower bay was similar to the previous year, with mean salinity = 17.18 psu (±0.05 SE) and a range of values of 32.98. In the upper bay, mean salinity was slightly higher in 2010 compared to 2009, with a mean value of 2.31 psu (±0.03 SE); however, a much larger range of salinity (11.26) was recorded in 2010 relative to 2009 (Figure 5).

Relating telemetered sharks to hydrographic variables

Bull sharks acoustically monitored in the lower and upper portions of Mobile Bay were not detected across the entire range of hydrographic conditions. In the lower bay, bull sharks were only detected in relatively warm waters, where they were present across 25% of the available range of temperatures in 2009. In 2010 the range of temperatures that tagged bull sharks were detected in increased to nearly 40% of the available temperatures, although this increase was due to the detection of a single individual. With respect to DO, acoustically tagged sharks in the lower bay were detected across two thirds of the available values of DO in 2009, yet only across 28% of the available DO regimes in 2010. Acoustically tagged bull sharks in this study were detected across nearly half of the available salinity values in 2009, decreasing to approximately 40% in 2010 (Figure 5). In 2009, tagged bull sharks in the upper bay were detected towards the upper portion of the available thermal conditions, detected across nearly a quarter (23%) of the available temperature values. This increased slightly in 2010 (29%), similar to what was seen with respect to temperature in the lower bay. For dissolved oxygen, the pattern was similar, with bull sharks detected at over a quarter (27%) of the available DO values in 2009 in the upper bay, increasing to 43% in 2010. The most striking difference between the available hydrographic regimes and the ones where bull sharks were detected was with respect to salinity. In 2009, tagged bull sharks were detected at only 13% of the available salinity values, where the mean value of salinity when bull sharks were present = 0.95 psu (±0.04 SE). In contrast, tagged bull sharks were detected across greater than three fourths (77%) of the available salinity values in 2010, where the mean value of salinity when bull sharks were present was 8.43 psu (±0.18 SE) (Figure 5).

Dynamic habitat use varied as a function of both region and year. In 2009, bull sharks monitored in the lower bay used a relatively narrow range of temperature in conjunction with a wider range of salinity and DO (Figure 6A and C). A similar pattern was observed in 2010 (Figure 6B and D). Whereas the prevailing hydrographic regimes in the lower bay were similar in 2009 and 2010, a higher degree of annual variability was seen in the upper bay, particularly with respect to salinity. The upper bay experienced a much narrower range of salinity in 2009 (Figure 7A and E) compared to 2010 (Figure 7B and F), such that the salinity occupied by bull sharks in 2010 did not occur in the upper bay in 2009. Conversely, ambient temperature values

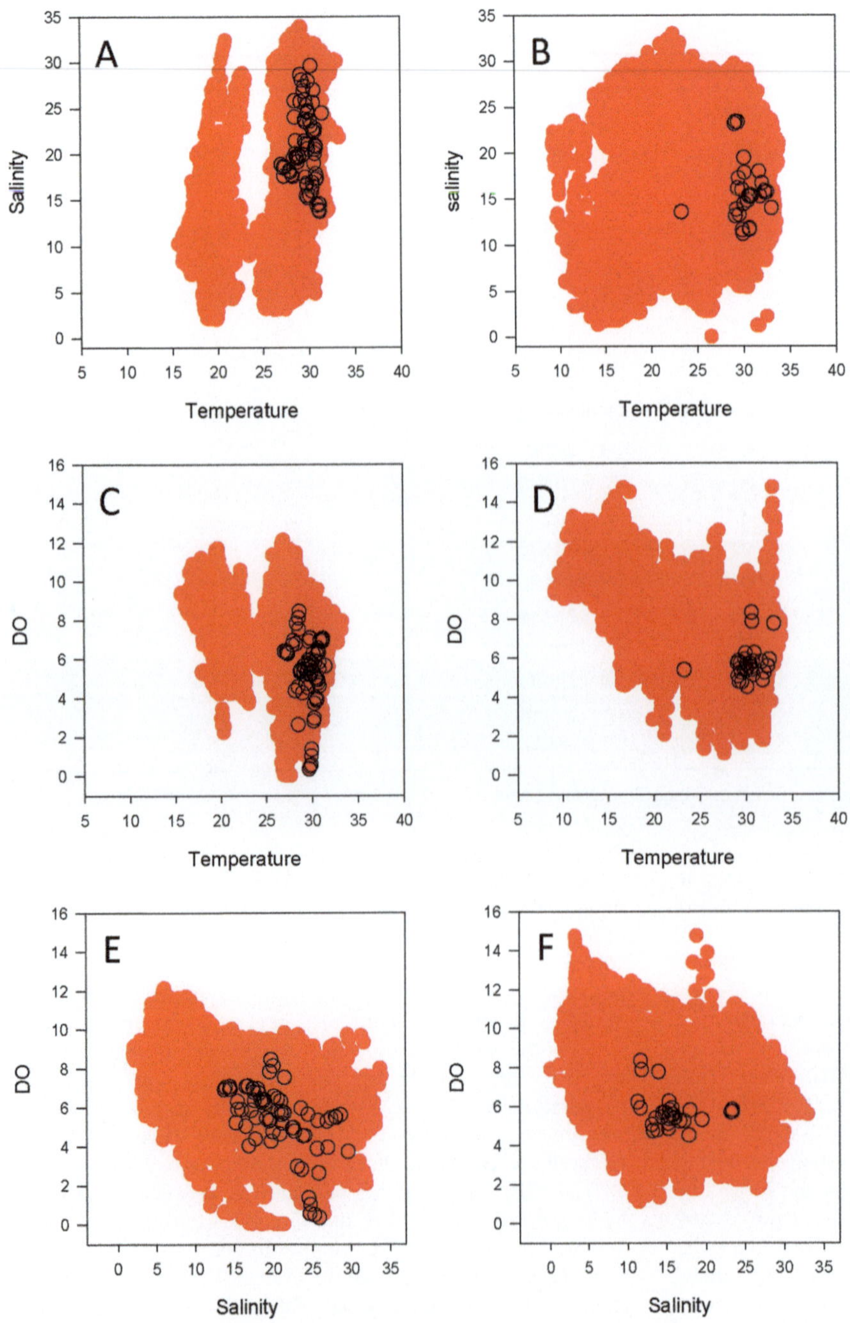

Figure 6. Bull shark dynamic habitat use in the lower bay. Dynamic habitat use of the lower Mobile Bay by bull sharks acoustically detected in 2009 (left column) and 2010 (right column): Temperature vs. Salinity (A, B), Temperature vs. Dissolved Oxygen (C, D), and Salinity vs. Dissolved Oxygen (E, F) are shown. Areas in red indicate the available dynamic habitat (data from the NEP mooring stations), and open circles indicate the presence of telemetered bull sharks.

were more similar between years, as were the temperatures occupied by telemetered bull sharks (Figure 7A–D).

Maximization of Firth's penalized likelihood function identified different combinations of the predictor variables across both region and year. In general, the models showed poorer fit in 2009 compared to 2010. In 2009, the best fit model for the lower bay included the terms temperature, DO, temperature*-salinity, temperature*DO and salinity*DO. The ROC value of 0.847 indicated very good model discrimination. Standardized regression coefficients indicated that temperature ($p<0.01$), DO ($p<0.01$) and their interaction ($p<0.01$) were the most influential parameters for the lower bay model in 2009. In 2010, the best fit model for the lower bay included the terms temperature, temperature*salinity, and temperature*DO. The ROC value of 0.864 indicated very good model discrimination. Standardized regression coefficients indicated that temperature ($p<0.01$), temperature*salinity ($p<0.01$) and temperature*DO ($p<0.01$) were equally influential parameters for the lower bay model in 2010. In the upper bay model for 2009, the best fit model included the terms temperature, DO, and temperature*-

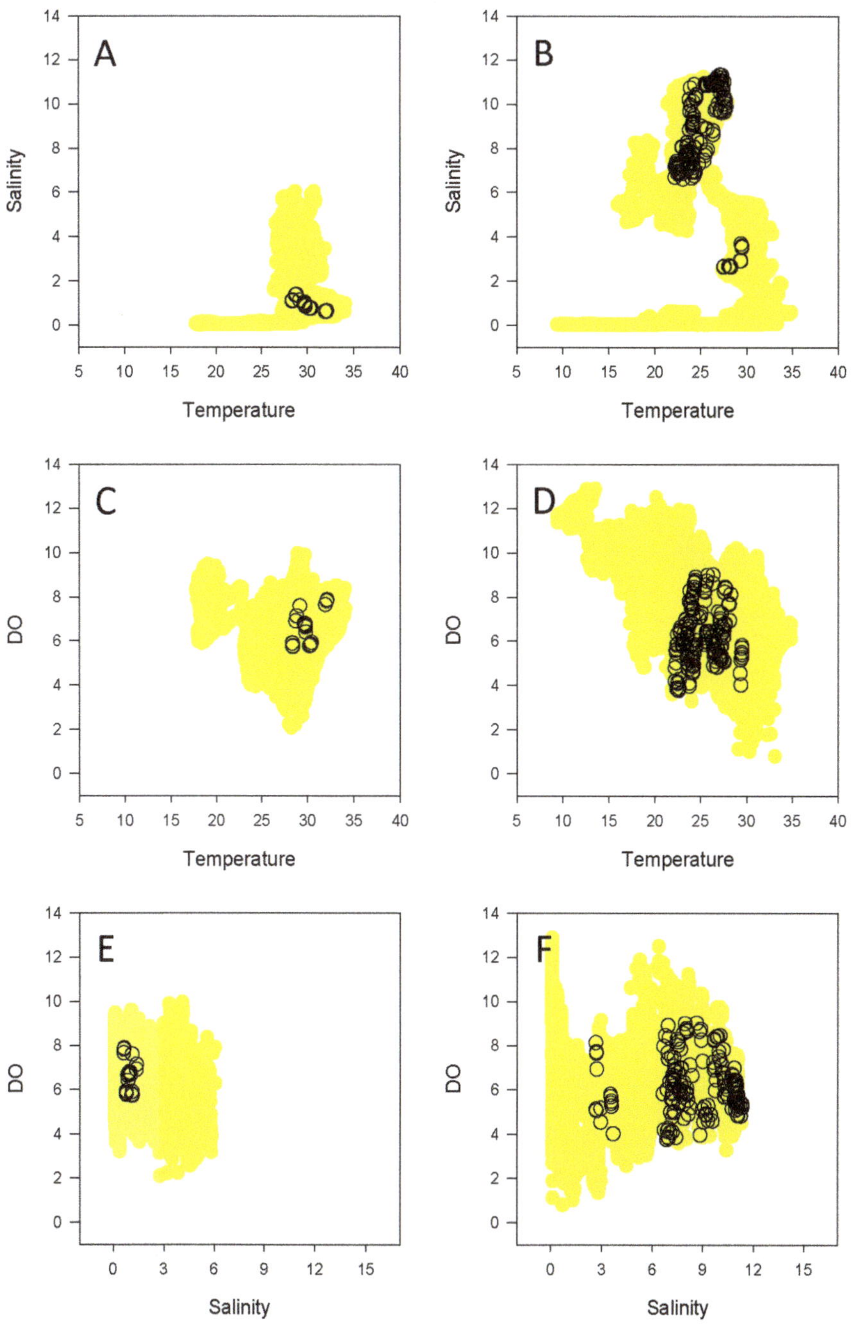

Figure 7. Bull shark dynamic habitat use in the upper bay. Dynamic habitat use of the upper Mobile Bay by bull sharks acoustically detected in 2009 (left column) and 2010 (right column): Temperature vs. Salinity (A, B), Temperature vs. Dissolved Oxygen (C, D), and Salinity vs. Dissolved Oxygen (E, F) are shown. Areas in yellow indicate the available dynamic habitat (data from the NEP mooring stations), and open circles indicate the presence of telemetered bull sharks.

salinity. The ROC value of 0.817 indicated a well-discriminated model. Standardized regression coefficients indicated that temperature ($p<0.01$) and the temperature*salinity interaction ($p<0.01$) were the most influential parameters in the upper bay model for 2009. In 2010, the best fit model included the terms temperature, salinity and DO, as well as all the first order interactions. The ROC value of 0.937 indicated excellent model discrimination. Standardized regression coefficients indicated that all predictors ($p<0.01$) were equally influential parameters for the upper bay model in 2010 (Table 2).

Electivity analyses were used to further investigate whether bull sharks showed an affinity for or avoidance of a specific range of hydrographic variables. In 2009, temperature electivity values were similar between the lower and upper bay regions, with affinity demonstrated for temperatures between 29 and 32°C. In 2010, the pattern in the lower bay was similar to the previous year, but bull sharks showed affinity for cooler temperatures (24–26°C) in the upper bay compared to 2009 (Figure 8A & D). With respect to salinity, in 2009 bull sharks demonstrated strong affinity for salinity regimes between 0 and 1 in the upper bay, with no

Table 2. Summary of predictive model results.

Model	Factor	p value
2009 Lower Bay	*Temperature*	<0.001
AIC = 706.524	*DO*	<0.001
ROC = 0.847	*Temp*Salinity*	0.02
	*Temp*DO*	<0.001
	Salinity*DO	0.111
2010 Lower Bay	*Temperature*	<0.001
AIC = 412.876	*Temp*Salinity*	<0.001
ROC = 0.864	*Temp*DO*	0.001
2009 Upper Bay	*Temperature*	<0.001
AIC = 281.561	DO	0.884
ROC = 0.817	*Temp*Salinity*	0.005
2010 Upper Bay	*Temperature*	<0.001
AIC = 1113.765	*Salinity*	<0.001
ROC = 0.937	*DO*	<0.001
	*Temp*Salinity*	<0.001
	*Temp*DO*	<0.001
	*Salinity*DO*	0.005

Summary of GLM results for predicting the probability of detecting at least one bull shark as a function of region (upper bay and lower bay) and year (2009 and 2010). Significant values ($\alpha < 0.005$) are indicated in italics.

selection for salinity regimes in the lower bay. In 2010, strong affinity was shown for salinity values of 10 and 11 in the upper bay, similar to what was seen in the lower bay (Figure 8B & E). In the lower bay in 2009, affinity for low DO waters was observed, where electivity increased with decreasing DO, reaching peak electivity when waters were hypoxic (0–1 mg/l). In contrast, telemetered bull sharks in the upper bay in 2009 showed strongest affinity for normoxic waters (5–7 mg/l). In comparison, bull sharks monitored in the lower bay in 2010 showed affinity for normoxic waters. A similar, but less pronounced pattern was seen with respect to DO affinity of bull sharks monitored in the upper bay (Figure 8C & F).

Discussion

Our data demonstrate the presence of juvenile bull sharks across a range of dynamic habitat conditions within Mobile Bay, Alabama. These findings add to a body of literature establishing the importance of Gulf of Mexico estuaries for bull sharks during this life stage [2,20–22], and although not explicitly tested in the current study, suggests Mobile Bay could function as the northern

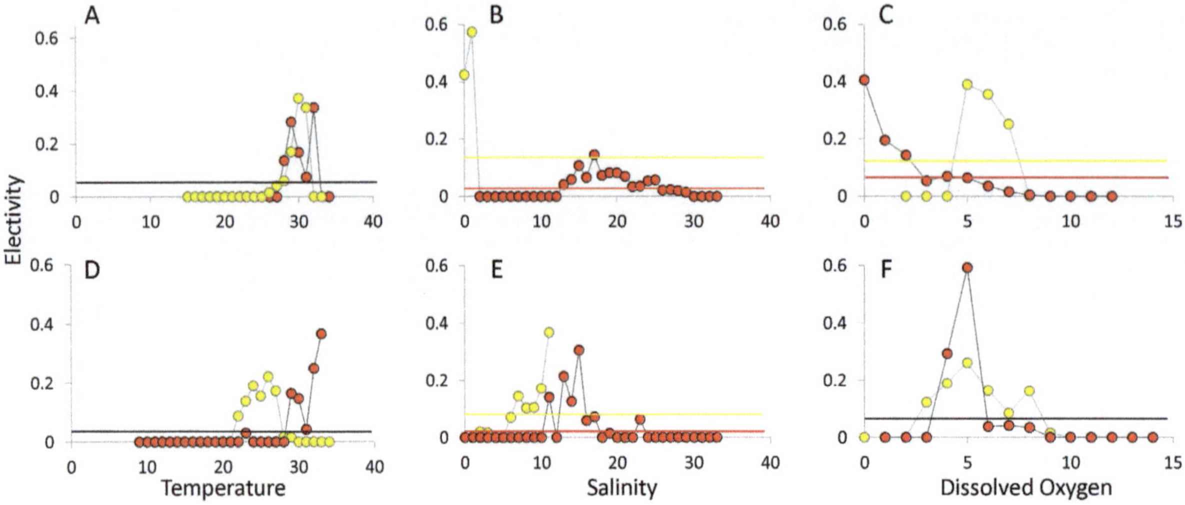

Figure 8. Bull shark electivity. Electivity plots for the lower (red) and upper (yellow) portions of Mobile Bay in 2009 (A–C) and 2010 (D–F). Electivity data are plotted for temperature (A, D) salinity (B, E) and dissolved oxygen (C, F). Horizontal lines represent neutral selection in the upper (yellow) and lower (red) bay. Horizontal black lines are used when the neutral selection value is the same between the upper and lower bay.

most potential nursery area for bull sharks in the Gulf of Mexico. While we were unable to establish acoustic coverage across the entirety of Mobile Bay, we constructed an array designed to monitor bull shark presence along various ingress and egress points of the system and its freshwater sources. Rather than true "presence/absence" data, we suggest our findings provide a framework by which future, more specific, habitat use hypotheses can be tested.

Many of the Gulf of Mexico estuaries containing bull sharks have single fresh water point sources; however, Mobile Bay represents an interesting system that contains several sources of freshwater input, not all of which were used equally. For example, a disproportionate number of bull sharks were originally tagged along the western portion of Mobile Bay, at Dog River, yet the frequency of acoustic detections at this location was low. In contrast, no bull sharks were tagged in the upper bay, yet this area had the greatest proportion of detections throughout the CAAMP array. The most notable difference between the freshwater sources in the upper bay relative to the eastern and western bay is the magnitude of the riverine discharge. With an average discharge of 1800 m^3sec^{-1}, the upper portion of Mobile Bay receives the sixth largest river discharge in North America [33] and thus represents a unique habitat for young bull sharks. Comparing yearly detection differences in the upper bay in light of discharge data supports this idea. Mean freshwater discharge into Mobile Bay varies seasonally, with distinct wet and dry periods. Mean wet season (late winter, early spring) discharge into Mobile Bay is 2637 m^3sec^{-1}, compared to mean dry season (late summer/early fall) discharge of 802 m^3sec^{-1} [34]. The upper bay experienced extreme discharge conditions between the two years we monitored bull sharks. Mean riverine discharge into Mobile Bay during the 2009 dry season was 2788 m^3/sec (i.e. higher than typical wet season values), compared to 2010, where dry season discharge was 279 m^3sec^{-1}, three times lower than average conditions (USGS discharge data, waterdata.usgs.gov, accessed January 7, 2014). Bull sharks are known to alter dynamic habitat use in response to extreme weather events [8]; while we lack both the long term data and continuous acoustic coverage, the dramatic difference in riverine discharge between 2009 and 2010 offers a potential explanation for the difference in acoustic detections between 2009 and 2010 in the upper bay.

Discharge is clearly correlated to salinity, the most often cited parameter influencing the distribution of young bull sharks. In the eastern Gulf of Mexico, salinity was shown to be the most important physical parameter affecting bull shark movements, a relationship that was stronger for young-of-the-year individuals compared to animals older than 1 year [19]. Similarly, salinity was the only measured physical parameter shown to influence habitat use for immature bull sharks in the Indian River Lagoon [16]. The unique physical attributes of Mobile Bay make identifying the important hydrographic variables influencing habitat use more complex. Comparing two systems in Florida, where juvenile bull sharks have been well studied, to Mobile Bay Alabama illustrates this point. The Caloosahatchee River and the Indian River Lagoon are both classified as vertically homogeneous systems [35]. Both systems share similarly sized drainage basins (3,678 and 3,226 km^2 for the Caloosahatchee and Indian River Lagoon, respectively) and receive similar average daily freshwater inputs (3,425 and 4,648 m^3 day^{-1} for the Caloosahatchee and Indian River Lagoon, respectively). In comparison, Mobile Bay is classified as a moderately to highly stratified system, with a drainage area and average daily freshwater input thirty and forty times greater, respectively, than the Caloosahatchee and Indian River systems [35]. Our electivity analyses suggest when moderate salinities are available in the upper bay, as was the case in 2010, bull sharks were selecting for regions within that salinity range, similar to the patterns shown by bull sharks in the Caloosahatchee River [19] and along the Texas coast [15]. However, given the expansive nature of Mobile Bay, coupled with its dynamic hydrographic qualities, it is likely that factors in conjunction with salinity are influencing dynamic habitat use for young bull sharks.

The interactive nature of multiple predictive hydrographic parameters can be best seen when examining the unique relationship between salinity and dissolved oxygen in Mobile Bay. The affinity demonstrated by bull sharks in the lower bay for hypoxic waters was unexpected. In the Florida Everglades, dissolved oxygen has been shown to have the greatest influence on the probability of bull shark capture, with bull shark abundance greatest in areas where DO >3.5 mg L^{-1} [11]. We suggest the strong affinity shown for bull sharks in the lower bay for hypoxic waters is best considered with respect to other predictors. Shallow estuaries, typical of those juvenile bull sharks have been studied in previously, are generally well mixed [36]; however, despite being considered a shallow estuary, this is not the case with Mobile Bay. The moderate to high levels of vertical stratification in Mobile Bay set up a strong bottom to surface salinity gradient (ΔS), which is strongly correlated to bottom dissolved oxygen levels [37]. When ΔS is less than 2 psu, hypoxia in the shallow monitored regions of Mobile Bay is rare; conversely, when ΔS is greater than 4 psu, roughly 75% of the same area becomes hypoxic (here defined as DO <2 g m^{-3}) [37]. Therefore, while periods of high freshwater discharge clearly lower salinity, they also establish a strong ΔS, which in turn can lead to wide-spread bottom hypoxia in the shallow regions of Mobile Bay. Our electivity data and others [19] demonstrate that while immature bull sharks are capable of inhabiting waters across a wide range of salinity, they tend to select waters from 7–11 psu. In Mobile Bay, these salinity regimes would be associated with a strong ΔS, and thus hypoxic conditions. This relationship offers a potential explanation for the apparent strong selection shown by young bull sharks in the lower bay in 2009 for hypoxic waters. The relationship between bull shark presence and hypoxia could also be explained by the limitations in our acoustic array. The electivity patterns we present need to be interpreted in light of our acoustic coverage, which we suggest provides "presence only" data. Interpreted in this respect, we may be simply detecting bull sharks as they exit Mobile Bay during hypoxic periods. Regardless of the mechanism, it seems clear that a combination of hydrographic variables influence the distribution of bull sharks monitored within Mobile Bay.

Temperature was the only main effect included in the final models for both lower and upper bays, in both 2009 and 2010. These findings complement previous studies investigating factors important in predicting the presence of young bull sharks. Analysis of the influence of salinity, temperature, depth, turbidity and DO on the distribution of juvenile bull sharks in the western Gulf of Mexico indicated salinity and temperature were the most influential factors, followed by freshwater inflow, turbidity and proximity to tidal inlets [15]. Interestingly, in Mobile Bay, the combination and strength of the factors that influence the distribution of young bull sharks varied over the two year period we examined. In the lower bay, the model fit the observations better in 2010, and with fewer parameters, all of which included temperature. Electivity patterns demonstrate that bull sharks appear to show preference for water greater than 30°C. We offer two potential explanations for these observations, the first being that bull sharks are simply selecting habitats with high water temperatures. Using multiple data sources, analysis of a thirty year data set from the Indian River Lagoon in Florida shows that the mean temperature of occurrence for juvenile bull sharks was

29.7°C [13], similar to the maximum electivity values shown for temperatures of 32 and 33°C for bull sharks in the lower bay in 2009 and 2010, respectively. In addition, acoustic tracking of an individual bull shark suggests long-term fidelity to thermal effluents in the Indian River Lagoon [16]. However, it could be that water temperatures in excess of 30°C represent a thermal maximum, and that our array detects the ensuing bull shark emigration from Mobile Bay, as suggested previously with dissolved oxygen.

While the interaction of hydrographic factors offers an explanation for the observed distribution of bull sharks in Mobile Bay, other factors not considered in this study, including the potential for increased prey availability and/or reduced mortality, are equally plausible. Several studies suggest that bull shark preference for freshwater is clearly not due to physiological constraints, and may be the result of increased resource abundance or decreased predation [38,39]. Using long-term acoustic monitoring data, Heupel and Simpfendorfer [7] demonstrated that young bull sharks in the Caloosahatchee River Estuary suffered lower mortality rates relative to similarly sized tiger (*Galeocerdo cuvier*), school (*Galeorhinus galeus*), blacktip (*Carcharhinus limbatus*) and lemon (*Negaprion brevirostris*) sharks which occupy polyhaline (18–30 psu) environments. Further investigation is required to understand if either or both of these explanations (increased resource availability, decreased mortality) interact with hydrographic factors to influence habitat use throughout Mobile Bay.

Given the unique osmoregulatory capability of bull sharks, and their subsequent propensity to occupy estuarine habitats early in their ontogeny, this species may be especially susceptible to anthropogenic alterations to coastal ecosystems. In the Indian River Lagoon, bull sharks show affinity for urbanized habitats, a trend that can be partially explained by an affinity to thermal outfall from powerplants [16]. The ability of bull sharks to occupy these altered habitats may also provide population level benefits; decreased mortality has been shown for young bull sharks in highly urbanized areas in the Caloosahatchee River [7]. With respect to local-scale habitat selection, this suggests bull sharks, in this region and during this life stage, show a moderate degree of habitat specialization and are thereby sensitive to environmental fluctuation [40]. That said, neonate bull sharks monitored with acoustic tags along the East coast of Australia showed a preference for natural compared to artificial habitat [41], suggesting spatial variation in the degree of habitat specialization exhibited in this species. Combined, our data suggest that future studies coupling telemetry and hydrographic variables should, when possible, consider the interactions of multiple environmental parameters when identifying the dynamic variables explaining the spatial distribution of coastal shark species.

Acknowledgments

The authors wish to thank Mike Dardeau and the technical support staff at the DISL for maintaining and servicing the NEP mooring stations from which hydrographic data were obtained for this study. Thanks are owed to Marla Valentine and Andrea Kroetz for field support, Aaron Pilnick and Stan Bosarge for assistance with maps, and John Froeschke and John McCreadie for statistical advice. This work would not have been possible without assistance from the interns and technicians in the Fisheries Ecology Lab. We thank Tobey Curtis and Steve Kessel for constructive comments that greatly improved this manuscript.

Author Contributions

Conceived and designed the experiments: JMD MJA SPP. Performed the experiments: JMD MJA. Analyzed the data: JMD MJA. Contributed reagents/materials/analysis tools: JMD MJA SPP. Wrote the paper: JMD MJA SPP.

References

1. McCandless CT, Pratt HL Jr, Kohler NE (2007) Shark nursery grounds of the Gulf of Mexico and east coast waters of the United States. American Fisheries Society Symposium 50. 402 p.
2. Froeschke JT, Stunz GW, Sterba-Boatwright B, Wildhaber ML (2010) An empirical test of the 'shark nursery area concept' in Texas using a long-term fisheries-independent data set. Aquat Biol 11: 65–76.
3. Knip DM, Heupel MR, Simpfendorfer CA (2010) Sharks in nearshore environments: models, importance and consequences. Mar Ecol Prog Ser 402: 1–11.
4. Simpfendorfer CA, Heupel MR (2004) Assessing habitat use and movement. In: Carrier JC, Heithaus MR, eds. Biology of sharks and their relatives. Boca Raton: CRC Press. 553–572.
5. Speed CW, Field IC, Meekan MG, Bradshaw JA (2010) Complexities of coastal shark movements and their implications for management. Mar Ecol Prog Ser 408: 275–293.
6. Wirsing AJ, Heithaus MR, Dill LM (2007) Can measures of prey availability improve our ability to predict the abundance of large marine predators? Oecologia 153: 563–568.
7. Heupel MR, Simpfendorfer CA (2011) Estuarine nursery areas provide a low-mortality environment for young bull sharks *Carcharhinus leucas*. Mar Ecol Prog Ser 433: 237–244.
8. Matich P, Heithaus MR (2012) Effects of an extreme temperature event on the behavior and age structure of an estuarine top predator, *Carcharhinus leucas*. Mar Ecol Prog Ser 447: 165–178.
9. Ubeda AJ, Simpfendorfer CA, Heupel MR (2009) Movements of bonnetheads, *Sphyrna tiburo*, as a response to salinity change in a Florida estuary. Environ Biol Fish 84: 293–303.
10. Simpfendorfer CA, Yeiser BG, Wiley TR, Poulakis GR, Stevens PW, et al. (2011) Environmental influences on the spatial ecology of juvenile smalltooth sawfish (*Pristis pectinata*): results from acoustic monitoring. PLoS ONE 6(2): e16918. doi: 10.1371/journal.pone.0016918.
11. Heithaus MR, Delius BK, Wirsing AJ, Dunphy-Daly MM (2009) Physical factors influencing the distribution of a top predator in a subtropical oligotrophic estuary. Limnol Oceanogr 54(2): 472–482.
12. Martin RA (2006) Conservation of freshwater and euryhaline elasmobranchs: a review. J Mar Biol Assoc UK 85: 1049–173.
13. Curtis TH, Adams DH, Burgess GH (2011) Seasonal distribution and habitat associations of bull sharks in the Indian River Lagoon, Florida: a 30-year synthesis. Trans Am Fish Soc 140(5): 1213–1226.
14. Simpfendorfer CA, Freitas GG, Wiley TR, Heupel MR (2005) Distribution and habitat partitioning of immature bull sharks (*Carcharhinus leucas*) in a southwest Florida estuary. Estuaries Coast 28(1): 78–85.
15. Froeschke JT, Stunz GW, Wildhaber ML (2010) Environmental influences on the occurrence of coastal sharks in estuarine waters. Mar Ecol Prog Ser 407: 279–292.
16. Curtis TH, Parkyn DC, Burgess GH (2013) Use of human-altered habitats by bull sharks in a Florida nursery area. Mar Coast Fish 5: 28–38.
17. Ortega LA, Heupel MR, Van Beynen P, Motta P (2009) Movement patterns and water quality preferences of juvenile bull sharks (*Carcharhinus leucas*) in a Florida estuary. Environ Biol Fishes 84: 361–373.
18. Yeiser BG, Heupel MR, Simpfendorfer CA (2008) Occurrence, home range and movement patterns of juvenile bull (*Carcharhinus leucas*) and lemon (*Negaprion brevirostris*) sharks within a Florida estuary. Mar Freshw Res 59: 489–501.
19. Heupel MR, Simpfendorfer CA (2008) Movement and distribution of young bull sharks *Carcharhinus leucas* in a variable estuarine environment. Aquat Biol 1: 277–289.
20. Heupel MR, Yeiser BG, Collins AB, Ortega L, Simpfendorfer CA (2010) Long-term presence and movement patterns of juvenile bull sharks, *Carcharhinus leucas*, in an estuarine river system. Mar Freshw Res 61: 1–10.
21. Blackburn JK, Neer JA, Thompson BA (2007) Delineation of bull shark nursery areas in the inland and coastal waters of Louisiana. In: McCandless CT, Pratt HL Jr, Kohler NE, editors. Shark nursery grounds of the Gulf of Mexico and east coast waters of the United States. American Fisheries Society Symposium 50. 331–344.
22. Parsons GP, Hoffmayer ER (2007) Identification and characterization of shark nursery grounds along the Mississippi and Alabama Gulf Coasts. In: McCandless CT, Pratt HL Jr, Kohler NE, editors. Shark nursery grounds of the Gulf of Mexico and east coast waters of the United States. American Fisheries Society Symposium 50. 301–316.
23. Sanchez-Rubio G, Perry HM, Patricia MB, Johnson DR, Lipcius RN (2011) Oceanic-atmospheric modes of variability and their influence on riverine input to coastal Louisiana and Mississippi. J Hydrol 396: 72–81.

24. Kim CK, Park K (2012) A modeling study of water and salt exchange for a micro-tidal, stratified northern Gulf of Mexico estuary. J Mar Syst 96–97: 103–115.
25. Heupel MR, Semmens JM, Hobday AJ (2006) Automated acoustic tracking of aquatic animals: scales, design and deployment of listening station arrays. Mar Freshw Res 57: 1–13.
26. Kessel ST, Cooke SJ, Heupel MR, Hussey NE, Simpfendorfer CA, et al. (2013) A review of detection range testing in aquatic passive acoustic telemetry studies. Rev Fish Biol Fish doi: 10.1007/s11160-013-9328-4.
27. Ajemian MJ, Powers SP (2012) Habitat-specific feeding by cownose rays (*Rhinoptera bonasus*) of the northern Gulf of Mexico. Environ Biol Fishes doi: 10.1007/s10641-011-9858-3.
28. Drymon JM, Powers SP, Carmichael RH (2012) Trophic plasticity in the Atlantic sharpnose shark (*Rhizoprionodon terraenovae*) from the north-central Gulf of Mexico. Environ Biol Fish 95: 21–35.
29. Ajemian MJ (2011) Foraging ecology of large benthic mesopredators: effects of myliobatid rays on shellfish resources. Marine Sciences, University of South Alabama, Mobile, Alabama, p 261.
30. Firth D (1993) Bias reduction of maximum likelihood estimates. Biometrika 80 (1): 27–38.
31. Chesson J (1983) The estimation and analysis of preference and its relationship to foraging models. Ecology 64(5): 1297–1304.
32. Neer JA, Thompson BA, Carlson JK (2005) Age and growth of *Carcharhinus leucas* in the northern Gulf of Mexico: incorporating variability in size and birth. J Fish Biol 67: 370–383.
33. Dzwonkowski B, Park K, Ha HK, Graham WM, Hernandez FJ, et al. (2011) Hydrographic variability on a coastal shelf directly influenced by estuarine outflow. Cont Shelf Res 31: 939–950.
34. Marr CD (2013) Hydrodynamic modeling of residence, exposure, and flushing time response to riverine discharge in Mobile Bay, Alabama. Civil Engineering, University of South Alabama, Mobile, Alabama, p102.
35. NOAA (National Oceanographic and Atmospheric Administration) (1985) Gulf of Mexico coastal and ocean zones strategic assessment: data atlas. Strategic Assessment Branch, NOS/NOAA. Rockville, MD.
36. Schroeder WW, Dinnel SP, Wiseman WJ (1992) Salinity structure of a shallow, tributary estuary. In: Prandle D, ed. Dynamics and exchanges in estuaries and the coastal zone, Coastal and Estuarine Studies 40. American Geophysical Union, Washington, D.C. 155–171.
37. Park K, Kim CK, Schroeder WW (2007) Temporal variability in summertime bottom hypoxia in shallow areas of Mobile Bay, Alabama. Estuaries Coast 30(1): 54–65.
38. Pillans RD, Franklin CE (2004) Plasma osmolyte concentrations and rectal gland mass of bull sharks (*Carcharhinus leucas*), captured along a salinity gradient. Comp Biochem and Physiol A Mol Integr Physiol 138: 363–371.
39. Pillans RD, Good JP, Anderson WG, Hazon N, Frankin CE (2005) Freshwater to seawater acclimation of juvenile bull sharks (*Carcharhinus leucas*): plasma osmolytes and Na^+/K^+-ATPase activity in gill, rectal, gland, kidney and intestine. J Comp Physiol B 175: 37–44.
40. Munroe SEM, Simpfendorfer CA, Heupel MR (2013) Defining shark ecological specialization: concepts, context, and examples. Rev Fish Biol Fish doi: 10.1007/s11160-013-9333-7.
41. Werry JM, Lee SY, Lemchert CJ, Otway NM (2012) Natural or Artificial? Habitat-use by the bull shark, *Carcharhinus leucas*. PLoS One 7(11) 1–17.

Shading as a Control Method for Invasive European Frogbit (*Hydrocharis morsus-ranae* L.)

Bin Zhu[1]*, Michael S. Ellis[2¤], Kelly L. Fancher[1], Lars G. Rudstam[3]

1 Department of Biology, University of Hartford, West Hartford, Connecticut, United States of America, **2** Department of Biology, Hobart and William Smith Colleges, Geneva, New York, United States of America, **3** Department of Natural Resources, Cornell University, Ithaca, New York, United States of America

Abstract

Invasive European frogbit (*Hydrocharis morsus-ranae* L.) has negative environmental and economic impacts in North American water bodies. It is therefore important to develop effective management tools to control this invasive species. This study investigated shading as a control method for European frogbit in both greenhouse and lake mesocosm experiments. A series of shade treatments (0%, 50%, 60%, 70%, 80%, and 100%) were tested in the greenhouse for three weeks. Results showed that the 100% shade was most effective at controlling European frogbit, and other shade treatments greater than 50% were less effective, reducing frogbit biomass up to 38.2%. There were no differences found in temperature between treatments, but dissolved oxygen decreased as shading increased. A lake mesocosm experiment utilizing 0% shade, 70% shade, and 100% shade treatments was performed in a sheltered inlet of Oneida Lake in New York State for over one month. Resulting European frogbit biomass was significantly (25 times) less in areas treated with the 70% shade and nearly zero with the 100% shade. Shading did not affect temperature but improved DO conditions. Results on the shading effects on submerged macrophytes were not conclusive: no significant differences in changes in species richness and abundance between the three groups at the end of studied period suggested no shading effects; significant differences between the beginning and end communities in the 70% shade and the 100% shade but not in the control group indicated significant impacts of shading. This study is the first one to investigate shading as a control method for European frogbit and it is concluded that a moderately high density shade can effective remove European frogbit likely with minor impacts on the environment. More experiments with larger scales and longer time periods are recommended for further investigation.

Editor: Shuijin Hu, North Carolina State University, United States of America

Funding: This project was supported by the New York Great Lakes Protection Fund and the Vincent B. Coffin Grant and the A&S Dean's Research Fund at the University of Hartford. The funders had no role in study design, data collection and analysis, decision to publish, or preparation of the manuscript.

Competing Interests: The authors have declared that no competing interests exist.

* E-mail: zhu@hartford.edu

¤ Current address: Department of Fisheries and Wildlife, Oregon State University, Corvallis, Oregon, United States of America

Introduction

Biological, physical, and chemical factors all interact to shape the growth, abundance, and distribution of aquatic plants [1,2,3]. Important factors affecting aquatic plants include light, depth, fetch, ice scour, latitude, temperature, and water levels. Among these factors, light is of paramount importance, because it exerts a major control on photosynthesis and declines with water depth due to attenuation, scattering, and absorption [4,5]. The general effect of an increase in light will be positive for aquatic plants, promoting plant growth by increasing photosynthesis [6,7]. For example, Zhu et al. [8] reported the submerged macrophytes increased their abundance, richness and depths in Oneida Lake, NY when the water clarity increased following the invasion of zebra mussels and nutrient reduction. Extension of submerged plant distribution to deeper depth was observed in bays of the Great Lakes as the water clarity increased [2,9]. Conversely, blocking light will lead to reduction in plant growth [5,6,10]. Therefore, shading as a control method is likely to inhibit growth of invasive aquatic plants.

European frogbit (*Hydrocharis morsus-ranae* L.) is an invasive floating plant in North American water bodies. The species escaped in 1939 from a Botanical Garden in Ottawa, Canada and was then spotted in the Rideau Canal, Ontario, Canada [11]. European frogbit has been travelling south since then and had reached the United States by 1974 [12]. This plant can be found in still, slow-moving shallow waters, such as ponds, ditches, wetlands, marshes and swamps, backwaters, beaver dams, canals, and sluggish creeks, as well as wind sheltered and wave protected areas of lakes and rivers [13,14]. European frogbit reproduces vegetatively through development of stolon buds and turions to form new plantlets and possibly through seeds as well [14]. It continues to spread and may invade further south as a result of range expansion. Global warming may facilitate spreading of this species [15]. European frogbit's dense leaves cover a large surface area that blocks or reduces sunlight penetrating the water below, thereby suppressing growth of submerged, native macrophytes [16]. Invasive European frogbit may harm the economy as well as the environment because of its ability to expand rapidly. It can block navigation channels, irrigation ditches, and water intake pipes and can reduce aesthetic and recreational value of water bodies, thus decreasing tourism and real estate values [14,17]. It is therefore important to develop effective management tools to control this invasive species.

Mechanical harvesting, chemicals, and biological agents are three common methods suggested or reported to control European frogbit. While a mechanical harvesting technique is a control method, it can have significant negative impacts on aquatic ecosystems [14]. Hand pulling has also been proven helpful in removing some frogbit from numerous environments [18]. However, this approach regularly requires the employment of costly labor forces [18] and frequent, repeated removal efforts in order to be effective (B. Zhu, unpublished data). Chemicals such as endothal and diquat have been used as effective controls in ditches against European frogbit [19,20]. However, chemical treatment sometimes is not target-specific and can eliminate other aquatic plants, including beneficial species, and possibly have negative impacts on other organisms [20]. Biological control agents were also suggested in some studies because European frogbit is a food resource for many animals including insects, rodents, water birds, freshwater snails, and fish [14,21,22,23]. Froemming [21] observed that consumption of *H. morsus-ranae* stimulates egg production of the freshwater snails *Lymnaea stagnalis* and *Rumina decollate*. Dabbling ducks (*Anas* spp.) have been documented to consume European frogbit in the eutrophic wetlands of central Finland [22]. Note though, these biological control candidates are not target-specific and could harm other native plants or animals [22]. To date, there are no classical biological control organisms in development or released for European frogbit.

Studies have shown that light is essential for the germination and growth of European frogbit, that light-deprivation may reduce frogbit root growth by 90% [24,25]. It follows that control methods utilizing shading are likely to inhibit the growth and spread of this invasive species. Shading has already been successfully used for controlling aquatic plants such as submerged cabomba (*Cabomba caroliniana*) [10]. It was reported that the 99% shade completely removed cabomba within four months and the 70% shade was effective at deeper depths [10]. However, shading may have negative impacts on beneficial submerged plants that grow underneath target species due to further reductions in the amount of light penetrating to deeper depths [16]. Therefore, shading that blocks too much light (e.g., >90%) would not be desirable despite its high potential to eradicate target species. A desirable outcome should result in the effective control of European frogbit and minor or no impacts on submerged macrophytes below. Consequently, the objectives of this study are: 1) to test the efficacy of different levels of shading as controls on European frogbit growth in both greenhouse and lake mesocosm experiments; 2) to assess the impacts of the shading method on aquatic ecosystems by examining temperature, dissolved oxygen, and submerged macrophytes underneath treated European frogbit mats.

Materials and Methods

Shading Experiment in the Greenhouse

An experiment with a random design was performed in a greenhouse using black shade cloths of various densities (50%, 60%, 70%, 80%, and 100%, International Greenhouse Company, Georgetown, IL) for three weeks from June 1 to June 21, 2010. We also included ambient light (0% shade) as the control. The greenhouse had air exchange with the outside and the temperature and irradiance were similar to ambient lake conditions. Individual full-grown plantlets were put in 5 gallon white buckets with different shade cloths (see Zhu et al. [15] for details), and all plantlets were similar in size at the beginning of the experiment. All buckets were placed in one large water bath at the ambient temperature to simulate lake conditions - the shallow portion of a lake where European frogbit is likely to grow. Three replicates were randomly selected for each shade level. Temperature and dissolved oxygen (DO) were recorded at the end of the experiment using an Orion 4 Star DO Portable multipurpose digital probe (Thermo Scientific, Waltham, MA). European frogbit was collected at the end and biomass was weighed after drying at 65°C for 72 hrs. Plant growth was evaluated by the number of plantlets in each bucket, the average biomass per plantlet, and total biomass at the end of the experiment.

Shading Experiment in the Lake Mesocosms

The natural ecosystem testing was performed in a sheltered inlet in Big Bay of Oneida Lake in New York State (43.25° N, 76.11° W) from June 23 to July 27, 2010. No specific permissions were required for this location for the purpose of this experiment and our field study did not involve endangered or protected species. A control group (European frogbit mats with no shading), 70% shade, and 100% shade were applied to 1×1 m^2 experiment plots with similar densities (about 80% coverage) of European frogbit, three replicates for each group. Experiment plots were defined by 1 m×1 m PVC pipe squares set floating on the surface, anchored with concrete blocks to the bottom of the lake, and labeled and kept afloat with buoys. Water depth in each plot was less than 1.5 m, and all plots were located in areas protected from wind and waves and with minimal boat traffic. The 70% shade treatment was chosen mainly based on the results from the greenhouse experiment (see the result section for details). Also in another study, the 70% shade was used to control submerged cabomba [10]. Therefore, we chose the 70% shade instead of a series of shade gradients in the lake experiment. Temperature and DO were measured below the cloths at noon of three separated dates (June 23, July 8, and July 27) during the experiment period. Total biomass of European frogbit was collected and measured at the end of the experiment to test the effectiveness of the shade method. Submerged aquatic macrophytes were collected at the beginning and the end of this experiment using a 0.25 m^2 (0.5 m×0.5 m) quadrat below each plot to evaluate the impacts of the shade method on submerged macrophytes. Dry weight was measured after drying at 65°C for 72 hrs and biomass was then calculated for analysis.

Statistical Analysis

All data except submerged macrophyte biomass were recorded as mean ±1 standard error. Standard errors were not shown for submerged macrophytes due to large variability of pre-existing submerged macrophyte communities between experimental sites within each group. We used non-metric multidimensional scaling (MDS) [26] to visualize differences in plant community structure before and after the experimental treatment. The differences among plots were based on the biomass (dry weight) of nine species of macrophytes. Significant differences between community structure before the experiment and community structure in the control group, the 70% shade and 100% shade treatments (four factors) were investigated with an analysis of similarities (ANOSIM) [26]. ANOSIM used all possible permutations of the rank similarity matrix to calculate the probability of the similarity within a factor to be larger than a random selection of samples using an R-statistics defined as the difference between the average of rank similarities of pairs of plots in different groups and the average of rank similarities of pairs of plots within a group, and dividing this difference by a measure of the number of samples under consideration [26]. Analyses were done with Primer v6.1.6 (Plymouth Routines in Multivariate Ecological Research). For other comparisons, all data was natural logarithm transformed (ln(x+1)) to reduce heteroscedasticity and analyzed using ANOVA (IBM SPSS Statistic 20) [27]. All ANOVAs were followed by the least significant difference (LSD) analysis to compare different treatments at the level $\alpha = 0.05$ [27].

Results

Greenhouse Experiment

Effectiveness of Shading. Plant growth was evaluated from three different variables: number of plantlets, average biomass per plantlet, and total biomass for all the plantlets in each group. We observed that the control group had the healthiest plants while no plants survived after the 100% shade treatment, and all treatments with 50% shading or higher affected frogbit growth (Figure 1). The control group with ambient light had an average of one plantlet after three weeks, as did the 50% shade treatment, while all other shading treatments but the 100% shade had more plantlets than the control group (Figure 1a). Average biomass per plantlet was highest in the control group with 0.35±0.05 g/plantlet while all shaded treatments were significantly lower (Figure 1b). The lowest was observed in the 100% shade treatment followed by the 60% and 70% shade groups. Like average biomass per plantlet, the treatment groups had much lower total biomass than the control group, with zero biomass under the 100% shade (Figure 1c). Combining the results from the three variables, we concluded that 100% shade was most effective for controlling European frogbit. Other shade treatments greater than 50% were relatively effective, reducing the biomass up to 38.2%.

Impacts of Shading on Temperature and DO. Water temperature under different shade treatments ranged from 21.4±0.44°C to 22.7±0.15°C and were not statistically different ($df = 5$, $F = 2.077$, $p = 0.139$). However, there were differences in dissolved oxygen content between the different groups ($df = 5$, $F = 3.783$, $p = 0.027$, Figure 2). Dissolved oxygen decreased when more light was blocked, from 5.9±1.3 mg/L in the control group to 4.3±1.2 mg/L in the 100% shade.

Lake Mesocosm Experiment

Effectiveness of Shading. Data from the lake mesocosm experiment suggested shading had effectively controlled European frogbit: its biomass was significantly reduced from 142.6±9.6 g/m^2 in the control group to 5.7±3.5 g/m^2 in the 70% shade, and to 0.01±0.01 g/m^2 in the 100% shade ($df = 2$, $F = 174.9$, $p < 0.001$).

Impacts of Shading on Temperature and DO. There was an obvious seasonal trend in temperature ($df = 2$, $F = 40.2$, $p < 0.001$), but no differences were found between the three different groups: control, 70% shade, or 100% shade in the lake ($df = 2$, $F = 0.564$, $p = 0.576$). DO levels were generally low (less than 4 mg/L) in densely vegetated water in this study (Figure 3). As the experiment proceeded, DO decreased from around 4 mg/L to less than 1.5 mg/L. However more oxygen was present in the two shade groups than the control group at the end of the experiment (1.66±0.19 and 0.74±0.20 mg/L vs. 0.23±0.04 mg/L, $df = 2$, $F = 20.57$, $p = 0.002$, Figure 3).

Effects of Shading on Submerged Macrophytes. Species richness of submerged macrophytes was found not significantly different between the control group and the shading treatment groups at the beginning of the experiments ($df = 2$, $F = 1.091$, $p = 0.394$) or at the end of the experiments ($df = 2$, $F = 1.50$, $p = 0.296$, Table 1). Species richness was similar for all treatments at the beginning of the experiment and had declined by the end for all treatments, including the control. Relatively more reduction in macrophyte richness occurred in plots with the 100% shade

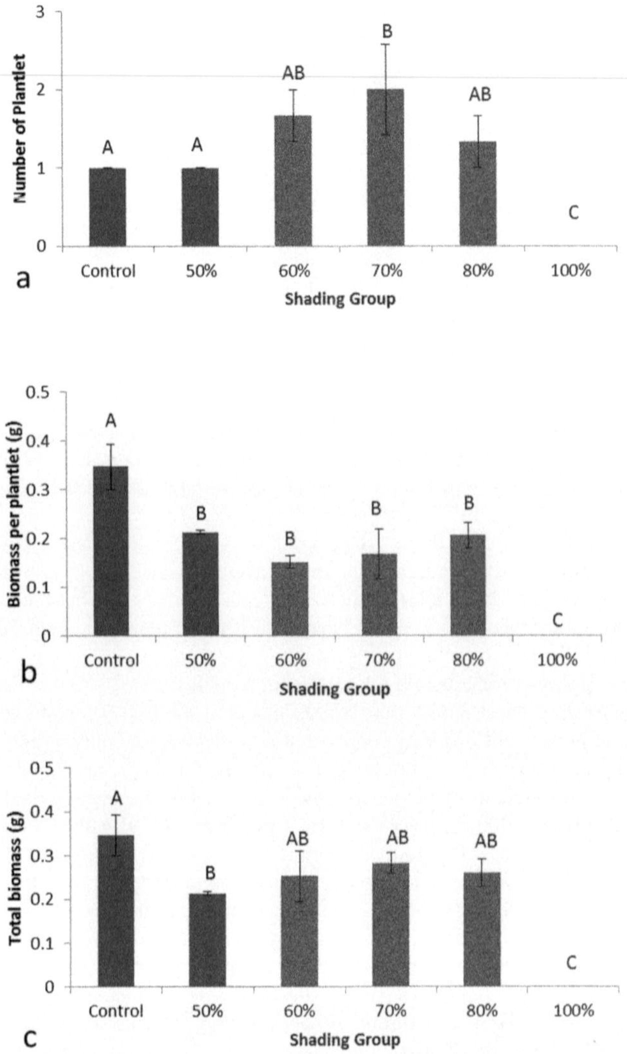

Figure 1. European frogbit growth in a series of shades in the greenhouse experiment: a. number of plantlets, b. average biomass per plantlet, and c. total biomass. Different letters indicate significant differences at $\alpha = 0.05$ following ANOVA LSD analysis.

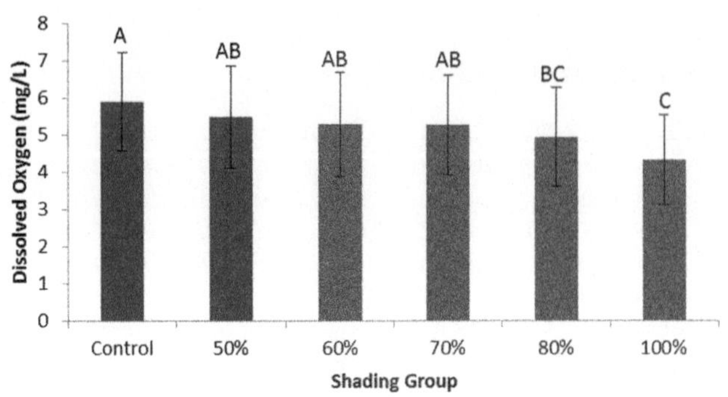

Figure 2. Dissolved oxygen in each shade treatment in the greenhouse experiment. Different letters indicate significant differences at $\alpha = 0.05$ following ANOVA LSD analysis.

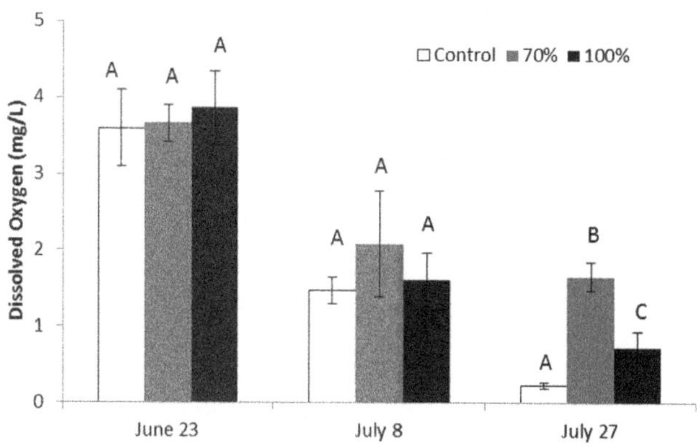

Figure 3. Dissolved oxygen in different treatment groups at different times during the lake mesocosm experiment. Different letters indicate significant differences at $\alpha = 0.05$ following ANOVA LSD analysis.

compared to those in the control group and under the 70% shade (Table 1).

Results of the impacts of shading on the submerged macrophyte community stucture in terms of species composition and abundance were not conclusive. The species composition varied among groups (Figure 4). For all groups, the dominant submerged macrophyte species were coontail (*Ceratophyllum demersum*), elodea (*Elodea canadensis*), Eurasian watermilfoil (*Myriophyllum spicatum*), flat-stem pondweed (*Potamogeton zosteriformis*), and small-leaf pondweed (*P. pusillus*), though their abundance (in terms of biomass) differed between groups (Figure 4). For example, at the beginning of the experiment, coontail was the most abundant species in the control group whereas elodea was most abundant in the 70% shade, and small-leaf pondweed was most abundant in the 100% shade. This also indicated the variability of initial submerged macrophyte communities. At the end of the experiment, most species had decreased in biomass, but the magnitude of this change varied between groups. Seven out of the nine species in the control group had 65.9%–100% (average of 85.4%) reduction in biomass whereas two species, stargrass (*Heteranthera dubia*) and water buttercup (*Ranunculus aquatilis*), increased in biomass depsite their overall abundance being relatively low. Similarly, eight out of the ten species in the 70% shade experienced 45.2%–100% (average of 68.6%) reduction in biomass, and two species, common naiad (*Najas flexilis*) and stargrass, slightly increased in biomass. However, the densities of all ten species were reduced by 57.1%–100% (average of 90.6%, Figure 4) in plots treated with the 100% shade. These changes led to the decrease in total abundance of submerged macrophyte (measured as biomass of all macrophytes) in all the three groups: 69.6% in the control group, 67.8% in the 70% shade group, and 89.8% in the 100% shade group (Table 1) and the changes did not differ between groups ($df = 2$, $F = 1.288$, $p = 0.342$). There were also no differences in the community structures at end of the experiment among the three groups ($R < 0.01$, $p > 0.55$). However, significant differences were found when the beginning and end community structures were compared for the three groups. There were no differences in the beginning and end community structures in the control group ($R = 0.36$, $p = 0.077$) whereas significant differences between the beginning and end communities were found in the 70% shade and the 100% shade ($R = 0.45$, $p = 0.03$, and $R = 0.84$, $p = 0.005$ respectively).

Discussion

Our results from both greenhouse and lake experiments revealed that the 100% shade was most effective at removing invasive European frogbit. However this level of shading should be considered unacceptable as a large-scale control option since its application significantly reduces native plant growth and would be detrimental to many other aquatic organisms [28].

A moderately high density of shade, such as 70%, did effectively control invasive European frogbit and could be a control option. In our lake mesocosm experiment, the 70% shade significantly reduced frogbit biomass by 25 times over the course of a one month trial, while in the greenhouse experiment, shade treatments greater than 50% were all relatively effective in the control of European frogbit, especially in reducing frogbit biomass. Interestingly, greater reproduction (indicated by the number of plantlets) was observed in the denser shade treatments. Because plants often show greater vegetative growth at the expense of reproduction under favorable conditions [29], this change in resource allocation from growth to reproduction is a predictable response to the low-light stress created by denser shading. For example, Miller et al. [29] reported that dioecious hydrilla (*Hydrilla verticillata*) decreased turion production when plant density increased (i.e., better growth). In stressful situations where survival probability is low and opportunities for reproduction may be limited to "now or never," plants tend to assign more resources to reproductive output [30]. However, if the stress persists, affected plants will eventually die.

Control or eradication efforts including physical, chemical and/or biological methods, often greatly disturb natural ecosystems, possibly increasing the vulnerability of an area to subsequent invaders [3,28]. An effective control method should be able to remove unwanted plants while minimizing environmental impacts. Shading had no impacts on temperature in the growing environment and may be able to improve DO condition, as was demonstrated by the 70% shade treatments in our lake mesocosm experiment, which had higher DO values than both the control and 100% shade treatments. Higher DO is likely improving environmental conditions for organisms living beneath the plants,

Table 1. Species richness and biomass of submerged macrophytes in different shading treatments.

Shade Group	Species Richness		Biomass (g/m²)	
	Start	End	Start	End
Control	7.0±0.6	6.0±0.6	33.2 (25.5, 43.6)	8.5 (1.2, 13.8)
70%	8.3±0.9	5.3±0.9	40.3 (15.8, 88.8)	17.5 (2.8, 45.1)
100%	7.7±0.3	3.7±1.3	31.5 (10.7, 62.8)	1.8 (0.4, 3.8)

Species richness was shown as the mean ±1 SE and biomass was shown as the mean followed by the range of biomass in parenthesis.

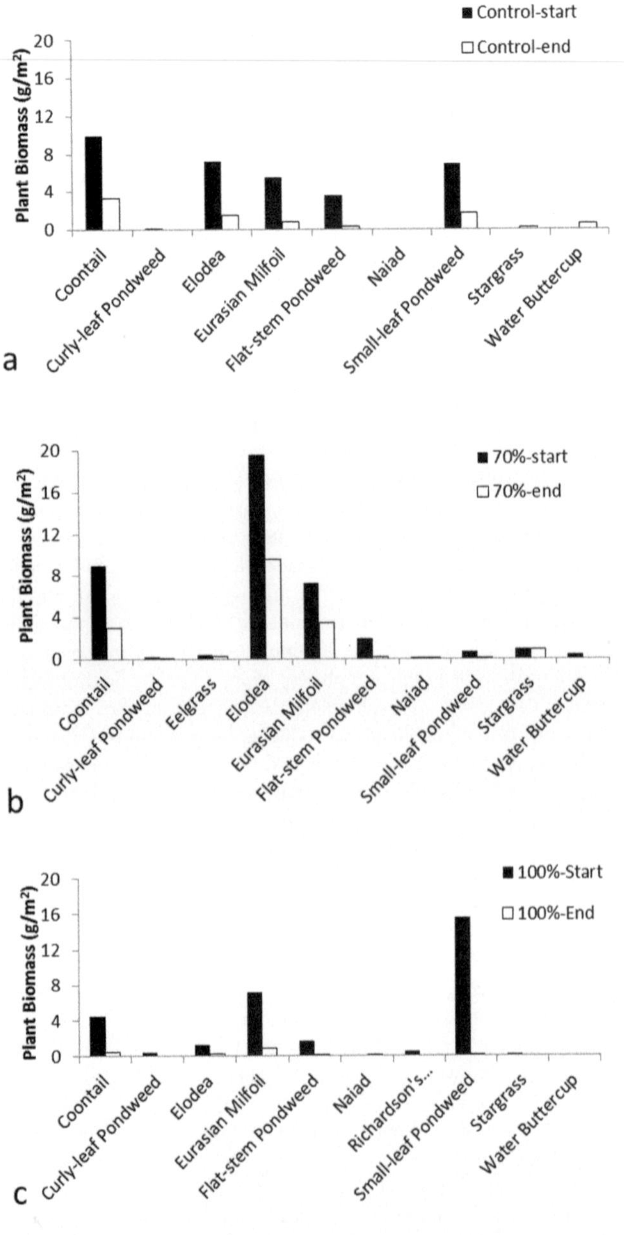

Figure 4. The species composition and abundance of submerged macrophytes at the start and the end of the lake experiment: a. the control group where European frogbit mats were present; b. the 70% shade; and c. the 100% shade. Due to large variability of pre-existing submerged macrophyte communities between experimental sites within each group, only means were shown here and standard errors were omitted for clear view.

including fish and benthic invertebrates. The lowest DO content was observed in the lake control group at the end of the experiment and may have been attributable to the control group's dense, floating mats of European frogbit cover (142.6 g/m^2 compared to 5.7 g/m^2 in the 70% shade). Conversely, in the greenhouse experiment, DO content was highest in the control group, which was likely a result of there being only one plant in each bucket, with plenty of gas exchange between air and water. DO content in the greenhouse experiment decreased as shade increased, likely because there was less photosynthesis under shade and greater rates of decomposition resulted from increased frogbit mortality alongside shade density. These same reasons also explain the low final DO observed in the 100% shade lake treatments. Highest DO in the 70% shade in the lake experiment may be due to less decomposition and more algal production after the European frogbit canopy was removed [31].

The lake experiment suggested possible negative impacts of the shade treatments on submerged macrophytes growing underneath the European frogbit mats. Like the control group, there was an overall decline in total macrophyte biomass along with a slight change in species richness in both the 70% shade and the 100% shade. However, there were no differences in the beginning and the end community structures in the control group whereas significant differences between the beginning and end communities were found in the 70% shade and 100% shading. This indicates there were likely significant impacts of shading on submerged macrophyte communities. Typically we would expect high levels of shading would have negative impacts on species richness and abundance of submerged plants because light is a key factor regulating submerged plant growth and depth distribution, both of which decrease as light is reduced [1,2,5,10]. Other factors might have weakened the negative effects of shading in this study such as large variation of pre-existing macrophyte communities and relatively small size of shading plots in the trial. Because the shading plot is small, submerged macrophytes can still have some light from the surrounding areas, which minimizes the shading effects. Therefore, experiments further investigating the effects of shading over larger spatial scales and longer time periods are recommended 1) because decreases in submerged macrophyte biomass were observed under all treatments in this study, 2) because a possible seasonal reduction in macrophyte biomass cannot be eliminated as a factor, and 3) because of the variability of pre-existing macrophyte communities between sites within the different shade groups at the beginning of the experiment.

Our lake mesocosm and greenhouse experiments have demonstrated that a moderately high density of shading, such as that achieved with the 70% shade cloth, can serve as an effective control for European frogbit. In some conditions, it may even improve some envrionmental conditions such as dissolved oxygen compared to the environment under dense beds of European frogbit. However, this might also have some negative impacts on submerged macrophytes. Shading has not been regularly used for controlling aquatic plants to date, with few invasive aquatic plant management cases utilizing this method [10]. The major reason is that shades are thought to interfere with recreation and are considered aesthetically displeasing because shades float on the water [32]. However, shading may be a feasible choice for controlling European frogbit because it grows in sheltered areas where there are few recreational activities. Additionally, shading has desirable qualities for invasive plant management – time and cost efficient. Shading requires only the preparation, placement, and retrieval of shade cloths, and the shade cloth is commercially available at low prices. The success and the feasibility of eradication are usually heavily dependent upon the amount of investment that can be made [33]. Consequently, it seems both effective and feasible to use shading to control European frogbit.

There are other options for controlling European frogbit, including manual or mechanical removal of plants, chemical control, and biological control [18,20,22]. All control mechanisms, including the shading method, have advantages and disadvantages. For example, shading can be effective, but it may affect other floating plants growing next to European frogbit. It is noteworthy to mention that prevention is the single best solution for European frogbit management, just as it is for many other invasive species [17,28,32]. Common methods of prevention include decontamination and cleaning of boats and equipment that could contain hitchhikers and restricting deliberate imports of potentially harmful species. However, once European frogbit is established, shading may be considered by aquatic plant managers as one of the possible control methods.

Acknowledgments

We thank many faculty and staff at the Finger Lakes Institute of Hobart and William Smith Colleges and the Cornell Biological Field Station for providing support for this project. We also thank anonymous reviewers and the Associate Editor for providing constructive comments to improve the manuscript.

Author Contributions

Conceived and designed the experiments: BZ LGR. Performed the experiments: MSE BZ. Analyzed the data: BZ KLF LGR. Wrote the paper: BZ MSE KLF LGR.

References

1. Spence DHN (1982) The zonation of plants in freshwater lakes. Adv Ecol Res 12:37–125.
2. Skubinna JP, Coon TG, Batterson TR (1995) Increased abundance and depth of submersed macrophytes in response to decreased turbidity in Saginaw Bay, Lake Huron. J Great Lakes Res 21:476–488.
3. Schindler DE, Scheuerell MD (2002) Habitat coupling in lake ecosystems. Oikos 98:177–189.
4. Sand-Jensen K, Borum J (1991) Interactions among phytoplankton, periphyton, and macrophytes in temperate freshwaters and estuaries. Aquat Bot 41:137–175.
5. Hudon C, Lalonde S, Gagnon P (2000) Ranking the effects of site exposure, plant growth form, water depth, and transparency on aquatic plant biomass. Can J Fish Aquat Sci (Suppl 1):31–42.
6. Wade PM (1990) Physical control of aquatic weeds. In: Pieterse AH, Murphy KJ, editors. Aquatic Weeds: The Ecology and Management of Nuisance Aquatic Vegetation. New York: Oxford University Press. pp. 93–135.
7. Karatayev AY, Burlakova LE, Padilla K (2002) Impacts of zebra mussels on aquatic communities and their roles as ecosystem engineers. In: Leppakoski E, Ollasch S, Olenin S, editors. Invasive aquatic species of Europe: distribution, impacts and management. Boston: Kluwer. pp. 433–446.
8. Zhu B, Fitzgerald DG, Mayer CM, Rudstam LG, Mills EL (2006) Alteration of ecosystem function by zebra mussels in Oneida Lake, NY: impacts on submerged macrophytes. Ecosystems 9:1017–1028.
9. Zhu B, Fitzgerald DG, Hoskins SB, Rudstam LG, Mayer CM, et al. (2007) Quantification of historical changes of submerged aquatic vegetation cover in two bays of Lake Ontario with three complementary methods. J Great Lakes Res 33:122–135.
10. Schooler SS (2008) Shade as a management tool for the invasive submerged macrophyte, *Cabomba caroliniana*. J Aquat Plant Manage 46:168–171.
11. Dore WG (1968) Progress of the European frogbit in Canada. Can Field-Nat 82:76–84.
12. Roberts ML, Stuckey RL, Mitchell RS (1981) *Hydrocharis morsus-ranae* (Hydrocharitaceae): new to the United States. Rhodora 83:147–148.
13. Cook CDK, Lüönd R (1982) A revision of the genus Hydrocharis (Hydrocharitaceae). Aquat Bot 14:177–204.
14. Catling PM, Miltrow G, Haber E, Posluszny U, Charlton WA (2003) The biology of Canadian weeds. 124. *Hydrocharis morsus-ranae* L. Can J Plant Sci 83:1001–1016.
15. Zhu B, Eppers ME, Rudstam LG (2008) Predicting invasion of European frogbit in the Finger Lakes of New York. J Aquat Plant Manage 46:186–189.
16. Catling PM, Spicer KW, Lefkovitch LP (1988) Effects of the floating *Hydrocharis morsus-ranae* (Hydrocharitaceae), on some North American aquatic macrophytes. Nat Can 115:131–137.
17. Pimentel D, Zuniga R, Morrison D (2005) Update on the environmental and economic costs associated with alien-invasive species in the United States. Ecol Econ 52:273–288.
18. Langdon S (2007) Eradication of European Frogbit (*Hydrocharis morsus-ranae*) from the Grasse River in the Town of Clare, St. Lawrence County, New York. Adirondack Park Invasive Plant Program 2007 Season Report. 23 p.
19. Holz W (1963) Chemical weed control in ditches; trials on the control of submerged plants. 5. Dtsch. Arbeitsbesprechung uber Fragen der Unkrautbiologie u. -bekampfung, Hohenheim, 4 p.
20. Renard C (1963) The use of diquat and paraquat to control aquatic plants. Compte rendu Conference du Comite Francais de Lutte contre les Mauvaises Herbes (COLUMA). 9 p.
21. Froemming E (1954) Problematic constituents in several marsh plants. Pharmazi 9:766–769.
22. Vaananen VM, Nummi P (2003) Diet of sympatric dabbling ducks in eutrophic wetlands. Suomen Riista 49:7–16.
23. Magomaev FM (1973) The daily diet of two- and three-year-old grass carp. Sbornik Nauchnykh Trudov, Vsesoyuznyi Nauchno-Issledovatel'skii Institut Prudovogo Rybnogo Khozyaistva 10:192–196.
24. Minshall WH (1959) Effect of light on the extension growth of roots of frog-bit. Can J Bot 37:1134–1136.
25. Richards AJ, Blakemore J (1975) Factors affecting the germination of turions in *Hydrocharis morsus-ranae* L. Watsonia 10:273–275.
26. Clarke KR, Warwick RM (2001) Change in marine communities: an approach to statistical analysis and interpretation. Primer-E Ltd. Plymouth, UK.
27. Kuehl RO (2000) Design of experiments: statistical principles of research design and analysis. 2nd edition. Pacific Grove, CA: Duxbury Press. 666 p.
28. Madsen JD (1997) Methods for management of nonindigenous aquatic plants. In: Luken JO, Thieret JW, editors. Assessment and Management of Plant Invasions. Springer, NY. pp.145–171.
29. Miller JD, Haller WT, Glenn MS (1993) Turion production by dioecious hydrilla in North Florida. J Aquat Plant Manage 31:101–105.
30. Chiariello NR, Gulmon SL (1991) Stress effects on plant reproduction. In: Winner WE, Pell EJ, Roy J, editors. Response of Plants to Multiple Stresses. New York: Academic Press. pp. 161–188.
31. Reed DC, Foster MS (1984) The effects of canopy shading on algal recruitment and growth in a giant kelp forest. Ecology 65:937–948.
32. New York State Department of Environmental Conservation (2005) A Primer on Aquatic Plant Management in New York State. Division of Water. 63 p.
33. Panetta FD (2009) Weed eradication - an economic perspective. Invasive Plant Sci Manage 2:360–368.

Cloud-Enabled Microscopy and Droplet Microfluidic Platform for Specific Detection of *Escherichia coli* in Water

Alexander Golberg[1]ⓢ, Gregory Linshiz[2,3,4]ⓢ, Ilia Kravets[5], Nina Stawski[2,3], Nathan J. Hillson[2,3,4], Martin L. Yarmush[1,6], Robert S. Marks[7,8,9], Tania Konry[10]*

1 Centre for Engineering in Medicine, Massachusetts General Hospital, Harvard Medical School, Shriners Burns Institute, Boston, Massachusetts, United States of America, **2** Fuels Synthesis Division, Joint BioEnergy Institute, Emeryville, California, United States of America, **3** Physical BioSciences Division, Lawrence Berkeley National Labs, Berkeley, California, United States of America, **4** DOE Joint Genome Institute, Walnut Creek, California, United States of America, **5** Department of Computer Science, Technion Institute of Technology, Haifa, Israel, **6** Department of Biomedical Engineering, Rutgers University, New Jersey, United States of America, **7** Department of Biotechnology Engineering, The National Institute of Biotechnology in Negev, Ben Gurion University, Beer-Sheva, Israel, **8** School of Materials Science and Engineering, Nanyang Technological University, Singapore, **9** NRF CREATE program for Nanomaterials in Energy and Water Management, Singapore, **10** Department of Pharmaceutical Sciences, School of Pharmacy Bouvé College of Health Sciences, Northeastern University, Boston, Massachusetts, United States of America

Abstract

We report an all-in-one platform – ScanDrop – for the rapid and specific capture, detection, and identification of bacteria in drinking water. The ScanDrop platform integrates droplet microfluidics, a portable imaging system, and cloud-based control software and data storage. The cloud-based control software and data storage enables robotic image acquisition, remote image processing, and rapid data sharing. These features form a "cloud" network for water quality monitoring. We have demonstrated the capability of ScanDrop to perform water quality monitoring via the detection of an indicator coliform bacterium, *Escherichia coli*, in drinking water contaminated with feces. Magnetic beads conjugated with antibodies to *E. coli* antigen were used to selectively capture and isolate specific bacteria from water samples. The bead-captured bacteria were co-encapsulated in pico-liter droplets with fluorescently-labeled anti-*E. coli* antibodies, and imaged with an automated custom designed fluorescence microscope. The entire water quality diagnostic process required 8 hours from sample collection to online-accessible results compared with 2–4 days for other currently available standard detection methods.

Editor: James P. Brody, University of California, Irvine, United States of America

Funding: AG and MY acknowledge Shriners Foundation Grant #85120-BOS for the support of this work. The portion of this work conducted by the Joint BioEnergy Institute, and the U.S. Department of Energy Joint Genome Institute, was supported by the Office of Science, Office of Biological and Environmental Research, of the U.S. Department of Energy (Contract No. DE-AC02-05CH11231). This research is funded in part by the Singapore National Research Foundation and the publication is supported under the Campus for Research Excellence And Technological Enterprise (CREATE) program for the project 'Nanomaterials for Energy and Water Management'. The funders had no role in study design, data collection and analysis, decision to publish, or preparation of the manuscript.

Competing Interests: The authors have declared that no competing interests exist.

* E-mail: t.konry @neu.edu

ⓢ These authors contributed equally to this work.

Introduction

Worldwide water-associated infectious diseases are a major cause of morbidity and mortality [1]. It is estimated that 4.0% of global deaths and 5.7% of the global disease burden are caused by waterborne diseases [1–4]. Common waterborne diseases include diarrhea (bacterial, viral and parasitic), schistosomiasis, trachoma, ascariasis, and trichuriasis [1–4]. Low income countries are particularly vulnerable to waterborne diseases because of their under-developed infrastructure and poor water management [5–14]. Water and sewage distribution systems in high income societies also require pollutant and microorganism monitoring [15].

Escherichia coli, found in mammalian feces [16], has been a biological indicator for water quality since the 19th century [16]. Testing for the presence of *E. coli* is obligatory for current water management systems [17–19]. Herein, we report a comprehensive system – ScanDrop – for the rapid and specific identification of *E. coli* in drinking water.

The identification of bacteria in a water sample includes two major steps: 1) the capture of target bacteria from the water sample, and 2) the identification of the captured bacteria. Traditional methods for *E. coli* detection include culture, fermentation, enzyme-linked immunosorbent (ELISA), and polymerase chain reaction (PCR) assays [20,21]. These traditional methods have disadvantages including long identification times (2–4 days), and/or high labor and reagent costs [20,21]. Despite high costs, rapid tests are necessary to enable quick responses to putative contamination threats. Recently, novel sensors and assays for rapid pathogen detection have been developed, including the capture of whole pathogen cells or molecular fragments for further amplification and identification [22–27], with detection methods utilizing a variety of transducing technologies (optical, electrochemical, surface plasmon resonance and piezoelectric) [27–40]. Many of these newer methods remain expensive and/or require sophisticated instrumentation, and most have yet to reach the market place. Therefore, there remains a need for alternative platforms for the detection of bacteria in water samples.

It remains challenging to inexpensively perform water quality control testing at multiple locations along a distribution system, and to rapidly process and share the test results. To address these challenges, we have developed the ScanDrop platform. ScanDrop is a self-contained detection platform that enables the online control of water testing at multiple locations along the distribution system. ScanDrop integrates live-bacteria capturing and detection, droplet microfluidics, automated fluorescence microscopy, and cloud-based data management and sharing. Droplet microfluidics, applied in ScanDrop, is an emerging application of microelectromechanical systems (MEMS) technology, where assay reagents and biological sample are confined to the pico-liter reactors, composed of water in oil emulsion [41–43]. Small volumes, rapid reagent mixing and non-complex droplet control make droplet microfluidics an attractive choice for the next-generation of high-throughput assays [41–43] and herein detection of bacteria in water samples.

In this work, we demonstrate ScanDrop's capability to detect live *E. coli* in water samples. Magnetic beads, conjugated with specific antibodies, were used to quickly and effectively capture *E. coli* from contaminated water. The captured bacteria were then encapsulated into pico-liter droplets containing fluorescently labeled antibodies, for subsequent detection using a proprietary automated optical fluorescence signal registration system. Imaging system control was facilitated by leveraging a cloud-based laboratory automation system, coined Programing a Robot, PR-PR [44]. We envision that multiple ScanDrop systems could be dispatched at multiple locations to form a cloud-enabled water quality assessment network. Each system could be managed in real-time from a remote control center. Such a network could potentially reduce the infrastructure, management, and labor costs required to perform multiple sample analysis and rapidly share results.

Results and Discussion

Bead-based *E. coli* capture and detection assay

Herein the isolation of bacteria and detection are conducted utilizing simple magnetic bead based immunoassay thus no bacteria agar plate cultivation step is necessary to identify a presumptive positive sample. This approach saves considerable time and resources. In our approach, magnetic beads conjugated with anti-*E. coli* antibodies are added to a water sample (**Fig. 1**). Within 10 min, the magnetic beads have captured the bacteria (if any) from the water sample. The beads are then concentrated with a simple magnet (**Fig. 1**), and a single immunoassay step labels the captured bacteria with a fluorescent antibody for subsequent detection (**Fig. 1**). Detection protocols are integrated into a droplet microfluidic device to reduce reagent volume and enhance reaction rates.

ScanDrop Sensor

The ScanDrop sensor consists of two major components: 1) a droplet microfluidic device for bacteria labeling, and 2) a portable fluorescent optical system for signal detection and sharing.

Droplet microfluidic device. To reduce reagent volumes and detection times, we designed a pico-liter droplet microfluidic chip. The design of poly(dimethylsiloxane) (PDMS) microfluidic device is shown in **Fig. 2A**, the generation of monodisperse droplets in a micro-channel through shearing flow at a flow-focusing zone in **Fig. 2B**, and the resulting droplet array in **Fig. 2C**. Three perpendicular inlet channels form a nozzle, (**Fig. 2A rectangle**), independent syringe pumps controlling flow rates for the oil, beads, and fluorescently labeled secondary antibodies streams. Each droplet in the array co-encapsulates fluorescently labeled anti-*E. coli* antibodies with captured bacteria (if any), to generate a localized fluorescent signal for subsequent detection. The chip enables the generation and incubation of 10^3 droplets with ~100 micron diameter (~520 pL). The advantages of this droplet-based array technique include the physical and chemical isolation of beads in droplets, and the rapid and efficient mixing of the reagents that occurs inside droplets providing fast reaction rates [45–48]. Importantly, this nano-liter microenvironment also enables gas exchange for bacterial viability if further studies are required [45–48]. Previous works in the field of droplet microfluidics showed that the chance to find a cell or a bead inside droplet follows Poisson distribution [49,50]. This puts certain theoretical limitations of the limit of detection of bacteria of droplet microfluidic system. Clausell-Tormos et al. showed that decreasing the number of cells in aqueous solution that is converted to droplets to less than 10^6 cell/mL reduced the probability to find droplets with encapsulated cells and increased the number of empty droplets [50]. Therefore, at the end of incubation time we need to get at least 10^6 CFU/mL of bacteria. The relation between the bacteria in the analyzed sample and the bacteria subjected to encapsulation after enrichment is as follows:

$$2^n N_0 = N_d$$

Where N_o (CFU/mL) is the initial load of bacteria, n is the number of generations in enrichment phase, N_d (CFU/mL) is the concentration of bacteria subjected to encapsulation after enrichment.

Given the generation time of 20 min for *E. coli* in the optimum cultivation condition, to get 10^6 CFU/mL at the end of 6 hours of incubation (18 generations), proposed in our assay, the initial concentration should be at least 3.5 CFU/mL.

To demonstrate the feasibility of our ScanDrop system for multiplex analysis, we co-encapsulated red florescent protein (RFP) and green florescent protein (GFP) expressing *E. coli* in the same droplet (**Fig. 2D**). Capturing on a bead and latter encapsulation for detection of two different bacteria in the same microenvironment will enable multiplex future studies using several types of beads conjugated with different antibodies that bind different target bacteria and different fluorescent tags. The probability of capturing two different objects in a single droplets were analyzed in [49].

Optical system

The schematic for the portable optical system for fluorescent signal detection in the droplet microfluidic device is presented in **Fig. 3A**. The system enables remote microscope control as well as simultaneous top and inverted image registration (**Fig. 3B**). The top camera allows for whole chip bright field imaging, while the bottom camera allows for fluorescence imaging with 10× magnification. This combination allows for high-throughput droplet imaging. A robotic stage is used to scan the array of multiple droplets, with an XY microscope scanning range of 45 mm×45 mm and a resolution of 5 μm/step (10 mm/sec). Z-axis focus capabilities include 15 mm travel with 1 μm/step, at 2 mm/sec. The ScanDrop optical system is controlled by Python scripts which can be automatically generated by PR-PR.

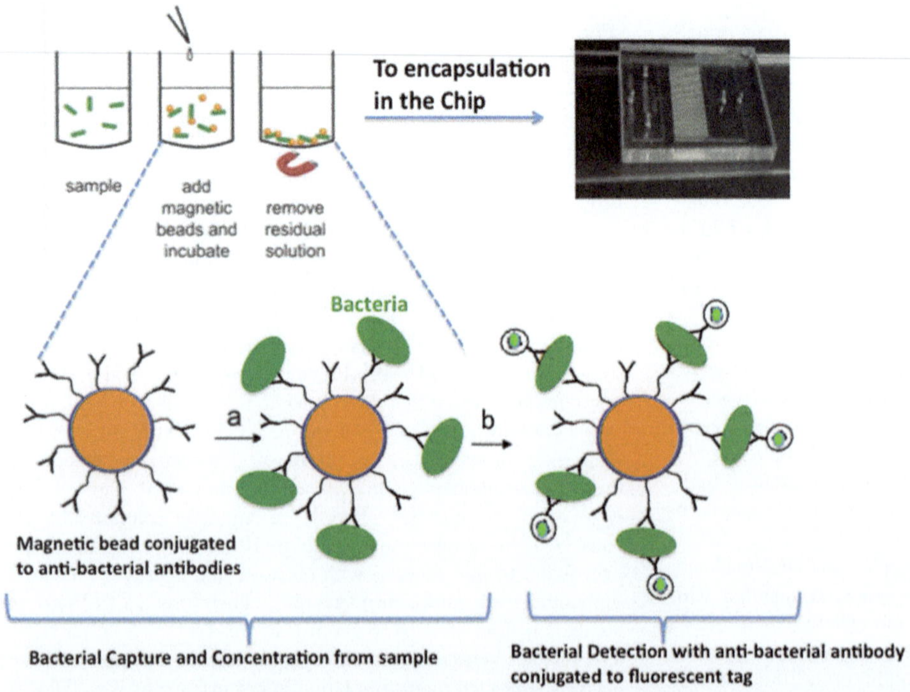

Figure 1. Bacteria capturing and detection assay. Magnetic bead capture of *E. coli* from enriched water samples, and downstream chip encapsulation for fluorescent labeling and detection. 1L of water is passed through a 0.22 μm filter, which is then incubated for 6 hr in LB media. Dynabeads® MAX anti-*E. coli* O157 are added to the resulting cell culture ("sample"), incubated for 20 min, and concentrated via magnet. The beads (potentially conjugated with bacteria) are then co-encapsulated with secondary fluorescently labeled anti-*E. coli* antibodies in the chip and incubated up to 1 hour before imaging.

PR-PR cloud-based Laboratory Automation System, and Data Management and Sharing

In this work, we have further developed PR-PR, a biology-friendly high-level language for laboratory automation [44], to control ScanDrop's automated microscopy system and enable ScanDrop to be promptly and easily adjusted to changes in experimental protocol. In PR-PR, transfer of a material (*e.g.*, a liquid) or system component (*e.g.*, a robotic arm) is described by a **S**ource, **D**estination, **Q**uantity, and **M**ethod. For ScanDrop, the **S**ource is the initial coordinates of the microscope stage (XY) and lens (Z), the **D**estination is the final target coordinates of interest, the **Q**uantity is the number of pictures that should be taken, and the **M**ethod specifies imaging parameters: light, filters, and delay between image capture. PR-PR inputs a script for ScanDrop automated microscope control (such as that presented in **Fig. S2**) and outputs a Python script that can directly operate the

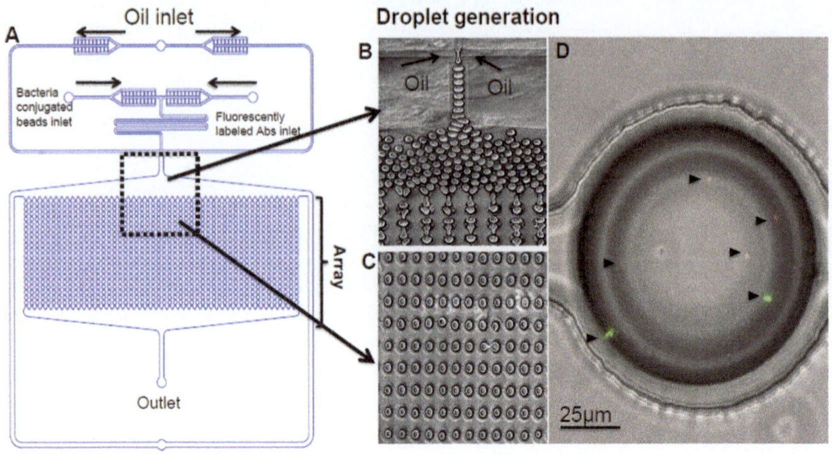

Figure 2. Droplet microfluidic device for bacteria monitoring: A) Schematic representation of ScanDrop droplet microfluidic chip and fluid control system. **B**) Droplet generator. **C**) Droplet incubation array (up to 10^3 droplets can be incubated simultaneously). **D**) Co-encapsulation of GFP- and RFP-expressing *E. coli* inside a single droplet (20× magnification, Zeiss microscopic imaging). Arrows indicate single bacteria cells.

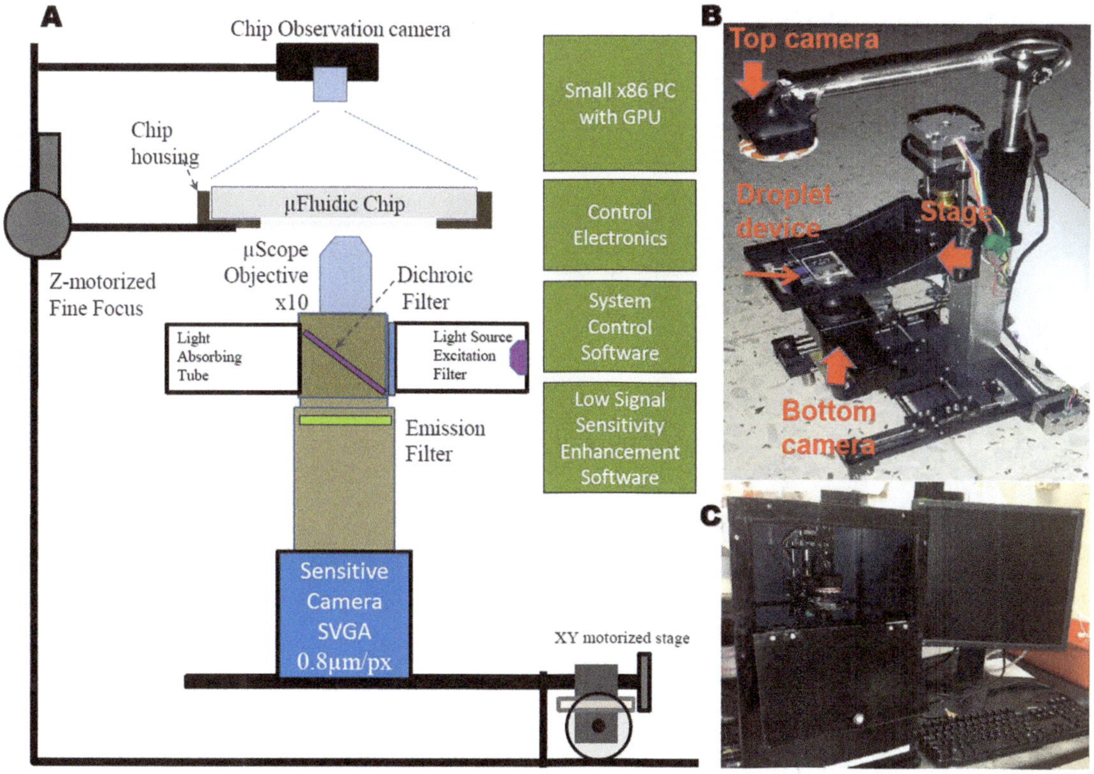

Figure 3. Portable fluorescence microscope system: A) Microscope design scheme. **B**) Robotic stage and two cameras (arrows) for chip observation and data acquisition. **C**) ScanDrop system set up, including 2 pressure pumps to create droplets, a microscope imaging system, internet access, and a monitor for local data viewing.

ScanDrop automated microscope system. The PR-PR script protocol and the resulting data for each experiment are stored in a local folder within the ScanDrop sensor and can shared between users via Dropbox.

Detection of RFP-expressing *E. coli* in drinking water

As a positive control, we tested ScanDrop for the detection of 150 CFU/mL of RFP-expressing *E. coli* in drinking water (**Fig. 4**). The overall assay for *E. coli* detection is divided into three steps: enrichment, capture, and detection. For enrichment, 1L of contaminated water sample was filtered and the filter with captured bacteria was incubated for 6 hours in the LB medium. For capturing, the enriched solution was mixed with Dynabeads® MAX anti-*E. coli* O157 for 10 min and separated by magnet.

Fig. 4B shows the droplet-based microfluidic chip used to perform the immunoassay described in **Fig. 1**. For the detection step, beads conjugated to bacteria captured from contaminated water sample were co-encapsulated with secondary FITC fluorescently labeled anti-*E. coli* antibodies in the droplet array and incubated up to 1 hr in the chip array at room temperature. **Fig. 4C** shows a representative droplet, and **Fig. 4D** shows the green florescence signal detected in a single droplet containing *E. coli* capture on the bead and tagged by secondary FITC labeled antibodies. The presence of RFP expressing *E. coli* in water was confirmed by PCR (**Fig. 5 C, D**).

Detection of fecal *E. coli* in drinking water

Next, we tested the ScanDrop system for the detection of fecal *E. coli* in drinking water. We contaminated the water with rat feces and applied the droplet detection assay described above. **Fig. 5A** presents the procedure flow for the detection of fecal *E. coli* in water. **Fig. 5B** shows a representative resulting image, with fluorescence signal indicating *E. coli* contamination. We confirmed the ScanDrop detection with PCR (**Fig. 5 C,D**), which clearly showed that the water samples were contaminated with *E. coli*. The results from the ScanDrop tests were uploaded to Dropbox cloud data storage.

The work presented here demonstrates the potential of automated microscale systems for water quality analysis. To detect *E. coli* in water samples, we developed and demonstrated a bead-based immuno-assay performed with a droplet microfluidic device to reduce reagent volume and enhance reaction rates. We integrated the microfluidic assay with a portable imaging system and remote control automation software. We demonstrated ScanDrop system capabilities through the detection of model coliform bacteria, *E. coli*, in feces-contaminated drinking water. Our successful multiplex detection assay results suggest that simultaneous multiple bacteria detection, using several types of beads conjugated with different antibodies that bind different target bacteria, will be possible with further development. The ScanDrop platform decreased reagent volumes, (the full chip uses 520 nL of reagents, while conventional assay require at least 10 μL of reagents) and allows for results within 8 hours from the time of water sampling. Our results demonstrate that a combination of droplet microfluidics with low cost optics and cloud network can provide a flexible and efficient alternative for pathogen detection in drinking water. The ScanDrop platform has the potential to significantly improve water diagnostics, particularly in low income countries where the infrastructure does not yet exist [51,52].

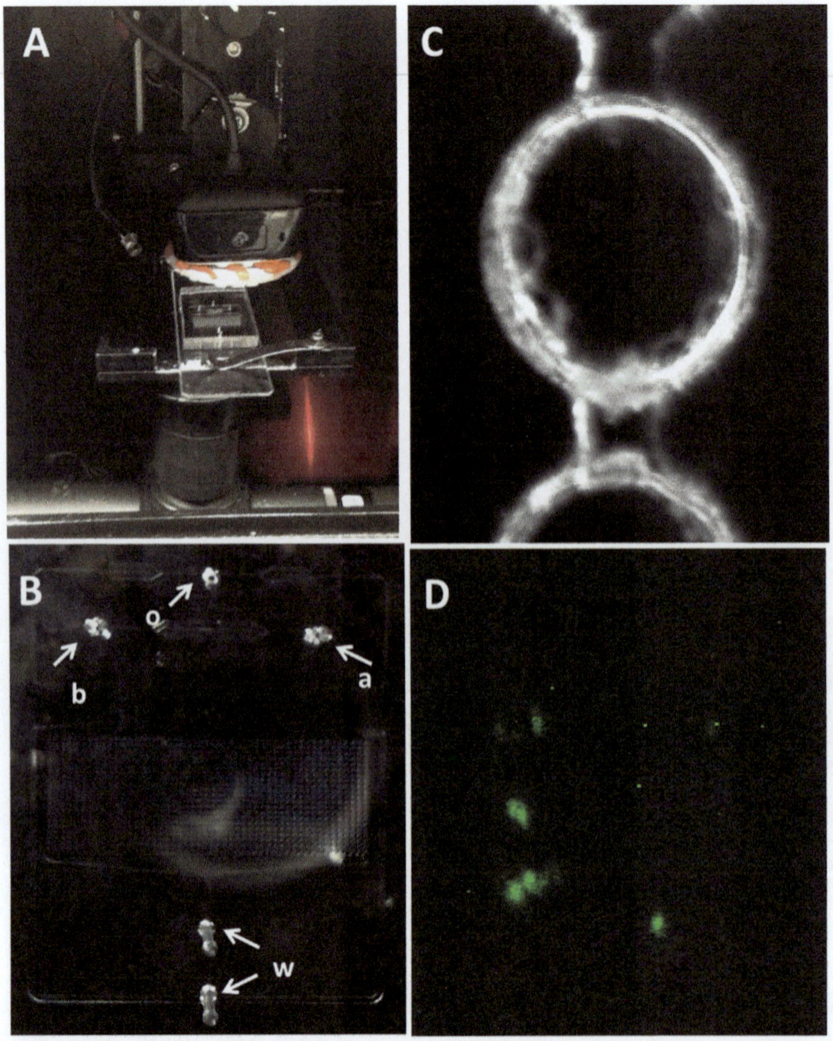

Figure 4. ScanDrop: *E. coli* detection: A) Microfluidic chip operating inside the imaging system. **B**) Droplet array as viewed from the top camera. Arrows indicate tubing/chip connection locations as follows: o- oil (inlet); b- beads conjugated with *E. coli* (inlet); a- fluorescently labeled antibody (inlet); w- waste (outlets). **C**) Droplet image as seen from the top camera with white LED illumination. **D**) Antibody green fluorescence indicates the presence of RFP-expressing *E. coli* (positive control).

Conclusions

We developed the ScanDrop platform for *E. coli* detection in water. The platform uses magnetic beads to capture bacteria, droplet microfluidics to encapsulate the captured bacteria with fluorescent antibodies, low cost portable optics for signal detection, PR-PR to facilitate microscopy control and data acquisition, and cloud-based storage for results sharing. The use of droplet microfluidics increases reaction kinetics and reduces reagent volumes (lowering the cost per test), the developed florescence microscopy system allows for data generation in multiple locations, and PR-PR facilitates ScanDrop control. A schematic illustration of an envisioned ScanDrop network for water quality analysis is shown in **Fig. 6**. The ScanDrop network would consist of 1) the PR-PR laboratory automation system and cloud-based data storage for remote control, image capturing, and result sharing; and 2) ScanDrop sensor stations deployed at multiple water distribution locations. The control station would perform image analysis for multiple sensors and shares the test results in the real time with multiple end users. This ScanDrop network could contribute to more rapid, cost-effective, and continuous water quality monitoring systems, with centralized facilities simultaneously monitoring multiple water sampling sites without complex imaging or data processing infrastructure.

Materials and Methods

ScanDrop optics system

A custom made, motorized, dual view, computerized portable microscopy system was designed for droplet microfluidic imaging (R&D Engineering Solutions, Netania, Israel). The dual view system was used for the simultaneous imaging of the whole chip (top view camera) and specific droplets (bottom view camera). The top view camera includes: 1280×768 resolution, color sensor, auto/computer-controlled focus, manually configurable [83×50 mm - 30×18 mm] field of view, 640×480 region of interest (ROI), and zoom functionality. The bottom view (microscope) camera includes: 752×582 resolution, monochrome 8.6 µm×8.3 µm pixels sensor, and a 10× objective. A single 3W 468 nm light emitting diode (LED) was used for florescence

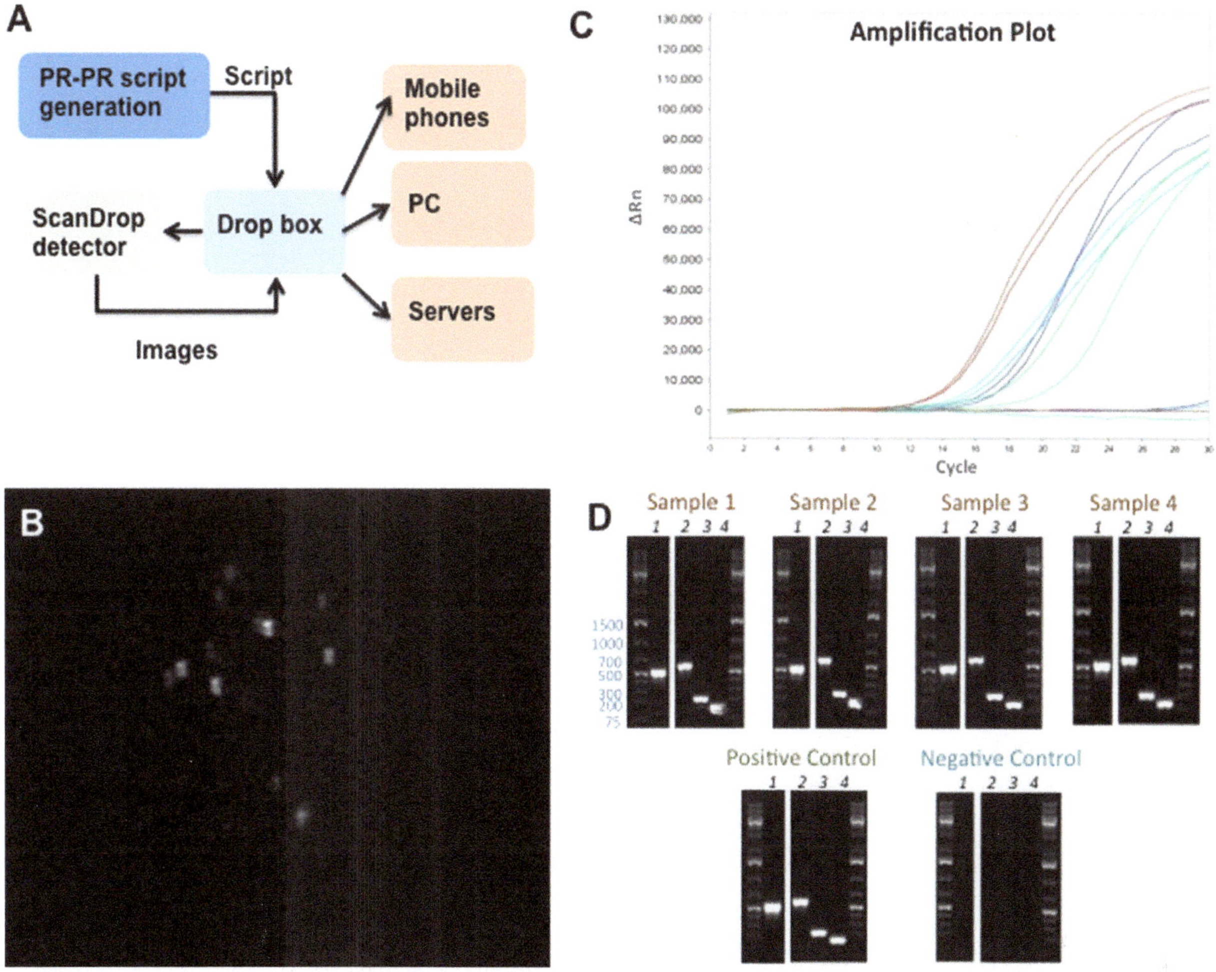

Figure 5. Detection of fecal *E. coli* in drinking water: A) ScanDrop detection network. PR-PR generates a Python script to control the ScanDrop detector. The Python script is uploaded to Dropbox, and then run on the ScanDrop detector. The captured images are uploaded to Dropbox, and then distributed to various devices. **B**) Representative ScanDrop image demonstrating fecal *E. coli* detection in drinking water. **C**) Real-time PCR amplification plot for contaminated sample 3, and positive and negative controls. Red curves indicate amplification of primary *16S rRNA* locus, cyan amplification of secondary *16S rRNA* locus, green *tuf* locus, and blue *uidA* locus. Sample 3 and the positive control amplified similarly. The negative control did not amplify. **D**) Gel electrophoresis analysis of PCR reaction products for the four contaminated samples, along with positive and negative controls. For all samples, gel lanes correspond to the amplification of loci as follows: lane 1 - *16S rRNA* primary locus, lane 2 - *16S rRNA* secondary locus, lane 3 - *tuf*, lane 4 – *uidA*.

excitation. A 41017 - Endow GFP/EGFP bandpass fluorescence filter set (Chroma Inc., VT) was used for florescence detection. Top illumination was made by a single 30 mW white LED for chip observation and microscope camera positioning. An embedded ×86 dual core computer with HDMI display port outputs (CompuLab, Israel) was used for the local control of the system. An embedded computer runs custom software, which allows full control of the microscope, including XY position, focus, illumination, image acquisition and enhancement. A system can be controlled manually by the human operator via a standard PC console (keyboard, mouse and monitor). Alternatively, a system can be controlled programmatically via a program written in Python. We further improved the programmatic control aspect of our system by leveraging PR-PR [44], whereby a PR-PR microscope control script (**Fig. S2**) is translated into a Python script that can control the microscope system, as described above. Python script deployment and image retrieval across distributed microscope systems was performed with the Dropbox cloud-based storage service.

Bacterial strains and plasmids

Plasmids pFAB_SchPMK36GFP and pFAB_SchPMK36RFP (unpublished results, Vivek Mutalik, Drew Endy, and Adam Arkin; see **Fig. S1**), both carrying a kanamycin resistance marker, were transformed into *E. coli* BW25113. These bacterial strains and plasmids, along with their associated information (*e.g.*, annotated Genbank-format DNA sequence files), have been deposited in the public instance of the JBEI Registry [53] (https://public-registry.jbei.org; corresponding Part IDs JPUB_001327-001329). For transformation, 1 μL pFAB_SchPMK36GFP or pFAB_SchPMK36RFP was mixed on ice with 40 μL chemically competent *E. coli* BW25113. The mixture was incubated on ice for 20 min, then placed at 42°C for 45 s (heat shock), and then returned to ice. 200 μL SOC media was then added to each tube of transformed cells and incubated with agitation at 37°C for 30 min. 100 μL of each transformation mixture was plated on solid LB media (Sigma-Aldrich, MO) supplemented with 30 μg/mL kanamycin (Sigma-Aldrich, MO) and then cultured at 37°C.

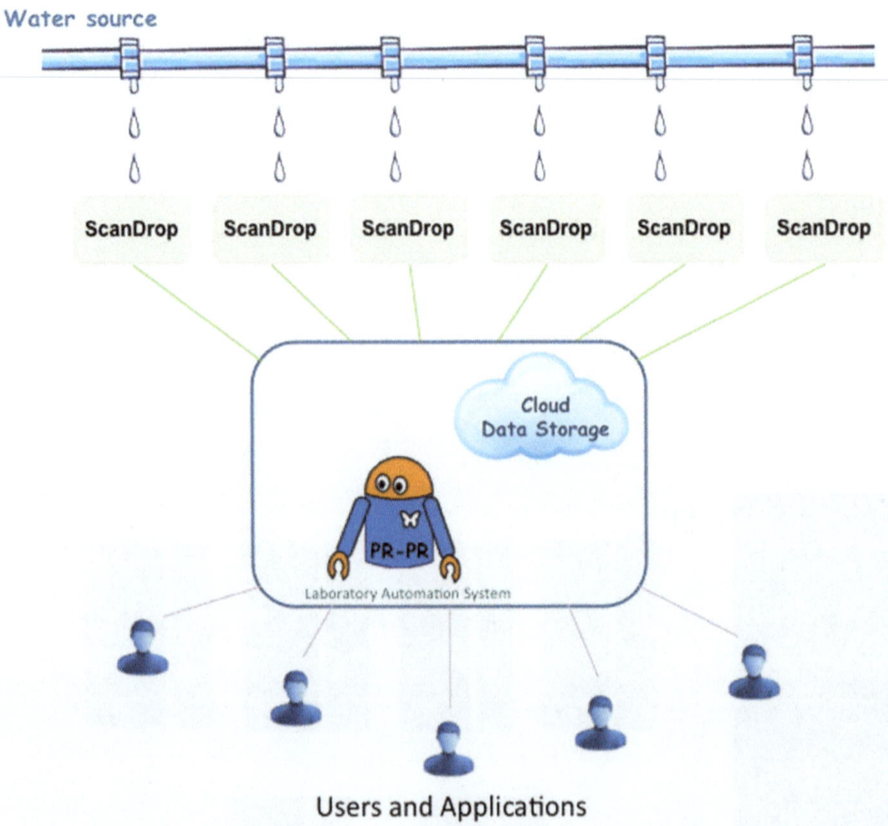

Figure 6. Schematic overview of a ScanDrop cloud-based water quality assessment system. The ScanDrop detector network is enabled by PR-PR and cloud-based data storage. Users send requests for water quality assessment at different locations in the distribution system. ScanDrop detectors perform the tests, the results are stored in the cloud, and the collected data is shared between users and applications.

Microfluidic device for droplet generation

The droplet microfluidic flow focusing device mask was fabricated by soft lithography. Negative photo resist SU-8 2100 (MicroChem, Newton, MA) was deposited onto clean silicon wafers to a thickness of 150 μm, and patterned by exposure to UV light through a transparency photomask (CAD/Art Services, Bandon, OR). To manufacture consumable devices, Sylgard 184 poly(dimethylsiloxane) (PDMS) (Dow Corning, Midland, MI) was mixed with cross-linker (ratio 10:1) and poured onto the photoresist pattern, degassed thoroughly and cured for 12 hours at 75°C. After curing, the PDMS devices were peeled off the wafer and bonded to glass slides after oxygen-plasma activation of both surfaces. The microfluidic device was composed of two parts: 1) a droplet forming nozzle (channel cross section $6.25 \cdot 10^{-8}$ m^2) and 2) a 10^3 droplets storage array (channel cross section $3.13 \cdot 10^{-7}$ m^2). The bonded microfluidic channels were treated with Pico-Sur^{TM} 2 (Dolomite Microfluidics, UK) by filling the channels with 10 μL of the solution as received and then flushing with air. This treatment was done to improve the wetting of the channels with mineral oil in the presence (1% w/w) of the surfactant (span80). 1 mL syringes were used to load the fluids into the devices through Tygon Micro Bore PVC Tubing 100f, 0.010″ ID, 0.030″ OD, 0.010″ Wall (Small Parts Inc, FL). Individual syringe pumps (Harvard Apparatus, USA) were used to control the flow rates of oil and other reagents. To form droplets, the flow-rate-ratio of water-to-oil was adjusted to $Q_w/Q_o = 1$.

Droplet microfluidics multiplex detection assay

E. coli expressing GFP or RFP were incubated for 12 hours at 37°C in LB media (Sigma-Aldrich, MO) to 10^6 CFU/mL and encapsulated into droplets. Fluorescence images were captured on a Zeiss 200 Axiovert microscope using an AxioCAM MRm digital camera and AxioVision 4.8 software at 20× magnification. Each experiment consisted of 4 repeats.

ScanDrop detection of *E. coli* in water

1 L drinking water was spiked with RFP-expressing *E. coli* to 150 CFU/mL. The spiked water was filtered through a 0.22 μm filter (Corning Inc., NY), and the filter was then inoculated in 10 mL LB media (Sigma-Aldrich, MO) and incubated for 6 hr at 37°C. 20 μL of Dynabeads® MAX anti-*E. coli* O157 beads (Life Technologies, CA) were added to 1.5 mL of the incubation media and further incubated for 20 min on a rotating stage at room temperature (RT). Beads with captured bacteria were separated by magnet and resuspended in 400 μL of Phosphate Buffered Saline (PBS). The resuspended solution was co-encapsulated 500:1 with green fluorescently labeled anti-*E. coli* antibody (FITC, ab30522, Abcam, MA) in droplet reactors inside the chip positioned on the ScanDrop robotic stage. After a further 1 hr of incubation at RT, images were taken from different locations on the chip. The objective position movements were controlled via PR-PR, and the generated images were automatically uploaded to Dropbox. Each experiment consisted of 4 repeats.

Table 1. Primers and loci used for PCR detection of *E. coli.*

Locus	Primers	T_m	Amplicon Length
16S rRNA	ECA75F - GGAAGAAGCTTGCTTCTTTGCTGAC	60°C	544 bp
(primary)	ECR619R- AGCCCGGGGATTTCACATCTGACTTA		
16S rRNA	16E1 - GGGAGTAAAGTTAATACCTTTGCTC	60°C	583 bp
(secondary)	16E2 - TTCCCGAAGGCACATTCT		
tuf gene	TEcol553 - TGGGAAGCGAAAATCCTG	58°C	258 bp
	TEcol754 - CAGTACAGGTAGACTTCTG		
UidA gene	UAL - TGGTAATTACCGACGAAAACGGC	62°C	147 bp
	UAR - ACGCGTGGTTACAGTCTTGCG		

Drinking water contaminated with rat feces

Fresh feces were collected from rat cages in the animal facility of Massachusetts General Hospital. 1.5 g feces was mechanicaly homogenized in 1 L of drinking water. The contaminated water was filtered twice through a 40 μm filter (BD Falcon™, BD Biosciences, CA). The detection of E.coli in permeate was done by ScanDrop assay (as described in the previous section) and by Real-Time PCR (as described in the following section). Each experiment consisted of 4 repeats.Detection of RFP expressing and fecal origin E. coli by PCR. We chose four primer sets (**Table 1**) to detect E. coli in the prepared drinking water contaminated with rat feces. The primers targeted specific sequences from different loci in the E. coli genome: two primer sets for 16S rRNA, one for *tuf*, and one for *uidA* [54–57]. For each primer set, we tested four contaminated samples and a positive and a negative control. Negative controls contained water only, and positive controls contained water supplemented with *E. coli* BW25113. After enrichment of the microbial population (described above), 5 μL of enriched culture was added to 45 μL H_2O. All samples were incubated 15 min at 98°C and then diluted in additional 100 μL H_2O. Each 30 μL PCR reaction contained 10 μL of the diluted cell lysate (as template), 10 μL of 3× qPCR master mix (H_2O 3.3 μL, 5× Phusion HF 6 μL, dNTP 100 mM 0.25 μL, Phusion DNA Polymerase (NEB) 0.3 μL, SYBR® Green II 200× (Molecular Probes) 0.15 μL), and a pair of primers at 5 pmol each. PCR reactions were subjected to thermal cycling (3 min at 95°C, and then 30 cycles of 30 s at 95°C, 30 s at 58°C, and 30 s at 72°C, with a final hold step at 10°C) in a StepOnePlus™ Real-Time PCR System (Life Technologies, CA). We tracked the amplification curves and stopped the PCR amplifications after most reactions plateaued (**Fig. 5c**). We analyzed PCR fragments using electrophoresis by running the PCR products in 1% agarose gels (**Fig. 5d**). Each experiment consisted of 4 repeats.

PR-PR software availability

PR-PR is open-source software under the BSD license and is freely available from GitHub (https://github.com/jbei/prpr), and is also available through its web interface on the public PR-PR webserver (http://prpr.jbei.org).

Supporting Information

Figure S1 Plasmid maps for pFAB_SchPMK36GFP and pFAB_SchPMK36RFP.

Figure S2 Representative PR-PR script for ScanDrop. LOCATION declarations define microscope stage (XY) and lens (Z) locations. TRANSFER commands specify the starting and destination locations, the number of pictures to capture, and the capture parameters. In a single TRANSFER statement, multiple sequential destinations can be defined by location offset and number of repetitions.

Acknowledgments

The authors thank Vivek Mutalik for the gift of plasmids pFAB_SchPMK36GFP and pFAB_SchPMK36RFP.

Author Contributions

Conceived and designed the experiments: AG GL TK. Performed the experiments: AG GL IK NS. Analyzed the data: AG GL TK. Contributed reagents/materials/analysis tools: AG GL IK NS NH TK MY. Wrote the paper: AG GL NH TK. Read the manuscript: MY RM Supervised the work: MY RM.

References

1. Yang K (2012) Global distribution of outbreaks of water-associated infectious diseases. PLoS Negl Trop Dis, 2012. 6(2): p. e1483.
2. Lewin S (2007), Estimating the burden of disease attributable to unsafe water and lack of sanitation and hygiene in South Africa in 2000. S Afr Med J, 2007. 97(8 Pt 2): p. 755–62.
3. Kosek M, Bern C, Guerrant RL(2000) The global burden of diarrhoeal disease, as estimated from studies published between 1992 and 2000. Bull World Health Organ, 2003. 81(3): p. 197–204.
4. Pruss A (2002) Estimating the burden of disease from water, sanitation, and hygiene at a global level. Environ Health Perspect, 2002. 110(5): p. 537–42.
5. Cann KF (2013) Extreme water-related weather events and waterborne disease. Epidemiol Infect, 141(4): p. 671–86.
6. Teschke K(2010) Water and sewage systems, socio-demographics, and duration of residence associated with endemic intestinal infectious diseases: a cohort study. BMC Public Health, 10: p. 767.
7. Sarkar R(2013) Burden of childhood diseases and malnutrition in a semi-urban slum in southern India. BMC Public Health, 2013. 13: p. 87.
8. Jafari N (2011) Prevention of communicable diseases after disaster: A review. J Res Med Sci, 2011. 16(7): p. 956–62.
9. Subbaraman R(2013) The social ecology of water in a Mumbai slum: failures in water quality, quantity, and reliability. BMC Public Health, 2013. 13: p. 173.
10. Levy K(2012) Rethinking indicators of microbial drinking water quality for health studies in tropical developing countries: case study in northern coastal Ecuador. Am J Trop Med Hyg, 2012. 86(3): p. 499–507.

11. Stoler J (2012) When urban taps run dry: sachet water consumption and health effects in low income neighborhoods of Accra, Ghana. Health Place, 2012. 18(2): p. 250–62.
12. Gwimbi P (2011) The microbial quality of drinking water in Manonyane community: Maseru District (Lesotho). Afr Health Sci, 2011. 11(3): p. 474–80.
13. Mandeville KL(2009) Gastroenterology in developing countries: issues and advances. World J Gastroenterol, 2009. 15(23): p. 2839–54.
14. Al-Bayatti KK, Al-Arajy KH, Al-Nuaemy SH(2012), Bacteriological and physicochemical studies on Tigris River near the water purification stations within Baghdad Province. J Environ Public Health, 2012. 2012: p. 695253.
15. Ashbolt N, Grabow W, Snozzi M(2001) Indicators of Microbial Water Quality, in Water Quality: Guidelines, Standards and Health, L. Fewtrell and J. Bartman, Editors. IWA Publishing: London,UK. p. 289–316.
16. Edberg SC(2000) Escherichia coli: the best biological drinking water indicator for public health protection. Symp Ser Soc Appl Microbiol, (29): p. 106S–116S.
17. (2006) EPA, Total Coliform Rule Requirements. Available: http://water.epa.gov/lawsregs/rulesregs/sdwa/tcr/regulation.cfm. Accessed Nov 25 2013.
18. WHO (2006), Guidlines for drinking-water Quality. Third Edition.
19. Leclerc, H., Relationships between common water bacteria and pathogens in drinking-water, in Heterotrophic Plate Counts and Drinking-water Safety, J. Bartman, et al., Editors. 2003, IWA Publishing: London, UK. p. 80–118.
20. (2005) StandardMethods, For the examination of water and wastewater. 9020 Quality Assurance/Quality control. Available: http://www.standardmethods.org/store/ProductView.cfm?ProductID = 538. Accessed Nov 25 2013
21. Mairhofer J, Roppert K, Ertl P (2009) Microfluidic systems for pathogen sensing: a review. Sensors (Basel), 2009. 9(6): p. 4804–23.
22. Kao MC, Durst RA (2010) Detection of Escherichia coli Using Nucleic Acid Sequence-Based Amplification and Oligonucleotide Probes for 16S Ribosomal RNA. Analytical Letters. 43(10–11): p. 1756–1769.
23. Xiao XL, (2013) Detection of viable but nonculturable Escherichia coli O157:H7 using propidium monoazide treatments and qPCR. Can J Microbiol. 59(3): p. 157–63.
24. Langer V, Niessner R, Seidel M (2013) Stopped-flow microarray immunoassay for detection of viable E. coli by use of chemiluminescence flow-through microarrays. Anal Bioanal Chem, 2013. 399(3): p. 1041–50.
25. Rompre A (2002) Detection and enumeration of coliforms in drinking water: current methods and emerging approaches. J Microbiol Methods, 2002. 49(1): p. 31–54.
26. Liu Y (2008) Detection of viable but nonculturable Escherichia coli O157:H7 bacteria in drinking water and river water. Appl Environ Microbiol. 74(5): p. 1502–7.
27. Noble RT, Weisberg SB (2005) A review of technologies for rapid detection of bacteria in recreational waters. J Water Health. 3(4): p. 381–92.
28. Setterington EB, Alocilja EC (2011) Electrochemical Biosensor for Rapid and Sensitive Detection of Magnetically Extracted Bacterial Pathogens. Biosensors. 2(1): p. 15–31.
29. Bogomolny E, Swift S, Vanholsbeeck F, (2013) Total viable bacterial count using a real time all-fibre spectroscopic system. Analyst. 138(14): p. 4112–9.
30. Allevi RP (2013) Quantitative analysis of microbial contamination in private drinking water supply systems. J Water Health. 11(2): p. 244–55.
31. Johnson PE (2006) Fountain Flow cytometry, a new technique for the rapid detection and enumeration of microorganisms in aqueous samples. Cytometry A. 69(12): p. 1212–21.
32. Sun H(2009) Nano-silver-modified PQC/DNA biosensor for detecting E. coli in environmental water. Biosens Bioelectron. 24(5): p. 1405–10.
33. Abu-Rabeah K (2009) Highly sensitive amperometric immunosensor for the detection of Escherichia coli. Biosensors and Bioelectronics. 24: p. 3461–3466.
34. Liebes Y (2009), Immobilization strategies of Brucella particles on optical fibers for use in chemiluminescence immunosensors. Talanta. 80: p. 338–345.
35. Liebes Y, Marks R, Banai M (2009) Chemiluminescent optical fiber immunosensor detection of Brucella cells presenting smooth A antigen. Sensors and Actuators B. 140: p. 568–576.
36. Abdulhalim I (2009)Surface enhanced fluorescence from metal sculptured thin films with application to biosensing in water. App Phys Lett 94: p. 063206.
37. Abdulhalim, I., Zourob M., and Lakhtakia A., Surface plasmon resonance sensors-a mini review. J Electromagnetism 2008. 28(3): p. 213–242.
38. Karabchevsky A (2012) Microspot biosensing based on surface enhanced fluorescence from Nano-STFs. J NanoPhotonics, 2012. 6(6): p. 061508–1.
39. Shalabney A, Abdulhalim I (2011) Sensitivity enhancement methods for surface plasmon sensors. Lasers and Photonics Reviews. 5: p. 571–606.
40. Zourob M, Elwary S, Turner A (2008) Principles of Bacterial Detection - Biosensors, Recognition Receptors and Microsystems. NY: Springer.
41. Niu X, deMello AJ (2012) Building droplet-based microfluidic systems for biological analysis. Biochem Soc Trans. 40(4): p. 615–23.
42. Choi K (2012) Digital microfluidics. Annu Rev Anal Chem (Palo Alto Calif). 5: p. 413–40.
43. Guo MT (2012) Droplet microfluidics for high-throughput biological assays. Lab Chip. 12(12): p. 2146–55.
44. Linshiz G (2012) PaR-PaR Laboratory Automation platform. ACS Synthetic Biology,. 2(5): p. 216–222.
45. Baroud CN, Gallaire F, Dangla R (2010) Dynamics of microfluidic droplets. Lab on Chip,. 10: p. 2032–2045.
46. Bringer MR (2004) *Microfluidic systems for chemical kinetics that rely on chaotic mixing in droplets.* Philosophical Transactions of the Royal Society of London. Series A: Mathematical, Physical and Engineering Sciences. 362(1818): p. 1087–1104.
47. Handique K, Burns MA (2001) Mathematical modeling of drop mixing in a slit-type microchannel. Journal of Micromechanics and Microengineering. 11: p. 548–554.
48. Sarrazin F(2007) Mixing characterization inside microdroplets engineered on a microcoalescer. Chemical Engineering Science. 62(4): p. 1042–1048.
49. Mazutis L(2013) Single-cell analysis and sorting using droplet-based microfluidics. Nature Protocols. 8: p. 870–891.
50. Clausell-Tormos J(2008) Droplet-based microfluidic platforms for the encapsulation and screening of Mammalian cells and multicellular organisms. Chem Biol, 2008. 15(5): p. 427–37.
51. Reyes DR (2002) Micro total analysis systems. 1. Introduction, theory, and technology. Anal Chem. 74(12): p. 2623–36.
52. Buffi N(2011) Development of a microfluidics biosensor for agarose-bead immobilized Escherichia coli bioreporter cells for arsenite detection in aqueous samples. Lab Chip, 2011. 11(14): p. 2369–77.
53. Ham TS(2012) Design, implementation and practice of JBEI-ICE: an open source biological part registry platform and tools. Nucleic Acids Res, 2012. 40(18): p. e141.
54. Maheux AF(2009) Analytical comparison of nine PCR primer sets designed to detect the presence of Escherichia coli/Shigella in water samples. Water Res, 2009. 43(12): p. 3019–28.
55. Tsen HY, Lin CK, Chi WR (1998) Development and use of 16S rRNA gene targeted PCR primers for the identification of Escherichia coli cells in water. J Appl Microbiol. 85(3): p. 554–60.
56. Tantawiwat S (2005) Development of multiplex PCR for the detection of total coliform bacteria for Escherichia coli and Clostridium perfringens in drinking water. Southeast Asian J Trop Med Public Health. 36(1): p. 162–9.
57. Sabat G (2000) Selective and sensitive method for PCR amplification of Escherichia coli 16S rRNA genes in soil. Appl Environ Microbiol, 2000. 66(2): p. 844–9.

9

Precipitation Effects on Microbial Pollution in a River: Lag Structures and Seasonal Effect Modification

Andreas Tornevi[1]*, Olof Bergstedt[2], Bertil Forsberg[1]

1 Occupational and Environmental Medicine, Department of Public Health and Clinical Medicine, Umeå University, Sweden, **2** Department of Civil and Environmental Engineering, Water Environment Technology, Chalmers University of Technology, Gothenburg, Sweden

Abstract

Background: The river Göta Älv is a source of freshwater for 0.7 million swedes. The river is subject to contamination from sewer systems discharge and runoff from agricultural lands. Climate models projects an increase in precipitation and heavy rainfall in this region. This study aimed to determine how daily rainfall causes variation in indicators of pathogen loads, to increase knowledge of variations in river water quality and discuss implications for risk management.

Methods: Data covering 7 years of daily monitoring of river water turbidity and concentrations of E. coli, Clostridium and coliforms were obtained, and their short-term variations in relation with precipitation were analyzed with time series regression and non-linear distributed lag models. We studied how precipitation effects varied with season and compared different weather stations for predictive ability.

Results: Generally, the lowest raw water quality occurs 2 days after rainfall, with poor raw water quality continuing for several more days. A rainfall event of >15 mm/24-h (local 95 percentile) was associated with a three-fold higher concentration of E. coli and 30% higher turbidity levels (lag 2). Rainfall was associated with exponential increases in concentrations of indicator bacteria while the effect on turbidity attenuated with very heavy rainfall. Clear associations were also observed between consecutive days of wet weather and decreased water quality. The precipitation effect on increased levels of indicator bacteria was significant in all seasons.

Conclusions: Rainfall elevates microbial risks year-round in this river and freshwater source and acts as the main driver of varying water quality. Heavy rainfall appears to be a better predictor of fecal pollution than water turbidity. An increase of wet weather and extreme events with climate change will lower river water quality even more, indicating greater challenges for drinking water producers, and suggesting better control of sources of pollution.

Editor: Jose Luis Balcazar, Catalan Institute for Water Research (ICRA), Spain

Funding: This research was founded by The Swedish Water & Wastewater Association and The Swedish Research Council for Environment, Agricultural Sciences and Spatial Planning (FORMAS), grant number 10-104. The funding is external to our organization. The project is entitled: Climate change, freshwater quality, treatment and distribution- risk assessment by health studies. (URL in Swedish: http://www.svensktvatten.se/FoU/SVU/Pagaende-projekt/Dricksvatten/SVU-projekt-10-104/). The funders had no role in study design, data collection and analysis, decision to publish, or preparation of the manuscript.

Competing Interests: The authors have declared that no competing interests exist.

* E-mail: andreas.tornevi@umu.se

Introduction

Drinking water producers are responsible for providing safe drinking water. One challenge faced by producers is that water supplies, especially surface water sources, may experience temporal variations in water quality. Therefore, the raw water quality needs to be repeatedly tested to validate if the current water treatment technique is sufficient for producing safe and clean drinking water. A common indicator of water quality is turbidity, a measure of water cloudiness, which is relatively easy to quantify with optical devices and is regularly used as a first indicator of levels of microbial contamination. However, turbidity reflects the load of organic and inorganic particles, so additional water samples are needed for more precise analysis of organic contaminants. Density analyses of indicator bacteria, such as coliforms or Escherichia coli, may return a better estimate of levels of human pathogens. Because precise analyses of indicator bacteria are performed in laboratories, there is a delay between sampling and results being available, which is why turbidity monitoring and indicator bacteria samples often complement each other.

Several studies have shown relationships with prior weather events, especially wet weather, and raw water quality parameters [1–3]. Heavy rainfall has also been linked to the majority of observed drinking water-related outbreaks of gastrointestinal diseases in developed nations worldwide [4–7]. However, the reported outbreaks may only represent a fraction of the total impact; the proportion of gastrointestinal infections caused by drinking water is argued to be much higher [8].

Within Sweden, the southwest is one of the regions with the highest annual precipitation, with an average of around 1 m per year, and climate change is projected to increase annual precipitation and heavy rainfall events. The river Göta Älv runs through this region, serving as a freshwater supply for 0.7 million people. The river is at risk of pathogen contamination from multiple sources, including runoff from agricultural land and point sources such as sewer system discharges [3]. More frequent events of heavy rainfall may increase the risk of contamination, increasing the challenges of providing safe water drinking water [9,10].

Aim

With the use of time series analysis we aim to describe how daily rainfall influences raw water quality parameters in this important freshwater supply in Sweden. We focus on water quality parameters commonly used as indicators of pathogen contamination and study their long term variations and determine precipitation effects on short term variations. We analyze the distribution of lagged effects of precipitation and determine the extent to which there are seasonal effect modifications. Such results will increase knowledge of fluctuations in raw water quality, and provide a causal link behind health effects and the evidence base to develop guidance for risk evaluation of drinking water production.

Materials and Methods

Study area

The river Göta Älv originates from Sweden's largest lake (Vänern) and runs 93 km to Kattegatt at the North Sea. Close to the sea, the river divides into a second branch (Nordre Älv), with Göta Älv continuing through the coastal city Gothenburg (latitude 57.708, longitude 11.975) (Figure 1). The City of Gothenburg is Sweden's second largest city with a population of about 500,000.

The main river arm has an average flow rate of 550 m^3/s and varies around 200 to 1000 m^3/s. Hydropower stations in the upper region (Vänersborg, Trollhättan and Lilla Edet) can cause rapid changes in stream flow. The flow time from Vänern to the sea varies from 1.5 to 5 days, with an average of about 3 days. The rivers total descent is about 44 meters, where the majority takes place at the hydropower stations.

Five drinking water utilities use the river as a freshwater supply, distributing water to around 700,000 people. Two of these utilities, Alelyckan and Lackarebäck, are distributing drinking water to the population in Gothenburg. Lackarebäck takes water from a lake reservoir (Delsjön) that under normal conditions is continually supplied with river water through a 9 km tunnel designed to maintain a constant water level in the lake. The river water intake at Alelyckan is located close to the utility and takes about 2 m^3/s, with about half the volume for Alelyckan and the other half for the reservoir. This river water intake is closed when information suggest the river water is inadequate for drinking water production. To maintain drinking water production at Alelyckan, the water in the tunnel is then taken back and the tunnel transports freshwater in the other direction. Closures of the river water intake are determined by analysis of indicator bacteria and turbidity levels, or high conductivity caused by inflows of seawater. Intake closures can also be based on events of heavy rainfall or information on upstream events indicating elevated contamination risks, such as releases of untreated sewage water. The river is the recipient for eight wastewater treatment plants upstream of the intake at Alelyckan.

Despite the northern latitude of the study area, the climate is comparably very mild. The latitude provides distinct winter and summer seasons and daylight spans between 7–17 h/day. On average, February is the coldest month with daily mean temperatures a few degrees Celsius below zero. The river is usually not covered with ice during winter seasons. Precipitation is fairly constant throughout the seasons although the second half of the year generally experiences more and heavier rainfall events.

Data

River water. Concentrations of the indicator bacteria Escherichia coli (E. coli), Clostridium perfringens and coliforms, together with water turbidity, are routinely monitored to indicate the quality of the river water. Seven stations along the river monitor quality parameters with varying frequencies. The municipal water department in Gothenburg, Department of Sustainable Waste and Water, provided laboratory analyses of concentrations of indicator bacteria, sampled outside the river water intake to Alelyckan during the time period 2004–2010. This location had the highest rate of sampling of indicator bacteria for laboratory analysis, three times weekly, although more frequent sampling may be performed, particularly when high concentrations are detected. Laboratory results of coliforms and E. coli were reported in units of Most Probable Number (MPN), water samples analyzed with Coliert-18/Quanti-Tray (ISO 9308-2:2012), enumerated after an incubation time of 18 hours at 35°C. Clostridium concentrations were reported in Colony Forming Unit (CFU), enumerated after incubation at 44°C for 21±3 hours (ISO/CD 6461-2 2002-12-20). These samples were not preheated with intention to kill vegetal cells. The analytic methods of enumeration of indicator bacteria have been the same during the study period. Turbidity has been continuously monitored at the river water intake and we obtained daily mean values in Formazin Nephelometric Unit (FNU) during the same period (2004–2010). Daily data on stream flow (m^3/s) measured near Lilla Edet were also provided. The time series data of water quality parameters are displayed in Figure 2.

Weather. The Swedish Metrological and Hydrological Institute (SMHI) provided daily weather data. Daily precipitation data were available for three stations along the river: Gothenburg, (south station) the village of Alvhem situated about 30 km upstream from the water intake (middle station) (Figure 2), and the City of Vänersborg situated around 90 km upstream (north station), where also data on snow depth were available. Daily mean temperature data measured in Gothenburg (Figure 2) were also obtained.

Statistical methods

The associations between daily precipitation and raw water quality were analyzed using time series regression. Raw water quality parameters were log-transformed (natural logarithm) and generalized additive regression models (GAM) were fitted [11]. We adjusted for long-term trends and seasonality patterns and analyzed how short-term effects of daily precipitation were distributed with Distributed Lag Non-linear Models (DLNM) [12], and unconstrained distributed lag models. We compared different weather stations for their predictive ability analyzed possible confounding factors such as change in upstream snow depth and temperature variations, and examined if weekday patterns were present. In addition we examined how consecutive wet or dry weather affect daily water quality. Finally, as the study area has distinct seasonality in temperature, daylight, ecology etc., we studied if short term effects of precipitations on raw water quality parameters vary with season and possible effect modifications due to river flow rate. This was done by only including specific time periods of data in regression analyses. Details on the statistical methods are described in a supplementary material (Text S1). The statistical software R (v 2.15.2) [13], together with MGCV [11] and DLNM [14] packages were used.

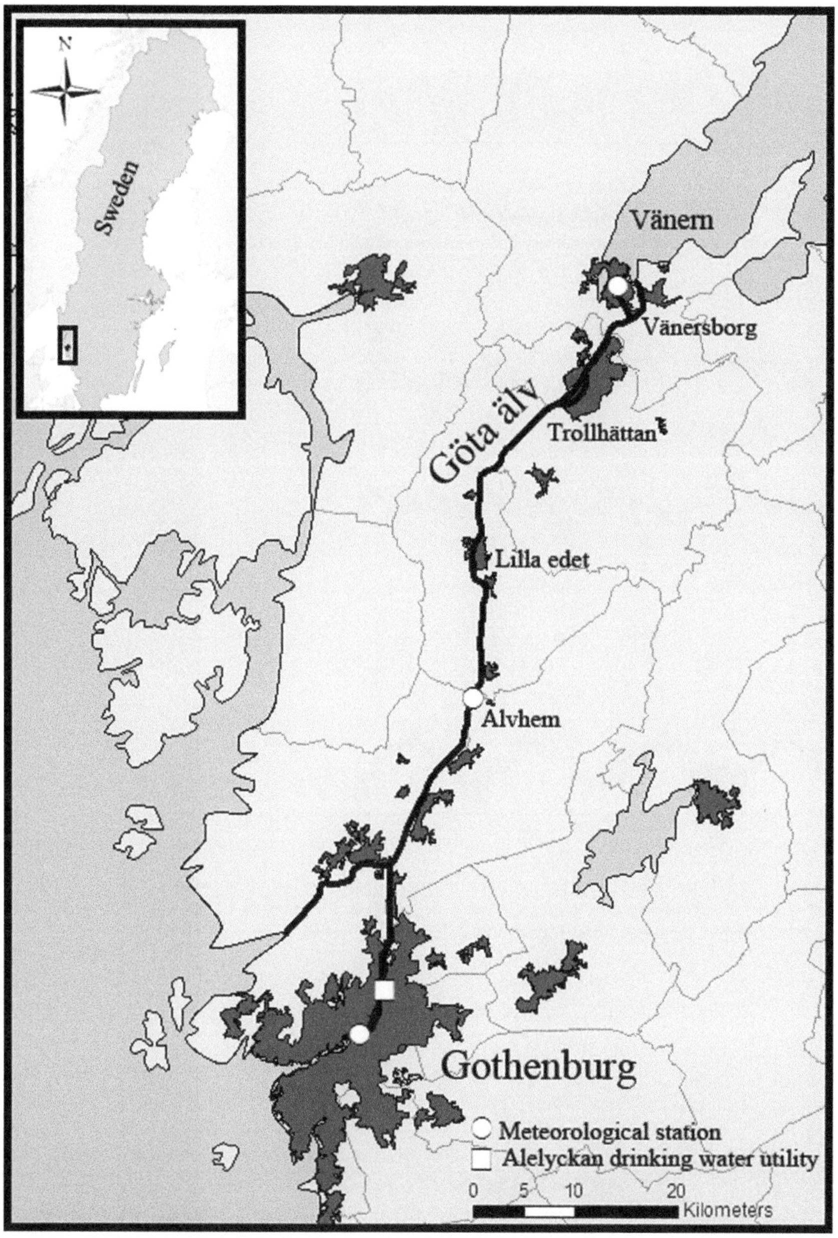

Figure 1. Map. Map of study area.

Results

Descriptive data and seasonal patterns

Daily mean values of river water turbidity ranged between 1.6 and 33.7 FNU with an overall mean of 7.0 FNU. The median value of E. coli was 96 MPN/100 mL and the highest concentration was enumerated to 2800 MPN/100 mL. Same statistics for coliforms and was 750 MPN/100 mL and 24000 MPN// 100 mL. Although these maximum concentrations were detected in June, the best average river water quality occurred in spring and summer (April–July). In general, average water quality was lowest in the darker seasons, as for example maximum observation of turbidity in May was below the average value for November. Coliforms showed less elevated concentrations in December through March relatively the other water quality parameters, and also showed a long-term declining trend throughout the study period.

Most precipitation was recorded at the middle station with an average daily precipitation of 3.1 mm, or 7.0 mm if only including days with precipitation, resulting in an average annual precipitation of 1138 mm. The average annual precipitation in the other stations was 872 mm/year (north) and 970 mm/year (south). The maximum daily precipitation was recorded at the south station (59.7 mm), which also had the highest total count of days with any observed precipitation. Figure S1 illustrates seasonal patterns of weather and water parameters, along with all data observations. Descriptive data on river water parameters and weather observations are shown in Table 1, and in Table S1 monthly statistics about water parameters is displayed.

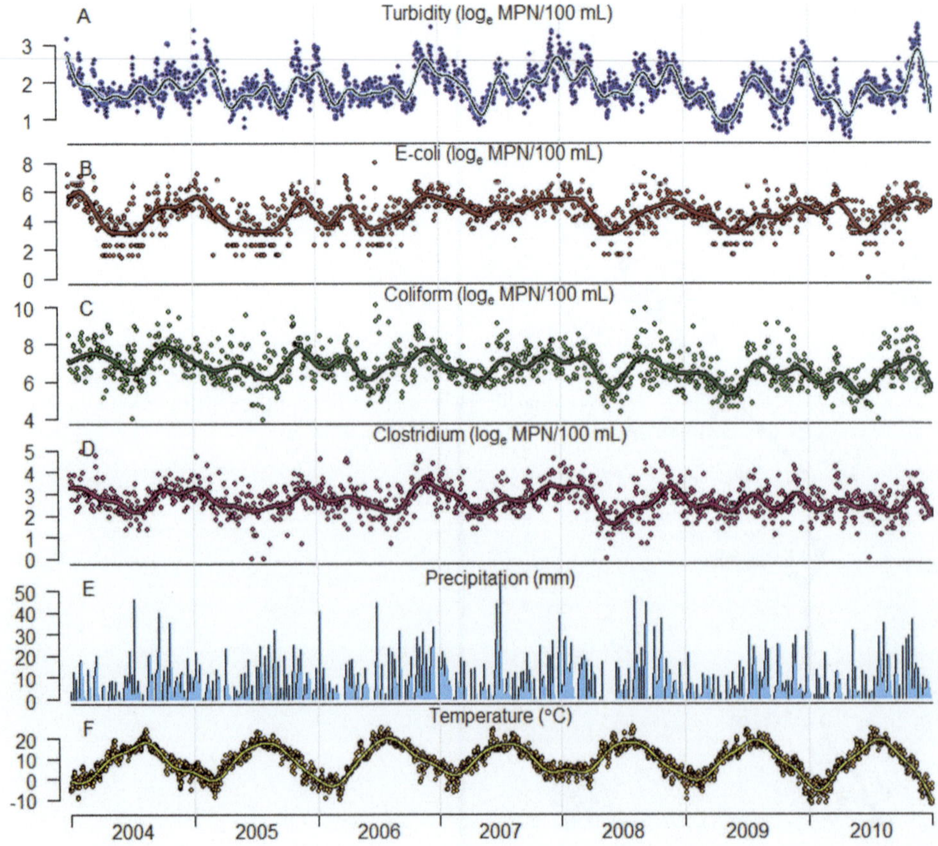

Figure 2. Data. Time series data for 2004 through 2010 for (from top) A: turbidity, B: E.coli, C: coliforms, D: Clostridium (river water intake Alelyckan), E: precipitation (Alvhem), and F: daily mean temperature (Gothenburg). A moving average is projected with a spline using 10 df/year for turbidity and 7 df/year for indicator bacteria and temperature. Data transformed by the natural logarithm (A–D).

Main effects of precipitation

Short-term variations in all raw water quality parameters were highly associated with prior precipitation regardless which weather stations provided data, although the middle station had the best predictive ability and the north station second best. An event of heavy precipitation decreased water quality over several days, with the effect peak in general two days later. Precipitation effects were significant over numerous lags, and the lag structure on turbidity appeared less peaky and to be affected a few days longer compared to indicator bacteria.

Precipitation events of 15 mm/24-h or more (n = 142) were associated with a three-fold increase in E. coli concentrations (+ 190%, 95% confidence interval (CI): 146–242%) two days later (lag 2) using a DLNM model. A similar model, with turbidity as outcome, estimated an average increase of 32% (CI: 27–37%) at lag 2. Figure 3 illustrates the relative effects on raw water quality parameters along lags 0 to 15, using both DLNMs and unconstrained distributed lag models.

A DLNM model with precipitation as continuous predictor, and allowing for non-linear associations, fitted increased concentrations with increased amount of precipitation analyzing indicator bacteria, while the effect on turbidity attenuated with extreme rainfall. An event of 40 mm/24-h was estimated to increase E. coli concentrations two days later by 580% (CI: 443–754%), and an extreme event of 50 mm/24-h was associated with increased concentrations by 11 times (+1000%, CI: 730–1360%). The precipitation effects at lag 2 are illustrated in Figure 4, and the associations along all lags (0–15) with quantity of precipitation (0–54 mm) are illustrated in Figure S2.

Analyzing the effect of consecutive wet or dry weather days and raw water quality also exposed clear relationships. On average, consecutive wet weather for more than a week increased E. coli concentrations four-fold compared to one week of dry weather (+ 299%, CI: 168–492). As estimates for singular precipitation events, this precipitation predictor also indicated a lagged effect after wet weather days. Figure 5 displays the estimated effects of consecutive dry and wet weather on raw water quality parameters using a categorical predictor together with a smooth association using a penalized spline function, which both result in similar associations.

Precipitation effect modifications

Analysis of seasonal effects indicated that precipitation influenced concentrations of indicator bacteria all year around and seasonal effect modifications were quite moderate. The effect of precipitation across the seasons on turbidity appeared to be modified more than for indicator bacteria, with the smallest relative effect during summer. Figure 6 illustrates the estimated effects two days after precipitation events (>15 mm/24-h) across seasons, and Figure S3 illustrates such events across lags 0–8 in a colored contour plot.

The extent to which short-term variations in raw water parameters were explained with prior precipitation varied over the year. The coefficient of determination (R^2) for precipitation predictors logically indicated a similar seasonal pattern as the seasonal moving average of precipitation, but the variability

Table 1. Summary statistics of observations in river water and weather during 2004–2010.

	N	Missing	Mean	St.Dev	Minimum	25%:tile	Median	75%:tile	Maximum
Temperature °C (south)	2556	1	8,8	7,48	−12,7	3,4	9,1	15.0	25,4
Precipitation mm (middle)	2557	0	3,1	6,1	0	0	0	3,8	54,3
Precipitation mm (south)	2557	0	2,7	5,5	0	0	0,1	3	59,7
Precipitation mm (north)	2557	0	2,4	4,9	0	0	0	2,5	49,5
Stream flow m^3/s (middle)	2555	2	566	216	148	416	571	730	1021
Coliforms MPN/100 mL	1156	1400	1374	1896	51	410	750	1600	24000
E. coli MPN/100 mL	1156	1400	156	200	1	41	96	190	2800
Clostridium CFU/100 mL	1054	1502	18,2	15,5	1	9	14	21	140
Turbidity FNU	2550	7	7,0	3,8	1,6	4,7	5,9	8,2	33,7

differed between water quality parameters. Precipitation explained around 50 % of the variation in coliform concentrations in river water during fall, and decreased to around 20 % in winter and spring. When analyzing turbidity in summer periods the R^2 values were low, especially when using precipitation data from the south station. R^2 values of precipitation across season, from different weather stations, are displayed in Figure S4.

Analyzing precipitation effects under different stream flows exposed effect modifications in the expected direction; precipitation during lower flow rates spreads the effect over more lags together with a decreased peak effect, while during higher rates a more concentrated lag distribution with a higher effect peak was estimated. Figure S5 illustrates the lag structures on turbidity after rainfall events (>15 mm/24-h) in episodes separated by stream flows quartiles, where effect peaks ranged between ~2 to 3.5 lag days.

Model details and covariates

Estimates of precipitation effects presented were generated with observations from the middle station (Alvhem). The DLNM models were fit with natural cubic splines in lag space and used 6 df, with knots distributed at a log scale. DLNM models with continuous precipitation predictor were fitted with a natural-cubic spline in exposure space and according to AIC scores, the log transformed turbidity data was best fit with 3 df in space of precipitation, log-Clostridium was associated with precipitation using 2 df, whereas a linear association with precipitation was sufficient when analyzing log-concentrations of E. coli and coliforms. Decreased snow cover (Vänersborg) was related to a short-term decrease in raw water quality, most clear regarding turbidity and just about significant regarding coliforms. Including such predictor in the models did not influence effects or lag structure of precipitation, and was not included in final models. Daily mean temperature showed none, or only small effects, on short-term variation and were not included in final models. Turbidity data contained a week day pattern which could be linked to scheduled maintenance of the measuring device (3 times weekly); an average increase of ~5% per day after filter cleaning was observed, and this effect was adjusted for in all turbidity models.

When comparing estimates between DLNM models and unconstrained distributed lag models (Figure 3) we concluded that the unconstrained designs produced reliable estimates (subject discussed in statistical details (Text S1)), and they were used when analyzing precipitation effect modifications.

Discussion

Findings

In this large data study clear relationships between precipitation and decreased river water quality were uncovered. Regression models showed that concentrations of indicator bacteria, but not turbidity, increase exponentially with the amount of observed precipitation (Figure 4, Figure S2), which suggests that prior precipitation in fact predicts fecal contaminations better than mean levels of turbidity at events of heavy rainfall. This conclusion also appears to be valid when creating models with turbidity as a predictor to indicator bacteria; Figure S6 illustrates the association between turbidity and indicator bacteria (lag 0), and for comparison associating indicator bacteria with precipitation two days prior.

When comparing the effect of precipitation across seasons, additional differences between turbidity and indicator bacteria were observed, with a weakened effect on turbidity during summer

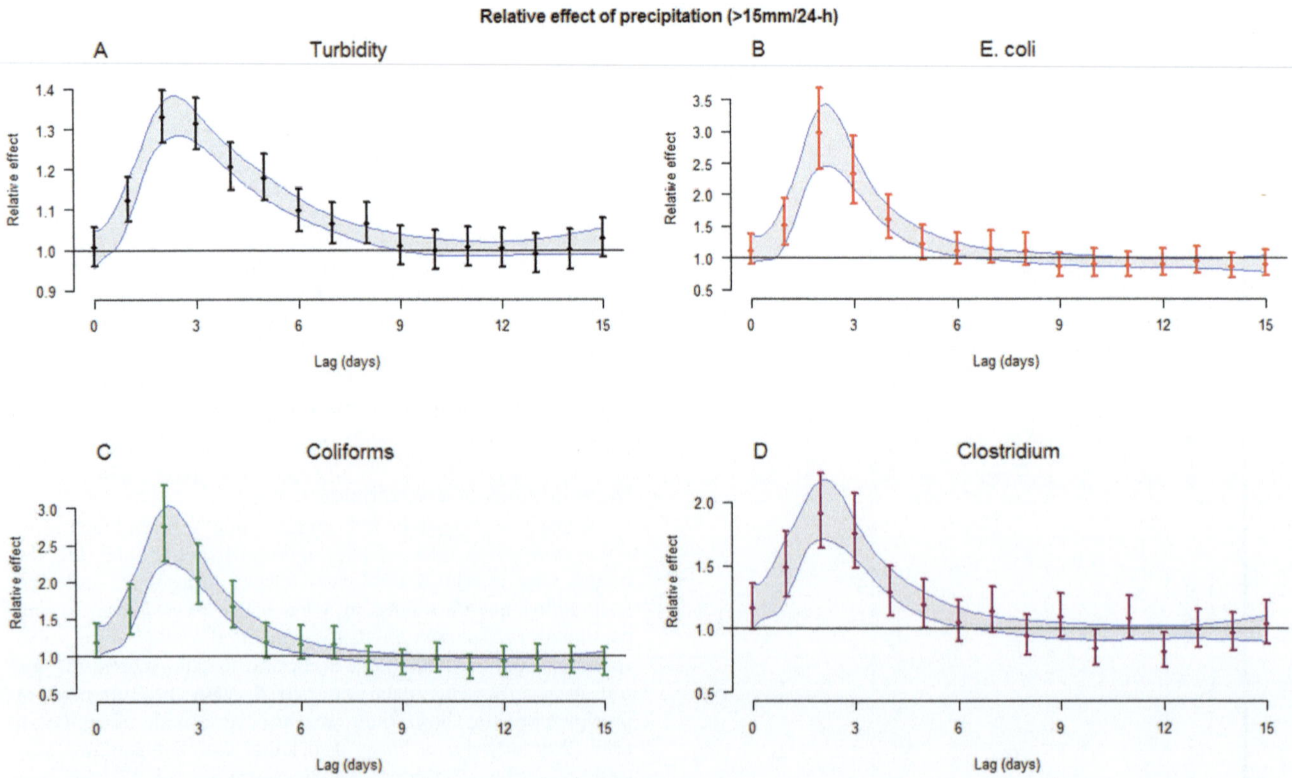

Figure 3. Effect of heavy precipitation across lags. Estimated relative effects of precipitation events (>15 mm/24-h) on river water quality parameters along 0 to 15 lag days: A: turbidity, B: E. Coli, C: coliforms, D: Clostridium perfringens. 95% confidence intervals are illustrated with bars (unconstrained distributed lag model) and with shaded area (DLNM model).

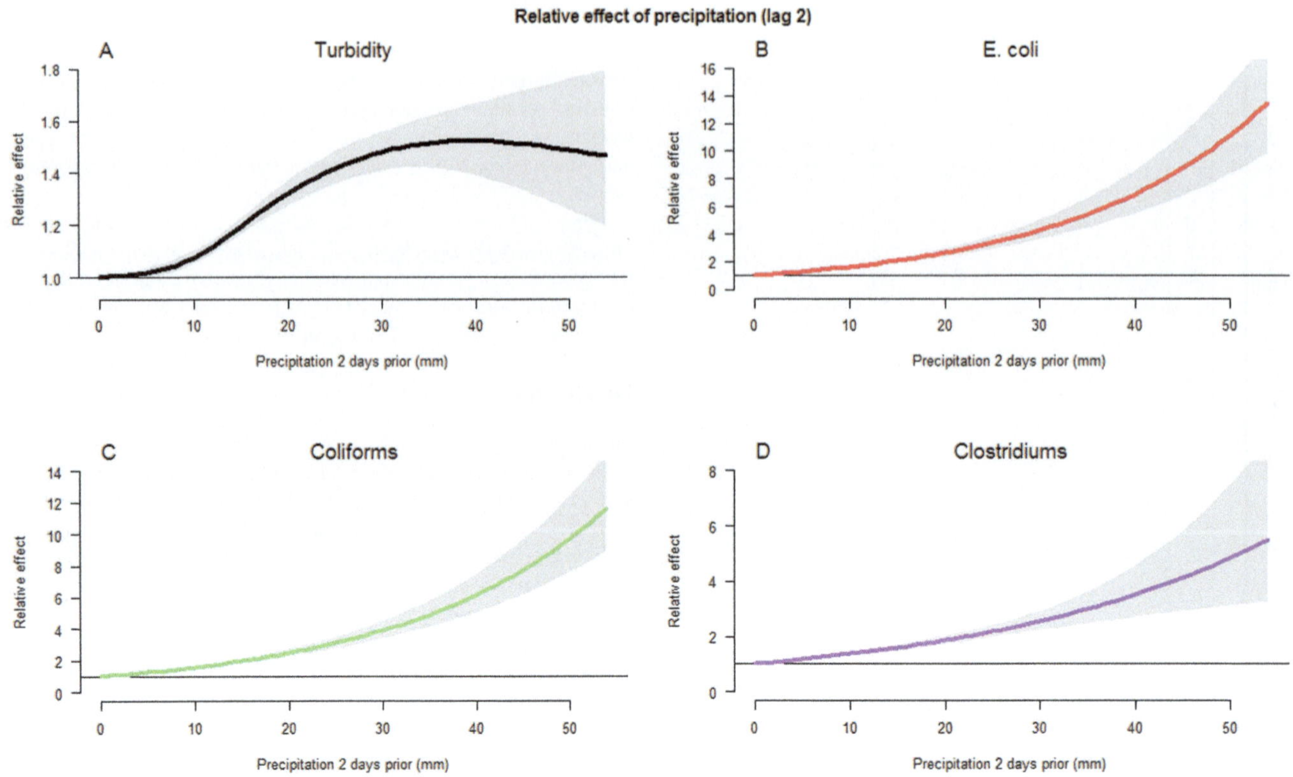

Figure 4. Precipitation effects lag 2. Relative effects of daily precipitation (0–54 mm) two days later on A: turbidity, B: E. Coli, C: coliforms, D: Clostridium perfringens. Gray areas represent 95% confidence intervals.

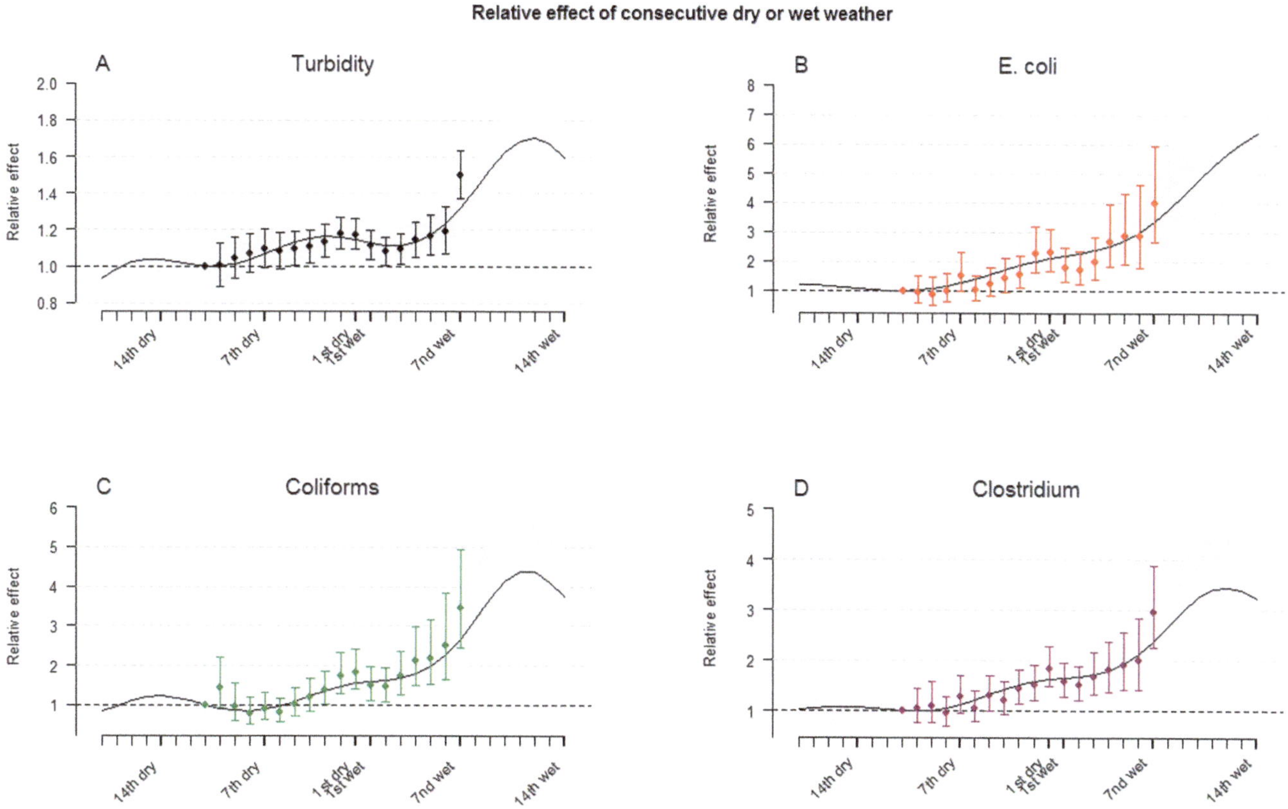

Figure 5. Effect of consecutive days with dry or wet weather. Relative effects (with 95% CIs) of consecutive wet or dry weather (middle station) on A: turbidity, B: E. Coli, C: coliforms, D: Clostridium perfringens. The reference is set to ten consecutive dry days using a continuous predictor (shaded), and at least ten dry days using a categorical predictor (bars). Seven wet days (bar) represent seven wet days or more.

(Figure 6). This effect decrease on turbidity can only partly be explained by a lower river flow rate in summer periods, which spreads the effects over more lags, and is likely to be a result of seasonal changes in the relation between runoff and rainfall as seasonal soil varies in its ability to absorb precipitation. However, significant precipitation effects on E. coli concentrations were observed across seasons, and we conclude that precipitation is associated with increased risks of introducing pathogens in to river water all year around.

The water quality parameters also showed varying seasonal averages, which to some extent could be explained by prolonged dry or wet weather, but it's likely a result of a combination of factors which might differ with water quality parameter. The inactivation rate of E. coli increases with higher temperatures, but the increased concentrations during winters could also be a consequence of less sunlight intensity [15]. Since E. coli are less long lived in warm temperatures [15], it could also be argued that if water samples were taken closer to the contamination sources a smaller seasonality pattern could have been observed. Additionally, as winter seasons usually holds higher flow rates E. coli survives over longer distances, and higher concentrations are therefore detected downstream. Coliforms showed however less elevated average concentrations during colder periods, and this could be a result of that they are capable of growing in nature and can originate from plant decay. This characteristic is also supported in our data since precipitation explained short-terms variations in coliforms with a larger seasonal variability than other indicator bacteria (Figure S4), and other indicator bacteria may better reflect the extent of fecal pollution. Coliforms were also observed with a declining long-term trend, and a speculative explanation could be that a large paper mill (Wargöns Bruk AB) situated 11 km north of Trollhättan decreased their production during the study period, to finally close in 2008. It is common that pulp and paper mills release high levels of coliforms in their waste [16].

Further model implications - potentials and limitations

As closures of the river water intake to Alelyckan are determined by concentrations of indicator bacteria (and other causes), the regression models can predict expected closures due to precipitation. Closure of intake due to E. coli counts is protocolled at 400 MPN/100 mL, which the DLNM model predicts to occur two days after precipitation events of 30 mm/24-h. The limit of accepted counts of coliforms are 7000 MPN/100 mL, which DLNM models predict to take place two days after events of 45 mm/24-h, while Clostridium counts of more than 50 CFU/100 mL are predicted to occur two days after events of 39 mm/24-h. These predictions use an average water quality as baseline (i.e. average concentrations are present before the precipitation event) and represent estimates of expected concentrations; peaks in indicator bacteria concentrations can be considerably higher and behave differently than suggested by the estimated lag structure. For example, some model outliers include the maximum observations of E. coli counts (2800 MPN/100 mL and 1400 MPN/100 mL) which were sampled the same day (lag 0) as two of the five heaviest precipitation observations (44.6 and 44.5 mm/24-h). This day had also a turbidity value below the overall mean value, and normal stream flows were registered (500–

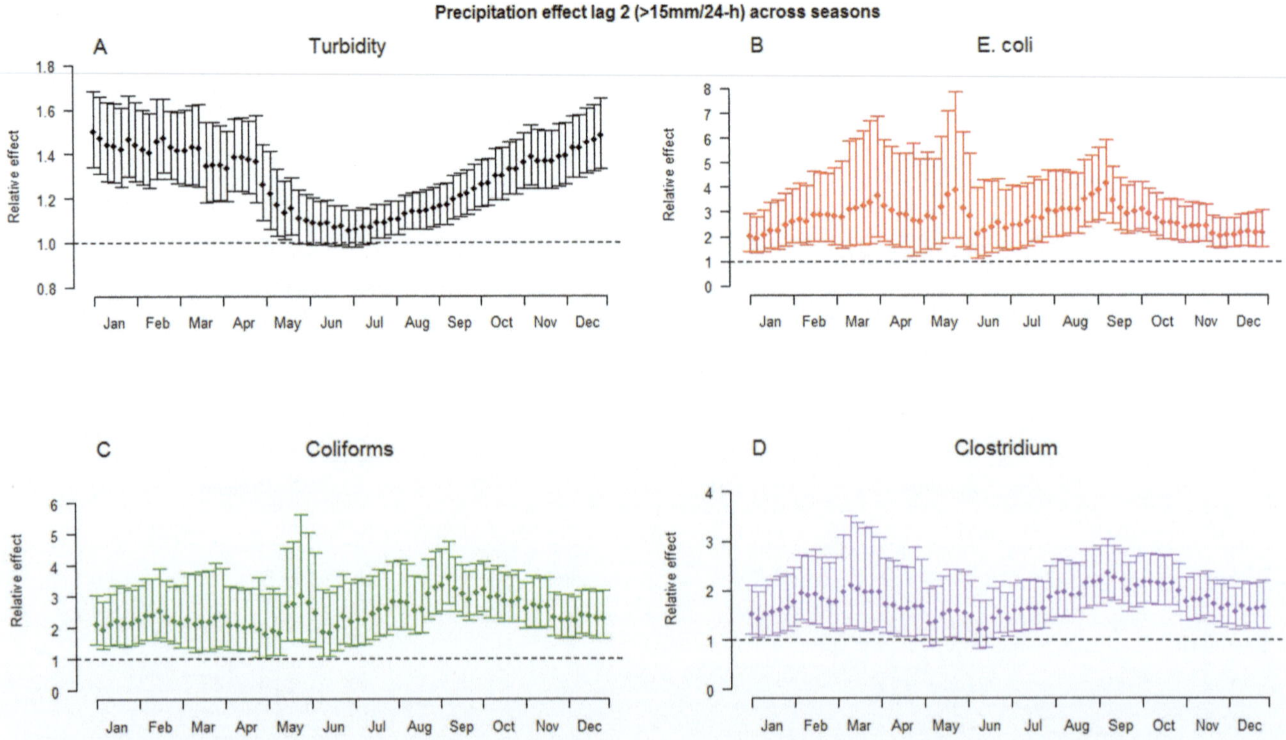

Figure 6. Effect modifications lag 2 across season. Relative effects of precipitation (>15 mm/24-h) 2 days later across seasons on A: turbidity, B: E. Coli, C: coliforms, D: Clostridium perfringens. Each estimate covers a range of 90 days: 45 days prior and 45 days following. Vertical bars represent 95% confidence intervals.

600 m^3/s) at these events. DLNM models on coliforms and Clostridium showed also higher effects after events of extreme rainfall when using a more relaxed association with precipitation (3 df), compared to the models presented and suggested by AIC scores. This indicates that the presented estimated effects of precipitation on levels of Clostridium and coliforms are actually conservative at extreme events. Models on E. coli were stable independently of the flexibility parameter, and E. coli is also argued to be the best indicator of bacteriological quality of water [17].

Related Studies and Perspectives

The findings in this study cannot be directly generalized to other fresh-water supplies because lag structures and effect of rainfall cannot be assumed to be similar. Therefore, assessments of how rainfall increases the risk of highly polluted raw water in other locations should be performed separately. Precipitation data are often well documented and easily accessible which means that water supply producers, once they understand the associations between precipitation and water quality, can better validate the risks, in magnitude and time. To our knowledge, no studies have addressed effects and lag structures of raw water quality with similar statistical methods and precision before, limiting comparison with other studies. Signor et al. [1] addressed variations in water quality (physical, chemical and bacteriological parameters) in an uncontrolled Australian river, where no point sources of contaminations were known. Mean concentrations were compared at runoff events with base flows, with bacteria and parasites counts increasing during increased flow. Kistemann et al. [2] studied three surface water reservoirs in Germany, comparing mean levels before and after runoffs, and detected elevated parasites and indicator bacteria concentrations with increasing water levels. Åström et al. [3] also studied Göta Älv with data from 2004 and found correlations between accumulated rainfall and elevated concentrations of E. coli, and also other pathogen indicators such as intestinal enterococci. Studies have also shown associations between precipitation and outbreaks of gastrointestinal illnesses (GI), indicating that rainfall has an important role in pathogen contamination in water supplies [4–7]. Other studies tried to assess the relation between water quality and gastrointestinal illnesses under non-outbreak situations, most commonly using turbidity as exposure variable [18–23]. This study suggests that precipitation can be an alternative exposure variable in such studies, since precipitation may better reflects peaks of fecal contamination than turbidity. Studies of rainfall and incidence of GI illnesses under non-outbreak situations are few in number. A study from Milwaukee (WI) reported an increase in emergency department visits 4 days after any amount of rainfall [24]. Associations between heavy rainfall and daily number of nurse advice calls relating to GI problems within the City of Gothenburg has recently been reported [25], and this study supports the causal pathway between rainfall and increased risks due to poor drinking water.

Climate change projections indicate annual precipitation will increase, with more heavy rainfall events in Sweden. Therefore, Göta Älv is expected to encounter more days in the future with inadequate raw water quality than today. Although the drinking water utility Alelyckan has the opportunity to close the water intake, this creates challenges because water in the reservoir lake cannot support the population in Gothenburg with drinking water over long time periods; river water is the only available source of freshwater in the quantities required. Elimination of combined storm and sewage water systems or other systems that result in emergency releases of untreated sewage water into the river should be an important step to increasing river water quality and lowering impacts of heavy precipitation, and to be more prepared for a future climate.

Supporting Information

Figure S1 Data–seasonality patterns. Observations from 2004–2010 plotted within season. Averages projected with a cyclic spline function (9 df). Top row shows observations of river water quality parameters measured near Alelyckan (Gothenburg). A: daily mean turbidity, B: E. coli, C: coliforms and F: Clostridium. Bottom row from left: E: maximum precipitation observation from the three weather stations, F: consecutive wet and dry (negative) days where a wet day was defined as any observed precipitation in any station, G: daily mean temperature observed in Gothenburg and H: stream flow measured in Lilla Edet.

Figure S2 Precipitation effects. Relative effect of daily precipitation (0–54 mm) along 0–15 lags on raw water quality. A: turbidity, B: E. Coli, C: coliforms, D: Clostridium perfringens.

Figure S3 Seasonal effect modifications. Relative effect of daily precipitation (>15 mm/24-h) along 0–8 lags on raw water quality across seasons. A: turbidity, B: E. Coli, C: coliforms, D: Clostridium perfringens.

Figure S4 R^2-seasonality patterns. Variation explained (R^2-values) across seasons by non-linear precipitation predictors 0–8 day prior observations of river water parameters A: turbidity, B: E. Coli, C: coliforms, D: Clostridium perfringens. Colors represent the three different precipitation stations and horizontal dotted lines represent the average R^2.

Figure S5 Stream flow effect modifications. Estimated relative effect on turbidity of a rainfall event of >15 mm along 0–15 lags days at different stream flows (quartiles). Vertical bars represent 95% CI. Vertical lines represent estimated effect peaks.

Figure S6 Turbidity, precipitation and indicator bacteria. Associations (penalized splines, max 5 df) between turbidity and indicator bacteria (lag 0) and precipitation and indicator bacteria (lag 2) (A and B: E. coli, C and D: coliforms, E and F: Clostridium), together with model residuals (dots). Blue dotted lines represent mean levels, and red dotted lines represent unaccepted levels for open raw water intake at Alelyckan drinking water utility.

Acknowledgments

Great acknowledgments to Kristie Ebi, and Annika Malm and the Department of Sustainable Waste and Water, City of Gothenburg, Sweden.

Author Contributions

Conceived and designed the experiments: AT OB BF. Analyzed the data: AT. Wrote the paper: AT. Choose statistical methods: AT. Evaluated results: AT OB BF. Contributed with data material: OB. Approved the final manuscript: AT OB BF.

References

1. Signor RS, Roser DJ, Ashbolt NJ, Ball JE (2005) Quantifying the impact of runoff events on microbiological contaminant concentrations entering surface drinking source waters. J Water Health 3: 453–468.
2. Kistemann T, Classen T, Koch C, Dangendorf F, Fischeder R, et al. (2002) Microbial load of drinking water reservoir tributaries during extreme rainfall and runoff. Appl Environ Microbiol 68: 2188–2197.
3. Astrom J, Petterson S, Bergstedt O, Pettersson TJR, Stenstrom TA (2007) Evaluation of the microbial risk reduction due to selective closure of the raw water intake before drinking water treatment. Journal of Water and Health 5: 81–97.
4. Curriero FC, Patz JA, Rose JB, Lele S (2001) The association between extreme precipitation and waterborne disease outbreaks in the United States, 1948–1994. Am J Public Health 91: 1194–1199.
5. Nichols G, Lane C, Asgari N, Verlander NQ, Charlett A (2009) Rainfall and outbreaks of drinking water related disease and in England and Wales. J Water Health 7: 1–8.
6. Thomas KM, Charron DF, Waltner-Toews D, Schuster C, Maarouf AR, et al. (2006) A role of high impact weather events in waterborne disease outbreaks in Canada, 1975–2001. Int J Environ Health Res 16: 167–180.
7. Cann KF, Thomas DR, Salmon RL, Wyn-Jones AP, Kay D (2013) Extreme water-related weather events and waterborne disease. Epidemiol Infect 141: 671–686.
8. Reynolds KA, Mena KD, Gerba CP (2008) Risk of waterborne illness via drinking water in the United States. Rev Environ Contam Toxicol 192: 117–158.
9. Delpla I, Jung AV, Baures E, Clement M, Thomas O (2009) Impacts of climate change on surface water quality in relation to drinking water production. Environ Int 35: 1225–1233.
10. Swedish Commission on Climate and Vulnerability (2007) Sweden facing climate change – threats and opportunities, Stockholm. Swedish Governmental Official Reports (SOU) 2007:60. Final report. Available: http://www.sweden.gov.se/sb/d/8704/a/89334 Sverige inför klimatförandringarna. Bilaga B 13: Dricksvattenförsörjning i förändrat klimat.
11. Wood SN (2011) Fast stable restricted maximum likelihood and marginal likelihood estimation of semiparametric generalized linear models. Journal of the Royal Statistical Society Series B-Statistical Methodology 73: 3–36.
12. Gasparrini A, Armstrong B, Kenward MG (2010) Distributed lag non-linear models. Stat Med 29: 2224–2234.
13. Team R (2010) R: A Language and Environment for Statistical Computing. R Foundation for Statistical Computing, Vienna, Austria, 2007. ISBN 3-900051-07-0.
14. Gasparrini A (2011) Distributed lag linear and non-linear models in R: the package dlnm. Journal of Statistical Software 43(8):1–20.
15. Blaustein RA, Pachepsky Y, Hill RL, Shelton DR, Whelan G (2013) Escherichia coli survival in waters: temperature dependence. Water Res 47: 569–578.
16. Gauthier F, Archibald F (2001) The ecology of "fecal indicator" bacteria commonly found in pulp and paper mill water systems. Water Research 35: 2207–2218.
17. Odonkor ST, Ampofo JK (2013) Escherichia coli as an indicator of bacteriological quality of water: an overview. Microbiology Research 4: e2.
18. Egorov AI, Naumova EN, Tereschenko AA, Kislitsin VA, Ford TE (2003) Daily variations in effluent water turbidity and diarrhoeal illness in a Russian city. Int J Environ Health Res 13: 81–94.
19. Gilbert ML, Levallois P, Rodriguez MJ (2006) Use of a health information telephone line, Info-sante CLSC, for the surveillance of waterborne gastroenteritis. J Water Health 4: 225–232.
20. Morris R, Naumova EN, Levin R, Munasinghe R (1996) Temporal variation in drinking water turbidity and diagnosed gastroenteritis in Milwaukee. American journal of public health 86: 237–239.
21. Schwartz J, Levin R, Goldstein R (2000) Drinking water turbidity and gastrointestinal illness in the elderly of Philadelphia. J Epidemiol Community Health 54: 45–51.
22. Schwartz J, Levin R, Hodge K (1997) Drinking water turbidity and pediatric hospital use for gastrointestinal illness in Philadelphia. Epidemiology 8: 615–620.
23. Tinker SC, Moe CL, Klein M, Flanders WD, Uber J, et al. (2010) Drinking water turbidity and emergency department visits for gastrointestinal illness in

Atlanta, 1993–2004. Journal of Exposure Science and Environmental Epidemiology 20: 19–28.
24. Drayna P, McLellan SL, Simpson P, Li SH, Gorelick MH (2010) Association between Rainfall and Pediatric Emergency Department Visits for Acute Gastrointestinal Illness. Environmental Health Perspectives 118: 1439–1443.
25. Tornevi A, Axelsson G, Forsberg B (2013) Association between precipitation upstream of a drinking water utility and nurse advice calls relating to acute gastrointestinal illnesses. PLoS One 8: e69918.

Assessing the Impact of Water Filters and Improved Cook Stoves on Drinking Water Quality and Household Air Pollution: A Randomised Controlled Trial in Rwanda

Ghislaine Rosa[1]*, Fiona Majorin[1], Sophie Boisson[1], Christina Barstow[2], Michael Johnson[3], Miles Kirby[1], Fidele Ngabo[4], Evan Thomas[5], Thomas Clasen[1,6]

1 Department of Disease Control, London School of Hygiene and Tropical Medicine, London, United Kingdom, **2** Department of Civil, Environmental and Architectural Engineering, University of Colorado at Boulder, Boulder, Colorado, United States of America, **3** Berkeley Air Monitoring Group, Berkeley, California, United States of America, **4** Ministry of Health, Government of Rwanda, Kigali, Rwanda, **5** Department of Mechanical and Materials Engineering, Portland State University, Portland, Oregon, United States of America, **6** Department of Environmental Health, Emory University, Atlanta, Georgia, United States of America

Abstract

Diarrhoea and respiratory infections remain the biggest killers of children under 5 years in developing countries. We conducted a 5-month household randomised controlled trial among 566 households in rural Rwanda to assess uptake, compliance and impact on environmental exposures of a combined intervention delivering high-performance water filters and improved stoves for free. Compliance was measured monthly by self-report and spot-check observations. Semi-continuous 24-h $PM_{2.5}$ monitoring of the cooking area was conducted in a random subsample of 121 households to assess household air pollution, while samples of drinking water from all households were collected monthly to assess the levels of thermotolerant coliforms. Adoption was generally high, with most householders reporting the filters as their primary source of drinking water and the intervention stoves as their primary cooking stove. However, some householders continued to drink untreated water and most continued to cook on traditional stoves. The intervention was associated with a 97.5% reduction in mean faecal indicator bacteria (Williams means 0.5 *vs.* 20.2 TTC/100 mL, $p<0.001$) and a median reduction of 48% of 24-h $PM_{2.5}$ concentrations in the cooking area ($p=0.005$). Further studies to increase compliance should be undertaken to better inform large-scale interventions.

Trial registration: Clinicaltrials.gov; NCT01882777; http://clinicaltrials.gov/ct2/results?term = NCT01882777&Search = Search

Editor: Robert K. Hills, Cardiff University, United Kingdom

Funding: This study was funded by DelAgua Health, a for-profit company that implements the intervention in Rwanda in conjunction with the Rwanda Ministry of Health. The funders had no role in study design, data collection and analysis, decision to publish, or preparation of the manuscript.

Competing Interests: We have the following interests: This study was funded by DelAgua Health, a for-profit company that implements the program in Rwanda in conjunction with the Rwanda Ministry of Health. Evan Thomas and Christina Barstow, co-authors of this article, are compensated consultants to DelAgua Health, and are in charge of overseeing the implementation of the program in Rwanda. Michael Johnson, co-authors of this article, is employed by the commercial company, Berkeley Air Monitoring Group, which was contracted to provide advice on household air pollution monitoring. This does not alter our adherence to all the PLoS ONE policies on sharing data and materials.

* E-mail: ghislaine.rosa@lshtm.ac.uk

Introduction

Environmental contamination at the household level is a major cause of death and disease, particularly among rural populations in low-income countries. Unsafe drinking water, together with poor sanitation, account for an estimated 0.9% of the global burden of disease and 0.3 million deaths [1]. Much of this disease burden is associated with diarrhoea, which alone accounts for 10.5% of deaths in children under 5 years in low-income countries [2]. Household air pollution (HAP) from biomass fuel smoke has been linked to increased risk of respiratory tract infections, low birth weight, exacerbations of inflammatory lung conditions, cardiac events, stroke, eye disease, tuberculosis, cancer and nutritional deficiencies [3]. The Global Burden of Disease (GBD) 2010 project found HAP from solid fuels to be responsible for 3.5 million premature deaths globally [1]. In this same assessment, smoke from household cooking fuels was also responsible for another half a million premature deaths due to contributions to outdoor air pollution [1]. These environmental hazards are aggravated among rural inhabitants of sub-Saharan Africa who are more likely to rely on unsafe water supplies and cook using biomass fuels on inefficient stoves [4–6].

Inefficient cookstoves also present substantial economic, developmental and environmental costs. At the household level, poverty is exacerbated and time spent at school is reduced by the burden of collecting more fuel for boiling drinking water and cooking [7]. Individuals, households and governments bear the cost of expenditures for seeking treatment of enteric and respiratory infections. Cookstove emissions also contribute to greenhouse gas and black carbon emissions, and in some cases the fuel harvesting can result in denuding of forests [8,9].

With a population of 10.5 million and a density of 412 persons per sq. km, Rwanda is the most densely populated country in East Africa [10]. Eighty per cent of the population of Rwanda lives in rural areas and is engaged in agriculture [11]. Despite significant progress over the last decade, 57% of the population is living below the poverty line, 37% of them living in extreme poverty [11]. While a large proportion of the rural population has access to improved water sources (71.2%), mainly through protected springs, only 2.2% of rural areas have water on their premises [12], resulting in an increased risk for drinking water contamination during transport and storage [13]. Almost all of rural Rwanda (99.0%) relies on biomass for their cooking needs [12]. Morbidity and mortality are largely dominated by communicable diseases, including HIV/AIDS, acute respiratory infections, diarrhoeal diseases, intestinal parisitoses, and malaria [14]. Among deaths of children under 5, pneumonia accounts for 20% and diarrhoea for 12% [15].

In an effort to reduce the disease burden in rural Rwanda, decrease poverty associated with expenditures for fuel, and minimize the impact of greenhouse gases from inefficient combustion of biomass in low-efficiency stoves, the Rwanda Ministry of Health (MiniSante) and the Rwanda Environmental Management Authority (REMA) have partnered with DelAgua Health (implementer) to design, deploy and evaluate the impact of a project that will deliver and promote the use of advanced water filters and high efficiency cookstoves to lower-income households in Rwanda. Prior to initiating the full campaign, the implementer with the Ministry of Health undertook a pilot distribution of filters and cookstoves to approximately 2200 households in 15 villages in 11 of the country's 30 districts. We conducted this study in three of those villages in order to assess the uptake of the intervention and its impact on drinking water quality and household air pollution.

Methods

Study setting

The study was conducted from September 2012 to April 2013 in three rural villages, Nyarutovu and Kabuga located in Muhanga district, Southern province; and Rubona, located in Gakenke district, Northern province. These villages were purposely selected from the 15 villages comprising the pilot distribution phase. The sites were changed from the original protocol, Karongi and Ngororero districts in Western province, to accommodate access to better microbiology laboratory facilities in Kigali.

Study design and sample size

The study employed a parallel, household-randomised, control trial design with a 1:1 ratio. This trial followed a non-blinded design because previous attempts to blind an earlier version of the LifeStraw Family filter in the Democratic Republic of Congo were unsuccessful [16]. The objectives of the study were to assess (i) uptake and use of the intervention by the target population when delivered programmatically, and (ii) the impact of the intervention on the microbiological quality of household drinking water and air quality near the self-reported cooking area over the 5-month follow-up period. Our primary outcomes were (i) to assess levels of faecal contamination (measured by thermotolerant coliforms, TTC) in stored water in the home that householders used for drinking, and (ii) to determine average 24-h concentrations of $PM_{2.5}$ in the main cooking area as identified by participants. Our secondary outcome was to assess use of the intervention filters and stoves based on self-report and spot-check observations.

The sample size calculation was based on $PM_{2.5}$ emissions reductions rather than TTC reductions in drinking water as the former was determined to require a larger sample. Assuming a 50% reduction in $PM_{2.5}$ emissions, 80% power, $\alpha = 0.05$ and a coefficient of variation (COV) of 1, we estimated a sample size of 63 households per arm.

The protocol of this trial and CONSORT checklist are available as supporting information; see Text S1 and S2.

Intervention

Each intervention household received one LifeStraw Family 2.0 filter and one EcoZoom Dura improved wood burning stove. The filter is the second-generation of a gravity-based water purifier that uses ultrafiltration in the form of a hollow-fibre cartridge to remove pathogens from drinking water. The first generation device has been shown in field studies to be highly effective in improving water quality and to achieve consistent (though not exclusive) use [16]. The second-generation version used in this study employs a table-top design and an integrated safe storage vessel. Untreated water is poured through a 20-µm pre-filter plastic mesh into a 6.0 L container; over time, gravity forces the water through the cartridge comprised of hollow-fibres with a 20-nm pore size. The water then passes into a 5.5 L storage vessel where it can be dispensed via a plastic tap. The device is cleaned daily by backwashing the cartridge using a squeeze-pump mounted on the back of the storage container. The device is designed to treat 18,000 L of water [17] with a flow rate of approximately 3 L per hour. In the laboratory, the filter cartridge was found to meet the USEPA standards for microbiological water purifiers by reducing bacteria by 6 logs, viruses by 5 logs and protozoa by 4 logs [18]. The filter meets the "highly protective" World Health Organization (WHO) rating for household water treatment technologies [19].

The intervention stove is based on the 'rocket' concept that uses an internal 'chimney' in the stove that directs air through the burning fuel (usually biomass), and encourages the mixing of gases and flame above it. Precise internal stove dimensions are used to achieve high combustion efficiency and transfer heat to the cooking pot. Two additional components are included with the stove, a "stick support" onto which fuel wood is placed to promote airflow and a "pot skirt" which increases fuel efficiency. A study comparing cookstoves in Uganda, Kenya and Tanzania reported that the EcoZoom (aka StoveTec) stove saved 39% to 54% of fuel compared to open fires, cooked meals faster, and was participants' most preferred stove during controlled cooking of local dishes [20–22]. In the intervention group, householders were encouraged to cook outdoors on the EcoZoom stove and to use dry wood only to increase the efficiency of the stove. Further details on the messaging used in the pilot distribution can be found elsewhere [23].

Houses that were allocated to the intervention group also received a poster with illustrations and instructions in *Kinyarwanda*, the local language, on filter and stove use, maintenance, and contact names and phone numbers for the implementer. Most households had easy access to a cellular phone for contacting the implementer. Intervention households received one-to-one training on use and maintenance in their homes by community health workers (CHWs) who were previously trained by trainers who themselves had been trained by the filter and stove manufacturers and implementer. Intervention households were then visited periodically at approximately one-month intervals by CHWs to refresh health messaging and encourage use. Households allocated to the control group were instructed to continue usual practices throughout the study. At the end of the study in April 2013, these control households received their own filters, stoves, posters and training.

Enrolment, baseline survey, randomisation and deployment of devices

Households were eligible to participate in the study if (i) they were registered as being members of the village, (ii) the head of the household was over 18 years, and (iii) no members of the household worked as a CHW. The last criterion was included after the original protocol was drafted as at the time of the design the researchers were not aware that the CHW that would deliver the intervention resided in the villages selected for the study. It was explained that while all participating households would receive filters and stoves, half would receive them at the outset of the study and the balance at the conclusion of the 5-month follow-up period. After obtaining consent from the heads of participating households, a baseline survey was undertaken in September-November 2012 to collect information on demographics, socio-economic characteristics, water, hygiene and sanitation practices as well as fuel and cooking practices. Data collection tools were translated into *Kinyarwanda* and piloted before use.

Following the baseline survey, a public lottery was organised by the implementer and research teams during a village meeting to randomly allocate an approximately equal number of households from each village to intervention or control groups. Local authorities and village chiefs were extensively engaged to assess the suitability of this randomisation approach. After the lottery, members of control households were invited to leave the venue while those of intervention households attended a demonstration on the use and maintenance of the filter and stove, collected their devices, and carried them to their homes.

Outcome assessment

Compliance. Monthly cross-sectional surveys were conducted by trained field investigators (the evaluation team) working independently of the implementation team at unannounced visits among each household. At each visit participants were asked to identify the main drinking water container in the household, whether it was the intervention filter or another container; the surveyor also recorded whether the filter contained water at the time of the visit, a possible objective indicator of filter use. The field investigators also observed the cookstove and if cooking was taking place at the time of the unannounced visit, recorded where and whether such cooking was on the intervention stove or the traditional stove. If no cooking was taking place, field investigators noted the presence of smoke marks on the intervention stove, a possible objective indicator of use. Reported measures of stove use were also collected by asking participants what stove had been used the last time cooking took place in their home.

Independent to our study, the implementers undertook a separate survey, conducted by Environmental Health Officers (EHOs), to assess use and acceptability of the intervention for their own monitoring and evaluation purposes. The details of this assessment have been presented elsewhere [23].

Additionally, to assess use of the intervention in a more objective manner, remotely reporting electronic sensors were mounted onto 23 intervention filters and 27 intervention stoves and deployed in a randomly selected sub-sample of intervention households for a two-week period. The details of the implementation of this nested study, data handling and analysis, and results are presented elsewhere [24].

Water quality. During each of the five monthly visits, field investigators took a sample from the water container identified by the householder as being used mainly for drinking by children under 5 years of age, or adults if no under 5 s resided in the household. If this was other than directly from the intervention filter, a second sample was taken directly from the filter if it contained water. All water samples were collected in sterile Whirl-Pak bags (Nasco, Fort Atkinson, WI) containing a tablet of sodium thiosulphate to neutralize any halogen disinfectant. Samples were placed on ice and processed within 6 h of collection to assess levels of TTC. Microbiological assessment was performed using the membrane filtration technique [25] on membrane lauryl sulphate medium (Oxoid Limited, Basingstoke, Hampshire, UK) using a DelAgua field incubator (Robens Institute, University of Surrey, Guilford, Surrey, UK).

Household air pollution. Monitoring of particulate matter with an aerodynamic diameter <2.5 µm ($PM_{2.5}$) in the main cooking area took place between November 2012 and March 2013. 126 households (63 control and 63 intervention households) were randomly selected for semi-continuous 24-h $PM_{2.5}$ monitoring. Households were numbered and selected by using a computerised random number generator. Upon arrival at the participant's home, the family member mainly responsible for cooking was identified and a short survey was employed to identify the area in the household where cooking primarily took place. "Stacking" of stoves (using different stoves, often in different locations) [26] was a common scenario, both in control and intervention households, though more common in the latter. In cases where the participant reported cooking equally in two locations or with two or more stoves, we sampled from indoor rather than outdoor locations and from traditional rather than intervention stove. UCB-PATS $PM_{2.5}$ monitors (described below) were placed 1.5 m above the ground and 1 m away from the stove and, whenever possible, at least 1.5 m from windows and doors by suspending the monitors from the roof beams. When cooking was reported to take place outdoors, the $PM_{2.5}$ monitor was mounted onto a vertical wooden stand and placed at the same distance and height from the stove. The location of the stand was marked on the floor and participants were advised not to touch or move the equipment.

$PM_{2.5}$ was measured using the University of California, Berkeley Particle and Temperature Sensor (UCB-PATS™), (Berkeley Air Monitoring Group, USA), a semicontinuous (1-min averages), light-scattering nephelometer [27,28]. Laboratory and field validations of the UCB-PATS have been described previously [27–30]. To take into account that nephelometer sensitivity is a function of an aerosol's specific optical properties such as size, colour, and shape [31], calibration of the UCB response with the target aerosol was undertaken by conducting 24-h $PM_{2.5}$ gravimetric co-location measurements in a sub-sample of homes ($n = 30$). Five field blanks were obtained, resulting in an adjustment of subtracting 5 µg to the final filter masses (<1% of the mean mass deposition). The UCB-PATS response was then linearly regressed against the gravimetric samples ($n = 27$, $R^2 = 0.86$), with the resulting equation then used to adjust the UCB-PATS response to the gravimetric measures (Figure S1 of supporting information). Three gravimetric samples were omitted due to incomplete sampling durations.

Gravimetric $PM_{2.5}$ samples were collected using standard air sampling pumps (PXR8, SKC Inc., USA) with $PM_{2.5}$ cyclones (SCC 1.062, BGI, USA) using a flow rate of 1.5 L/min. Flow rates were measured before and after installation of the sampling equipment in the home with a rotameter (Matheson Trigas, Montgomeryville, PA, USA) that had been calibrated using a TSI Flow Calibrator 4146 (TSI, Inc., USA). $PM_{2.5}$ was collected on 37-mm Teflon filters (Pall, USA). Filters were stored at 4°C until shipment to Berkeley Air Monitoring Group in California, USA for weighing. Filters were equilibrated for 24 h at 22±3°C and 40±5% relative humidity before being weighed on a 0.1

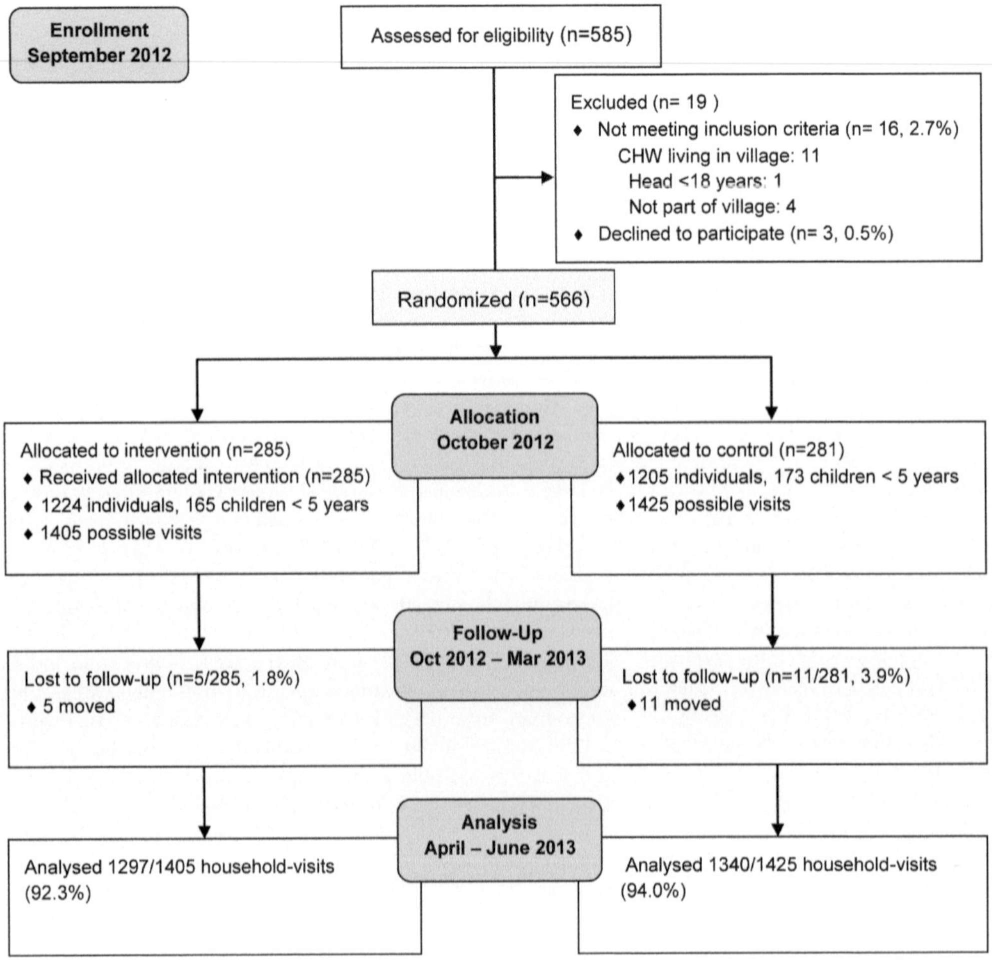

Figure 1. CONSORT diagram showing the flow of participants through the trial.

microgram resolution electro microbalance (XP2U, Mettler Toledo, USA).

Data analysis

All data were analysed using Stata 12 (Stata Corporation, College Station, TX, USA). Because both $PM_{2.5}$ concentrations and TTC counts in drinking water followed non-normal distributions, medians, geometric means and Williams means are presented together with arithmetic means. The Williams mean is calculated by adding 1 to all the data values, then taking the geometric mean, then subtracting 1 again [32]. Categorical data were compared using a Chi square or a Fisher's exact test where appropriate. The non-parametric Wilcoxon rank sum test was used to compare $PM_{2.5}$ concentrations in the main cooking area between intervention and control groups. To assess the effect of the intervention on water quality, TTC counts during follow-up were compared using random effects negative binomial regression as describe elsewhere [33] to account for (i) repeated observations within households and, (ii) the skewed distribution of the TTC counts. Model comparison was assessed by using the Bayesian information criterion (BIC), which is a well-established measure of goodness of fit that also applies to non-nested models [33,34]. For the purpose of analysis, plates that yielded coliform forming units (CFUs) that were too numerous to count (TNTC) were assigned a value of 300 TTC/100 mL. Data were analysed in an intention-to-treat basis in order to estimate the effect of the intervention regardless of compliance. Only those households with complete follow-up data were analysed.

Ethics

The study was reviewed and approved by the ethics committee at the London School of Hygiene and Tropical Medicine (No. 6239, as amended) and the Rwanda National Ethics Committee (No. 328 RNEC/2012). Written informed consent to participate in the research was obtained from the male or female head or the wife of each participating household.

Results

Study population

The three villages participating in the study comprised 585 households, all of which were screened to participate in the study, 16 (2.7%) were ineligible and 3 (0.5%) refused to participate (Figure 1). A total of 566 households with 2429 individuals were enrolled in the study. Of those 281 (49.7%) were assigned to the control group and 285 (50.4%) were assigned to receive the intervention filter and cookstove. Household loss-to-follow-up was 2.8%, primarily due to participants moving out of the study area. A total of 2737 household-visits were completed during the follow-

Table 1. Filter and stove use among intervention households: Evaluator's survey.

	All visits	
	N	%
Filter use[1]		
Reported drinking container		
Intervention filter	1210	89.2
Other container	146	10.8
Water stored in other container treated	57	39.0
Method of treatment: Intervention filter	56	98.2
No water in intervention filter among households identifying filter as drinking container	12	1.0
No water in intervention filter among households not identifying filter as drinking container	83	56.8
Stove use[2]		
Observation data on use		
Intervention household cooking at time of visit	280	26.9
Stove in current use		
Intervention stove only	152	54.3
Both stoves simultaneously	12	4.3
Traditional stove only	116	41.4
Currently cooking outdoors	59	21.1
Intervention stove users cooking outdoors[3]	49	32.2
Reported data on use		
Reported last stove used[4]		
Intervention stove only	593	78.0
Both stoves simultaneously	15	19.3
Traditional stove only	147	2.0
Reported using intervention stove in last three follow-up visits	130	47.5

[1]Based on households that completed the visits and allowed enumerators to observe the container, 1356/1393 = 97.3%.
[2]Data only available from mid follow-up 2 onwards (1040/1393 = 74.7%).
[3]Among those households cooking only on the intervention stove.
[4]Excludes those households cooking at time of home visit.

up period (96.7%) and data on one of the primary outcomes (water quality) was collected for 2637 households-visits (93.2%).

Baseline characteristics

Baseline characteristics were distributed evenly between the trial arms, with the exception of availability of soap among households with a designated hand washing area and boiling or chlorination of drinking water (see Table S1 of supporting information). At baseline, drinking water samples were obtained from 551 (97.3%) households. The median and Williams mean of drinking water was 14 and 20.2 TTC/100 mL (95% CI: 15.0–27.0 TTC/100 mL) and 22 and 30.3 TTC/100 mL (95% CI: 22.8–40.2 TTC/100 mL) for control and intervention groups, respectively.

Filter and improved stove use and compliance

Most households used the filter throughout the study period (Table 1). Intervention households identified the filter as the main drinking source in 89.2% of all household visits where drinking water was available. Visual inspection at the time of the unannounced visit was consistent with reported use, with 99% of the filters containing water. Of the 10.8% of intervention households that stored their drinking water elsewhere, overall only 39.0% of them reported that the water had been treated with the intervention filter. Over the course of the study, however, only 62.9% of intervention households identified the filter as the main drinking water storage container in all five follow-up visits with available water (n = 240, 84.2%). Of the remainder, 11.2% reported treating it and storing it elsewhere at least once during the 5-month follow-up, 25.0% reported drinking untreated water at least once during follow-up and 0.8% did not know the status of their water in at least one of the visits. During the last follow-up visit, the major reasons for not having filtered water at the time of the visit were (i) forgetting to fill the filter (48.1%), (ii) drinking mainly locally produced beer instead of water (22.2%), or (iii) having a broken or not properly functioning filter (18.5%).

The intervention stoves were also used throughout the study, though most householders also continued to use their traditional stoves. Field investigators observed actual cooking on about a quarter (26.9%) of their unannounced visits. Of these, 54.3% were cooking only with the intervention stove and 4.3% were using both the intervention and traditional stoves (Table 1). Reported use was higher, with householders claiming they last cooked solely on the intervention stove on 78.0% of visits. Use of the intervention stove was not consistent, with 47.5% of intervention households reporting to have used the intervention stove during the last cooking event at all three home visits (data not collected during initial phases of follow-up). Likewise, of the households that were cooking at all three unannounced visits (n = 8), or at two of the

Table 2. Filter and stove use among intervention households: Implementer's survey.

	N	%
Filter use		
Filter presence confirmed in households[1]	283	99.7
Tap accessible to <5 s	267	94.4
Water present in filter	269	95.1
Stove use		
Observation data on use		
Intervention household cooking at time of visit[2]	78	27.7
Stove in current use		
Intervention stove only	50	64.1
Both stoves simultaneously	4	5.1
Traditional stove only	24	30.8
Currently cooking outdoors	30	38.5
Intervention stove users cooking outdoors[3]	28	17.2
Reported data on use		
Reported last stove used[4]		
Intervention stove only	163	79.9
Both stoves simultaneously	2	1.0
Traditional stove only	39	19.1
Primary stove in current use is intervention stove[4]	253	89.1
Use intervention stove ≥7/week	236	93.3
Use intervention stove ≥14/week	137	54.2
Continue using traditional stove	217	76.4
Use traditional stove ≥7/week	58	26.7
Reported cooking less indoors	175	61.6
Reported main cooking is outdoors	163	57.4
Tend more the fire with the intervention stove[5]	212	83.8

[1]Observation not allowed in one household.
[2]Of those households that allowed the observation (n = 282, 99.3%).
[3]Among those households cooking only on the intervention stove.
[4]Excludes those households cooking at time of home visit.
[5]Among those households identifying intervention stove as main cooking stove.

three unannounced visits (n = 52), only 50.0% and 34.6% were using the intervention stove at all three or two visits, respectively. During the last follow-up visit, the major reasons for not using the intervention stove during the last cooking event were (i) having no time to tend the fire (34.1%), (ii) not having dry (30.7%) or the right-size wood (10.2%) or, (iii) cooking beans for which a traditional stoves was regarded most appropriate (10.2%).

Data on use of the intervention from the implementer's survey was very similar to our assessment (Table 2). A similar percentage of intervention households (27.7%) were cooking at the time of the visit. Of these, just over two thirds (64.1%) were exclusively cooking on the intervention stove, but only 17.2% of these were cooking outdoors, a figure just slightly lower than the one observed on our independent follow-up. Data from the implementer's more extensive survey confirmed the stacking of stoves in intervention households, with 76.4% of intervention households reporting to continue using their traditional stoves. Of these, 26.7% reported using it ≥7 times per week. Of interest was the fact that 83.8% of intervention households identifying the intervention stove as their primary cookstove reported that the intervention stove required more active tending of the fire as compared to the traditional stove.

In the last round of our follow-up, 5.4% and 5.1% of intervention households reported having problems with their filter or stove at the time of the visit, respectively. Data collected from the implementer's repair team indicates that 24.9% of filters and 6.7% of stoves had to be repaired during the study. No devices had to be fully replaced, though some repairs involved the replacement of individual components. The main reasons for filters being repaired were (i) filters being clogged (48.6%) and, (ii) tubes being damaged by rodents (27.0%). The main reasons for the intervention stoves being repaired included (i) pot skirts melting (65%), and (ii) stick supports breaking (10%).

Overall, only 1.0% of water samples collected from control households were reported to have been treated with a neighbour's intervention filter, showing low levels of cross-contamination between groups.

Water quality

The microbiological quality of the stored drinking water was significantly higher in intervention households than control households (Williams means 0.5 vs. 20.2 TTC/100 mL, respectively, $p<0.001$). Overall, 86.8% (95% CI: 84.9%–88.6%) of drinking water samples from intervention households were free of TTC compared to 22.4% (95% CI: 20.1%–24.6%) of control household samples ($p<0.001$) (Figure 2). The proportion of samples that had >100 TTC/100 mL was 3.6% (95% CI: 2.6%–4.6%) for intervention households and 31.9% (95% CI: 29.4%–34.5%) for control households. Overall, 96.6% of drinking water samples collected directly from filters were free of TTC. In intervention households, water quality was significantly higher in water samples collected directly from the filter (Williams mean 0.14 TTC/100 mL; 95% CI: 0.10–0.18) than water stored in another container (Williams mean 13.8 TTC/100 mL; 95% CI: 9.0–20.7) (see Table S2 of supporting information). The quality of the drinking water stored in other containers did not differ significantly between control and intervention households ($p = 0.07$). However, among intervention households, water that was stored in another container and was reportedly treated with the intervention filter was significantly of higher quality than reportedly non-treated stored water (Williams means 5.4 vs. 23.2 TTC/100 mL, respectively, $p<0.001$). Throughout the duration of the study, only 2.5% of control households had drinking water free of TTC on all follow-up visits as opposed to 56.5% of intervention households. Overall 15.2% of samples from control households and 5.1% of samples from intervention households yielded plates that were TNTC.

Air quality

A total of 121 households (60 intervention and 61 control) completed the 24-h $PM_{2.5}$ monitoring of the main cooking area. 66.7% of intervention households identified the intervention stove as their main cooking stove. However, only 23.3% of intervention households reported that their main cooking area was outdoors as promoted by the intervention. Of these, all households reported cooking with the intervention stove. Among the control households, the three stone fire was identified as the main cooking stove in 65.6% of cases, followed by the locally made *rondereza* stove (24.6%). Only one control household reported cooking outdoors.

Table 3 shows the $PM_{2.5}$ concentrations of the main cooking area for control and intervention households on an aggregate level and stratified by reported main area of cooking. Overall, mean and median 24-h $PM_{2.5}$ concentrations in intervention households were 0.485 mg/m^3 and 0.267 mg/m^3, respectively, compared to

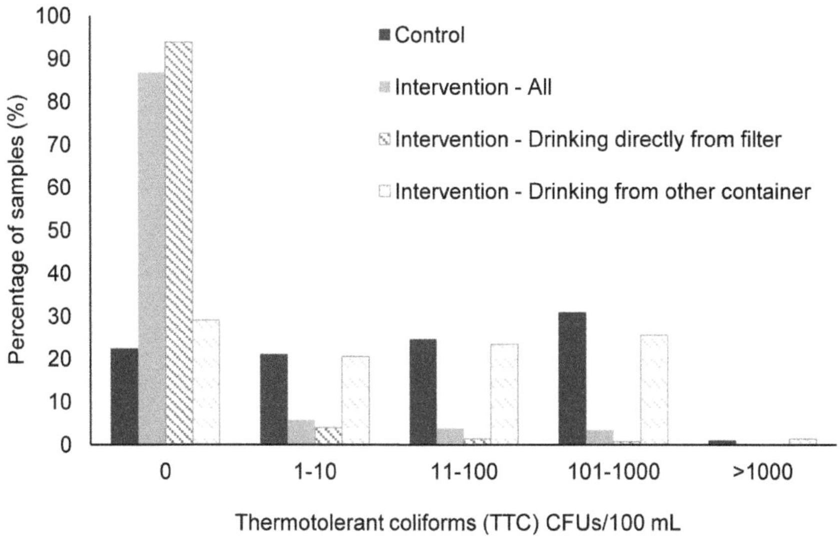

Figure 2. Percentage of water samples by level of contamination (TTC/100 mL).

0.905 mg/m^3 and 0.509 mg/m^3 for control households. This represents a 48% reduction in median 24-h concentrations ($p = 0.005$). Compared to control households that predominantly cooked indoors, intervention homes that reported indoor cooking showed a reduction in median concentrations of 37%, which was only borderline significant, possibly due to the smaller sample size ($p = 0.08$). Outdoor cooking in the intervention was associated with a median reduction of 73% when compared to control households ($p < 0.001$) and 57% reduction when compared to indoor-cooking intervention homes ($p = 0.02$).

Discussion

We report on a randomised controlled trial to independently evaluate a pilot implementation program distributing free water filters and improved cooking stoves to rural homes in Rwanda. We found high reported use of the intervention filter, which was associated with significantly higher microbiological quality of drinking water when consumed directly from the filter. Nevertheless, such use was not exclusive; a sizable proportion of householders continued to drink untreated water. We also found improved household air quality among intervention households despite continued use of the traditional stove.

Filter uptake among the intervention population was high, with filters being reportedly used in 89.2% of all household visits. Similar levels of uptake of filter-based interventions have been reported elsewhere [16,35,36]. Nevertheless, we found that 25% of intervention householders were reporting untreated water in at least one of the five follow-up visits. The nested study within this RCT using remotely reporting electronic sensors that collected objective data on use of the intervention devices (mainly times and volumes of water filtered for the intervention filter and times and duration of use for the intervention stove) corroborated our findings, showing that the filters and stoves were not used in a consistent and exclusive manner [24]. Epidemiological modelling based on quantitative microbial risk assessment suggests that even occasional consumption of untreated water can vitiate the health benefits associated with improved water quality interventions [37–39]. However, the intervention did significantly improve the microbiological quality of the drinking water when the filter was used as the main storage container. Since 96.6% of drinking water samples collected directly from filters were free of TTC, the conditions for achieving health gains may be achieved with better messaging.

Exclusive use was more problematic for the intervention stove. Only half of the intervention households reported that the last

Table 3. Summary statistics for 24-h $PM_{2.5}$ concentrations in the reported main cooking area.

$PM_{2.5}$ (mg/m^3)	N	Mean	SD	Min	Median	Max	Geometric mean	% Mean reduction	% Median reduction	Wilcoxon RST[1] p-value
Control	61	0.905	1.05	0.06	0.509	4.69	0.51	-	-	
Intervention	60	0.485	0.53	0.04	0.267	2.28	0.28	46%	48%	0.005
Reported cooking location										
Control- Indoor cooking	60	0.910	1.06	0.06	0.506	4.69	0.51	-	-	
Intervention- Indoor cooking	46	0.558	0.56	0.04	0.321	2.28	0.33	39%	37%	0.08
Intervention- Outdoor cooking	14	0.243	0.34	0.05	0.139	1.40	0.16	73%	73%	<0.001

[1]Wilcoxon rank-sum (Mann-Whitney) test

cooking event was performed with the intervention stove in the last three monthly follow-up visits. Likewise, only a third of those households that were visited twice at times that cooking was taking place were using the intervention stove at both instances, showing that among the intervention arm, households continued to rely on their traditional stove. Results from the implementers' survey showed similar results, with 76.4% of households reporting the continued use of their traditional stove, 26.7% of them using it more than 7 times per week. This is consistent with other studies that have shown that the introduction of a new stove often results in "stacking" rather than an immediate complete substitution [40–43].

Households reported continuing the use of their traditional stove because the intervention stove required more tending, unavailability of the adequate fuel or personal preferences for cooking traditional dishes. Context-specific issues regarding a community's cooking needs and preferences have been commonly cited in the literature as reasons for not achieving higher uptake and/or exclusive sustained use of improved cookstoves [42,44]. Thus re-considerations of the promoted stove or more active messaging addressing each of the main barriers may be required if a switching of the stove as opposed to an addition of the intervention stove to the current cooking system is to be achieved. This is not only going to affect the potential health impact of the intervention but also its environmental impact.

The assessment of HAP among control and intervention households showed an overall reduction of 48% of 24-h $PM_{2.5}$ among intervention households, which was comparable to reductions in household air pollution for rocket stove interventions in Ghana (52%) and Kenya (33%) [22,45]. Indoor cooking with the intervention stoves as opposed to the traditional stove was associated with a 37% reduction in 24-h $PM_{2.5}$, which was of borderline significance. However, we cannot rule out that this association may be due to residual bias by comparing sub-groups. Likewise, cooking outdoors, as recommended by the implementer, doubled the reduction in 24-h $PM_{2.5}$ from 37% to 73% as compared to indoor cooking on traditional stoves. Future studies, randomising participants not only to stove technology but also to cooking location (indoors *vs.* outdoors) would be advisable. More effective messaging may increase the levels of outdoor cooking expected by the intervention, as only 57.4% of households reported that their main cooking area was outdoors. Nevertheless, both the indoor and outdoor concentrations in the cooking area were well over even the initial interim 24-h WHO target for $PM_{2.5}$ (75 $\mu g/m^3$) [46]. At the same time, it will be important to monitor personal exposure directly, as most householders that identified the intervention stove as their primary cooking stove (83.8%) reported that the intervention stove required more tending than their traditional one, which could mitigate some of the impact from the household level reductions in $PM_{2.5}$. Indeed, many studies have found that reductions in personal exposure tend to be lower than reductions of emissions in the cooking area [47,48]. Given that a recent RCT study suggested that personal exposure reductions exceeding 50% may be required to achieve meaningful health impacts [47], further assessments of the intervention stove maybe be needed to determine whether the use of the intervention stove translates into meaningful health benefits.

This study has certain limitations. First, the villages included in the RCT were not selected randomly and should not be viewed as representative of any larger population. Second, we cannot rule out the potential for reactivity due to repeated monthly follow-up visits [49]. Third, while we attempted to collect objective indicators of use, by both undertaking visual observations of the filter and stoves and cooking events, the study relied heavily on reported data, which is susceptible to reporting bias. Furthermore, in this study we failed to collect data on reported supplementation of treated water with untreated water, which would have further implications for the health impact of the study. Previous studies with the earlier version of the LifeStraw Family filter have found quite varied results. A study in the Democratic Republic of Congo showed substantial supplementation despite high levels of filter use [16]. On the other hand, a study among HIV-positive mothers, who may be more aware of their health and their children's health, reported almost no supplementation [36]. However, in the latter storage containers were provided. Fourth, budget constraints allowed only the main cooking area, as identified by the participant, to be monitored for HAP. Given the potential for reporting bias and that stacking was commonly reported among the study population, it is very likely that cooking events may have taken place during the monitoring period in areas other than the one being monitored, thus giving a misleading and probable underestimate of the actual total HAP. Likewise, budget constraints did not permit personal $PM_{2.5}$ assessment, a more reliable metric for exposures associated with health outcomes [50]. Fifth, we did not collect any self-reported or other measures of diarrhoea or respiratory infections in our study communities. Finally, the follow-up period of this evaluation was limited to 5 months. This represents under a fraction of the lifespan of both the filter and stove and provided little opportunity to assess the impact of seasonal variations that are common in water quality and HAP. It also provided no opportunity to assess long-term patterns of use, which have been shown to diminish or vary over time for both water filters and improved cookstoves [35,43]. We are endeavouring to address some of these shortcomings in a longer-term follow-up study, currently underway, that will focus on health outcomes and sustained use.

Notwithstanding these limitations, this study suggests that a combined filter/stove intervention accompanied by consistent follow-up to promote use has the potential to significantly improve drinking water quality and household air pollution among a vulnerable population in Rwanda. If the longer-term follow-up study demonstrates sustained use with more exclusive reliance on the intervention hardware and lower personal exposure to HAP, then a large-scale roll out in Rwanda could significantly reduce exposures linked to much of the country's disease burden.

Acknowledgments

We thank all of the participants who contributed to this study. We also thank the staff for their diligent efforts in collecting the data. We thank too Wolf-Peter Schmidt and Alexandra Huttinger for their contribution to the design of the study. Michael Johnson, senior scientist at Berkeley Air Monitoring Group was responsible for providing advice on HAP monitoring and acts as lead author for the group, mjohnson@berkeleyair.com.

Author Contributions

Conceived and designed the experiments: GR SB FN TC. Performed the experiments: GR FM MK. Analyzed the data: GR MJ. Contributed reagents/materials/analysis tools: CB ET. Wrote the paper: GR TC.

References

1. Lim SS, Vos T, Flaxman AD, Danaei G, Shibuya K, et al. (2012) A comparative risk assessment of burden of disease and injury attributable to 67 risk factors and risk factor clusters in 21 regions, 1990–2010: a systematic analysis for the Global Burden of Disease Study 2010. Lancet 380: 2224–2260.
2. Liu L, Johnson HL, Cousens S, Perin J, Scott S, et al. (2012) Global, regional, and national causes of child mortality: an updated systematic analysis for 2010 with time trends since 2000. Lancet 379: 2151–2161.
3. Ezzati M, Kammen DM (2001) Quantifying the effects of exposure to indoor air pollution from biomass combustion on acute respiratory infections in developing countries. Environ Health Perspect 109: 481.
4. Rehfuess E, Mehta S, Pruss-Ustun A (2006) Assessing Household Solid Fuel Use: Multiple Implications for the Milleniem Development Goals. Environ Health Perspect 114.
5. WHO/UNICEF (2012) Progress on drinking water and sanitation: 2012 update. Geneva: WHO.
6. Bonjour S, Adair-Rohani H, Wolf J, Bruce NG, Mehta S, et al. (2013) Solid Fuel Use for Household Cooking: Country and Regional Estimates for 1980–2010. Environ Health Perspect 121: 784–790.
7. Barnes DF, Openshaw K, Smith KR, van der Plas R (1994) What Makes People Cook with Improved Biomass Stoves? A Comparative International Review of Stove Programs. World Bank.
8. Johnson M, Edwards R, Ghilardi An, Berrueta V, Gillen D, et al. (2009) Quantification of Carbon Savings from Improved Biomass Cookstove Projects. Environ Sci Technol 43: 2456–2462.
9. Bond TC, Doherty SJ, Fahey DW, Forster PM, Berntsen T, et al. (2013) Bounding the role of black carbon in the climate system: A scientific assessment. J. Geophys Res: Atmos 118: 5380–5552.
10. NISR (2012) Rwanda Ministry of Health Annual Health Statistics Booklet 2011. National Institute of Statistics of Rwanda.
11. NISR (2008) Rwanda in Statistics and Figures 2008. National Institute of Statistics of Rwand.
12. National Institute of Statistics of Rwanda (NISR) [Rwanda], Ministry of Health (MOH) [Rwanda], International I (2012) Rwanda Demographic and Health Survey 2010. Calverton, Maryland, USA: NISR, MOH, and ICF International.
13. Wright J, Gundry S, Conroy R (2004) Household drinking water in developing countries: a systematic review of microbiological contamination between source and point-of-use. Trop Med Int Health 9: 106–117.
14. RMOH (2011) Rwanda Health Statistics Booklet 2011. Rwanda Ministry of Health.
15. UNICEF (2012) Committing to Child Survival: A Promise Renewed. Progress Report 2012. New York: United Nations Children's Fund.
16. Boisson S, Kiyombo M, Sthreshley L, Tumba S, Makambo J, et al. (2010) Field Assessment of a Novel Household-Based Water Filtration Device: A Randomised, Placebo-Controlled Trial in the Democratic Republic of Congo. PLoS One 5: e12613.
17. Clasen T, Naranjo J, Frauchiger D, Gerba C (2009) Laboratory Assessment of a Gravity-Fed Ultrafiltration Water Treatment Device Designed for Household Use in Low-Income Settings. Am J Trop Med Hyg 80: 819–823.
18. Naranjo J, Gerba C (2011) Assessment of the LifeStraw Family Unit using the World Health Organization Guidelines for "Evaluating Household Water Treatment Options: Health-based Targets and Performance Specifications". University of Arizona, Department of Soil, Water and Environmental Science.
19. WHO (2011) Evaluating household water treatment options: health-based targets and performance specifications. Geneva: WHO.
20. Adkins E, Tyler E, Wang J, Siriri D, Modi V (2010) Field testing and survey evaluation of household biomass cookstoves in rural sub-Saharan Africa. Energy Sustain Dev 14: 172–185.
21. USAID (2011) In-Home Emissions of Greenhouse Pollutants from Rocket and Traditional Biomass Cooking Stoves in Uganda.
22. Pennise D, Brant S, Agbeve SM, Quaye W, Mengesha F, et al. (2009) Indoor air quality impacts of an improved wood stove in Ghana and an ethanol stove in Ethiopia. Energy Sustain Dev 13: 71–76.
23. Barstow CK, Thomas EA, Ngabo Fidele, Ghislaine R, Majorin F, et al. (2013) Environmental health product delivery and education in rural Rwanda through a public-private partnership. Unpublished.
24. Thomas EA, Barstow CK, Rosa G, Majorin F, Clasen T (2013) Use of Remotely Reporting Electronic Sensors for Assessing Use of Water Filters and Cookstoves in Rwanda. Environ Sci Technol 47: 13602–13610.
25. APHA (2001) Standard Methods for the Examination of Water and Wastewater. 21st Edition. Washington, (DC): APHA.
26. Masera OR, Saatkamp BD, Kammen DM (2000) From Linear Fuel Switching to Multiple Cooking Strategies: A Critique and Alternative to the Energy Ladder Model. World Development 28: 2083–2103.
27. Edwards R, Smith K, Kirby B, Allen T, Litton C, et al. (2006) An inexpensive dual-chamber particle monitor: laboratory characterization. J Air Waste Manag Assoc 56: 789–799.
28. Edwards RD, Liu Y, He G, Yin Z, Sinton J, et al. (2007) Household CO and PM measured as part of a review of China's National Improved Stove Program. Indoor Air 17: 189–203.
29. Chowdhury Z, Edwards RD, Johnson M, Naumoff Shields K, Allen T, et al. (2007) An inexpensive light-scattering particle monitor: field validation. J Environ Monit 9: 1099–1106.
30. Dutta K, Shields KN, Edwards R, Smith KR (2007) Impact of improved biomass cookstoves on indoor air quality near Pune, India. Energy Sustain Dev 11: 19–32.
31. Cynthia AA, Edwards RD, Johnson M, Zuk M, Rojas L, et al. (2008) Reduction in personal exposures to particulate matter and carbon monoxide as a result of the installation of a Patsari improved cook stove in Michoacan Mexico. Indoor Air 18: 93–105.
32. Alexander N (2012) Review: analysis of parasite and other skewed counts. Trop Med Int Health 17: 684–693.
33. McElduff F, Cortina-Borja M, Chan S-K, Wade A (2010) When t-tests or Wilcoxon-Mann-Whitney tests won't do. Adv Physiol Educ 34: 128–133.
34. Kuha J (2004) AIC and BIC: Comparisons of Assumptions and Performance. Sociol Methods Res 33: 188–229.
35. Hunter PR (2009) Household Water Treatment in Developing Countries: Comparing Different Intervention Types Using Meta-Regression. Environ Sci Technol 43: 8991–8997.
36. Peletz R, Simunyama M, Sarenje K, Baisley K, Filteau S, et al. (2012) Assessing Water Filtration and Safe Storage in Households with Young Children of HIV-Positive Mothers: A Randomized, Controlled Trial in Zambia. PLoS One 7: e46548.
37. Hunter PR, Zmirou-Navier D, Hartemann P (2009) Estimating the impact on health of poor reliability of drinking water interventions in developing countries. Sci Total Environ 407: 2621–2624.
38. Brown J, Clasen T (2012) High Adherence Is Necessary to Realize Health Gains from Water Quality Interventions. PLoS One 7: e36735.
39. Enger KS, Nelson KL, Rose JB, Eisenberg JNS (2013) The joint effects of efficacy and compliance: A study of household water treatment effectiveness against childhood diarrhea. Water Res 47: 1181–1190.
40. Ruiz-Mercado I, Masera O, Zamora H, Smith KR (2011) Adoption and sustained use of improved cookstoves. Energy Policy 39: 7557–7566.
41. Burwen J, Levine DI (2012) A rapid assessment randomized-controlled trial of improved cookstoves in rural Ghana. Energy Sustain Dev 16: 328–338.
42. Hanna R, Duflo E, Greenstone M (2012) Up in smoke: The influence of household behaviour on the long-run impact of improved cooking stoves. NBER Working Paper No. 18033. Cambridge, MA: National Bureau of Economic Research.
43. Pine K, Edwards R, Masera O, Schilmann A, Marrón-Mares A, et al. (2011) Adoption and use of improved biomass stoves in Rural Mexico. Energy Sustain Dev 15: 176–183.
44. Manibog FR (1984) Improved Cooking Stoves in Developing Countries: Problems and Opportunities. Annu Rev Energy 9: 199–227.
45. Ochieng CA, Vardoulakis S, Tonne C (2013) Are rocket mud stoves associated with lower indoor carbon monoxide and personal exposure in rural Kenya? Indoor Air 23: 14–24.
46. WHO (2006) Global update 2005: Particulate matter, ozone, nitrogen dioxide and sulfur dioxide, air quality guidelines. Copenhagen: WHO.
47. Smith KR, McCracken JP, Weber MW, Hubbard A, Jenny A, et al. (2011) Effect of reduction in household air pollution on childhood pneumonia in Guatemala (RESPIRE): a randomised controlled trial. Lancet 378: 1717–1726.
48. Fitzgerald C, Aguilar-Villalobos M, Eppler AR, Dorner SC, Rathbun SL, et al. (2012) Testing the effectiveness of two improved cookstove interventions in the Santiago de Chuco Province of Peru. Sci Total Environ 420: 54–64.
49. Zwane AP, Zinman J, Van Dusen E, Pariente W, Null C, et al. (2011) Being surveyed can change later behavior and related parameter estimates. Proc Natl Acad Sci U S A.
50. Naeher LP, Brauer M, Lipsett M, Zelikoff JT, Simpson CD, et al. (2007) Woodsmoke health effects: A review. Inhal Toxicol 19: 67–106.

11

Temporary Storage or Permanent Removal? The Division of Nitrogen between Biotic Assimilation and Denitrification in Stormwater Biofiltration Systems

Emily G. I. Payne[1]*, Tim D. Fletcher[2], Douglas G. Russell[3], Michael R. Grace[3], Timothy R. Cavagnaro[4], Victor Evrard[3], Ana Deletic[1], Belinda E. Hatt[1], Perran L. M. Cook[3]

1 Monash Water for Liveability, Department of Civil Engineering, Monash University, Victoria, Australia, **2** Department of Resource Management and Geography, Melbourne School of Land and Environment, The University of Melbourne, Victoria, Australia, **3** Water Studies Centre, School of Chemistry, Monash University, Victoria, Australia, **4** School of Biological Sciences, Monash University, Victoria, Australia

Abstract

The long-term efficacy of stormwater treatment systems requires continuous pollutant removal without substantial re-release. Hence, the division of incoming pollutants between temporary and permanent removal pathways is fundamental. This is pertinent to nitrogen, a critical water body pollutant, which on a broad level may be assimilated by plants or microbes and temporarily stored, or transformed by bacteria to gaseous forms and permanently lost via denitrification. Biofiltration systems have demonstrated effective removal of nitrogen from urban stormwater runoff, but to date studies have been limited to a 'black-box' approach. The lack of understanding on internal nitrogen processes constrains future design and threatens the reliability of long-term system performance. While nitrogen processes have been thoroughly studied in other environments, including wastewater treatment wetlands, biofiltration systems differ fundamentally in design and the composition and hydrology of stormwater inflows, with intermittent inundation and prolonged dry periods. Two mesocosm experiments were conducted to investigate biofilter nitrogen processes using the stable isotope tracer $^{15}NO_3^-$ (nitrate) over the course of one inflow event. The immediate partitioning of $^{15}NO_3^-$ between biotic assimilation and denitrification were investigated for a range of different inflow concentrations and plant species. Assimilation was the primary fate for NO_3^- under typical stormwater concentrations (~1–2 mg N/L), contributing an average 89–99% of $^{15}NO_3^-$ processing in biofilter columns containing the most effective plant species, while only 0–3% was denitrified and 0–8% remained in the pore water. Denitrification played a greater role for columns containing less effective species, processing up to 8% of $^{15}NO_3^-$, and increased further with nitrate loading. This study uniquely applied isotope tracing to biofiltration systems and revealed the dominance of assimilation in stormwater biofilters. The findings raise important questions about nitrogen release upon plant senescence, seasonally and in the long term, which have implications on the management and design of biofiltration systems.

Editor: Marie-Joelle Virolle, University Paris South, France

Funding: This research was supported by funding from the Australian Research Council (LP0990153)(http://www.arc.gov.au/), Melbourne Water (http://www.melbournewater.com.au/Pages/home.aspx), and Western Australia Department of Water (http://www.water.wa.gov.au/). Fletcher is supported by an ARC Future Fellowship (FT100100144). The funders assisted with the study design and discussions of overall project results but had no role in data collection and analysis, decision to publish, or preparation of the manuscript.

Competing Interests: The authors have declared that no competing interests exist.

* E-mail: emily.payne@monash.edu

Introduction

The performance of stormwater biofilters (also known as bioretention systems or raingardens) has traditionally been expressed in terms of simple pollutant removal. Few studies consider the permanency of this removal, yet many processes in such systems may be better described as attenuation - when retention is only temporary and the pollutant is at some point re-released, either in its original or transformed state. The fate of a pollutant between temporary and permanent removal pathways is fundamental to long-term performance. Nitrate is a critical waterway pollutant with possible transformations in both of these categories – biotic assimilation provides temporary immobilization, or denitrification offers permanent removal in gaseous form. While nitrogen transformation and cycling processes have been characterised across wide natural and engineered environments, they have not been explicitly quantified in the unique conditions of stormwater biofilters. This leaves the long-term efficiency of biofilter nitrogen treatment open to question and constrains the potential for future design improvements.

Biofiltration typically consists of a vegetated layer of sandy loam overlying sand and gravel layers, designed to capture, infiltrate and treat urban stormwater runoff before discharge downstream or into the surrounding environment or collection for harvesting [1,2]. Like wastewater treatment wetlands, biofilters are engineered systems which harness natural biogeochemical processes. However, biofilters differ fundamentally from wetlands as a result of stormwater inflows and infiltration. While biofilters share some common design features with vertical flow wetlands, they are distinguished by being ephemeral, fed by urban intermittent stormwater runoff, which differs substantially from wastewater in composition and inflow hydrology [3,4]. This leads to large, irregular variances in inundation, soil moisture and potentially nutrient, carbon and oxygen availability. As a result, biofilters are typically vegetated with terrestrial and semi-terrestrial plant species. Such differences likely alter the dominant nitrogen processes and drivers between treatment wetlands and stormwater biofilters. Characterising pollutant fate within stormwater biofilters is necessary not only for the optimal design of systems, but also to understand their long-term performance and determine suitable maintenance regimes.

Nitrogen is an essential nutrient in all biomass, but its natural cycling has been substantially altered by anthropogenic inputs and as a result forms a major contaminant of surface and ground waters [5]. Consequently, nitrogen processing has been extensively studied across terrestrial, semi-terrestrial and aquatic environments. This knowledge can be applied to infer possible nitrogen removal pathways in stormwater biofilters. Incoming nitrogen associated with urban stormwater runoff may undergo a range of potential fates, including assimilation, transformation by microbial processes (including nitrification, denitrification, dissimilatory nitrate reduction to ammonium (DNRA)), abiotic processes (including filtration and adsorption), or leaching from the system [6]. Based on research in other environments, the key fates for nitrate, a mobile inorganic form of nitrogen, are expected to be biotic assimilation (uptake by plants, bacteria, fungi or other microbes) and denitrification (conversion into gaseous forms primarily N_2 or N_2O) [7]. Assimilated nitrate is subsequently converted into an array of organic compounds and stored for some period, before return upon cell death or exudation. Decomposition processes act to either lock nitrogen up for longer term storage in recalcitrant components of the soil organic matter or re-release, when it is again available for uptake, transformation or leaching [6]. Hence, in many environments temporary storage from assimilation can occur over days, years, decades and beyond [6,8]. However, in biofiltration systems concentrations of organic matter are initially very minimal in the engineered sandy substrate [9], and it is unknown if a pool will develop to provide significant long-term nitrogen storage. Other key nitrogen processes include nitrification and denitrification, which are both mediated by microbes, but require contrasting redox conditions. Many biofilter designs incorporate an upper drained layer underlain by a saturated layer, maintained using a raised outlet, which may theoretically provide zones for nitrification and denitrification respectively. A supplementary carbon source (e.g. wood chips, straw) is often mixed throughout the saturated zone to provide electrons for denitrification [10]. Despite increased nitrate removal associated with these design features [11,12], to date no study has confirmed that this is due to denitrification. Without definitively relating design features to nitrogen processes, designers may be blind to opportunities for future performance enhancement. While biofilter design has progressed substantially over the past decade [9,13], further improvements may be confined by the current lack of process knowledge. This is particularly the case for nitrogen, where the cyclical nature of assimilation and mineralisation – internal recycling – threatens to eventually overwhelm the demand for incoming nitrogen.

The removal performance quoted by most biofilter studies, which use a simple black-box input-output approach, is a reflection of predominantly short-term processes, ignoring possible long-term changes. While authors acknowledge the potential for re-mobilisation and leaching of previously attenuated nitrogen [14,15], and several have investigated pollutant profiles [2,16] or estimated plant accumulation [15,17], none have explicitly characterised nitrogen fate to account for re-release. An initial step towards this understanding is determining the immediate fate of incoming nitrogen. While such an assessment focuses on rapid processing in a short timeframe, it has implications on longer term dynamics, given that a system cannot indefinitely accumulate nitrogen [18].

Assimilation and denitrification have been quantified in a range of engineered and natural systems. Denitrification has been reported to dominate processing in wastewater treatment wetlands [7,19], wetlands treating high nitrate groundwater [20] and riparian soils receiving agricultural runoff [21]. In contrast, other studies have noted, sometimes surprisingly, the key role played by plant uptake in environments including streams [22], peat bogs [23], flooded soils planted with wetland species [24] and grassed buffer strips [25]. Factors driving the division may include carbon and nitrogen availability, nitrogen speciation, sediment characteristics, plant species morphology and physiology, plant density and hydrological regime [26,27]. The dominant pathways have not been identified in biofilters, yet these influential factors may differ substantially under the characteristics of stormwater biofilters.

Given the 'black-box' approach of most biofiltration research to date, this experiment was designed to provide an initial investigation into nitrogen processes. The experiment aimed to quantify the immediate nitrate transformation pathways in stormwater biofilters, focusing upon the initial division between assimilation and denitrification, with the specific objectives to investigate the:

(1) Effect of TN influent concentration on nitrate removal pathways in biofilter mesocosms planted with *Carex appressa*
(2) Effect of plant species on nitrate removal pathways within laboratory-scale biofilters

This study uniquely applied isotope tracing techniques to stormwater biofilters in order to quantify processes, an approach which, to the authors' knowledge, has not previously been reported. In addition, few studies in other fields have applied isotope tracer across plant species, despite considerable interaction between species and nitrogen cycling [28]. Quantification across a loading gradient provides a basis to understand how results might vary across systems and inform comparisons. Overall, the study aimed to indicate if nitrogen may accumulate within components of the biofilter over time, which has critical implications for system design, maintenance and life span.

Methods

Experimental overview

This study is made up of two components. The first tested the effect of influent concentration on nitrate removal pathways in mesocosms planted with *Carex appressa*, selected as a high performing species in biofilters [14,29]. In the second, nitrogen pathways were tracked in biofilter columns planted with various species. Both experiments use a laboratory-scale approach to provide insight into biofilter processes under controlled conditions, and include non-vegetated controls.

Key differences between the experiments include their establishment period, depth and configuration of the saturated zone, replicate number, and the mixture of nitrogen species added (Table 1 and Figure 1). The *Carex appressa* influent concentration experiment was conducted under fully saturated conditions (Figure 1(a)), providing a simplified design to specifically investigate the effect of nitrate loading on biofilter saturated zone processes, which have been hypothesized as of primary impor-

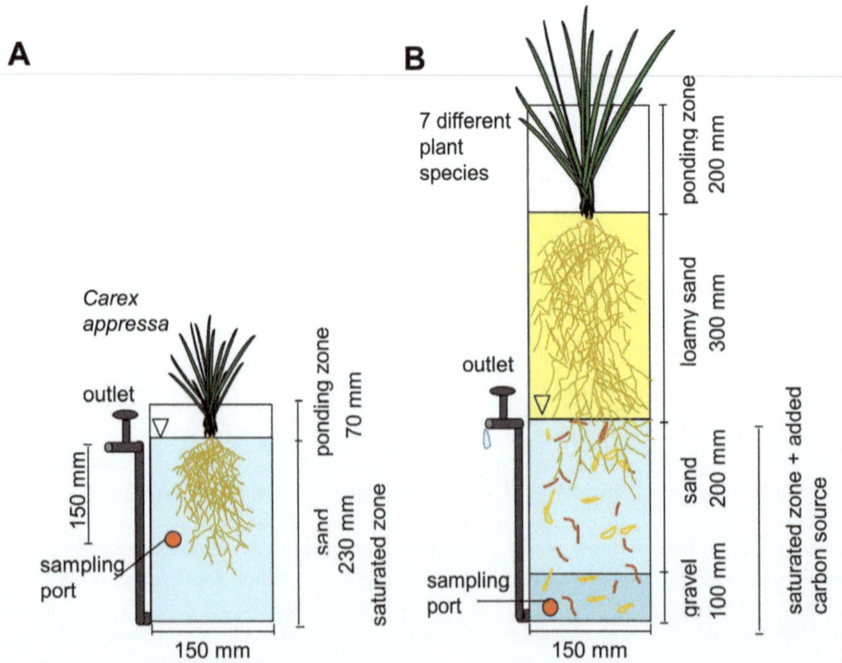

Figure 1. Experiment Configuration. A.) Influent concentration experiment (under fully saturated conditions) and B.) Multiple plant species experiment (with saturated zone overlaid by a non-saturated zone). Note diagrams are not drawn to scale.

tance in biofilter nitrogen removal performance [10,12]. The experiment across plant species was established over a much longer period of plant growth and stormwater application, and incorporated an unsaturated layer above the saturated zone (typical of most vertical-flow stormwater biofilters, Figure 1 (b)).

Experimental setup

(i) Influent concentrations experiment. Twenty mesocosms (4 replicates for each tested nitrogen inflow concentration) were constructed using 150 mm diameter polyvinyl chloride (PVC) pipe and 230 mm of washed landscaping sand. Access holes were drilled and blocked using 12 mm butyl septa to allow sampling. A constant height of water was maintained to saturate the entire sediment. Two week old *Carex appressa* seedlings were planted into 16 of the mesocosms and grown for 8 weeks in a greenhouse alongside 4 non-vegetated controls.

(ii) Multiple plant species experiment. Twenty-four single-plant biofilter columns (3 replicates for each of the 7 tested plant species and non-vegetated control) were grown in an open-air roofed greenhouse for a period of 11 months. The plants were originally sourced as tubestock and established in planter bags (300 mm by 150 mm) using tapwater for 6 months. The columns were constructed using PVC pipe with a clear Perspex ponding zone and filled with loamy sand overlying sand and gravel layers. A raised outlet tap allowed a 300 mm saturated zone with a carbon source of pine chips and sugar cane mulch mixed throughout to form 5% by volume (Figure 1), typical of common stormwater biofilter design guidelines [9]. Seven plant species (Table 1) were selected to cover a range of plant forms and include both high and poor performing species for nitrogen removal in biofilters, determined from previous sampling (not presented here). These columns formed part of a broader 18 month study, which investigated nitrogen cycling across plant species and design configurations.

Dosing and sampling

(i) Influent concentrations experiment. The mesocosms were dosed with varying concentrations of the modified Long-Ashton nutrient solution containing a 1:1 ratio of $Na^{14}NO_3$:-$Na^{15}NO_3$ ($Na^{15}NO_3$>98% isotopic purity, Cambridge Isotope Laboratories), as described in Cavagnaro et al. [30]. The molar amount of NO_3^- was matched with NH_4^+. Four vegetated mesocosms and four non-vegetated control mesocosms received the equivalent of 1 mg N/L (NO_3^- and NH_4^+) and 0.3 mg P/L. This comprised the 'Low' inflow concentration. The remaining vegetated mesocosms were dosed with 'Medium' = 2 mg N/L, 0.6 mg P/L, 'High' = 10 mg N/L, 2.8 mg P/L and 'Very high' = 20 mg N/L, 5.6 mg P/L concentrations. This range of nutrient concentrations covers the typical range of urban stormwater runoff (1–3 mg N/L) [3] and beyond, towards concentrations more typical of wastewater. A weekly dose of 1.63 L was applied, which flushed the mesocosm by one pore volume based on a sand porosity of 0.4.

Before tracer addition, samples were collected from each inflow solution and the pore water, and analysed for background concentrations of NH_4^+, NO_3^-, N_2 and N_2O. Pore water samples were removed using a 25 mL plastic syringe with an 18-gauge needle. Samples for NH_4^+ and NO_3^- were filtered through a 0.45 μm membrane filter (MicroAnalytix 30PS045AN) into a 12 mL container and frozen for subsequent analysis. Water samples for analysis of N_2 and N_2O concentrations were transferred into a 12.5 mL glass gas-tight vial (Exetainer, Labco). Zinc chloride (250 μL, 50% w/v) was added as a preservative [31]. This sampling regime was repeated 6 hourly for a period of 30 hours. Two pulses of isotope tracer were added, one in July and another in August 2012.

(ii) Multiple species experiment. The columns were dosed with semi-natural stormwater twice weekly, following the method outlined in Bratieres et al. [14] with a target Total Suspended Solids concentration of 150 mg/L and nutrient concentrations

Table 1. Summary of experiment details.

Experiment	1. Influent concentration	2. Multiple plant species
Description	Tested effect of 4 different influent N and P concentrations on NO_3^- partitioning between denitrification, pore water and vegetation, using single plant species *Carex appressa* and non-vegetated controls.	Tested effect of 7 different plant species and non-vegetated control on NO_3^- partitioning between denitrification, pore water and vegetation using constant influent composition of 'typical' stormwater.
	150 mm diameter PVC pipe containing 230 mm washed sand with constant saturation maintained.	150 mm diameter PVC pipe with layers of loamy sand, sand and gravel (Figure 1) across 600 mm. Saturated conditions maintained in lower 300 mm with sugar cane mulch and pine chips mixed throughout.
Stormwater application	Weekly dose of 1.63 L (~one total pore volume)	Twice-weekly dose of 3.7 L (to Vic plant species, see below) and 4.2 L (to WA plant species) in accordance with local rainfall (~one total pore volume)
	1 mg N/L, 0.3 mg P/L (non-vegetated control and 'low' dose)	~2.2 mg N/L, 0.36 mg P/L (all columns)
	2 mg N/L, 0.6 mg P/L ('medium')	
	10 mg N/L, 2.8 mg P/L ('high')	
	20 mg N/L, 5.6 mg P/L ('very high')	
	Modified Long-Ashton nutrient solution [30] – (~50% NO_3^-,50% NH_4^+ by weight)	Semi-natural urban stormwater with 'typical' components [3,32] – (~45% NO_3^-, 18% NH_4^+, 37% organic N by weight), ~150 mg/L TSS
	Tracer added twice – July and August	Tracer added once - October
Plant species	*Carex appressa* (sedge)	*Carex appressa* (sedge; Vic)
		Palmetto Soft Leaf Buffalo (lawn grass/Vic)
		Dianella revoluta (sedge; Vic)
		Juncus kraussii (reed; WA)
		Allocasurina littoralis (tree; Vic)
		Leptospermum continentale (tree; Vic)
		Hypocalymma angustifolium (tree; WA)
Initial plant growth period	8 weeks	>17 months (includes 11 months in columns with twice-weekly of stormwater application)
Location	Controlled greenhouse	Open-air roofed greenhouse
Number of replicates	4 replicates	3 replicates

based on typical stormwater composition [3,32]. The average nitrogen composition delivered is shown in Table 1 and comprised approximately 2.2 mg N/L and 0.36 mg P/L. The dose volume reflected a biofilter sized to 2.5% of its impervious catchment area, a twice weekly watering frequency and the annual average effective rainfall for Melbourne (Victoria) and Perth (Western Australia). Columns with Victorian (Vic) species were dosed with 3.7 L and Western Australian (WA) species with 4.2 L (Table 1). Buffalo, a common lawn grass, was cut regularly using scissors to simulate mowing. Following 11 months of stormwater dosing, the influent was enriched with $Na^{15}NO_3^-$ in October 2011. Inflow samples were collected before and after tracer addition. Pore water samples were collected approximately 1.5 cm from the base of the saturated zone and processed as described for the previous experiment. An O_2 minisensor (2.5 mm tip diameter) connected to a FireSting O_2 oxygen meter (PyroScience GmbH, Germany) immediately measured dissolved oxygen concentrations in the sample. This sampling procedure was validated in the laboratory by collecting samples from anoxic water (created by 20 minutes of Argon gas bubbling). The anoxic water recorded an average of 1.4% air saturation (±0.1 standard deviation), but following sample collection using a syringe, plastic tubing and exetainer, the average dissolved oxygen concentration was 5.4% air saturation (±0.4), indicating the sampling procedure introduced a small amount of O_2. Therefore, samples recording around 7% air saturation or lower can be considered anoxic. Sampling using a continuous oxygen probe was attempted but the fragile probes were repeatedly damaged when inserted into the biofilters. It should also be noted measurement of pH was not deemed necessary as stormwater influent and effluent from similar laboratory biofilters has been previously measured near neutral [33], CO_2 production acts as a buffer against potentially low pH and denitrifiers operate under a wide range of conditions [34].

Six sets of samples were collected across 5 days from 25th–29th October 2011. Samples were collected initially as each column finished draining, the next day in the morning and afternoon, then daily in the afternoon and following the next stormwater dosing (4 days after tracer addition).

The effluent from each column was sampled on 2nd November 2011 to determine concentrations of total nitrogen (TN), total phosphorus (TP), filterable reactive phosphorus (FRP), total dissolved nitrogen (TDN), NH_4^+ and NO_x, as per the method outlined in Payne et al. [26].

Laboratory analyses - NH_4^+, NO_3^-, N_2 and N_2O

Concentrations of NH_4^+ and NO_x in the pore water were analysed using standard flow injection analysis methods in a NATA (National Association of Testing Authorities) accredited lab. Dissolved N_2O, $^{28}N_2$, $^{29}N_2$ and $^{30}N_2$ were quantified in the water samples after a 4 mL He (99.9%) headspace was placed in the vials and equilibrated by vigorously shaking for 5 minutes. The concentration of N_2O was analysed by injecting a 100 μL sample of the headspace by gas chromatography (Hewlett Packard 5710A Gas Chromatograph). The total amount of N_2O in the vials was calculated using Henry's law [35]. N_2 was analysed on an ANCA GSL2 elemental analyser coupled to a Hydra 20–22 isotope ratio mass-spectrometer (IRMS; Sercon Ltd., UK).

Data analysis

$^{28}N_2$, $^{29}N_2$ and $^{30}N_2$ production and denitrification rates. A linear regression was fitted to the amount of excess $^{29}N_2$ and $^{30}N_2$ over time to calculate a production rate. The rate was adjusted to compensate for the loss rate, determined from a linear regression across the decline in labelled N_2 in the later portion of the time series. The rate of ^{15}N denitrification (D_{15}), ^{14}N denitrification (D_{14}), the proportion of denitrification coupled to nitrification and total denitrification (D_{total}) were calculated from the rates of $^{30}N_2$ and $^{29}N_2$ production, p $^{30}N_2$ and p $^{29}N_2$, respectively and the ratio of $^{14}NO_3^-/^{15}NO_3^-$ [31].

Mass balance. The mass balance was calculated over a period of 12 hours for all treatments, consistent with the period in which denitrification occurred in the *Carex appressa* mesocosm study (illustrated later by the rapid rise and peak in $^{30}N_2$ and $^{29}N_2$ in Figure 2). The total amount of ^{15}N denitrified was calculated by integrating the rate of denitrification. This estimate of denitrification is expected to be conservative (i.e. an overestimate) given denitrification does not commence until anaerobic conditions establish, but achieved our objective of determining the maximum denitrification occurring within the systems. Both experiments were dosed with approximately one pore volume, resulting in negligible loss of ^{15}N as outflow. The amount of $^{15}NO_3^-$ remaining in the pore water was calculated using the input ratio ^{15}N:$^{14+15}N$ and the pore water NO_3^- concentration at 12 hours. If no sample was collected at this point, a concentration was linearly interpolated across the surrounding sampling time points. The proportion of $^{15}NO_3^-$ assimilated was calculated as the difference between the total amount of $^{15}NO_3^-$ added and the amount denitrified and remaining pore water $^{15}NO_3^-$.

Correlation and Significance Analysis. Pearson's product-moment correlation was used to determine if there was a relationship between the proportions assimilated and denitrified, and the proportions denitrified and remaining in the pore water. As the data were non-normally distributed in some cases, the use of Spearman rank correlation was used to confirm these. We tested for significant differences in assimilation between species, and in denitrification rate between nitrogen loadings using the non-parametric Kruskal-Wallis test. A critical value of $\alpha = 0.05$ was used for hypothesis tests. Analyses were performed using the car Package [36] within the R Software Environment [37]. Michaelis-Menten curves were fitted using the drc Package [38] in R.

Results

Effect of TN influent concentration on nitrate removal pathways in *Carex appressa* mesocosms

Concentrations of NH_4^+ and NO_3^- in the pore water decreased rapidly within 24 hours in all treatments, except the non-vegetated controls which produced NO_3^-, as a result of significant nitrification. Consistent with the decline in NO_3^- in all vegetated treatments, $^{29}N_2$ and $^{30}N_2$ were produced over the first 12 hours after dosing, followed by a decrease in their concentrations thereafter (Figure 2 demonstrates these patterns for the very high nutrient treatment). Coupled nitrification-denitrification comprised approximately one third of denitrification in the high treatment. The production of N_2O in the pore water represented an insignificant fraction of the nitrogen budget (<1%) and was therefore not considered further.

Rates of denitrification were extremely low in the control and low nutrient dosed columns (<25 μmol m^{-2} h^{-1}), but increased sharply with higher nutrient loading, reaching a maximum rate in the high treatments (600 and 1800 μmol m^{-2} h^{-1}, during July and August respectively) before decreasing in the very high treatment (Figure 3). There were significant differences in denitrification rate between the treatments in both July and August ($p<0.05$).

There was negligible assimilation or denitrification of nitrate in the non-vegetated control columns; most nitrogen was recovered

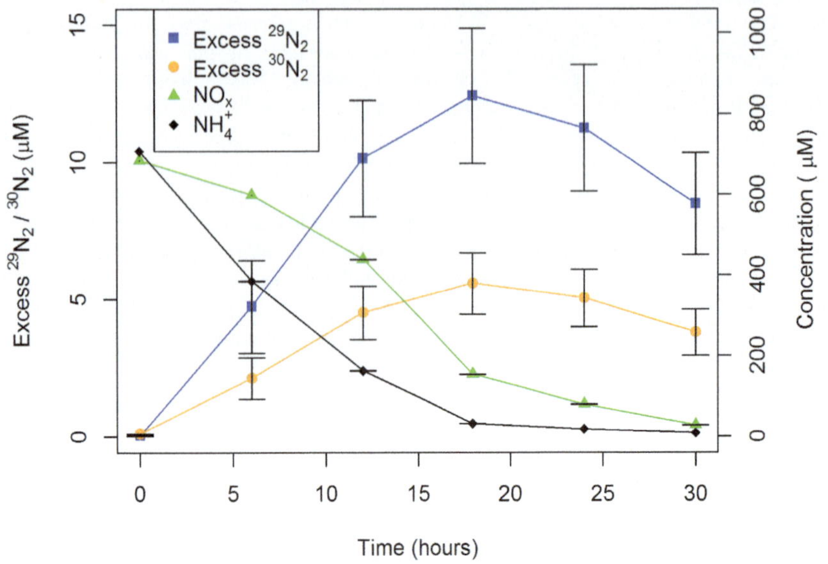

Figure 2. Nitrogen species concentrations. Examples of time series NH_4^+, NO_x, excess $^{29}N_2$ and $^{30}N_2$ concentrations (± standard error (n = 4)) following dosing in the influent concentration experiment under very high nutrient dosing (20 mg N/L) measured in July 2012.

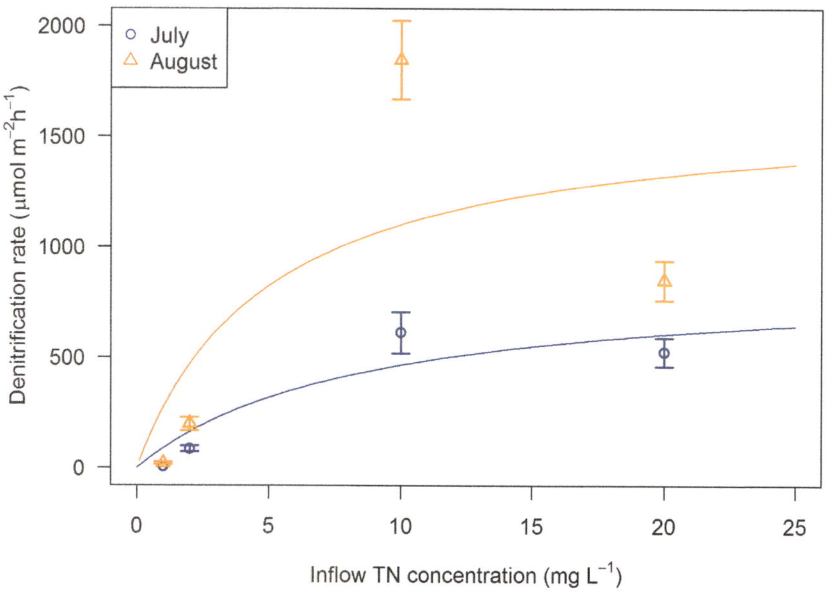

Figure 3. Rates of denitrification (^{14}N+^{15}N) against inflow TN concentration. Measured in the influent concentration experiment (± standard error (n = 4)) during July and August. Michaelis-Menten curves were fitted to give V_{max} = 861 μmol m^{-2} h^{-1} and K_m = 8.46 mg L^{-1} in July and V_{max} = 1653 μmol m^{-2} h^{-1} and K_m = 5.01 mg L^{-1} in August.

as NO_3^- after 12 hours (Figure 4). In the vegetated columns, assimilation dominated in the low dose mesocosms, but generally decreased with higher loading. The fraction of $^{15}NO_3^-$ denitrified increased with nitrogen loading alongside the proportion remaining as nitrate. However, in the very high treatment denitrification declined and assimilation increased again.

Effect of plant species on nitrate removal pathways within biofilters

The vegetated biofilter columns effectively reduced concentrations of TN and TP in the stormwater from 2.1 mg N/L and 0.31 mg P/L to averages of 0.27 mg N/L and 0.01 mg P/L respectively. NH_4^+ concentration reductions were high irrespective of plant species or the presence of vegetation, reduced from 0.4 mg N/L to <0.05 mg N/L. NO_x removal was also high but more variable; effluent concentrations ranged from 0.001–0.27 mg N/L from an influent concentration of 1.0 mg N/L.

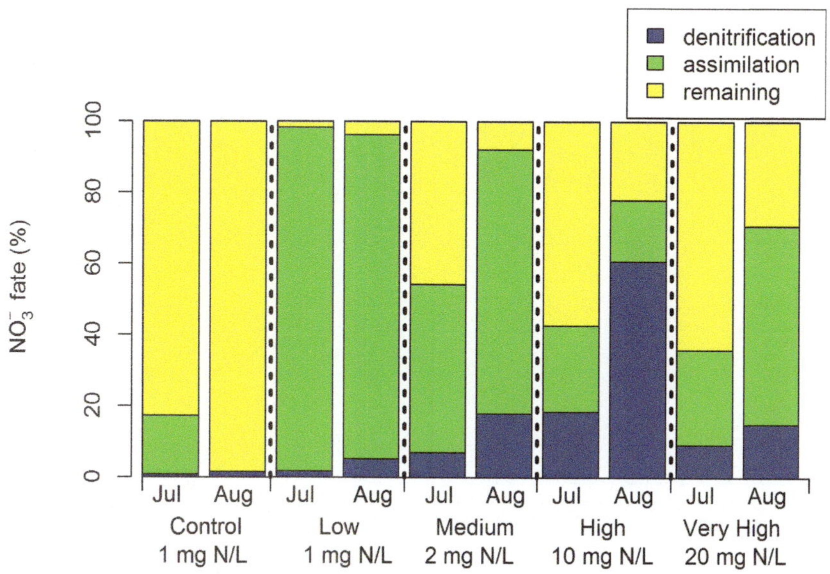

Figure 4. Division of $^{15}NO_3^-$ between denitrification, plant or microbial assimilation and remaining as $^{15}NO_3^-$ within the pore water. Measured 12 hours after dosing during July and August in the control (non-vegetated) and low, medium, high and very high (vegetated) nutrient dosing rates (n = 4).

Removal of organic nitrogen was also effective with dissolved and particulate forms on average reduced from 0.4 mg N/L and 0.3 mg N/L to 0.12 mg N/L and 0.05 mg N/L respectively. The non-vegetated controls were less effective, with outflow concentrations averaging 1.47 mg N/L TN and a net production of NO_x to 1.27 mg N/L.

Similar patterns for NO_3^-, $^{29}N_2$ and $^{30}N_2$ illustrated by Figure 2 for the pore water were also evident for the multiple species column experiment. Concentrations of NO_3^- declined rapidly within 24 hours of tracer addition in all vegetated treatments, but remained elevated with some production in the non-vegetated controls, again indicating nitrification in these systems similarly to the mesocosm experiment. Concentrations of $^{29}N_2$ and $^{30}N_2$ increased rapidly, generally peaking between 25 and 45 hours, but continued to increase in the non-vegetated controls. N_2O production was again minimal (<1% of $^{15}NO_3^-$) and was thus ignored in the mass balance. Concentrations of dissolved oxygen within the saturated zone rapidly declined towards anoxia, with a sharp decline measured across the first 22 hours for all species excluding Buffalo grass and the non-vegetated control which demonstrated a slower decline (Figure 5). Given the introduction of a small amount of oxygen during sample collection (up to ~7% air saturation) and the expected commencement of denitrification <0.5 mg/L (or approximately <5% air saturation) [39], appropriate conditions for denitrification were considered to have occurred for most columns within 22 hours. In addition, the concentrations of dissolved oxygen in vegetated columns were significantly lower than in the non-vegetated columns across the first and second sampling times (collected an average of 5 hours and 22 hours after tracer addition respectively).

Biotic assimilation was the primary fate of $^{15}NO_3^-$ in all vegetated columns; ranging between 58% and 100% of $^{15}NO_3^-$ (Figure 6). While an average of 88% (±7% standard error, n = 3) of $^{15}NO_3^-$ was assimilated, individual species differed significantly ($p<0.05$). The lowest uptake was associated with columns planted with *Dianella* and *Hypocalymma* which assimilated 58–80% and 69–85% respectively.

Assimilation and denitrification were inversely related (Supplementary information, Figure S1, $r = -0.79$, $p<0.05$). Denitrification was only a minor removal mechanism in the vegetated columns, providing a sink for, on average, only 3% (±2%) of $^{15}NO_3^-$ and ranging to a maximum of 5 to 8% across columns planted with *Dianella* and *Hypocalymma* species. These same treatments also had a greater proportion of $^{15}NO_3^-$ remaining in the pore water, demonstrated by a positive relationship between denitrification and nitrate remaining for vegetated columns (Figure S2, $r = 0.66$, $p<0.05$). In the non-vegetated controls assimilation was low, accounting for 19 to 26% of $^{15}NO_3^-$ fate. Instead $^{15}NO_3^-$ primarily remained in the pore water, and 3 to 4% was denitrified.

Discussion

Assimilation as a key biofilter pathway

Denitrification only formed a minor removal mechanism at typical stormwater concentrations in the biofilter columns and mesocosms. Biotic assimilation (uptake by plants and microbes) functioned as the major sink for incoming nitrate. The minimal contribution from denitrification is somewhat unexpected, given the focus on designing biofilters to promote denitrification [10,12] and the dominance of denitrification in many treatment wetlands and some aquatic systems [7,40]. Denitrification has been reported to account for 60–95% of removal in wastewater treatment wetlands [19], 89–96% in wetlands treating high nitrate groundwater [20], and up to 61–63% in riparian wetland soils treating agricultural runoff [21]. Denitrification can also be a critical process in semi-aquatic and terrestrial systems, including the soils of urban retention basins and parks [41].

However, the critical role of assimilation has also been noted in many studies, where plant and microbial assimilation make a significant, if not dominant, contribution to the removal of nitrate. This has been observed across riparian zones [42], flooded soils planted with wetland plants [24] and vegetated streams [22]. Assimilation accounted for 75% of nitrate retention in headwater

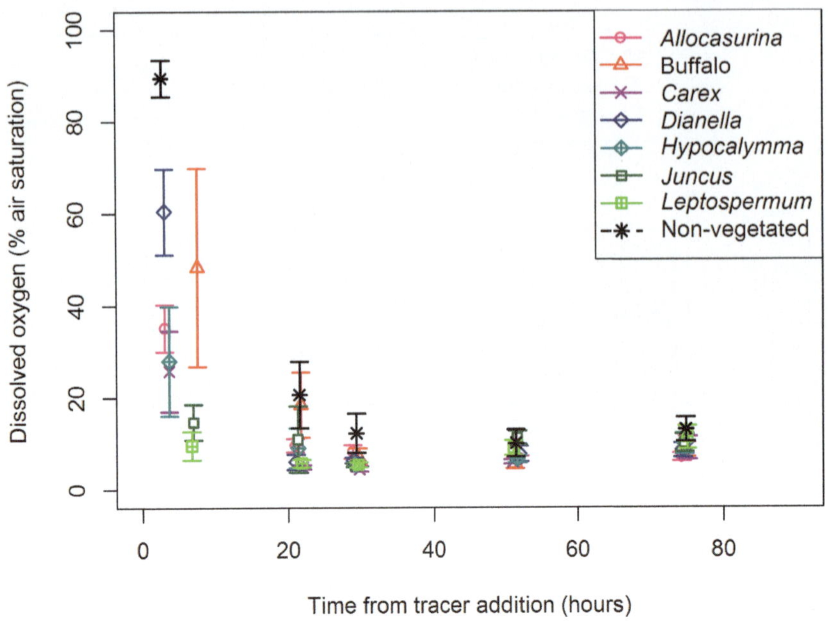

Figure 5. Change in pore water dissolved oxygen. Dissolved oxygen (% air saturated) (± standard error (n = 3) at base of columns across sampling period. Note the sample collection method introduced up to ~7% air saturation.

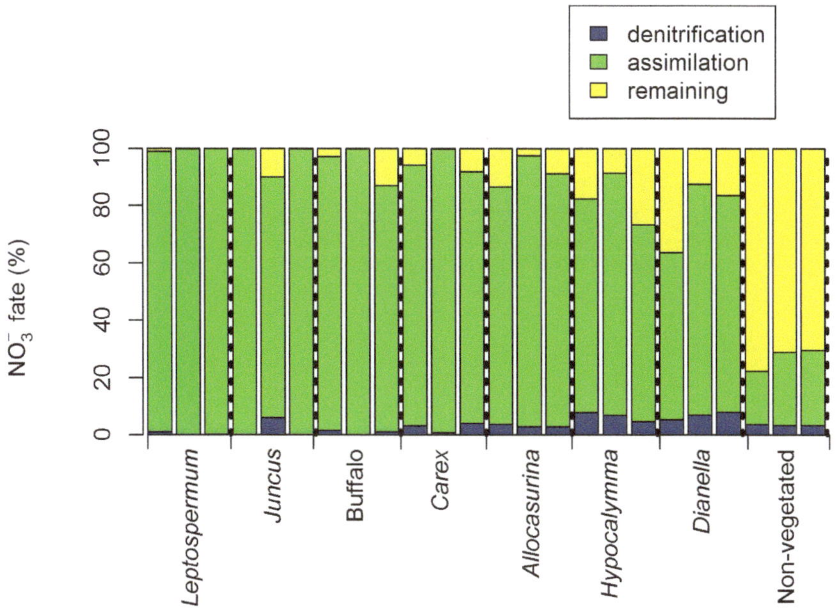

Figure 6. Division of $^{15}NO_3^-$ between denitrification, plant or microbial assimilation or remaining in the pore water. Results for each biofilter column in the multiple species experiment (3 replicates per species, n = 1).

streams [22], almost all nitrate deposited atmospherically on peat bogs [23] and was the primary removal mechanism in grassed buffers treating agricultural irrigation runoff [25].

The variation in nitrate fate between wastewater treatment wetlands and biofilters may result from their fundamental differences. While both are high nutrient, engineered and vegetated systems, biofilters generally experience greater moisture fluctuations, including prolonged drying. Hence, even with an underlying saturated zone, biofilter redox potential is dynamic. For optimal hydraulic and nutrient performance biofilter media is designed to be relatively free-draining with low organic matter or clay content [2]. As a result the system characteristics differ greatly from organic-rich and anoxic wetland sediments. Vertical subsurface flow wetlands (which operate on similar principles to biofilters) can show particularly low denitrification relative to other treatment wetlands due to oxygenated conditions in the sediment [7], and this may similarly limit denitrification in the saturated zone of biofilters. While anoxic conditions develop (Figure 5), the influent is oxygenated.

It is questionable if assimilation will remain the dominant pathway throughout the biofilter lifespan. Denitrification may increase as organic matter accumulates [41,43] and the uptake capacity of the plant biomass may diminish over time [27,28]. The division between assimilation and denitrification will depend upon the magnitude of nitrogen immobilised in organic material and the availability of carbon, oxygen and nitrate. These dynamics may be sensitive to plant species and nitrate loading, as discussed further in the following sections.

In contrast to nitrate, ammonia reduction from the stormwater was effective regardless of the presence of vegetation or plant species. Removal processes included uptake, nitrification, and there was evidence of some coupled nitrification-denitrification in vegetated treatments.

Adaptability of denitrifying bacteria

The importance of denitrification as a removal mechanism increased as nitrogen concentrations rose towards those of wastewater (~10 mg N/L). The results suggest that the denitrifying bacteria are largely utilising the nitrate remaining after assimilation and that this portion grows as loading increases (Figure 7). In contrast, plant assimilation diminished in proportion – likely becoming constrained by other growth requirements or uptake rate after a critical point [7,44]. Hence, denitrifying bacteria appear to have the adaptability to increase nitrate processing to a greater extent than plant and microbial assimilation, but remain dependent upon plant-derived carbon. As nitrate loading increases, the key role of plants may shift from assimilation to facilitation of denitrification.

The *Carex apressa* mesocosms represented a simplified experiment designed to investigate the influence of loading rate on the biofilter saturated zone; this experiment lacked the upper unsaturated layer or longer term loading of the multiple species biofilter column experiment, However, the consistency between the two experiments in terms of the proportion denitrified (7–18% mesocosms and 0–8% vegetated biofilter columns) and denitrification rate (88–202 μmol m^{-2} hr^{-1} and <1–635 μmol m^{-2} hr^{-1} respectively), when the same nitrate concentration was applied, suggests the results across the loading gradient can be meaningfully applied to biofiltration systems.

Other studies have also noted increases in denitrification across a loading gradient, but findings differ in the nature of this increase and relative change in assimilation. Denitrification and assimilation may both increase with loading, either linearly [22], or assimilation may increase at a much lower rate than denitrification [43], or denitrification rates may plateau (e.g. at loadings of around 5 mg N/L) [40,44]. In the current study the denitrification rate appeared to increase rapidly, but reach a plateau, before declining at the highest loading rate (20 mg N/L comprising 10 mg NO_3^--N/L). This may be due to a lower investment in root biomass by plants under the nutrient-rich conditions, which may result in less carbon to drive denitrification. Alternatively, the population of denitrifying bacteria may be inhibited by their generally facultative nature or other factors such as plant allelopathy, preventing optimal functioning in response to higher

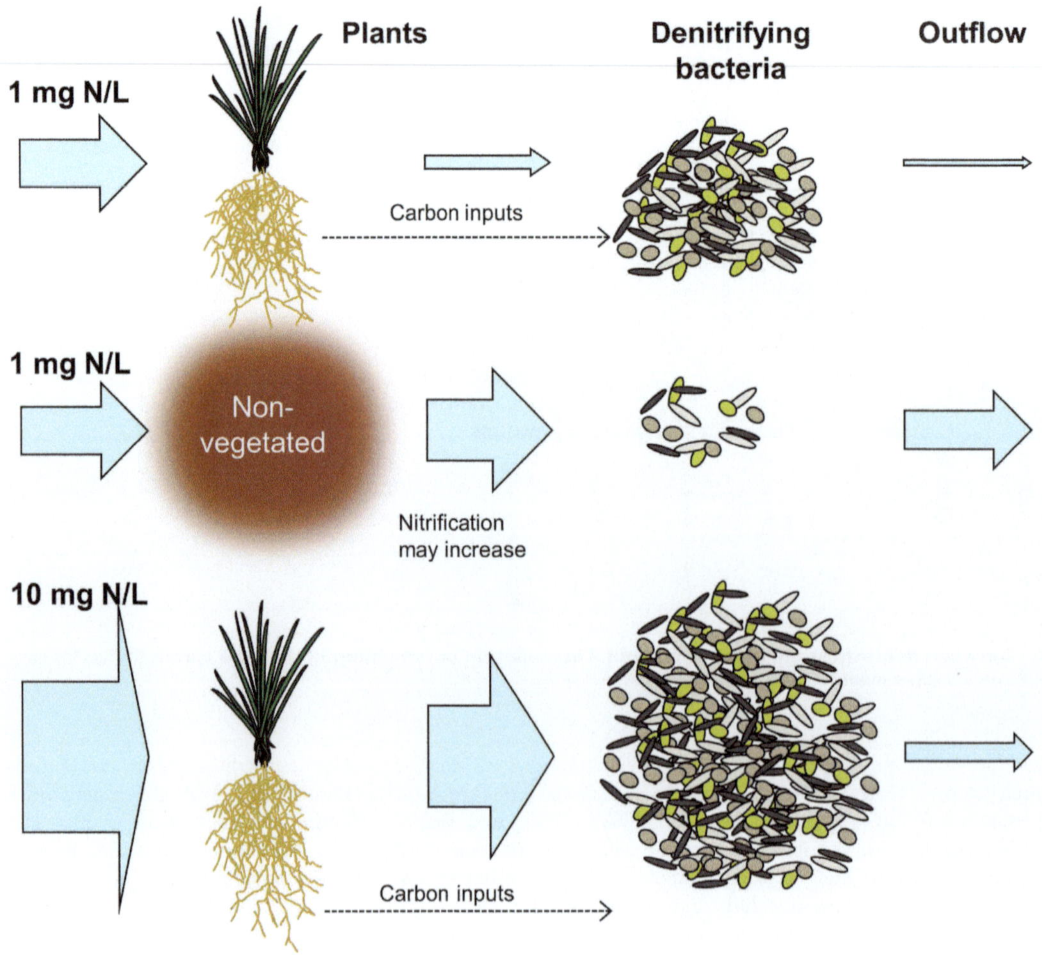

Figure 7. Conceptual diagram illustrating nitrate processing. Removal by assimilation and denitrification at different nitrogen loadings in vegetated and non-vegetated systems. The dependence of denitrifying bacteria on plant-derived carbon is also represented.

nitrate concentrations. However, data from the current study were inconclusive and testing these hypotheses will require further research. Nonetheless, the findings of the current study and others generally suggest that high loading leads to increased processing, although the proportion of nitrate retained by assimilation and denitrification is likely to decline across the loading spectrum as efficiency decreases [20], pathways saturate and mineralisation rates increase [44].

Synergistic relationships

While the discussion has so far focused on the division between biotic assimilation and denitrification, the two processes are not independent. Plants can facilitate denitrification, either directly or coupled to nitrification, by carbon provision (root exudates or sloughed cells) and promoting heterogeneity in redox potential; anoxic (driven by intense heterotrophic decomposition) and oxic (due to root oxygen release by some species in waterlogged conditions) [45,46]. This facilitative role of plants is highlighted by the performance of the non-vegetated control columns. In these treatments net nitrate production was commonly observed, indicating nitrification, which is facilitated by higher nitrate availability in the absence of plant-derived carbon (fuelling heterotrophic respiration) and plant assimilation (the small portion attributed to assimilation (Figure 6) was likely associated with bacteria consuming high C:N ratio organic matter from the carbon source). Despite the availability of nitrate penetrating the saturated zone and the provision of a carbon source in the non-vegetated treatment, denitrification still failed to dominate processing and was higher in columns planted with *Hypocalymma* or *Dianella*. In the influent concentration experiment, which lacked a carbon source, denitrification was negligible in the non-vegetated controls. This illustrates the importance of root-derived carbon, which acts as a continuous and dynamic source, in driving denitrification, despite the competition from plant and microbial nitrate assimilation [21,24]. This facilitation highlights the contradiction in the close relationship between plants and microbes; it is both synergistic and competitive, but essential and inter-dependent. Root exudation also hastens the onset of anaerobic conditions [46], as observed by the significantly lower concentrations of dissolved oxygen in the vegetated biofilter columns relative to the non-vegetated controls. However, a comparison of species rates of oxygen consumption and denitrification (Figures 5 and 6 respectively) yields no clear relationship (e.g. *Leptospermum*, *Juncus* and *Carex* demonstrate relatively rapid decline in dissolved oxygen but minimal denitrification, while the reverse is apparent for *Dianella*), suggesting available carbon is not exerting primary control on denitrification.

Do differences in kinetics or opportunity dictate the division between assimilation and denitrification? In the multiple species column experiment (and similar to biofilters in the field [9]) plants had the first opportunity to access incoming nitrate in the surface unsaturated layer before it reached the underlying saturated zone with high denitrification potential. In peat bogs this mechanism allowed a 5–10 cm layer of sphagnum moss to assimilate virtually all atmospheric nitrate additions despite favourable conditions for denitrification in the underlying soil [23]. In addition, the stormwater influent oxygenated the saturated zone, delaying the onset of suitable anaerobic conditions for denitrification (Figure 5), which may have advantaged biotic assimilation irrespective of rates. Hence, the kinetic rates of assimilation and denitrification cannot be conclusively compared in the current study. Regardless, biotic assimilation of ammonium appeared more rapid than nitrate (Figure 2), as expected due to general plant preference for ammonium [24].

Plant species variation

The proportion of nitrate assimilated displayed differences among plant species, from an average 99% uptake by *Leptospermum* to 74% for *Hypocalymma* and *Dianella* treatment replicates (Supplementary information, Figure S3). Interestingly, species demonstrating higher nitrate assimilation (83–99% $^{15}NO_3^-$ processing) also tended to be more effective for use in biofilters (i.e. less nitrate remaining in the pore water) (*Leptospermum, Carex,* Buffalo lawn grass, *Juncus* and *Allocasurina*), despite minimal denitrification (0–6%) (Figures S4 and S5). Alongside less effective species (*Dianella* and *Hypocalymma*) assimilation was reduced (58–85%), and while denitrification increased (5–8%), it did not compensate entirely, leading to increased nitrate concentrations in the pore water (9–36%). The minimal contribution of denitrification to performance was unexpected, but is compatible with the hypothesis that denitrifying bacteria primarily receive any nitrate left over after assimilation. Many characteristics may contribute to plant species variation including morphological (such as root architecture, biomass and diameter profile) and physiological traits (including root oxygenation, biomass composition, plant strategy, seasonality, assimilation rate, nitrogen preference, mycorrhizal symbiosis and photosynthate partitioning) [28,45,47]. Studies have already identified the importance of the root zone (rhizosphere) [48] or correlated biofilter efficiency for nitrogen with long, deep roots, a high root biomass and rapid growth [29], and further relationships are the subject of a current study.

Organic nitrogen stores – beneficial or false economy?

Plant nitrogen uptake and release follow seasonal cycles and the effect in wetlands has been likened to a 'spiral' [49]. The process acts to attenuate nitrogen within the system [49] and convert inorganic forms to an array of organic compounds. The significant impact of these plant services are recognised across wetlands, aquatic systems and vegetated buffers [7,22,25]. These functions may be particularly beneficial in biofilters, given that inflow events are relatively intermittent and transient. Assimilation will slow nitrogen passage through the biofilter and re-release will occur over a relatively longer timescale, which may facilitate more effective microbial processing. The benefits of temporary nitrogen immobilisation in plants are also well acknowledged in agriculture through the use of cover crops, which are used to minimise nitrate leaching between main crops and increasingly applied to protect groundwater quality [8].

However, the benefits of storage in the biomass may extend well beyond short seasonal or annual cycles if incorporated into recalcitrant compounds. These may be stored over the long term in woody biomass [44] or incorporated into the soil organic matter (SOM), which may exceed the nitrogen storage capacity of the plant biomass by a factor of 10 [50]. Such stores could potentially endure beyond the biofilter lifespan (generally 15–20 years [9]). In treatment wetlands, the accumulation of organic material, alongside denitrification, can both form significant pathways for nitrogen removal over the long term, even under high loading [43]. However, due to low anaerobic decomposition rates, wetlands are natural sinks for organic material [6] – it is less certain if significant accumulation will occur in the ephemeral environment of biofilters.

On the other hand, the conversion and attenuation function of plants leads to the production of nitrogen forms that require multiple processing steps before permanent removal is possible via denitrification. In particular, loss of dissolved organic nitrogen from the system is a high risk unless efficient mineralisation occurs [25]. In addition, both plants and microbes may over time increasingly source nitrogen from internal cycling processes. Harvesting the plant biomass to remove nitrogen could extend plant nitrogen demand, as has been observed in grassed systems [51]. In biofilters, however, this may be constrained by cost, the likelihood of filter media compaction [2] and concerns over pollution translocation [52].

What does this mean for long term biofilter function?

This experiment was limited by constraints inherent in studies at the laboratory scale, including small biofilter surface area, single-plant columns, regular inflows and the short time frame of the tracer experiment. In light of this, do the findings have any implications on processes in mature field-scale biofilters?

Concentrations of TN in the effluent from vegetated columns in the current study (0.11–0.45 mg N/L) are much lower than previous laboratory and field studies (typically 1 mg N/L at best [1,53]). This difference may in part reflect a design change towards media with a high sand and low nutrient content, which minimises nutrient leaching [9,26]. However, the current results do require validation under field conditions and extension by tracing nitrogen fate over longer periods. Nevertheless, the findings form an initial step in identifying and quantifying biofilter nitrogen processes, and thus represent an important advance on the predominantly "black-box" approach of studies to date.

The experiment quantified denitrification in the first hours following an inflow event. Within the multiple species column experiment the vegetated biofilters functioned effectively and little nitrate remained in the pore water after 24 hours (Figure 6), suggesting rapid initial processing, particularly alongside the most effective plant species. Given the transient nature of biofilter inflows, rapid initial processing may be an inherent characteristic of biofilters, possibly more so than it is for treatment wetlands. Hence, the results of the experiment may have some realistic implications on longer-term nitrogen fate in biofilters.

If denitrification does not form a significant long-term nitrogen sink in biofilters, ongoing performance is heavily dependent upon the capacity and duration of the biomass storage. Studies of ecosystem succession suggest both mature and early-stage systems have limited retentive capacity, and intermediate systems have the greatest potential to capture nitrogen [54]. The storage capacity increases over time as plants and a pool of soil organic matter establish [55,56]. However, growth of these storages will eventually plateau as the system moves towards a steady state [54] and nitrogen saturation [18]. At this point nitrogen returns (e.g. net nitrification) counteract the net biotic uptake, such that system inputs again equal outputs (e.g. leaching) [18]. Sustained high nutrient loading will exacerbate the saturation process and impair long-term functionality [44].

It might be expected that biofilters will similarly display an increase in performance towards this intermediate state, followed by a gradual decline in performance. However, do biofilters follow these same successional patterns and if so, is the point of zero net retention reached within the biofilter lifespan? These succession theories were developed for terrestrial forests where denitrification is negligible. If extrapolated to ephemeral biofiltration systems, the peak in performance may be sustained over a longer period of time if denitrification increases to counteract the decline in net biotic uptake.

Further work

The minor role of denitrification in early biofilter life lends greater urgency to the need to quantify nitrogen processing across the entire biofilter lifespan. If assimilation continues to be a major pathway, the temporal and quantitative dynamics of storage in the biomass or soil organic matter need to be delineated. The return flux of nitrogen from mineralisation should be incorporated into assessments of biofilter lifetime performance. In addition to the techniques employed in this paper, further understanding can also be gained using molecular biology techniques (such as qPCR), which are capable of characterising the bacterial community. As this study has demonstrated, plant species and loading can be critical influences on processes. Hence, the interconnectedness between assimilation, denitrification, plant species and cumulative loading over extended time periods require further research.

Conclusions

This study is the first known to apply a nitrate isotope tracer to the quantification of internal stormwater biofilter processes. Nitrate processing varied significantly with plant species and influent nitrogen concentration. Denitrification was only responsible for processing 0–8% of incoming nitrate in the laboratory-scale stormwater biofilters, suggesting biotic assimilation is the primary sink. Species identified as effective for reducing effluent concentrations (e.g. *Leptospermum*, *Carex*, Buffalo, *Juncus*, *Allocasurina*) tended to be associated with higher nitrate assimilation and minimal denitrification. This is surprising, given past efforts in biofilter design to promote denitrification. Instead, the denitrifying bacteria in the underlying saturated zone of biofilters appear to receive only the nitrate remaining after assimilation, such that nitrate plays a more important role in biofilters planted with species shown to be less effective in nitrogen removal. Higher nitrate loads increased the relative contribution of denitrification, implying denitrifying bacteria have greater adaptability to process high concentrations, whereas biotic assimilation becomes overwhelmed. While the results contrast with wastewater treatment wetlands, where microbial processing commonly dominates under higher loading, they are compatible with other studies highlighting the importance of plant assimilation as a nitrogen conversion mechanism and either a temporary or long-term (in soil organic matter or woody biomass) storage. This distinction from wetland functioning may reflect the unique characteristics of biofilters as quasi-terrestrial, engineered, ephemeral and highly-dynamic systems. The results have significant implications for biofilter design, maintenance and lifespan. With biotic assimilation dominating processing in early biofilter life, the need to characterise long-term organic matter accumulation and decomposition, the influence of plant species and prevalence of denitrification in mature systems becomes far more critical.

Supporting Information

Figure S1 Relationship between percentage of $^{15}NO_3^-$ denitrified and assimilated. Results for each biofilter column in the multiple species experiment (3 replicates per species, n = 1). Linear regression line fitted to the vegetated treatments only.

Figure S2 Relationship between percentage of $^{15}NO_3^-$ remaining in the porewater and denitrified. Results for each biofilter column in the multiple species experiment (3 replicates per species, n = 1). Linear regression line fitted to the vegetated treatments only.

Figure S3 Boxplot comparison of percentage of $^{15}NO_3^-$ assimilated across the vegetated treatments. Results for replicated columns in the multiple species experiment (3 replicates per species, n = 3).

Figure S4 Boxplot comparison of percentage of $^{15}NO_3^-$ denitrified across the vegetated treatments. Results for replicated columns in the multiple species experiment (3 replicates per species, n = 3).

Figure S5 Boxplot comparison of percentage of $^{15}NO_3^-$ remaining in the porewater across the vegetated treatments. Results for replicated columns in the multiple species experiment (3 replicates per species, n = 3).

Acknowledgments

The support of the project sponsors and invaluable assistance from Tracey Pham, Richard Williamson, Frank Winston, Anthony Brosinsky, Jennifer Read, Antonietta Torre, Marion Urrutiaguer, Tania Liaghati, Keralee Browne and Tina Hines are gratefully acknowledged. The authors would also like to thank the anonymous peer reviewers for their helpful comments and suggestions to improve the quality of this paper.

Author Contributions

Conceived and designed the experiments: PLMC TDF MRG AD BEH TRC VE EGIP DGR. Performed the experiments: EGIP DGR VE PLMC. Analyzed the data: DGR EGIP VE PLMC TDF MRG TRC AD BEH. Contributed reagents/materials/analysis tools: PLMC TDF MRG AD BEH TRC VE EGIP DGR. Wrote the paper: EGIP TDF PLMC AD BEH MRG TRC DGR VE.

References

1. Hunt WF, Smith JT, Jadlocki SJ, Hathaway JM, Eubanks PR (2008) Pollutant Removal and Peak Flow Mitigation by a Bioretention Cell in Urban Charlotte, N.C. Journal of Environmental Engineering 134: 403–408.
2. Hatt BE, Fletcher TD, Deletic A (2008) Hydraulic and pollutant removal performance of fine media stormwater filtration systems. Environmental Science and Technology 42: 2535–2541.
3. Taylor GD, Fletcher TD, Wong THF, Breen PF, Duncan HP (2005) Nitrogen composition in urban runoff–implications for stormwater management. Water Research 39: 1982–1989.
4. Gray SR, Becker NSC (2002) Contaminant flows in urban residential water systems. Urban Water 4: 331–346.
5. Vitousek PM, Aber JD, Howarth RW, Likens GE, Matson PA, et al. (1997) Human alteration of the global nitrogen cycle: Sources and consequences. Ecological Applications 7: 737–750.
6. Reddy KR, DeLaune RD (2008) Biogeochemistry of wetlands science and applications. Boca Raton: Taylor & Francis, 2008.
7. Vymazal J (2007) Removal of nutrients in various types of constructed wetlands. Science of the Total Environment 380: 48–65.
8. Powlson DS (1993) Understanding the soil nitrogen cycle. Soil Use and Management 9: 86–93.
9. FAWB (2009) Adoption Guidelines for Stormwater Biofiltration Systems: Facility for Advancing Water Biofiltration, Monash University, June 2009.

10. Kim H, Seagren EA, Davis AP (2003) Engineered bioretention for removal of nitrate from stormwater runoff. Water Environment Research 75: 355.
11. Zinger T, Fletcher TD, Deletic A, Blecken GT, Viklander M (2007) Optimisation of the nitrogen retention capacity of stormwater biofiltration systems. Lyon, France.
12. Zinger Y, Blecken GT, Fletcher TD, Viklander M, Deletic A (2013) Optimising nitrogen removal in existing stormwater biofilters: Benefits and tradeoffs of a retrofitted saturated zone. Ecological Engineering 51: 75–82.
13. Davis AP, Hunt WF, Traver RG, Clar M (2009) Bioretention Technology: Overview of Current Practice and Future Needs. Journal of Environmental Engineering 135: 109–117.
14. Bratières K, Fletcher TD, Deletic A, Zinger Y (2008) Nutrient and sediment removal by stormwater biofilters: A large-scale design optimisation study. Water Research 42: 3930–3940.
15. Davis AP, Shokouhian M, Sharma H, Minami C (2006) Water Quality Improvement through Bioretention Media: Nitrogen and Phosphorus Removal. Water Environment Research 78: 284.
16. Davis AP, Shokouhian M, Sharma H, Minami C (2001) Laboratory study of biological retention for urban stormwater management. Water Environment Research 73: 5.
17. Lucas WC, Greenway M (2011) Hydraulic Response and Nitrogen Retention in Bioretention Mesocosms with Regulated Outlets: Part I - Hydraulic Response. Water Environment Research 83: 692–702.
18. Aber J, McDowell W, Nadelhoffer K, Alison M, Berntson G, et al. (1998) Nitrogen Saturation in Temperate Forest Ecosystems. Bioscience 48: 921–934.
19. Lee C-g, Fletcher TD, Sun G (2009) Nitrogen removal in constructed wetland systems. Engineering in Life Sciences 9: 11–22.
20. Lin Y-F, Jing S-R, Wang T-W, Lee D-Y (2002) Effects of macrophytes and external carbon sources on nitrate removal from groundwater in constructed wetlands. Environmental Pollution 119: 413–420.
21. Matheson FE, Nguyen ML, Cooper AB, Burt TP, Bull DC (2002) Fate of ^{15}N-nitrate in unplanted, planted and harvested riparian wetland soil microcosms. Ecological Engineering 19: 249–264.
22. Cooper AB, Cooke JG (1984) Nitrate loss and transformation in 2 vegetated headwater streams. New Zealand Journal of Marine and Freshwater Research 18: 441–450.
23. Urban NR, Eisenreich SJ, Bayley SE (1988) The Relative Importance of Denitrification and Nitrate Assimilation in Midcontinental Bogs. Limnology and Oceanography 33: 1611–1617.
24. Kirk GJD, Kronzucker HJ (2005) The Potential for Nitrification and Nitrate Uptake in the Rhizosphere of Wetland Plants: A Modelling Study. Annals of Botany 96: 639–646.
25. Bedard-Haughn A, Tate KW, Kessel Cv (2004) Using Nitrogen-15 to Quantify Vegetative Buffer Effectiveness for Sequestering Nitrogen in Runoff. Journal of Environmental Quality 33: 2252–2262.
26. Payne EGI, Pham T, Cook PLM, Fletcher TD, Hatt BE, et al (in press) Biofilter design for effective nitrogen removal from stormwater - influence of plant species, inflow hydrology and use of a saturated zone. Water Science and Technology.
27. Mitsch WJ, Gosselink JG (2000) Wetlands. New York, United States of America: John Wiley & Sons, Inc.
28. Payne EGI, Fletcher TD, Cook PLM, Deletic A, Hatt BE (in press) Processes and drivers of nitrogen removal in stormwater biofiltration. Critical Reviews in Environmental Science and Technology.
29. Read J, Fletcher TD, Wevill T, Deletic A (2010) Plant Traits that Enhance Pollutant Removal from Stormwater in Biofiltration Systems. International Journal of Phytoremediation 12: 34–53.
30. Cavagnaro TR, Smith FA, Lorimer MF, Haskard KA, Ayling SM, et al. (2001) Quantitative Development of *Paris*-Type Arbuscular Mycorrhizas Formed between *Asphodelus fistulosus* and *Glomus coronatum*. New Phytologist 149: 105–113.
31. Nielsen LP (1992) Denitrification in sediment determined from nitrogen isotope pairing. FEMS Microbiology Letters 86: 357–362.
32. Duncan HP (1999) Urban Stormwater Quality: A Statistical Overview. Melbourne, Australia: Cooperative Research Centre for Catchment Hydrology.
33. Blecken G-T, Zinger Y, Deletic A, Fletcher TD, Viklander M (2009) Impact of a submerged zone and a carbon source on heavy metal removal in stormwater biofilters. Ecological Engineering 35: 769–778.
34. Šimek M, Jíšová L, Hopkins DW (2002) What is the so-called optimum pH for denitrification in soil? Soil Biology and Biochemistry 34: 1227–1234.
35. Weiss RF, Price BA (1980) Nitrous oxide solubility in water and seawater. Marine Chemistry 8: 347–359.
36. Fox J, Weisberg S (2011) An {R} Companion to Applied Regression. Thousand Oaks, CA: Sage.
37. R Core Development Team (2012) R: A language and environment for statistical computing. Vienna, Austria: R Foundation for Stastistical Computing
38. Ritz C, Streibig JC (2005) Bioassay Analysis using R. Journal of Statistical Software 12.
39. Seitzinger SP (1988) Denitrification in Freshwater and Coastal Marine Ecosystems: Ecological and Geochemical Significance. Limnology and Oceanography 33: 702–724.
40. Kreiling R, Richardson W, Cavanaugh J, Bartsch L (2011) Summer nitrate uptake and denitrification in an upper Mississippi River backwater lake: the role of rooted aquatic vegetation. Biogeochemistry 104: 309–324.
41. Zhu W-X, Dillard N, Grimm N (2004) Urban nitrogen biogeochemistry: status and processes in green retention basins. Biogeochemistry 71: 177–196.
42. Pinay G, Ruffinoni C, Wondzell S, Gazelle F (1998) Change in Groundwater Nitrate Concentration in a Large River Floodplain: Denitrification, Uptake, or Mixing? Journal of the North American Benthological Society 17: 179–189.
43. Craft CB (1997) Dynamics of nitrogen and phosphorus retention during wetland ecosystem succession. Wetlands Ecology and Management 4: 177–187.
44. Bernot M, Dodds W (2005) Nitrogen Retention, Removal, and Saturation in Lotic Ecosystems. Ecosystems 8: 442–453.
45. Reddy KR, Patrick WH Jr, Lindau CW (1989) Nitrification-denitrification at the plant root-sediment interface in wetlands. Limnology & Oceanography 34: 1004–1013.
46. Minett DA, Cook PL, Kessler AJ, Cavagnaro TR (2013) Root effects on the spatial and temporal dynamics of oxygen in sand-based laboratory-scale constructed biofilters. Ecological Engineering 58: 414–422.
47. Tanner CC (1996) Plants for constructed wetland treatment systems — A comparison of the growth and nutrient uptake of eight emergent species. Ecological Engineering 7: 59–83.
48. Lucas W, Greenway M (2008) Nutrient Retention in Vegetated and Nonvegetated Bioretention Mesocosms. Journal of Irrigation and Drainage Engineering 134: 613–623.
49. Kadlec RH, Tanner CC, Hally VM, Gibbs MM (2005) Nitrogen spiraling in subsurface-flow constructed wetlands: Implications for treatment response. Ecological Engineering 25: 365–381.
50. Jenkinson DS (1990) An introduction to the global nitrogen cycle. Soil Use and Management 6: 56–61.
51. Bedard-Haughn A, Tate KW, Kessel Cv (2005) Quantifying the Impact of Regular Cutting on Vegetative Buffer Efficacy for Nitrogen-15 Sequestration. Journal of Environmental Quality 34: 1651–1664.
52. Collins KA, Lawrence TJ, Stander EK, Jontos RJ, Kaushal SS, et al. (2010) Opportunities and challenges for managing nitrogen in urban stormwater: A review and synthesis. Ecological Engineering 36: 1507–1519.
53. Roberts SJ, Fletcher TD, Garnett L (2012) Bioretention saturated zones; do they work at the large-scale? 7th International Conference on Water Sensitive Urban Design 21–23 February 2012. Melbourne, Australia.
54. Vitousek PM, Reiners WA (1975) Ecosystem Succession and Nutrient Retention: A Hypothesis. Bioscience 25: 376–381.
55. Odum EP (1969) The strategy of ecosystem development. Science 164: 262–270.
56. Mitsch WJ, Day JW, Zhang L, Lane RR (2005) Nitrate-nitrogen retention in wetlands in the Mississippi River Basin. Ecological Engineering 24: 267–278.

12

Modelling Coral Reef Futures to Inform Management: Can Reducing Local-Scale Stressors Conserve Reefs under Climate Change?

Georgina G. Gurney[1,2*], Jessica Melbourne-Thomas[3,4], Rollan C. Geronimo[5], Perry M. Aliño[6], Craig R. Johnson[7]

1 Australian Research Council Centre of Excellence for Coral Reef Studies, James Cook University, Townsville, Queensland, Australia, **2** School of Zoology, University of Tasmania, Hobart, Tasmania, Australia, **3** Australian Antarctic Division, Department of Sustainability, Environment, Water, Population and Communities, Kingston, Tasmania, Australia, **4** Antarctic Climate & Ecosystems Cooperative Research Centre, University of Tasmania, Hobart, Tasmania, Australia, **5** Conservation International Philippines, Quezon City, Philippines, **6** Marine Science Institute, University of the Philippines, Quezon City, Philippines, **7** Institute for Marine and Antarctic Studies, University of Tasmania, Hobart, Tasmania, Australia

Abstract

Climate change has emerged as a principal threat to coral reefs, and is expected to exacerbate coral reef degradation caused by more localised stressors. Management of local stressors is widely advocated to bolster coral reef resilience, but the extent to which management of local stressors might affect future trajectories of reef state remains unclear. This is in part because of limited understanding of the cumulative impact of multiple stressors. Models are ideal tools to aid understanding of future reef state under alternative management and climatic scenarios, but to date few have been sufficiently developed to be useful as decision support tools for local management of coral reefs subject to multiple stressors. We used a simulation model of coral reefs to investigate the extent to which the management of local stressors (namely poor water quality and fishing) might influence future reef state under varying climatic scenarios relating to coral bleaching. We parameterised the model for Bolinao, the Philippines, and explored how simulation modelling can be used to provide decision support for local management. We found that management of water quality, and to a lesser extent fishing, can have a significant impact on future reef state, including coral recovery following bleaching-induced mortality. The stressors we examined interacted antagonistically to affect reef state, highlighting the importance of considering the combined impact of multiple stressors rather than considering them individually. Further, by providing explicit guidance for management of Bolinao's reef system, such as which course of management action will most likely to be effective over what time scales and at which sites, we demonstrated the utility of simulation models for supporting management. Aside from providing explicit guidance for management of Bolinao's reef system, our study offers insights which could inform reef management more broadly, as well as general understanding of reef systems.

Editor: Sebastian C. A. Ferse, Leibniz Center for Tropical Marine Ecology, Germany

Funding: This work was supported by the Modelling and Decision (MDS) working group of the Coral Reef Targeted Research and Capacity Building for Management Program (CRTR-CBMP). The funders had no role in study design, data collection and analysis, decision to publish, or preparation of the manuscript.

Competing Interests: The authors have declared that no competing interests exist.

* E-mail: georgina.gurney@gmail.com

Introduction

Coral reefs are of immense natural and anthropocentric value. Aside from boasting the highest diversity of all marine ecosystems [1], coral reefs provide a wealth of ecosystem goods and services to millions of people in more than 100 tropical countries [2,3]. However, coral reefs are in decline worldwide [4,5], and given the present momentum of human population growth, resource use and the imminent threat of climate change, this decline is likely to be exacerbated [6]. Coral reef degradation is driven by a myriad of stressors, which are largely a result of marine- and land-based anthropogenic activities such as fishing, coastal development, and aquaculture [7]. More recently, climate change has emerged as a principal threat to coral reefs [8,9]. There is growing evidence that the vulnerability of coral reefs to the effects of climate change is increased by chronic, local-scale stressors [10,11] and that the cumulative effect of these stressors may be synergistic [12,13]. Here, a stressor is defined as any environmental or biotic factor that exceeds natural levels of variation to cause a harmful effect on a system or organism [14–16].

Managing coral reefs under climate change

Given that climate change impacts on coral reefs cannot be mitigated directly, the question arises whether reduction of stressors that originate and can be managed at a local scale (i.e. local-scale stressors) provides a tractable opportunity to increase the potential of coral reefs to cope with inevitable changes in the climate [2,11]. Critical to undertaking such management actions is an understanding of how reef systems will respond to alternative actions under different climatic scenarios. Managers need to be able to identify appropriate management actions from a suite of options, and have an understanding of where they will most likely be effective and over what time scales. Although the management of local-scale stressors, typically fisheries and poor water quality (i.e. nutrification and sedimentation), is widely advocated to conserve coral reefs in the context of climate change [2,7], the extent to which reduction of local-scale stressors confers resilience to climate change and how to identify the particular stressors that should be targeted by management to most cost-effectively optimise resilience, is uncertain [17,18]. These issues have been flagged as key research questions for reef scientists [13].

Inherent to assessing the importance of local management action is knowledge of the combined impact of multiple stressors [19,20], including the nature of their interactions; i.e. whether the effect of multiple stressors is simply the sum of the effects of each stressor (additive), or if the cumulative effect is more than (synergism) or less than (antagonism) the additive effect [20,21]. Although stressors almost always co-occur [6] and the nature of their interaction can have a profound influence on management outcomes [19], research and management has tended to consider stressors in isolation [14,20]. Thus there is limited understanding of the effect of concurrent stressors on ecological systems in general [20,22], and marine systems in particular [23,24], which hampers managers' ability to diagnose drivers of environmental degradation and consequently the appropriate course of management action. The dearth of research on stressor interactions is likely because isolating and quantifying individual effects and interactive effects is a difficult task [6,15]. This is particularly pertinent in natural systems where most stressors occur simultaneously; thus, the few existing studies of multiple stressors tend to be manipulative experiments at small spatial scales [14,20].

Modelling coral reef systems

Models are ideal tools for tackling questions regarding cumulative stressors at appropriate spatial and temporal scales, and for supporting environmental decision-making in general, through elucidating the consequences of alternative management actions in the context of global climate change [25]. Stochastic simulation models are particularly relevant because they can be used to explore the possible outcomes of alternative management scenarios in the face of imperfect knowledge of the future, and with a limited understanding of complex ecological systems [26,27]. This is valuable given that decisions about when, where and how management actions should be implemented are founded on our predictions of the future [28]. In addition, models provide a useful way in which to distil the intrinsic complexity of coral reef systems, and they allow the ready application of complex experimental designs to address critical management questions at different temporal and spatial scales, including those related to the effects of simultaneous stressors.

Models have been used widely to aid understanding of the impact of stressors on coral reefs [29,30], including analysis of reef state under alternative climatic [8] and/or management scenarios [31,32]. However, there are a number of key limitations of existing models with respect to their ability to address questions relevant to management. First, a whole-of-system approach is not always adopted. For example, some studies consider the dynamics of corals [30,33] or fish [32] only, which does not allow for investigation of whole-of-system processes, such as phase-shifts from coral- to macroalgal-domination. Other studies do not explicitly model consumer dynamics [8,31], which prevents investigation of feedbacks between the benthos and consumers. A second important limitation is that existing models are often not instantiated using spatially explicit data on reef state for particular locations, instead they tend to be designed to apply to a general system within a broad bio-geographical area, most commonly the Caribbean [29,31,34]. Third, modelling studies tend to be restricted to examining the impact of one or two stressors. For example, scenario analyses have focused solely on fisheries [32,35], the effects of climate change [8], or water quality [36], and only a handful examine the effects of two concurrent stressors [34,37]. Thus there is a scarcity of spatially explicit models able to provide decision support for management of the multiple simultaneous stressors that many coral reefs are subject to, including how management of local stressors might affect future reef state under the expected effects of climate change.

Recent work by Melbourne-Thomas et al. [38,39] is an exception; these studies used models instantiated at a regional scale for particular sites to examine the effect of multiple stressors on coral reef systems. While consideration of regional-scale dynamics (e.g. connectivity between reefs through dispersal of fish larvae) is critical for the effective management of coral reefs, equally important is an understanding of reef systems at a local scale. Reef state and the magnitude of stressors will differ spatially even at a local scale, requiring management actions to be tailored to the local context. Further, governance and management of marine resources is undertaken locally in many developing countries [40]. However, studies using simulation models of coral reefs as a decision support tool for local-scale management of multiple stressors are lacking.

Aims

Here we use a simulation model to investigate the extent to which management of local-scale stressors might influence future trajectories of coral reefs under varying climatic scenarios. We use a model originally developed by Fung et al. [41], which captures the dynamics of key coral reef consumers and benthic organisms. We validate it for four sites in Bolinao, the Philippines, and simulate future reef state for each site 40 years into the future under scenarios involving the cumulative impact of fishing, poor water quality and thermal bleaching-induced mortality related to climate change. These stressors are major threats to reef systems in Bolinao [42,43], and the Philippines more broadly [40]. Further, both water quality and fishing pressure are suggested to be important determinants of reef resilience to climate change [10,11], and management of both stressors is tractable, with marine reserves already being implemented extensively in the Philippines [44].

Thus our study has two distinct aims: first, to ascertain how management of local stressors might affect future trajectories of coral reefs under varying bleaching scenarios. We address this by investigating (1) the cumulative impact of fishing, poor water quality (i.e. nutrification and sedimentation) and bleaching-induced coral mortality on reef systems, and how their interactions manifest over time; and (2) patterns of reef recovery following bleaching. Our second aim is to assess the utility of simulation modelling for providing decision support for local-scale management of coral reef systems subject to multiple stressors, which involves providing guidance on which courses of management are most likely to be effective in different locations of the Bolinao reef system and over what time scales. This assessment also involves identifying the circumstances under which projections of future reef state are likely to be most certain.

Methods

Ethics Statement

The University of The Philippines Marine Science Institute (UPMSI) has been given Prior Informed Consent (PIC) by the Mayor of Bolinao, Pangasinan, to undertake the research from which data was drawn for this modelling analysis. The PIC is continually updated by the Deputy Director for the Bolinao Marine Laboratory. Presently the director is Dr. Ronald Villanueva, and previously, during the research cited, the director was Dr. Antonette Juinio-Menez. Since there were no materials transferred outside the country, there was no need for a gratuitous permit for materials transfer from the Bureau of Fisheries and Aquatic Resources (BFAR) of the Department of Agriculture (DA). Since the sites are not part of the National Protected Areas System, the PIC from the Mayor was sufficient. Data from 1988 & 1995 was collected by Dr. Cleto Nanola, under the UPMSI Community Ecology (ComEco) Lab with permission from Dr. Porfirio M. Aliño as the ComEco Lab director. During this period (1988 – 1995), no PIC was required to undertake field observations in non-protected areas such as those cited in the publication. For all studies only visual censuses of fish and benthic communities were conducted; no fauna or flora were collected or manipulated.

The model

Ecological dynamics are simulated using a local-scale mean-field model developed by Fung [41] and modified by Melbourne-Thomas et al. [45] and during the course of this work. Fung et al. [41] provides a description of the model development and its general behaviours, and thus only the fundamental structure of the model is described here (also see Text S1, Figure S1, Table S1 and Table S2 for details of model equations and parameters). While Fung et al.'s [41] model uses ordinary differential equations, we represented ecological processes of shallow (~5 – 20 m depth) coral reef systems by difference equations (see Text S1; Melbourne-Thomas et al. [45]), which simulate the biomass dynamics of three consumer functional groups (herbivorous fish, piscivorous fish and sea urchins; Table 1), and the proportional covers of five benthic functional groups (hard coral, grazed epilithic algal communities (EAC), macroturf and macroalgae; Table 1). We modified the structure of the model to include feedback between benthic structure and fish dynamics by implementing a scaling term for fish recruitment which is dependent on coral cover (based on evidence from Feary et al. [46] and Holbrook et al. [47]; see Text S2 for further details).

A range of natural and anthropogenic stressors can be applied as external stressor functions (hereafter 'forcings'), that include nutrification and sedimentation (modelled as parameter scalings), and fishing and coral bleaching (which act to reduce biomass or cover; Table 2). All parameter values (both for the underlying model and external forcings) were from the literature. Where possible we used values from the literature relating to a particular functional group, otherwise we generalised the value from one or a few species across a functional group (see Fung et al. [41] and Melbourne-Thomas et al. [39] for further details).Where we had multiple values for responses to stressors, we specified these as a range, from which the model selected a value at random to introduce a stochastic element. Thus the forcing scheme represents a generalised response of coral reefs to stressors.

The model has been evaluated to ensure that it accurately captures the chief dynamic processes of coral reef systems [48,49], such as 'phase shifts' from coral to algal dominance [41], which are well described for degraded coral reefs [50]. Sensitivity analysis has been undertaken for all parameters and equilibrium behaviour has been examined [51]. The model was implemented in the object-oriented Python Programming Language (Python Software Foundation, Hampton, New Hampshire, USA) version 2.5.

Study site

The model was instantiated for four sites in the coral reef system near Bolinao in the Lingayen Gulf area, the Philippines (Figure 1). Bolinao's reef system is located around Santiago Island and covers approximately 66 km^2, including an extensive reef slope accounting for 65% of the area [52].

The Bolinao reefs are highly degraded, with the most significant stressors being poor water quality and fishing [42,43]. Water quality has decreased significantly in Bolinao over the last two decades [53–55], primarily due to the development of aquaculture of milkfish (*Chanos chanos*) in the Guiguiwanen Channel since 1995 [56,57]. The closest major river discharge is Alaminos River, located approximately 16 km to the south-east of Bolinao, but allochthonous inputs from this river are considered insignificant compared to excess fish feed and faecal matter from the fish farms, which causes nutrification and sedimentation of Boliano's reefs [56]. Artisanal fishing pressure is intense in Bolinao, with the area having one of the highest densities of fishers in the Philippines [58]. Therefore, the local fishery is highly overexploited [59]. Management of Bolinao's marine and coastal resources is the responsibility of the municipal government, as is the case in all of the Philippines since 1991 when many national government responsibilities were devolved to municipalities [60]. Existing marine management in Bolinao includes a no-take marine reserve at Lucero (one of our four sites), which was established in 2004.

The Bolinao reef system provided an ideal case study to examine local-scale management of multiple stressors because: (1) reefs are heavily affected by numerous stressors originating from anthropogenic activities and climate change; (2) marine governance is undertaken at the local-scale; and (3) data were available to parameterise and validate the model.

Table 1. Definition of functional groups modelled, with example taxa from the Indo-Pacific region.

Functional group	Description
Coral	Hermatypic (i.e. contain zooxanthellae) scleractinian coral species (e.g. *Pocillopora damicornis, Acropora* spp.).
Epilithic algal communities (EAC)	Components of the benthos such as dead coral skeletons, non-geniculate coralline algae and rocks, on which cropped filamentous algae (i.e. fine turf) of ≤~2 - 4 mm in height grow. Fine turf exerts a constant growth pressure to form macroturf, but fine turf may be maintained by grazing of herbivorous fish and sea urchins.
Macroturf	Filamentous algal species that form turfs >~2 - 4 mm in height (e.g. *Spyridia filamentosa, Hersiphonia secunda*).
Macroalgae	Brown, red and green algae which is more structurally complex and has a greater thallus size than macroturf (e.g. *Sargassum* spp., *Lobophora variegata*).
Herbivorous fish	Reef fish which feed predominantly on macroturf, macroalgae and EAC (e.g. families Siganidae, Scaridae, Acanthuridae).
Piscivorous fish	Reef associated fish which predate on herbivorous fish (e.g. families Lutjanidae, Serranidae).
Sea urchins	Grazing sea urchins which feed on EAC, macroturf and macroalgae (e.g. *Echinothrix* spp., *Diadema setosum*).

Table 2. The effect of forcings (i.e. stressors) on ecological processes and functional groups in the model.

Forcing	Implementation
Nutrification	Increases macroturf and macroalgal growth by a scaling factor of 1.3 – 2 (1, 2, 3) and 2 – 7 (4, 5, 6, 7, 8), respectively.
Sedimentation	Decreases coral growth by a scaling factor of 0 – 0.3 yr^{-1} (9, 10) and coral recruitment by 0 – 0.6 (11); increases coral mortality by 0 – 0.2 yr^{-1} (12), and prevents recruitment of coral onto macroturf algae (13).
Coral bleaching	Coral bleaching events decreases coral cover. The number of events and the percentage of coral lost per event are specified explicitly for each scenario.
Fishing	The rate of fish extraction (t km^{-2} $year^{-1}$) for herbivorous and piscivorous fish is specified explicitly for each scenario.

References: (1) Miller et al. [103], (2) McClanahan et al. [104], (3) McClanahan et al [105], (4) Lapointe [106], (5) Lapointe and O'Connell [107], (6) Larned [108], (7) Larned and Stimson [109], (8) Schaffelke and Klump [110], (9) Dodge et al.[111], (10) Cortes and Risk [112], (11) Babcock and Smith [113], (12) Nugues and Roberts [114], (13) Birrell et al. [115].

Parameterisation of the model for Bolinao

The model was parameterised for four distinctly different sites on Bolinao's reef slope to capture spatial differences in local reef state, hence providing guidance on the locations where different management actions are most likely to be effective. To represent these differences we parameterised the model for four sites, *viz.* Tomasa, Lucero, Cangaluyan and Malilnep (Figure 1c). The coral reefs at Tomasa and Cangaluyan are more degraded than at Lucero and Malilnep [61].

Sea urchins are present at Lucero and Malilnep, but not at Tomasa and Cangaluyan [62], so the model was modified to exclude sea urchins for the latter two sites. The area modelled at each site was 500 m × 500 m because this is the scale at which marine reserves are often implemented in the area (pers. observation). The updating interval was daily. General parameter

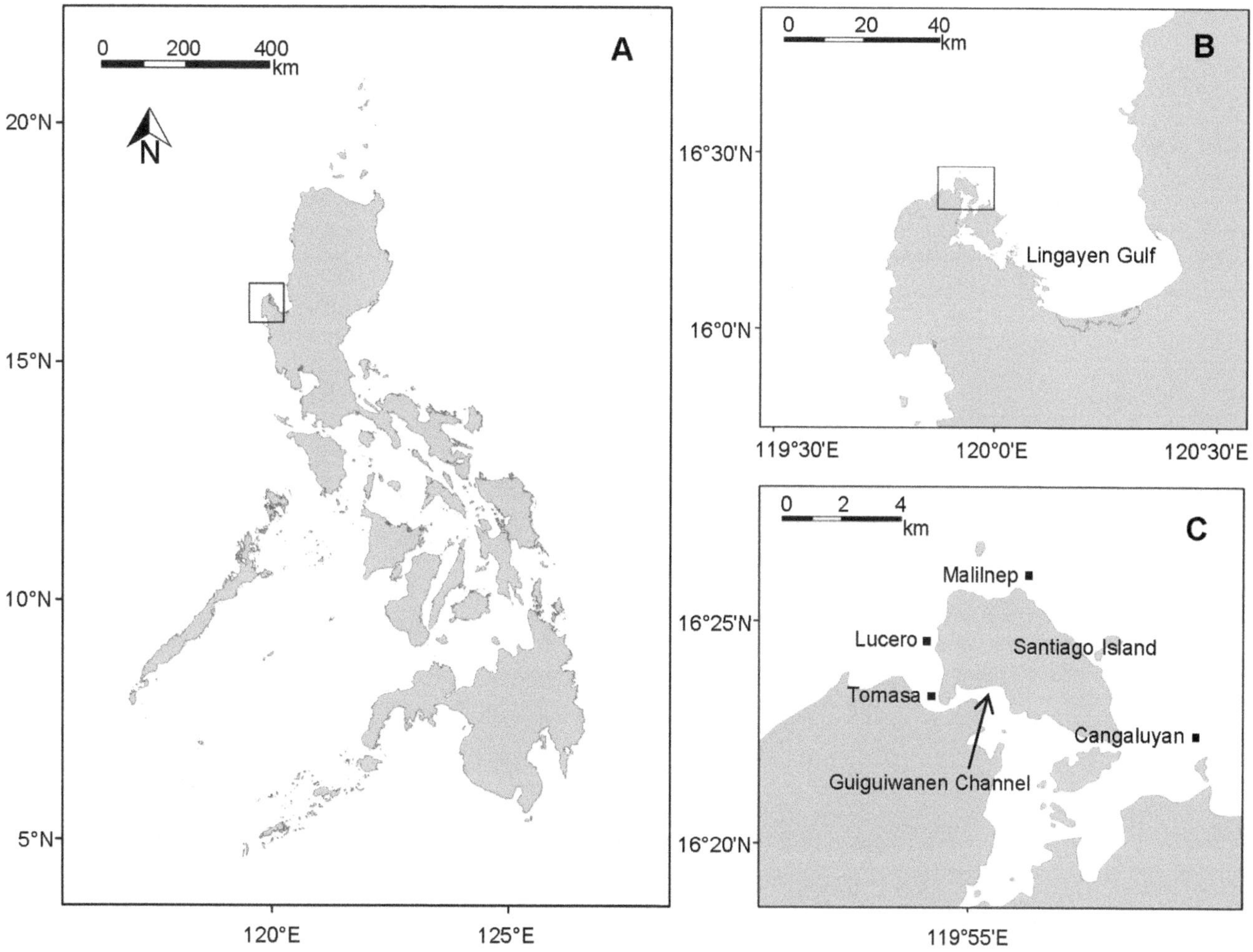

Figure 1. Location of study site. Location of (A) the Lingayen Gulf in the Philippines, (B) the municipality of Bolinao in the Lingayen Gulf area, and (C) the four sites (i.e. Lucero, Tomasa, Malilnep and Cangaluyan) on Bolinao's reef system for which the model was instantiated.

values (those not specific to each site) for the model were derived specifically for Bolinao when possible, with the remaining values derived using values measured elsewhere in the Indo-Pacific region (Table S1, Table S2). The different instantiations of the model were validated for each of the four sites following the method of Melbourne-Thomas et al. [45], i.e. we assessed whether the model could reproduce reef dynamics of the four sites from 1987 – 2008, given a known series of stressors at these sites, and data on reef state in 1987 (Figure S2, Figure S3, Figure S4, Table S3, Table S4, Text S3).

Future management and climatic scenarios

We simulated future reef state for each of the four sites in Bolinao under 18 scenarios, which comprised of all possible combinations of different levels of fishing pressure (three levels), coral mortality due to bleaching (three levels) and water quality (two levels; Table 3).

We examined three scenarios of future fisheries management (Table 3). First, no fisheries management (i.e. high impact fishing pressure), with fishing pressure set at the current yield at Bolinao (i.e. 2.6–2.8 t km^{-2} $year^{-1}$; see Text S3 for derivation). Second, a 50% reduction in fishing pressure (i.e. moderate impact fishing pressure) which could be achieved through fisheries management tools such as gear restrictions, quotas or development of alternative livelihoods for fishers. For these two scenarios fishing pressure acted on each fish functional group following Aliño et al. [63], where 68.3% of the total catch on Bolinao's reefs was herbivorous fish and 31.7% was piscivorous fish. The third scenario was implementation of no-take marine reserves which equates with zero fishing pressure (we assumed effective compliance). We do not include destructive fishing in our scenarios because fishers have ceased to employ those practices in Bolinao [64].

We represented water quality in the model through the combined effects of the nutrification and sedimentation forcings, because these two stressors co-occur in Bolinao, with both originating predominantly from aquaculture [65]. Thus management aimed at either stressor would necessarily lead to changes in the level of the other stressor. We examined two scenarios of future water quality, namely, unregulated (i.e. high impact nutrification and sedimentation) and highly regulated water quality (i.e. no impact of nutrification and sedimentation; Table 3). Unregulated water quality refers to the current situation in Bolinao, where aquaculture development is largely uncontrolled and water quality is consequently poor, leading to nutrification and sedimentation of Bolinao's reefs [66,67]. Under the regulated water quality scenario, reefs were not subject to nutrification and sedimentation. This reduction in nutrient and sediment input could be achieved through regulation of the aquaculture industry, including restricting both the number of aquaculture cages and stocking density.

We considered climate change in terms of coral bleaching only because the effects of ocean acidification in the short- to medium-term future are uncertain [68], and the likelihood of increased frequency and/or severity of damaging typhoons due to climate change continues to be debated [69,70]. Three scenarios for future bleaching frequency were explored, namely zero (low), one (medium) and two (high) events per decade; medium and high frequencies of bleaching were derived from Donner et al. [71] and Donner [9]. Bleaching acted to reduce coral cover by 10–60%; the upper end of this range was the average mortality of corals in Bolinao during the severe 1998 bleaching event [72,73].

We assumed uniform magnitude of stressors for all sites. While the magnitude of bleaching, nutrification and sedimentation is likely to vary with location in the future, we did not have information on the spatial distribution of bleaching susceptibility and water quality. Further, following Fung et al. [51] and Melbourne-Thomas et al. [45], the nutrification and sedimentation forcings were based on the linear threshold model [74], and under current conditions these thresholds are exceeded for all sites (Text S3). The four instantiations of the model were parameterised to reflect reef state in each location in 2008 [61], and each scenario was modelled over 40 years for each site using Monte Carlo simulations of 20 replicate runs.

Scenario analysis

Simulated future reef states were analysed using multivariate statistical techniques, and by examining model trajectories under these scenarios. Both approaches provide useful complementary insights into the effect of stressors on reef function. We used multivariate analyses to analyse modelled reef states for each site at four points in time selected *a priori*, *viz.* 5, 10, 20 and 40 years into the future. These are time-scales relevant to managers, and allow us to assess how interactions between stressors manifest over the long-term, including recovery patterns after bleaching. We used permutational multivariate analysis of variance (PERMANOVA) [75] to determine the significance of differences in modelled reef state, in terms of 'location' (i.e. mean reef state) and 'dispersion' (i.e. variability in reef state). The 20 model runs formed the independent replicates for the analyses.

Given that terms in the PERMANOVA can be significant due to differences in mean reef state (i.e. location) only, or differences in reef state *and* variability in reef state (i.e. dispersion), we used permutational analysis of multivariate dispersions (PERMDISP) [75] to test for homogeneity of dispersions among sets of the 20 independent model runs for each scenario. Thus, PERMDISP provided a means to: (1) clarify the nature of the significant PERMANOVA terms; and (2) quantify variability in projected trajectories within a set of Monte Carlo simulations for each scenario. This provides insights into the circumstances under which projected reef state is most certain (i.e. greater dispersion implies lower predictability). For all significant terms in the PERMANOVA we used *a posteriori* pairwise comparisons to identify the treatment groups that differed. Following the method

Table 3. Details of the three stressors (and levels of impact) examined in future management and climatic scenarios.

Stressor	No impact	Moderate impact	High impact
Fishing pressure (F)	Reef not affected	1.3–1.4 t $km^{-2}year^{-1}$ (50% of current yield)	2.6–2.8 t $km^{-2}year^{-1}$ (current yield)
Poor water quality (i.e. nutrification and sedimentation; NS)	Reef not affected	Not modelled	Nutrification and sedimentation present (effect of forcings detailed in Table 2)
Coral bleaching (B)	Reef not affected	Reduces coral cover by 10 –60%. Occurs with a long-term frequency of once per decade	Reduces coral cover by 10 –60%. Occurs with a long-term frequency of twice per decade

of Anderson [76], where pairwise comparisons are undertaken, no adjustment of the significance level was made for multiple tests.

We used canonical analysis of principal coordinates (CAP) [75] and non-metric multi-dimensional scaling (MDS) [77] to visualise differences in multivariate location and dispersion, respectively, and hence to explore the nature of significant effects detected using PERMANOVA and PERMDISP. CAP is a constrained ordination technique in which between-group variability is scaled by within-group dispersion, and is thus useful to visualise results of PERMANOVA [75]. Conversely, MDS is an unconstrained ordination technique, which preserves distances between groups and thus reflects raw dispersion patterns. All multivariate analyses were based on Euclidean distances and conducted on normalised data (as consumer and benthic variables use different units, i.e. gm^{-2} and %). 4,999 permutations of the residuals under a reduced model were used to calculate *p* values for PERMANOVA, PERMDISP and CAP analyses.

We examined trajectories of reef state over 40 years for six of the 18 scenarios, representing all combinations of the different levels of fishing and water quality under the climatic scenario of decadal bleaching. These scenarios were chosen based on results from the statistical analysis and because fishing and water quality, but not bleaching, can potentially be managed at a local scale. Analysis of the trajectories allowed us to visualise changes in the functional groups driving differences in reef state detected in the PERMANOVA, and to assess whether these changes were meaningful in a management context. Further, trajectory analysis elucidates patterns of recovery following bleaching events under alternative management scenarios. Here we define coral recovery as the rate at which coral cover increases after simulated bleaching. To allow assessment of coral recovery, the magnitude and timing of bleaching events was fixed for each scenario and site for the trajectory analysis. Bleaching events were set to occur 20 and 38 years into the future and to reduce coral cover by 30% and 15% respectively.

Results

Interactions and impacts of multiple stressors

Modelled reef communities differed significantly under the alternative management and climatic scenarios, in terms of multivariate location (i.e. mean composition of the reef) and, in some cases, dispersion (i.e. variability of modelled communities; Table 4). There were significant interaction effects between water quality (i.e. nutrification and sedimentation, hereafter 'nutrification-sedimentation' or NS), fishing (F) and bleaching (B).

There were instances of significant interactions between all pairs of the three stressors (i.e. NS×F, NS×B, and F×B) in both the PERMANOVA and PERMDISP analyses. However, the F×B interaction occurred only where a significant three-way interaction was also evident, and thus will be analysed only in the context of the three-way interaction. In considering results, first we characterise the NS×F and NS×B interactions in terms of reef state only, through examining those terms which were significant in the PERMANOVA but did not differ in dispersion between scenarios (i.e. non-significant PERMDISP results). Second, we investigate NS×F and NS×B interactions terms which were significant in both the PERMANOVA and PERMDISP, which affords some insights into the predictability of future reef condition under alternative management and bleaching scenarios. Finally, we examine the nature of significant NS×F×B interactions, guided by our analysis of the two-way interactions.

Characterising two-way interactions between stressors: reef state. The precise nature of the NS×B and NS×F interactions differed among scenarios, but in each case were consistent across all sites and time periods. For both interactions, reef state under scenarios in which nutrification-sedimentation was present (at any level of the other stressor), was characterized by high macroalgae cover and low piscivorous and herbivorous biomass (Figure 2, 3). Given that the NS×B and NS×F interactions did not differ temporally or spatially, we present only a single example for each interaction in detail, both for Lucero after 20 years (Figure 2, 3).

For the NS×B interaction, while there was a significant distinction between levels of bleaching when this stressor acted alone, when bleaching occurred together with nutrification-sedimentation, there was no difference between the levels of bleaching (Figure 2). This was because nutrification-sedimentation was dominant to bleaching.

Conversely, for the NS×F interaction, all levels of one factor were significant at each level of the other factor, but the combined effect of both impacts was less than the sum of the individual effects (Figure 3). Following Anderson [75], we interpreted effect size as average Euclidean distance between groups. Thus the effect size of each level of NS or fishing (singularly or combined) can be assessed through looking at the average Euclidean distance between reef state under the no impact scenario and scenarios in which NS and/or fishing is present. Therefore, fishing and nutrification-sedimentation may both be described as dominant stressors; the effect of either stressor on its own degrades the reef to such an extent that there was limited scope for other stressors to cause further damage. The effect size for nutrification-sedimentation was greater than for both levels of fishing (Figure 3b), indicating that nutrification-sedimentation degraded the reef to a much greater extent than fishing. When the two impacts occurred together, poor water quality resulted in algal overgrowth of corals, with a concomitant reduction in fish biomass. In these circumstances the reef was so degraded that further loss of herbivores (i.e. grazing pressure) and piscivores due to fishing did not have a large effect on benthic composition. For NS×B and NS×F interaction terms which were significant in terms of both reef state and dispersion (i.e. significant in the PERMANOVA and PERMDISP analyses; Table 4), pairwise and CAP analyses indicated that the nature of the interactions were consistent with those described above.

Characterising two-way interactions between stressors: variability of modelled reef state. Patterns of dispersion differed between the NS×B and NS×F interaction, but in each case were consistent across all sites and time periods. For both kinds of interactions modelled reef state was significantly less variable for scenarios in which nutrification-sedimentation was present (Figure 4 and Figure 5), and this trend was more pronounced for the more degraded sites (i.e. Tomasa and Cangaluyan; Figure 4a and Figure 5a). Hence for both interactions we examine an example of a degraded site (Figure 4a and Figure 5a) and less degraded site (Figure 4b and Figure 5b).

For the NS×B interaction dispersion of modelled reef state did not differ significantly between levels of bleaching for either level of nutrification-sedimentation (Figure 4). However, for the NS×F interaction, variability in reef state differed significantly between levels of fishing under either level of nutrification-sedimentation, with dispersion increasing as fishing decreases (Figure 5). Again this trend was more pronounced for the more degraded sites (Figure 5a). These results suggest that future reef state is more predictable for degraded sites which are subject to high impact stressors (i.e. nutrification-sedimentation and to a less extent fishing) because these sites converge on a reef state dominated by algae with few fish and coral, and have low probability of recovery.

Table 4. Summary of results from PERMANOVA and PERMDISP analyses of model output under 18 scenarios of different combinations of nutrification and sedimentation (NS), fishing (F) and bleaching (B).

Site	Effect	PERMANOVA *p*(perm)					PERMDISP *p* (perm)				
		years from 2008					years from 2008				
		df	5	10	20	40	df	5	10	20	40
Malilnep	NS	1	**0.0002**	**0.0002**	**0.0002**	**0.0002**	1	**0.0002**	0.6754	**0.0003**	0.1129
	F	2	**0.0002**	**0.0002**	**0.0002**	**0.0002**	2	0.7176	**0.0077**	**0.0005**	**0.0004**
	B	2	**0.0010**	**0.0064**	**0.0002**	**0.0002**	2	0.8134	0.1817	**0.0269**	0.0875
	NS x F	2	**0.0004**	**0.0002**	**0.0002**	**0.0002**	5	**0.0002**	0.0516	0.0518	0.0632
	NS x B	2	0.0870	**0.0084**	**0.0002**	**0.0002**	5	-	0.8328	0.2008	0.2047
	F x B	4	0.0862	0.4390	0.4660	0.6748	8	-	-	-	-
	NS x F x B	4	0.0800	0.0514	0.9196	0.1606	17	-	-	-	-
Lucero	NS	1	**0.0002**	**0.0002**	**0.0002**	**0.0002**	1	**0.001**	**0.0066**	**0.0208**	**0.011**
	F	2	**0.0002**	**0.0002**	**0.0002**	**0.0002**	2	0.0946	0.0890	**0.0062**	0.7218
	B	2	**0.0004**	**0.0036**	**0.0002**	**0.0002**	2	0.8224	0.4520	**0.0002**	**0.0114**
	NS x F	2	0.3388	**0.0118**	**0.0002**	**0.0046**	5	-	**0.0176**	0.4333	0.1418
	NS x B	2	0.2612	**0.0086**	**0.0002**	**0.0008**	5	-	**0.0252**	0.3537	**0.0091**
	F x B	4	0.0740	0.3660	0.1588	0.3954	8	-	-	-	-
	NS x F x B	4	0.1108	0.6486	0.0716	0.2574	17	-	-	-	-
Cangaluyan	NS	1	**0.0002**	**0.0002**	**0.0002**	**0.0002**	1	**0.0002**	**0.0002**	**0.0002**	**0.0002**
	F	2	**0.0002**	**0.0002**	**0.0002**	**0.0002**	2	**0.0158**	0.6322	**0.0352**	0.1668
	B	2	**0.0002**	**0.0002**	**0.0002**	**0.0002**	2	**0.0194**	**0.0002**	**0.0002**	**0.0002**
	NS x F	2	**0.0002**	**0.0002**	**0.0002**	**0.0002**	5	**0.0002**	**0.0002**	**0.0002**	**0.0002**
	NS x B	2	**0.0002**	**0.0002**	**0.0002**	**0.0002**	5	**0.0002**	**0.0002**	**0.0002**	**0.0002**
	F x B	4	0.4844	**0.0062**	**0.0016**	**0.0002**	8	-	**0.0002**	**0.0002**	**0.0002**
	NS x F x B	4	0.3200	**0.0388**	**0.0160**	**0.0002**	17	-	**0.0002**	**0.0002**	**0.0002**
Tomasa	NS	1	**0.0002**	**0.0002**	**0.0002**	**0.0002**	1	0.1792	0.358	0.0582	**0.0008**
	F	2	**0.0002**	**0.0002**	**0.0002**	**0.0002**	2	**0.0002**	**0.0002**	**0.0002**	**0.0002**
	B	2	**0.0052**	**0.0002**	**0.0002**	**0.0036**	2	0.9362	0.1470	**0.0326**	0.0814
	NS x F	2	**0.0002**	**0.0002**	**0.0002**	**0.0002**	5	**0.0001**	**0.0001**	**0.0001**	**0.0001**
	NS x B	2	0.5398	**0.0002**	**0.0002**	**0.0010**	5	-	0.4491	0.1022	**0.0003**
	F x B	4	0.5980	0.0534	**0.0002**	**0.0004**	8	-	-	**0.0001**	**0.0001**
	NS x F x B	4	0.0816	0.1506	**0.0084**	**0.0008**	17	-	-	**0.0001**	**0.0001**

For each site, a PERMANOVA and a PERMDISP were conducted on simulated reef state for each of four points in time selected *a priori*, i.e. 5, 10, 20 and 40 years into the future. All analyses were conducted using normalised data, Euclidean distance and 4,999 permutations of residuals under a reduced model. Bolded values are significant at $\alpha = 0.05$.

Characterising three-way interactions between stressors. Three-way interactions between nutrification-sedimentation, fishing and bleaching, were significant only for Tomasa and Cangaluyan (Table 4). All of these high order interaction terms were also significant for dispersion (i.e. significant in the PERMDISP) and therefore it is not possible to interpret the nature of the interactions solely in terms of mean reef state. However, pairwise PERMANOVA tests and PERMDISP analyses indicated that three-way interaction terms tend to manifest in the same way for both sites and all time periods. Pairwise PERMANOVA tests between each of the 18 scenarios (153 tests) indicated that differences in reef state identified for the two-way effects above were evident in the three-way interactions; namely reef state differs significantly between the different levels of fishing, bleaching, and nutrification-sedimentation, except between levels of bleaching when nutrification-sedimentation was also present. For the F×B interaction (where nutrification-sedimentation was absent) all levels of fishing were significant over all levels of bleaching. However, while all levels of bleaching were significant when fishing was not present, when fishing was present (at either level), there was no difference between scenarios in which bleaching was absent and those in which bleaching occurred at a low frequency. Note that due to the large number of *a posteriori* pairwise tests, there was a risk of compounding of Type I errors (following Anderson [76] p values were not adjusted to yield constant experiment-wise error rates). However, since significant terms were highly significant ($p<0.0004$ in all cases) and non-significant terms were highly non-significant ($p>0.1259$ in all cases), there is no ambiguity in the results.

Patterns in dispersion of reef state identified for the two-way interactions were similarly evident for third order interactions. Following the method of Anderson et al. [76], we plotted the mean distance-to-centroid in Euclidean distance to visualize patterns of dispersion (see Figure S5 for an example of Cangaluyan after 10

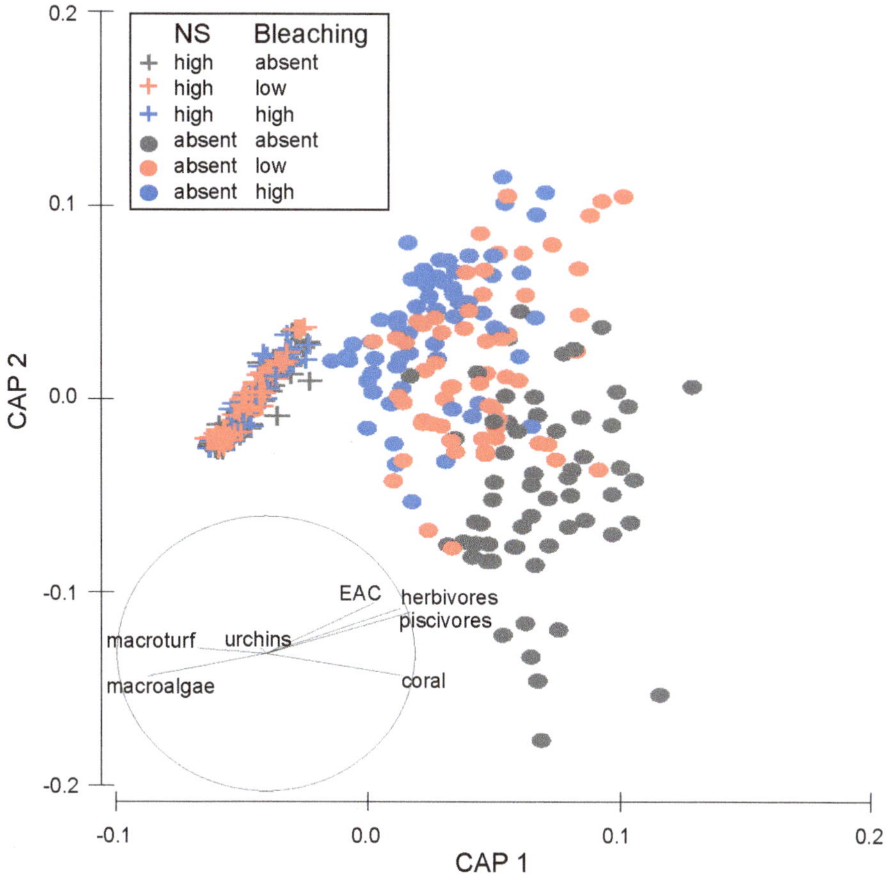

Figure 2. CAP ordination of modelled reef state showing the interaction between nutrification-sedimentation and bleaching. CAP ordination of modelled reef state at Lucero after 20 years indicating the nature of the two-way interaction between nutrification-sedimentation (NS) × bleaching (B), under 18 scenarios, with 20 model runs per scenario. NS had a significant effect on reef state across all levels of B (pairwise $p<0.0001$ in all cases). However, while there was a clear distinction between levels of bleaching when the disturbance acted alone (pairwise $p<0.0007$ in all cases), when it acted in conjunction with NS, there was no difference between levels of bleaching (pairwise $p>0.1887$ in all cases). A vector overlay of Spearman rank correlations ($|r|>0.20$) between functional groups and ordination axes is displayed on the plot, and indicates that modelled reef state represented by points on the left on the plot are dominated by macroturf and macroalgae, with low coral cover and fish biomass relative to reef state represented to the right hand side.

years). As we found for the two-way interactions, dispersion was lower when nutrification-sedimentation was present compared to when it was not. Within levels of nutrification-sedimentation, dispersion increased as fishing level decreased, while variability in reef state did not differ between the different levels of bleaching.

Trajectory analysis

Our analysis of reef state trajectories for various fishing and water quality scenarios under decadal bleaching revealed that differences in reef state under various scenarios detected in the PERMANOVA were driven by changes in all functional groups (Figure 6). The dominant effect of nutrification-sedimentation was evident for all sites (Figure 6). Under scenarios in which nutrification-sedimentation was absent, fish biomass and coral cover was higher, and algae cover was lower, than under scenarios in which water quality was poor. Importantly, coral cover declined to 0% within 12 years at all sites under scenarios in which reefs were subject to nutrification-sedimentation. Thus differences in reef state after 12 years under any fishing intensity (when nutrification-sedimentation was present), were driven largely by fish biomass and to a lesser extent algal cover. However, under poor water quality scenarios, the biomass of both herbivorous and piscivorous fish was less than 1400 kg km^{-2} for all sites. For reference, Nañola et al. [3] classed fish biomass at less than 5000 kg km^{-2} as 'very low' in their report on the status of Philippine coral reefs. Thus fish biomass was too small to be meaningful in a management context, suggesting that regulating fishing will have negligible effect when water quality is poor.

Under scenarios in which nutrification-sedimentation was not present, the effect of different levels of fishing on reef state was evident for all functional groups at all sites (except Tomasa in regards to fish). Coral cover increased as fishing pressure decreased for all sites; after 40 years coral cover was 15% higher at Lucero, 24% higher at Malilnep, 300% higher at Tomasa and 400% higher at Cangaluyan, under the zero fishing scenario compared with the high impact fishing scenario. While the differences in coral cover between fishing scenarios were highest for Tomasa and Cangaluyan, absolute coral cover was less than 3% at Tomasa and less than 12% at Cangaluyan for all scenarios. Further, at Tomasa, under all scenarios of fishing and water quality, modelled fish biomass was almost identical and was less than 500 kg km^{-2}. This is largely because fish recruitment at Tomasa was constrained by low coral cover (i.e. less than 3% under all scenarios). While statistical analysis indicated that there

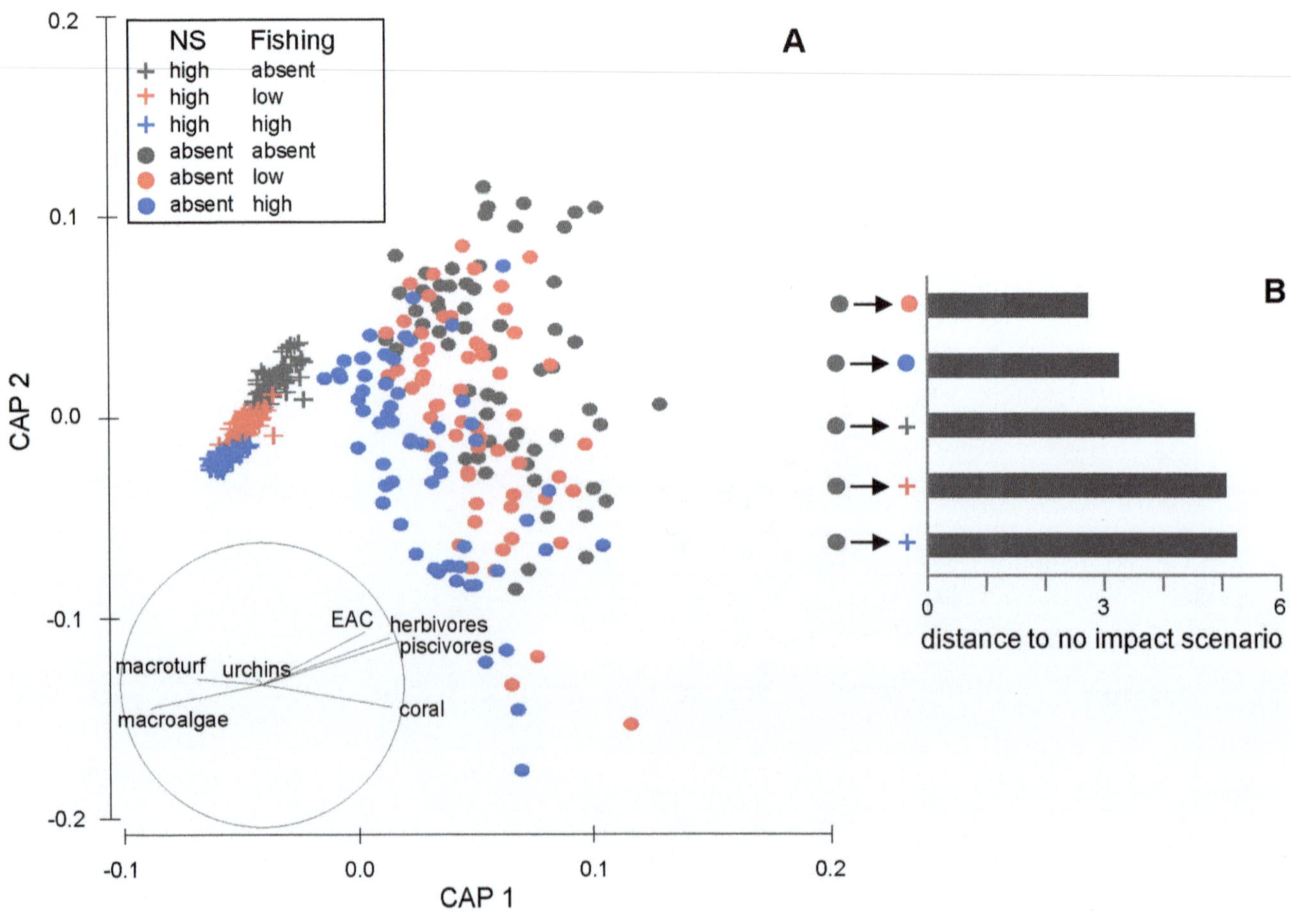

Figure 3. CAP ordination of modelled reef state showing the interaction between nutrification-sedimentation and fishing. CAP ordination of modelled reef state at Lucero after 20 years (a), reflecting the nature of the two-way interaction between nutrification and sedimentation (NS) × fishing (F), under 18 scenarios and 20 model runs per scenario. The average Euclidean distance between reef state under the no impact scenario and scenarios in which NS and/or fishing is present (b) indicates effect sizes. All levels of one factor were significant (pairwise $p<0.0354$ in all cases) at each level of the other factor, but the combined effect of both stressors was less than the sum of the individual effects in (b), indicating an antagonistic interaction. A vector overlay of Spearman rank correlations ($|r|>0.20$) between functional groups and ordination axes is displayed on the plot, and indicates that modelled reef state represented by points to the left on the plot are dominated by macroalgae and sea urchins, with low coral cover and fish biomass relative to reef state represented on the right hand side.

was a significant difference between all fishing and water quality scenarios, trajectory analysis showed that these differences are not ecologically meaningful and thus are irrelevant in a management context for Tomasa. Therefore, projected responses to management were strongly non-linear and depended on the initial state of the reef.

Coral recovery post bleaching

Analysis of the trajectories for scenarios in which nutrification-sedimentation was absent (i.e. where corals were present) highlighted that management of local stressors influenced patterns of recovery after bleaching events, although this was dependent on initial reef state (Table 5). For all sites bleaching events were modelled to occur in 2028, where coral cover was reduced by 30% (relative to cover in 2027), and in 2044, where coral cover was reduced by 15%. Loss of coral cover after each bleaching event was accompanied by an increase in algae cover and a decrease in fish biomass (which occurred because fish recruitment was impacted by reduced coral cover) for all sites. Coral recovery post-bleaching in 2028, and prior to the 2044 event, differed by site and fishing scenario. There was a clear trend in coral recovery between fishing scenarios for the least degraded sites (Lucero and Malilnep), that showed that recovery in absolute and relative terms increased as the intensity of fishing decreased (Table 5). Although this trend was apparent for Cangaluyan for scenarios in which fishing was low or absent, under high fishing pressure there was no recovery of coral cover post bleaching in 2028. This was also the case for the other highly degraded site, Tomasa, where there was no recovery under high and low fishing scenarios.

The increase in coral cover varied between sites (in relative and absolute terms) reflecting initial reef state; i.e. changes in cover was highest at Lucero (which increased by more than 50%), followed by Malilnep and Cangaluyan. The exception to this trend was Tomasa, where the relative increase in coral cover (when fishing was present) was 50% during 2028–2044, although the absolute increase in cover was only 1%. In Tomasa and Cangaluyan during 2028–2044 changes in coral cover – i.e. very small increases in coral cover of 1–2% or maintenance of very low coral cover – were mainly due to external coral recruitment. Thus coral recovered more quickly after bleaching events at those sites which were subject to fewest additional stressors and disturbances, and were least degraded (and thus had higher endogenous recruit-

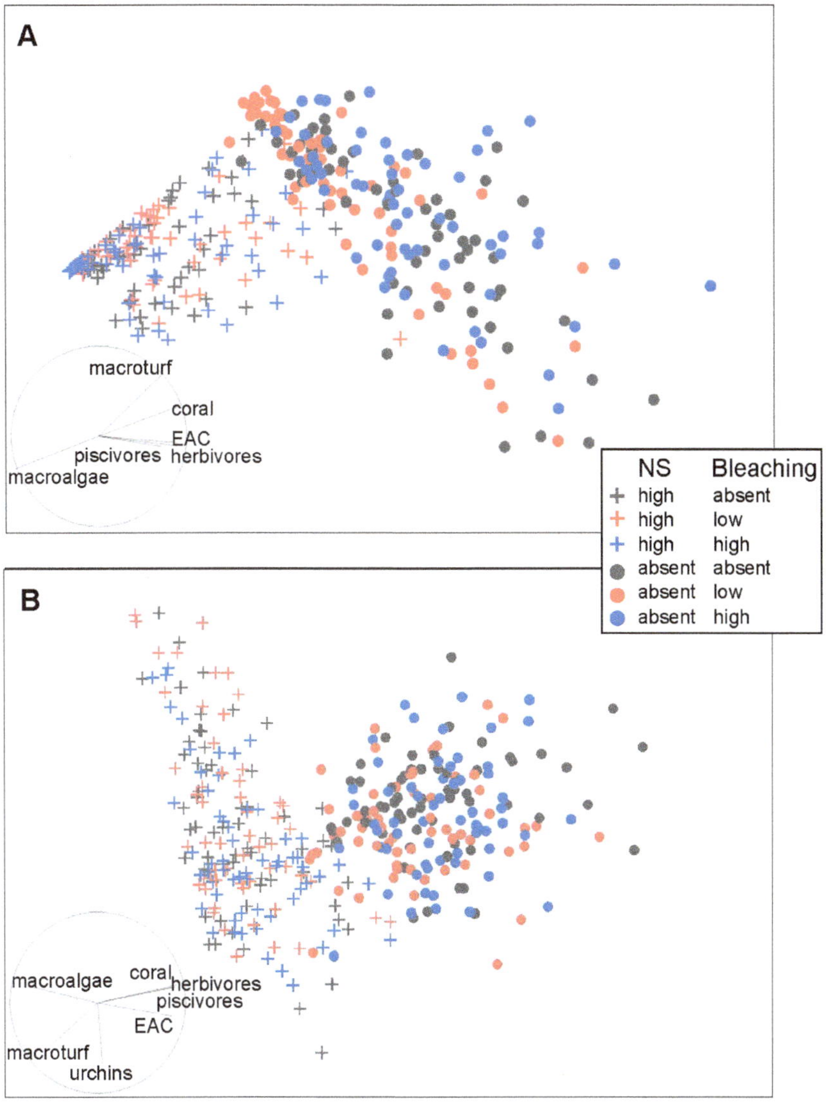

Figure 4. MDS ordinations of modelled reef state showing the interaction between nutrification-sedimentation and bleaching. MDS ordinations of modelled reef state showing the nature of the interaction between nutrification-sedimentation (NS) and bleaching (B) at Cangaluyan after 5 years (a) and Lucero after 10 years (b). For scenarios in which NS was not present (dots), variability in reef state increased significantly at Cangaluyan after 5 years (a; pairwise $p<0.0018$ in all cases), and was marginally significant for Lucero after 10 years (b; pairwise $p<0.0385$ in all cases). Dispersion in reef state was not significantly different between any level of B within either level of NS for either site (pairwise $p>0.1753$ in all cases). A vector overlay of Spearman rank correlations ($|r|>0.20$) between functional groups and ordination axes is displayed on the plot, and indicates that modelled reef state represented by points to the left on the plot are dominated by macroalgae, with low coral cover and fish biomass relative to reef state represented on the right hand side. 2D stress $=0.09$ (a) and 0.14 (b).

ment). These results suggest that if coral becomes too degraded, such as at Tomasa, coral will not recover from bleaching-induced mortality irrespective of the absence of other stressors.

Discussion

While reducing the impact of local stressors is widely advocated to bolster coral reef resilience under anticipated future changes in the climate [2,7], the degree to which management of local stressors might affect future trajectories of reef state remains a critical knowledge gap [13]. Models are useful for understanding future reef state under alternative management and climatic scenarios, but to date have not been developed as decision support tools for local-scale management of coral reefs subject to multiple stressors. We used a simulation model of coral reefs to estimate future reef state for Bolinao, the Philippines, under alternative management and climatic scenarios. We found that management of local-scale stressors (i.e. water quality and to a lesser extent fishing) can have significant impact on future reef state. In discussing this result we first consider interactions between multiple stressors and then factors influencing coral recovery post bleaching. Next we discuss the second aim of our study, namely assessing the utility of simulation modelling as a decision support tool for management of coral reefs.

Interactions and impacts of multiple stressors

Integral to assessing how management of local-scale stressors might influence future trajectories of reef state is investigating the cumulative impact of those stressors, including whether the effect

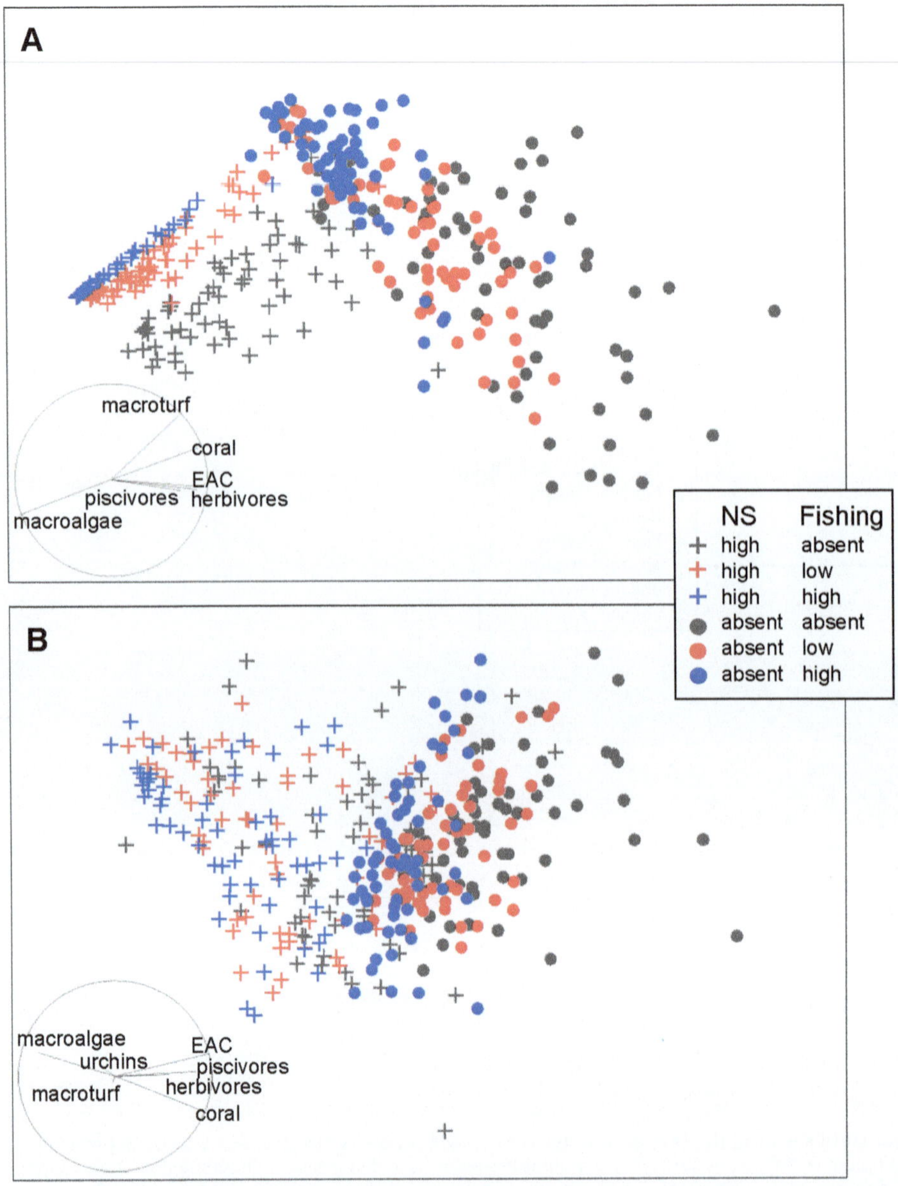

Figure 5. MDS ordinations of modelled reef state showing the interaction between nutrification-sedimentation and fishing. MDS ordinations of modelled reef state showing the nature of the interaction between nutrification-sedimentation (NS) and fishing (F). At Cangaluyan after 5 years (a) variability in reef state increased significantly as the magnitude of F and NS decreased (pairwise $p<0.0385$ all cases). This pattern of dispersion was also evident at Malilnep after 5 years (b) but was less clear and differences in dispersion were marginally significant (pairwise $p<0.0463$ in all cases). A vector overlay of Spearman rank correlations ($|r|>0.20$) between functional groups and ordination axes is displayed on the plot, and indicates that modelled reef state represented by points to the left on the plot are dominated by macroalgae, with low coral cover and fish biomass relative to reef state represented on the right hand side. 2D stress $=0.09$ (a) and 0.15 (b).

of multiple stressors is the sum of single-stressor effects (additive), or an amplification (synergism) or a reduction (antagonism) relative to an additive effect [21]. We identified antagonistic interactions between water quality, fishing and bleaching, that were driven by a dominant stressor mechanism rather than through active opposition between the stressors. The antagonistic interaction occurred because modelled reef state was initially degraded and had limited capacity to degrade further in response to multiple stressors. For example poor water quality resulted in such a degraded reef state that there was little or no coral remaining to be bleached. The interaction identified between water quality and fishing also reflected antagonism through a dominant stressor mechanism because either stressor acted individually to degrade the reef to such an extent that there was limited scope for the other to cause further damage.

The antagonistic interactions between stressors identified in this study may be driven by the asymmetric effects of the three stressors. The individual effect of poor water quality on overall reef state was much greater than the individual effect of the two other stressors, particularly bleaching. This suggests that the interactions between poor water quality and the other two stressors are caused by the dominant impact of nutrification-sedimentation. Antagonistic interactions driven by a dominant stressor have been identified in other studies [15,78]. In particular, Darling et al. [15]

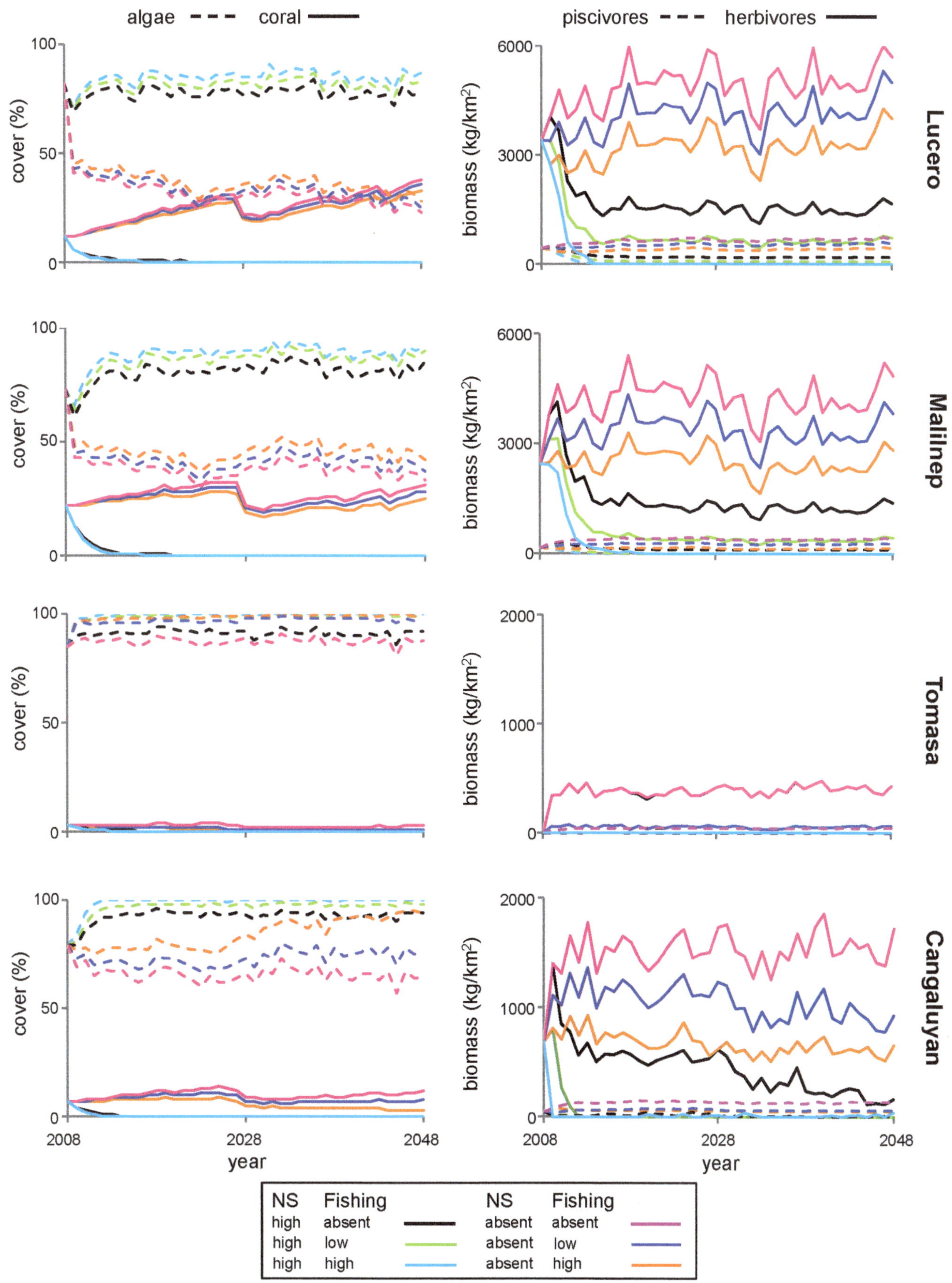

algae
coral
piscivores
herbivores
cover (%)
biomass (kg/km²)
Lucero
Malilnep
Tomasa
Cangaluyan
100
50
0
6000
3000
2000
1000
2008
2028
2048
year
NS Fishing
high absent
high low
high high
absent absent
absent low
absent high

Figure 6. Model trajectories of benthic cover and fish biomass simulated 40 years into the future. Model trajectories of mean benthic cover and fish biomass from 20 model runs under six scenarios for all sites. The six scenarios were simulated for 40 years and were combinations of two levels of nutrification and sedimentation (NS) and three levels of fishing intensity (F) under decadal bleaching. 'Algae' refers to the combined covers of macroturf and macroalgal. Trajectories for sea urchins are not shown as they were stable for the entire period for all scenarios. Coral recovery after the major bleaching event in 2028 and prior to another in 2044, differed between sites and fishing scenarios.

suggested that coral bleaching was driving an antagonistic interaction between bleaching and fishing, which they identified using 20 years of data from Kenyan coral reefs. Antagonistic interactions have also been identified between the effects of fishing and nutrification on macroalgae in coral reefs [24], while other studies have found this pair of stressors interact synergistically [79,80]. This disparity of interactions identified for the same stressors highlights that interactions between stressors are likely to be dependent on context, including the initial condition of the system and the magnitude of the stressors.

Management of multiple stressors requires an understanding of how interactions between stressors manifest in different contexts. This is particularly pertinent to coral reef management given that reefs are almost always subject to more than one stressor [6], and that the nature of interactions between stressors can fundamentally affect management outcomes [19]. However, while coral reef decline has been widely attributed to the cumulative effects of multiple stressors acting simultaneously [12,18,81], misunderstanding of the combined impact of stressors persists [13,23]. The effect of multiple stressors is often assumed to be the additive accumulation of impacts associated with single stressors [13,14], and Dunne [21] asserts that in the few cases where synergies are referred to in the context of coral reefs, the term 'synergy' is misused. This is in part due to the inherent difficulty of teasing apart the independent and combined effects of multiple stressors [82], especially at large spatial scales. Here we demonstrate the utility of modelling for understanding the impact of cumulative stressors and suggest its application in future research on cumulative impacts.

Coral recovery post bleaching

A critical component of understanding the importance of managing local-scale stressors to conserve coral reefs in the context of future climate change is knowledge of the factors influencing coral recovery post bleaching. We found that simulated coral recovery following bleaching was greatest at sites that were least degraded and subject to the least impact from local stressors. For the most degraded site (i.e. Tomasa), coral did not recover (under high and medium levels of fishing) or showed very little recovery (under no fishing), with a only 1% increase in absolute coral cover over 18 years. This indicates a hysteresis effect, and suggests that if coral cover becomes too depleted recovery is not possible even in the absence of stressors. This is in part because endogenous supply of coral recruits is less for degraded reefs, with a number of studies showing that coral recovery is highly dependent on the extent of remnant coral survival [83,84]. Note that although it was beyond the scope of our study to examine the nature of phase-shifts, this model can be used to this end [41,85].

These results support consensual understanding of conditions which foster coral resilience following bleaching-induced coral mortality [7,8], and are consistent with experimental studies showing that reducing herbivorous fish biomass decreases coral recovery [11]. However, results from site-specific studies which assessed coral recovery inside and outside marine reserves are mixed; McClanahan [86] found no differences in recovery inside and outside reserves, while Mumby and Harborne [87] found greater recovery within marine reserves. This highlights that the extent to which marine reserves will assist coral recovery following bleaching, and other climate-induced coral mortality, will vary with context, including factors such as the presence of other concurrent stressors, coral community structure and absolute level of live coral cover post bleaching [84,87,88]. Indeed, in the case of

Table 5. Increase in coral cover between the 2028 and 2044 bleaching events for all sites under scenarios in which fishing occurred but nutrification-sedimentation was absent.

Site	Fishing scenario	Coral cover post 2028 bleaching event	Increase in coral cover between 2028 - 2044	
			absolute (% increase)	relative (% increase)
Lucero	high	20	11	55
	low	21	12	57
	absent	22	13	59
Malilnep	high	19	4	21
	low	21	5	24
	absent	22	7	32
Cangaluyan	high	5	-	-
	low	7	1	14
	absent	9	2	22
Tomasa	high	0	-	-
	low	1	-	-
	absent	2	1	50

Dashes indicate no change in coral cover.

Bolinao, our study suggests that if water quality is not regulated, marine reserves are unlikely to confer any significant benefits to coral health in the future.

Managing local stressors in the face of climate change

Our research supports the paradigm that managing local-scale stressors is critical to the persistence of coral reefs in the context of global climate change, a concept that is widely advocated [2,8] but still subject to debate [17,18]. Our analysis of the cumulative impact of bleaching, poor water quality and fishing indicate that management of the two local stressors will significantly influence future reef state under climate change. We found that coral recovery post bleaching can be bolstered through management of fishing by building reef health prior to bleaching and aiding recovery after bleaching-induced mortality. However, the extent to which reducing fishing or improving water quality will influence future coral reef health will also vary with context [15,87].

For our study sites at Bolinao, poor water quality emerged as the pre-eminent driver of deterioration of coral reefs, and overrode the effects of fishing and bleaching. The combined effect of excess nutrient and sediment input is well known to degrade real coral reef systems [89–91], and the two stressors have been suggested to interact synergistically to affect reef health [92]. For our modelled reef system the deleterious impact of poor water quality was worsened by the low level of fish grazing pressure exerted on the modelled reef, even under no fishing scenarios.

The utility of modelling for supporting environmental management

The utility of simulation modelling to environmental management lies in the capacity of models to estimate future ecosystem state under alternative management and climatic scenarios. Through this process, models can greatly aid identification of drivers of degradation and consequently provide guidance on appropriate management actions, where they will most likely to be effective and over what time scales. Given that resources for management are limited and that management is often constrained by human use of marine resources, this information also usefully informs a 'triage' approach to management, whereby locations for undertaking management actions are prioritised according to where actions are most likely to be effective given existing limitations such as resources, time and knowledge.

We have demonstrated how simulation modelling can be used to support coral reef management at local scales. Our results clearly indicate that reefs in Bolinao would benefit from improving water quality above all other stressors, because even in the absence of fishing and bleaching, the present levels of nutrient enrichment and sediment deposition are likely to lead to loss of corals within a short time scale (within 12 years). Improvement in water quality could be achieved by regulating the aquaculture industry, including reducing the stocking density and number of fish cages. If water quality is improved, our results suggest that under decadal bleaching the present level of fishing can be supported at Malilnep and Lucero, and half of the present level at Cangaluyan, without further decline in coral cover. Conversely, any management intervention at Tomasa, including action to improve water quality, is unlikely to make a meaningful difference to reef health over at least a 40 year time frame. Thus, responses to management are strongly non-linear depending on the state of the reef, and management focused initially on the Lucero and Malilnep will likely yield greatest recovery for the least investment. This example highlights the importance of site-specific information in management.

Critical to using models to forecast future reef state under alternative scenarios is consideration of the predictability of those trajectories. Variability in modelled reef state (i.e. dispersion of modelled reef state between model runs) affords insights into the predictability of future reef state. We found reef state more predictable for degraded reefs, especially if the reef was subject to high impact stressors (i.e. poor water quality and to a lesser extent fishing), because these sites converge on a reef state dominated by algae with few fish and coral with low probability of recovery. Note however, that predicting the exact future state of a reef under a given scenario is not our intention, nor is it necessary to achieve good management outcomes. Rather, managers are more interested in broad-scale trends that indicate whether reefs will be broadly healthy or degraded.

Further, in considering the utility of simulation modelling to environmental management, it is important to acknowledge the host of other biological and socioeconomic factors which were not included in our modelling but which will greatly influence decision making in regards to management of fisheries, bleaching and water quality in Bolinao and elsewhere. For example, marine reserve design may be contingent upon factors such as the presence of threatened species, distribution of fishing effort by gear type or cultural practices associated with fisheries. Thus, we suggest that simulation modelling be used in combination with other decision support tools such as fisheries stock assessment models [93] or reserve zoning software [94]. Critically, simulation models, such as the one used here, are intended to provide a subset of management options, which should then be evaluated in light of the relevant socioeconomic context [39].

Critiques and caveats

Uncertainty is inherent to modelling complex systems [95] and arises at all stages of the modelling process. When using complex systems models it is necessary to determine and describe model uncertainties, especially if the model is to be used as a decision support tool for environmental management [96]. Uncertainty in relation to model structure and function is discussed in detail in Fung et al. [41] and Melbourne-Thomas et al. [39,45], so here we focus on uncertainties in relation to the validation of the model for Bolinao and the scenario analysis.

Uncertainty is related to our model validation process mainly due to the paucity of empirical observations of reef state in Bolinao in the past. We validated the model for each of the four sites through assessing whether the model could reproduce reef dynamics from 1987–2008 given empirical data on reef state in 1987 and a known series of stressors at these sites (Figure S2, Figure S3, Figure S4, Table S3, Table S4,Text S3). Although there was satisfactory correspondence between model trajectories and empirical data of reef state for 1987–2008, uncertainty is related to this validation process mainly because available empirical data were limited to two or three time periods for each functional group at each site. This highlights the importance of empirical data for simulation modelling of complex systems.

Given imperfect knowledge of both future impacts and how a system will react to these factors [97], scenario analysis necessarily involves simplifications and assumptions which leads to uncertainty in model projections. One of the foremost assumptions associated with predicting futures is that biological processes and relationships will not change in the future, which may not always hold [98]. For example, evidence is emerging that corals may have scope for acclimatization to predicted increases in sea surface temperature associated with climate change [99,100]. Further, a key assumption which we made in relation to biological processes relates to recruitment of coral and fish larvae; we assumed a constant supply of external larvae. In reality, the condition of reefs close to Bolinao is likely to change in the future, which will affect exogenous recruitment to Bolinao's reefs and consequently reef state.

Uncertainty is also related to the scenarios under which projections about the future are made. For instance, the frequency of bleaching events assumed in the future scenarios modelled here were predicted using climate models [9,71], which have their own uncertainty. Also, given uncertainty related to future frequency and nature of other climate change related stressors, such as ocean acidification and damaging storms [69,101], we modelled climate change in terms of bleaching only. We made further simplifications given uncertainty associated with the impact of climate change on biological processes, e.g. coral growth rates remained constant in our scenarios despite some evidence that coral growth may decline under bleaching [102].

In summary, there are inevitably sources of uncertainty in scenario analysis which translates to uncertainty in the associated projections. In this sense the kind of stochastic simulation model used here is realistic because uncertainty about the future is inevitable and is captured through the model providing a probabilistic spectrum of possible futures for a given scenario. Future scenarios are intended to illustrate the range of possibilities for the future, rather than predict actual trajectories [97].

Conclusions

Our analysis of potential reef state under alternative management and climatic scenarios highlights two key principles critical to management of coral reefs in the face of climate change. First, local-scale reef management to improve water quality and prevent overfishing can play a significant role in conserving coral reefs under the expected effects of global climate change. Second, to develop effective management actions it is critical to consider the combined impact of multiple stressors, and potential interactions between them. However, the relative importance of different stressors and how they interact is likely to vary with context, ensuring that generalization and extrapolation of specific findings is difficult. Therefore, managers require decision support tools to identify drivers of degradation for particular reefs, allowing the development of site-specific management. Simulation models, such as the one used here, are ideal tools in this endeavour, and can further support management decisions by providing insights into the efficacy of different management approaches under alternative climate (or other) scenarios. Thus we suggest that simulation modelling offers a critical and much needed opportunity for aiding local-scale management of coral reefs.

Supporting Information

Figure S1 The ecological processes between the seven functional groups in the local-scale mean-field model.

Figure S2 Historical timeline of stressors to Bolinao's coral reef system during 1987 – 2008.

Figure S3 Comparison of empirical observations and model trajectories of benthic covers in the four sites at Bolinao from 1987 – 2008.

Figure S4 Comparison of empirical observations and model trajectories of consumer biomass in the four sites at Bolinao from 1987 – 2008.

Figure S5 Mean distance-to-centroid in Euclidan space of the 20 predicted community states (i.e. results from Monte Carlo simulations) for each of the 18 scenarios for Cangaluyan after 10 years.

Table S1 Parameter definitions and values derived for Bolinao, the Philippines.

Table S2 Exogenous recruitment parameters for consumer groups used for historical trajectories.

Table S3 Additional forcings (i.e. stressors) used in model validation.

Table S4 Validation data for the historical reconstruction trajectories for Bolinao.

Text S1 Model equations.

Text S2 Additions to the model for Bolinao instantiation.

Text S3 Model validation.

Acknowledgments

This work is part of that undertaken by the Modelling and Decision (MDS) working group of the Coral Reef Targeted Research and Capacity Building for Management Program (CRTR-CBMP). We thank Mark Vergara and Cleto Nañola for supplying data, and Jorge Alvarez Romero, Natalie Ban, Joshua Cinner and three anonymous reviewers for helpful comments on the manuscript.

Author Contributions

Conceived and designed the experiments: GGG JMT CRJ RCG PMA. Performed the experiments: GGG JMT. Analyzed the data: GGG JMT CRJ. Contributed reagents/materials/analysis tools: GGG CRJ JMT RCG. Wrote the paper: GGG JMT CRJ RCG PMA.

References

1. Roberts CM, McClean C, Veron J, Hawkins J, Allen G, et al. (2002) Marine biodiversity hotspots and conservation priorities for tropical reefs. Science 295: 1280–1284.
2. Pandolfi JM, Connolly SR, Marshall DJ, Cohen AL (2011) Projecting coral reef futures under global warming and ocean acidification. Science 333: 418–422.
3. Burke LM, Reytar K, Spalding M, Perry A (2011) Reefs at risk revisited: World Resources Institute Washington, DC, USA.
4. Wilkinson C (2008) Status of coral reefs of the world: 2008 Townsville, Australia: Australian Institute of Marine Science. 304 p.
5. De'ath G, Fabricius KE, Sweatman H, Puotinen M (2012) The 27–year decline of coral cover on the Great Barrier Reef and its causes. Proceedings of the National Academy of Sciences 109: 17995–17999.
6. Halpern BS, Walbridge S, Selkoe KA, Kappel CV, Micheli F, et al. (2008) A global map of human impact on marine ecosystems. Science 319: 948–952.
7. Hughes TP, Baird A, Bellwood DR, Card M, Connolly S, et al. (2003) Climate change, human impacts and the resilience of coral reefs. Science 301: 929–933.
8. Hoegh-Guldberg O, Mumby PJ, Hooten AJ, Steneck RS, Greenfield P, et al. (2007) Coral reefs under rapid climate change and ocean acidification. Science 318: 1737–1742.

9. Donner S (2009) Coping with commitment: Projected thermal stress on coral reefs under different future scenarios. PLoS ONE 4:e5712.
10. Carilli J, Norris R, Black B, Walsh S, McField MD (2009) Local stressors reduce coral resilience to bleaching. PLoS ONE 4:e6324.
11. Hughes TP, Rodrigues MJ, Bellwood DR, Ceccarelli D, Hoegh-Guldberg O, et al. (2007) Phase shifts, herbivory, and the resilience of coral reefs to climate change. Current Biology 17: 360–365.
12. Veron J, Hoegh-Guldberg O, Lenton T, Lough J, Obura D, et al. (2009) The coral reef crisis: The critical importance of $<$359ppm CO_2. Marine Pollution Bulletin 58: 1428–1436.
13. Knowlton N, Jackson J (2008) Shifting baselines, local impacts and global change on coral reefs. PLoS Biology 6: 215–220.
14. Crain C, Kroeker K, Halpern B (2008) Interactive and cumulative effects of multiple human stressors in marine systems. Ecology Letters 11: 1304–1315.
15. Darling ES, McClanahan TR, Côté IM (2009) Combined effects of two stressors on Kenyan coral reefs are additive or antagonistic, not synergistic. Conservation Letters 3: 122–130.
16. Breitburg D, Riedel G (2005) Multiple stressors in marine systems. In: Norse E, Crowder L, editors. Marine conservation biology. Washington DC, USA: Island Press. pp. 416–431.
17. Keller BD, Gleason DF, McLeod E, Woodley CM, Airamé S, et al. (2009) Climate change, coral reef ecosystems, and management options for marine protected areas. Environmental Management 44: 1069–1088.
18. Baker A, Glynn P, Riegl B (2008) Climate change and coral reef bleaching: an ecological assessment of long-term impacts, recovery trends and future outlook. Estuarine, Coastal and Shelf Science 80: 435–471.
19. Hughes T, Connell J (1999) Multiple stressors on coral reefs: a long-term perspective. Limnology and Oceanography: 932–940.
20. Darling ES, Côté IM (2008) Quantifying the evidence for ecological synergies. Ecology Letters 11: 1278–1286.
21. Dunne R (2010) Synergy or antagonism—interactions between stressors on coral reefs. Coral Reefs 29: 145–152.
22. Didham R, Tylianakis J, Gemmell N, Rand T, Ewers R (2007) Interactive effects of habitat modification and species invasion on native species decline. Trends in Ecology & Evolution 22: 489–496.
23. Wernberg T, Smale DA, Thomsen MS (2012) A decade of climate change experiments on marine organisms: procedures, patterns and problems. Global Change Biology 18:1491–1498.
24. Crain C, Halpern B, Beck M, Kappel C (2009) Understanding and managing human threats to the coastal marine environment. Annals of the New York Academy of Sciences 1162: 39–62.
25. Van Kouwen F, Dieperink C, Schot P, Wassen M (2007) Applicability of decision support systems for integrated coastal zone management. Coastal Management 36: 19–34.
26. Grimm V, Railsback SF (2005) Individual-based modelling and ecology. Princeton, USA: Princeton University Press.
27. Westmacott S (2001) Developing decision support systems for integrated coastal management in the tropics: Is the ICM decision-making environment too complex for the development of a useable and useful DSS? Journal of Environmental Management 62: 55–74.
28. Peterson G, Cumming G, Carpenter S (2003) Scenario planning: A tool for conservation in an uncertain world. The Journal of the Society for Conservation Biology 17: 358–366.
29. Mumby PJ, Hedley JD, Zychaluk K, Harborne AR, Blackwell PG (2006) Revisiting the catastrophic die-off of the urchin *Diadema antillarum* on Caribbean coral reefs: Fresh insights on resilience from a simulation model. Ecological Modelling 196: 131–148.
30. Langmead O, Sheppard C (2004) Coral reef community dynamics and disturbance: a simulation model. Ecological Modelling 175: 271–290.
31. Edwards HJ, Elliott IA, Eakin M, Irikawa A, Madin JS, et al. (2011) How much time can herbivore protection buy for coral reefs under realistic regimes of hurricanes and coral bleaching? Global Change Biology 17: 2033–2048.
32. Little LR, Punta AE, Mapstone BD, Pantus F, Smith ADM, et al. (2007) ELFSim—A model for evaluating management options for spatially structured reef fish populations: An illustration of the "larval subsidy" effect Ecological Modelling 205: 381–396.
33. Lirman D (2003) A simulation model of the population dynamics of the branching coral *Acropora palmata*: Effects of storm intensity and frequency. Ecological Modelling 161: 169–182.
34. Blackwood JC, Hastings A, Mumby PJ (2011) A model-based approach to determine the long-term effects of multiple interacting stressors on coral reefs. Ecological Applications 21: 2722–2733.
35. McClanahan TR (1995) A coral reef ecosystem-fisheries model: impacts of fishing intensity and catch selection on reef structure and processes. Ecological Modelling 80: 1–19.
36. Wolanski E, Richmond RH, McCook L (2004) A model of the effects of land-based, human activities on the health of coral reefs in the Great Barrier Reef and in Fouha Bay, Guam, Micronesia. Journal of Marine Systems 46: 133–144.
37. Kubicek A, Muhando C, Reuter H (2012) Simulations of long-term community dynamics in coral reefs - how perturbations shape trajectories. PLOS Computational Biology 8.
38. Melbourne-Thomas J, Johnson C, Fulton E (2011) Regional-scale scenario analysis for the Meso-American Reef system: Modelling coral reef futures under multiple stressors. Ecological Modelling 222: 1756–1770.
39. Melbourne-Thomas J, Johnson CR, Aliño PM, Geronimo RC, Villanoy CL, et al. (2011) A multi-scale biophysical model to inform regional management of coral reefs in the western Philippines and South China Sea. Environmental Modelling & Software 26: 66–82.
40. Burke L, Reytar K, Spalding M, Perry A (2012) Reefs at Risk Revisited in the Coral Triangle. World Resources Institute.
41. Fung T, Seymour RM, Johnson CR (2011) Alternative stable states and phase shifts in coral reefs under anthropogenic stress. Ecology 92: 967–982.
42. Sumalde Z, Francisco K, Penales M (2003) Pollution-induced fish kills in Bolinao: Effects of excessive aquaculture structures and overstocking. In: Francisco H, de los Angeles M, editors. Economy and environment: Selected reading in the Philippines. Manila, the Philippines: EDM Press.
43. PhilReefs (2008) Reefs Through Time 2008: Initiating the State of the Coasts Reports. Quezon Citythe Philippines: Coral Reef Information Network of the Philippines (PhilReefs), MPA Support Network, Marine Environment & Resources Foundation, Inc. and the Marine Science Institute, University of the Philippines. 152 p.
44. Horigue V, Aliño PM, White AT, Pressey RL (2012) Marine protected area networks in the Philippines: Trends and challenges for establishment and governance. Ocean & Coastal Management 64:15–26.
45. Melbourne-Thomas J, Johnson CR, Fung T, Seymour RM, Chérubin LM, et al. (2011) Regional-scale scenario modeling for coral reefs: a decision support tool to inform management of a complex system. Ecological Applications 21: 1380–1398.
46. Feary D, Almany G, McCormick MI, Jones GB (2007) Habitat choice, recruitment and the response of coral reef fishes to coral degradation. Oecologia 153: 727–737.
47. Holbrook SJ, Schmitt RJ, Brooks AJ (2008) Resistance and resilience of a coral reef fish community to changes in coral cover. Marine Ecology Progress Series 371: 263–271.
48. Travers H, Clements T, Keane A, Milner-Gulland EJ (2011) Incentives for cooperation: The effects of institutional controls on common pool resource extraction in Cambodia. Ecological Economics 71: 151–161.
49. Volk S, Thöni C, Ruigrok W (2012) Temporal stability and psychological foundations of cooperation preferences. Journal of Economic Behavior & Organization 81: 664–676.
50. Bellwood DR, Hughes TP, Folke C, Nystrom M (2004) Confronting the coral reef crisis. Nature 429: 827–833.
51. Fung T (2009) Local scale models of coral reef ecosystems for scenario testing and decision support. Phd thesis: University College London.
52. McManus J, Nañola C, Reyes R, Kesner K (1992) Resource ecology of the Bolinao coral reef system. Manila, the Philippines: International Center for Living Aquatic Resources Management (ICLARM) Studies and Reviews 22.
53. David C, Maria Y, Siringan F, Reotita J, Zamora P, et al. (2008) Coastal pollution due to increasing nutrient flux in aquaculture sites. Environmental Geology 58: 447–454.
54. Sumagaysay-Chavoso N, San Diego-McGlone M (2003) Water quality and holding capacity of intensive and semi-intensive milkfish (*Chanos chanos*) ponds. Aquaculture 219: 413–429.
55. Castillo D, Saysel AK (2005) Simulation of common pool resource field experiments: a behavioral model of collective action. Ecological Economics 55: 420–436.
56. Villanueva RD, Yap HT, Montaño MNE (2005) Survivorship of coral juveniles in a fish farm environment. Marine Pollution Bulletin 51: 580–589.
57. Holmer M, Marbá N, Terrados J, Duarte CM, Fortes MD (2002) Impacts of milkfish (*Chanos chanos*) aquaculture on carbon and nutrient fluxes in the Bolinao area, Philippines. Marine Pollution Bulletin 44: 685–696.
58. Fisheries Resource Management Project (FRMP) (2006) Sustainable Philippines Fisheries Agenda. Volume 1: National Report.
59. Hilomen VV, Jimenez LF (2001) Fisheries of Lingayen Gulf. Manilathe Philippines: Marine Environment Resource Foundation-University of the Philippines-Marine Science Institute. 125 p.
60. White AT, Deguit E, Jatulan W, Esima-Osorio L (2006) Integrated coastal management in Philippine local governance: Evolution and benefits. Coastal Management 34: 287–302.
61. Vergara MWR (2009) Coral community structure of the Bolinao Reef System, Northwestern Philippines. MSc Thesis: University of the Philippines, Manila.
62. Geronimo RC (2009) Trophic dynamics of the Santiago Island Reef Slope (Pangasinan, Philippines). MSc Thesis: University of the Philippines, Manila.
63. Alino P, McManus L, McManus J, Nañola C, Fortes G, et al. (1993) Initial parameter estimations of a coral reef flat ecosystem in Bolinao, Pangasinan, northwestern Philippines. ICLARM Conference Proceedings.
64. Deocadez M, Alino P, Bautista A, Gaite P (2003) Lingayen Gulf, Northwestern Philippines. Manila: Coral Reef Information Network of the Philippines (Philreefs).
65. Villanueva R, Yap H, Montano M(2006) Intensive fish farming in the Philippines is detrimental to the reef-building coral *Pocillopora damicornis*. Marine Ecology Progress Series 316: 165–174.
66. Cruz-Trinidad A, Geronimo RC, Aliño PM (2009) Development trajectories and impacts on coral reef use in Lingayen Gulf, Philippines. Ocean & Coastal Management 52: 173–180.
67. Cruz-Trinidad A, Geronimo RC, Cabral RB, Aliño PM (2011) How much are the Bolinao-Anda coral reefs worth? Ocean & Coastal Management 54: 696–705.

68. Anthony KRN, Kline DI, Diaz-Pulido G, Dove S, Hoegh-Guldberg O (2008) Ocean acidification causes bleaching and productivity loss in coral reef builders. Proceedings of the National Academy of Sciences 105: 17442–17446.
69. Elsner JB, Kossin JP, Jagger TH (2008) The increasing intensity of the strongest tropical cyclones. Nature 455: 92–95.
70. Tu J, Chou C, Chu P (2009) The abrupt shift of typhoon activity in the vicinity of Taiwan and its association with western north Pacific-east Asian climate change. American Meteorological Society 22: 3617–3628.
71. Donner S, Skirving W, Little C, Oppenheimer M, Hoegh-Guldberg O (2005) Global assessment of coral bleaching and required rates of adaption under climate change. Global Change Biology 11: 2251–2265.
72. Cesar H, Pet-Soede L, Quibilan MCC, Alino P, Arceo H, et al. (2001) First evaluation of the impacts of the 1998 coral bleaching event to fisheries and tourism in the Philippines. In: Schuttenberg H, editor. Coral bleaching: Causes, consequences and responses (Selected papers presented at the 9th International Coral Reef Symposium, October 2000). Rhode Island, USA: Coastal Resources Center.
73. Arceo H, Quibilan MCC, Alino P, Lim G, Licuanan WY (2001) Coral bleaching in Philippine reefs: coincident evidences with mesoscale thermal anomalies. Bulletin of Marine Science 69: 579–593.
74. Bell PR, Lapointe BE, Elmetri I (2007) Reevaluation of ENCORE: Support for the Eutrophication Threshold Model for coral reefs. Ambio 36: 416–424.
75. Anderson MJ, Gorley RN, Clarke KR (2008) PERMANOVA+ for Primer: Guide to software and statistical methods: PRIMER-E: Plymouth, UK.
76. Anderson MJ (2006) Distance-based tests for homogeneity of multivariate dispersions. Biometrics 62: 245–253.
77. Clarke KR (1993) Non-parametric multivariate analyses of changes in community structure. Australian Journal of Ecology 18: 117–143.
78. Folt C, Chen C, Moore M, Burnaford J (1999) Synergism and antagonism among multiple stressors. Limnology and Oceanography 44: 864–877.
79. Gruner DS, Smith JE, Seabloom EW, Sandin SA, Ngai JT, et al. (2008) A cross-system synthesis of consumer and nutrient resource control on producer biomass. Ecology Letters 11: 740–755.
80. Burkepile DE, Hay ME (2006) Herbivore vs. nutrient control of marine primary producers: context-dependent effects. Ecology 87: 3128–3139.
81. Selkoe K, Halpern B, Ebert C, Franklin E, Selig E, et al. (2009) A map of human impacts to a "pristine" coral ecosystem, the Papahanaumokuakea Marine National Monument. Coral Reefs 28: 635–650.
82. Mora C, Metzger R, Rollo A, Myers RA (2007) Experimental simulations about the effects of overexploitation and habitat fragmentation on populations facing environmental warming. Proceedings of the Royal Society B: Biological Sciences 274: 1023–1028.
83. Adjeroud M, Michonneau F, Edmunds P, Chancerelle Y, De Loma TL, et al. (2009) Recurrent disturbances, recovery trajectories, and resilience of coral assemblages on a South Central Pacific reef. Coral Reefs 28: 775–780.
84. Golbuu Y, Victor S, Penland L, Idip D, Emaurois C, et al. (2007) Palau's coral reefs show differential habitat recovery following the 1998-bleaching event. Coral Reefs 26: 319–332.
85. Fung T, Seymour RM, Johnson CR (2013) Warning signals of regime shifts as intrinsic properties of endogenous dynamics. The American Naturalist 182: 208–222.
86. McClanahan T (2008) Response of the coral reef benthos and herbivory to fishery closure management and the 1998 ENSO disturbance. Oecologia 155: 169–177.
87. Mumby PJ, Harborne AR (2010) Marine reserves enhance the recovery of corals on Caribbean reefs. PLoS ONE 5: e8657.
88. Marshall P, Baird A (2000) Bleaching of corals on the Great Barrier Reef: differential susceptibilities among taxa. Coral Reefs 19: 155–163.
89. Holmes K, Edinger E, Hariyadi E, Limmon G, Risk MJ (2000) Bioerosion of live massive corals and branching coral rubble on Indonesian coral reefs. Marine Pollution Bulletin 40: 606–617.
90. Van Woesik R, Tomascik T, Blake S (1999) Coral assemblages and physico-chemical characteristics of the Witsunday Islands: Evidence of recent community changes. Marine and Freshwater Research 50: 427–440.
91. Edinger E, Limmon G, Jompa J, Widjatmoko W, Heikoop J, et al. (2000) Normal coral growth rates on dying reefs: Are coral growth rates good indicators of reef health? Marine Pollution Bulletin 40: 404–425.
92. Knowlton N (2001) The future of coral reefs. Proceedings of the National Academy of Sciences (USA) 98: 5419–5425.
93. Licuanan WY, Aliño PM, Campos WL, Castillo GB, Juinio-Meñez MA (2006) A decision support model for determining sizes of marine protected areas: biophysical considerations. Philippine Agricultural Scientist 89: 34.
94. Ball IR, Possingham H, Watts M (2009) Marxan and relatives: Software for spatial conservation prioritisation. In: Moilanen A, Wilson KA, Possingham H, editors. Spatial conservation prioritisation: Quantitative methods and computational tools. Oxford, UK: Oxford University Press. pp. 185–195.
95. Reichert P, Borsuk M (2005) Does high forecast uncertainty preclude effective decision support? Environmental Modelling & Software 20: 991–1001.
96. Ascough II JC, Maier HR, Ravalico JK, Strudley MW (2008) Future research challenges for incorporation of uncertainty in environmental and ecological decision-making. Ecological Modelling 219: 383–399.
97. Coreau A, Pinay G, Thompson J, Cheptou P, Mermet L (2009) The rise of research on futures in ecology: Rebalancing scenarios and predictions. Ecology Letters 12: 1277–1286.
98. Carpenter S (2002) Ecological futures: Building an ecology of the long now. Ecology 83: 2069–2083.
99. Jones A, Berkelmans R (2010) Potential costs of acclimatization to a warmer climate: growth of a reef coral with heat tolerant vs. sensitive symbiont types. PLoS ONE 5: e10437.
100. Baker AC, Starger CJ, McClanahan TR, Glynn PW (2004) Coral reefs: corals' adaptive response to climate change. Nature 430: 741–741.
101. Anthony KRN, Kline DI, Diaz-Pulido G, Dove S, Hoegh-Guldberg O (2008) Ocean acidification causes bleaching and productivity loss in coral reef builders. Proceedings of the National Academy of Sciences (USA) 105: 17442–17446.
102. Suzuki A, Gagan M, Fabricius K, Isdale P, Yukino I, et al. (2003) Skeletal isotope microprofiles of growth perturbations in *Poirites* corals during the 1997–1998 mass bleaching event. Coral Reefs 22: 357–369.
103. Miller MW, Hay ME, Miller SL, Malone D, Sotka EE, et al. (1999) Effects of nutrients versus herbivores on reef algae: A new method for manipulating nutrients on coral reefs. Limnology and Oceanography 44: 1847–1861.
104. McClanahan TR, Carreiro-Silva M, DiLorenzo M (2007) Effect of nitrogen, phosphorous, and their interaction on coral reef algal succession in Glover's Reef, Belize. Marine Pollution Bulletin 54: 1947–1957.
105. McClanahan TR, Steneck RS, Pietri D, Cokos B, Jones S (2005) Interaction between inorganic nutrients and organic matter in controlling coral reef communities in Glovers Reef Belize. Marine Pollution Bulletin 50: 566–575.
106. Lapointe BE (1989) Macroalgal production and nutrient relations in oligotrophic areas of Florida Bay. Bulletin of Marine Science 44: 312–323.
107. Lapointe BE, O'Connell J (1989) Nutrient-enhanced growth of *Cladophora prolifera* in Harrington Sound, Bermuda: Eutrophication of a confined, phosphorus-limited marine ecosystem. Estuarine Coastal and Shelf Science 28: 347–360.
108. Larned ST (1998) Nitrogen- versus phosphorus-limited growth and sources of nutrients for coral reef macroalgae. Marine Biology 132: 409–421.
109. Larned ST, Stimson J (1996) Nitrogen-limited growth in the coral reef chlorophyte *Dictyosphaeria cavernosa*, and the effect of exposure to sediment-derived nitrogen on growth. Marine Ecology Progress Series 145: 95–108.
110. Schaffelke B, Klumpp DW (1998) Nutrient-limited growth of the coral reef macroalga *Sargassum baccularia* and experimental growth enhancement by nutrient addition in continuous flow culture. Marine Ecology Progress Series 164: 199–211.
111. Dodge RE, Aller RC, Thomson J (1974) Coral growth related to resuspension of bottom sediments. Nature 247: 574–576.
112. Cortes J, Risk MJ (1985) A reef under siltation stress: Cahuita, Costa-Rica. Bulletin of Marine Science 36: 339–356.
113. Babcock R, Smith L (2000) Effects of sedimentation on coral settlement and survivorship. Proceedings of the Nith International Coral Reef Symposium.
114. Nugues MM, Roberts CM (2003) Coral mortality and interaction with algae in relation to sedimentation. Coral Reefs 22: 507–516.
115. Birrell CL, McCook LJ, Willis BL (2005) Effects of algal turfs and sediment on coral settlement. Marine Pollution Bulletin 51: 408–414.

13

Seasonal and Genotypic Changes in *Escherichia coli* Phylogenetic Groups in the Yeongsan River Basin of South Korea

Jeonghwan Jang[1], Doris Y. W. Di[1], Anna Lee[1], Tatsuya Unno[3], Michael J. Sadowsky[4,5], Hor-Gil Hur[1,2]*

1 School of Environmental Science and Engineering, Gwangju Institute of Science and Technology, Gwangju, Republic of Korea, **2** International Environmental Analysis and Education Center, Gwangju Institute of Science and Technology, Gwangju, Republic of Korea, **3** College of Molecular Life Sciences, Jeju National University, Jeju, Republic of Korea, **4** Department of Soil, Water and Climate, University of Minnesota, St. Paul, Minnesota, United States of America, **5** BioTechnology Institute, University of Minnesota, St. Paul, Minnesota, United States of America

Abstract

With 3,480 *E. coli* strains isolated from the Yeongsan River basin, South Korea, correlations between phylogenetic groups and horizontal fluorophore enhanced rep-PCR (HFERP) genotypes were examined, and environmental factors affecting *E. coli* phylogenetic groups in the river water were determined. Interestingly, multidimentional scaling (MDS) analyses based on HFERP DNA fingerprint data indicated that *E. coli* in phylogenetic groups A and B1 were uniquely clustered. Results of self-organized maps (SOMs) analyses also indicated that *E. coli* phylogenetic groups were seasonally affected by water temperature, with greater occurrences of phylogenetic groups A and B1 in low and high temperature seasons, respectively. The presence of *E. coli* in phylogenetic groups A and B1 were inversely related. Furthermore, redundancy analysis (RDA) revealed that phylogenetic group B1 correlated positively with temperature, strain diversity, and biochemical oxygen demand (BOD) but negatively with phylogenetic group A. Results of this study indicated that while *E. coli* strains could be clustered based on their genotypes and environment conditions, their phylogenetic groups did not change in relation to the same conditions. The distributional differences of phylogenetic groups among *E. coli* populations in different environments may be caused by different genomic adaptability and plasticity of *E. coli* strains belonging to each phylogenetic group. Although several previous studies have reported different *E. coli* ecological structures depending on their origins, this study is a first description of the specific environmental factors affecting *E. coli* phylogenetic groups in river water.

Editor: A. Mark Ibekwe, U. S. Salinity Lab, United States of America

Funding: The funders had no role in study design, data collection and analysis, decision to publish, or preparation of the manuscript. This work was supported by the National Research Foundation of Korea Grant funded by the Korean Government (MEST) (NRF: 2011-0029860) and the GSR(GIST Specialized Research) Project through a grant provided by GIST in 2014. The authors also wish to acknowledge a grant-in-aid for research from the Gwangju Regional Environmental Technology Development Center (10-2-70-73).

Competing Interests: The authors have declared that no competing interests exist.

* Email: hghur@gist.ac.kr

Introduction

Escherichia coli is generally considered to be a commensal bacterium found in the intestinal tracts of warm-blooded animals. Although this bacterium has been adopted as one of the best indicator organisms to examine fecal pollution in freshwater environments [1], its effectiveness as a fecal indicator has been questioned since several recent studies have reported naturalized *E. coli* populations which can survive for long period of time and reproduce in the environments outside the host bodies such as sand, soil and sediments [2–5]. Surrounding environmental conditions have been suggested to influence the genotypic traits of *E. coli* in the extraintestinal environments [6]. In a previous study, it was observed that genotypic richness of *E. coli* strains obtained from surface water of the Yeongsan River of South Korea was fluctuated with environmental conditions, especially water temperature, and season-specific *E. coli* genotypes were found in cold months [7]. Taken together, in order to fully understand ecology of this bacterium and its use as a fecal indicator, ecological approaches of *E. coli* community dynamics in the environments seem to be essential.

E. coli strains can be mainly assigned to four major phylogenetic groups: A, B1, B2 and D [8]. Recently, the extensive multi-locus sequence typing (MLST) and genome data sets for *E. coli* strains have refined our knowledge of *E. coli* phylogenetic group structure, and new phylogenetic groups of *E. coli* are now recognized such as phylogenetic groups C, E, F, and *Escherichia* clade I, which are comparatively rare to the major phylogenetic groups [9–14].

Strains belonging to each phylogenetic group possess different phenotypic and genotypic traits relative to each other [15,16]. Furthermore, the four major phylogenetic groups have been reported to differ in their ecological habitats. While phylogenetic groups B2 and D have been less frequently found than A or B1 in the environment [17], strains in group B2 have been observed to persist longer than the other groups in infants [18]. Moreover, *E. coli* strains in phylogenetic group B2 or D were more frequently isolated from extraintestinal sites within host bodies than group A or B1 strains [16]. Some strains belonging to phylogenetic group

B1 were reported to persist in water environments [17,19]. Many studies have also reported the relationship between virulence and phylogenetic groups of *E. coli*. Phylogenetic group B2 strains tend to be more virulent than other groups [20–23], and virulence genes were most frequently present in phylogenetic group B1 strains in the absence of B2 strains [24]. Thus, phylogenetic group identification of unknown *E. coli* strains may provide important information of their physiological and ecological aspects.

While initial studies used multilocus enzyme electrophoresis (MLEE) to classify *E. coli* strains into phylogenetic groups [25], Clermont and colleagues (2000) developed a multiplex PCR-based method to determine the four major phylogenetic groups of *E. coli* strains [26]. This method was further validated by Gordon et al. (2008) and showed that the PCR-based and multi-locus sequence typing (MLST) methods classified 80–85% of 662 *E. coli* isolates into the same phylogenetic groups [27]. These methods have been often used for the phylogenetic grouping of *E. coli* strains from various sources [22,24,28,29] since it reduces the time and cost of experimental procedures needed to examine a large number of strains. Recently, a new quadruplex PCR method enables an *E. coli* strain to be assigned to one of the seven phylogenetic groups (A, B1, B2, C, D, E, and F) and *Escherichia* clade I, which is also considered to be a phylogenetic group of *E. coli* [30].

The horizontal fluorophore-enhanced rep-PCR (HFERP) DNA fingerprinting technique enables accurate genotyping of *E. coli* strains [31,32]. However, no direct correlation between HFERP DNA fingerprint patterns and phylogenetic groups of *E. coli* strains was described previously [24,31]. Nevertheless, since rep-PCR DNA fingerprint patterns are susceptible to alteration of microbial genome structure, the technique can be used to study plasticity, molecular phylogeny, and evolution of microbial genomes [31]. The HFERP DNA fingerprinting technique, which revealed 1749 genotypes among 3480 *E. coli* isolates in a previous study [7], provides the much higher discriminatory power than the MLST or multiplex PCR methods assigning a *E. coli* strain to one of the eight phylogenetic groups as described above.

The aim of this study was to: 1) examine the relationship between phylogenetic groups and the HFERP DNA fingerprints of 3,480 *E. coli* strains isolated from the Yeongsan River basin, 2) determine correlations between *E. coli* phylogenetic groups and environmental factors in the environment, and 3) relations between virulence gene profiles of shiga toxigenic *E. coli* (STEC) and the phylogenetic groups of *E. coli* in the environment.

Materials and Methods

Environmental samples and bacterial strains

The sampling approach used and the isolation of *E. coli* strains from surface water samples was done as described previously [7]. A total of 3,480 *E. coli* strains were the same as those used in a previous study [7], and were subjected to further experiments here. Briefly, surface water samples were collected monthly from the 7 sites of the Yeongsan River basin (Table 1), including 3 agricultural sites, 3 urban sites, and 1 site affected by agricultural and urban, from April to December 2009, then sixty *E. coli* strains were isolated from each water sample but samples less than 60 isolates were excluded from further analyses. The membrane filtration technique with mTEC agar (Difco, Detroit, MI) was employed to obtain *E. coli* strains from freshwater samples [1]. ChromAgar ECC (Chromagar Microbiology, Paris, France) was used to reconfirm the phenotypical identity of potential *E. coli* isolates [6]. As practical point of view considering the large number of *E. coli* isolates, no further confirm tests for *E. coli* identification were proceeded since 97.3% of presumptive *E. coli* colonies obtained from the membrane filtration method with mTEC agar medium have been identified as *E. coli* strains in a previous study [33]. All *E. coli* isolates were preserved in LB freezing medium at −70°C [34], and their genomic DNA was prepared by boiling in 0.05 N NaOH. Colonies formed on LB agar plates were suspended in 100 μl of 0.05 N NaOH. The cells were lysed at 95°C for 15 min, and cell debris was precipitated by brief centrifugation. A dilution (1:10) of the supernatant was used as a DNA template of the further PCR assays.

Moreover, water quality parameters such as temperature, pH, biochemical oxygen demand (BOD), and *E. coli* MPN values were also obtained from all water samples by using the YSI 6600 Sonde (YSI Incorporated, Yellow Springs, OH) and the Colilert (IDEXX Laboratories, Inc., Westbrook, MN) system as described previously [7].

All the sampling sites were located in the Yeongsan River basin which is one of public rivers managed by Korean government. While the Youngsanriver Environmental Management Office is the government organization which manages the study area, no specific permission was required for any research activities performed in this study.

Horizontal fluorophore-enhanced rep-PCR (HFERP) DNA fingerprinting technique and defining HFERP genotypes of *E. coli* strains. Horizontal fluorophore-enhanced rep-PCR (HFERP) genotyping of the *E. coli* strains used in this study was done previously as described in our previous study [7]. Briefly, the BOX A1R (5′-CTACGGCAAGGCGACGCTGACG-3′) primer labeled with 6-FAM (6-carboxyfluorescein; Genotech Co. Ltd., Korea) was employed for the rep-PCR genotyping, then amplified DNA fragments and the Genescan-2500 ROX (6-carboxy-X-rhodamine) (Applied Biosystems, Foster City, CA) internal standard were separated by gel electrophoresis (70 V for 16 hours at 4°C) in 0.5 x TAE buffer using 1% Seachem LE agarose gels (FMC Bioproducts, Rockland, ME). Gel images were obtained by using by using a Typhoon 9400 variable mode imager (Molecular Dynamics/Amersham Biosciences, Sunnyvale, CA), and the image files were analyzed using BioNumerics v.6.01 software (Applied Maths, Sint-Martens-Latem, Belgium). Detailed protocols for HFERP DNA fingerprinting are accessible at http://www.ecolirep.umn.edu.

Phylogenetic grouping and virulence detection of *E. coli* strains. The phylogenetic grouping of the 3,480 *E. coli* strains was done using the Clermont multiplex PCR methods as described previously [26]. Briefly, the multiplex PCR assay was performed in 20 μl total volume containing 2 μl of 10X PCR buffer supplied with Top DNA polymerase (Bioneer, Korea), 1.5 μl of 20 mM $MgCl_2$, 20 pmol of each primer shown in top part of Table 2, 2 μM of each deoxynucleoside triphosphate, 2.5 U of Top DNA polymerase (Bioneer, Korea), and 3 μl of bacterial lysate. The PCR steps were composed of denaturation for 4 min at 94°C, 30 cycles at 94°C for 5 s and 10 s at 59°C, and a final extension step of 5 min at 72°C.

The presence of virulence genes among the *E. coli* strains related to shiga toxin producing *E. coli* (STEC), enterohemorrhagic *E. coli* (EHEC), and enteropathogenic *E. coli* (EPEC) (*eaeA*, stx_1, and stx_2) was determined by using multiplex PCR as previously described [35]. Briefly, the target genes were amplified in 25 μl reaction mixture containing 200 μM concentrations of deoxynuceloside triphosphates, 250 nM concentration of each primer shown in bottom part of Table 2, 1 U of Top DNA polymerase (Bioneer, Korea), 2.5 μl of 10X PCR buffer supplied with Top DNA polymerase (Bioneer, Korea), 2 mM $MgCl_2$, and 2 μl of bacterial lysate. The reaction mixtures were subjected to first 10 cycles of 1 min of denaturation at 95°C, 2 min of annealing at 65°C, and

Table 1. Information of each sampling location.

Sampling sites	Sub-basin	Major land use affecting sub-basin	Latitude and longitude
YS1	Manbong tributary	Agricultural	34°59′39.66"N, 126°42′5.92"E
YS2	Jangseong tributary	Agricultural	35° 3′10.05"N, 126°44′29.45"E
YS3	Main stream of the Yeongsan River	Agricultural and Urban	35° 6′34.02"N, 126°49′8.87"E
YS4	Orye tributary	Agricultural	35°16′57.44"N, 126°57′29.88"E
GJ1	Gwangju tributary	Urban	35° 9′11.54"N, 126°50′3.90"E
GJ2	Gwangju tributary	Urban	35°10′7.19"N, 126°53′3.13"E
GJ3	Gwangju tributary	Urban	35° 7′45.20"N, 126°55′41.71"E

1.5 min of elongation at 72°C; a second 15 cycles of 1 min of denaturation at 95°C, 2 min of annealing at 60°C, and 1.5 min of elongation at 72°C; and a third 10 cycles of 1 min of denaturation at 95°C, 2 min of annealing at 65°C, and 2.5 min of elongation at 72°C. All of the PCR products were assessed by using an automated capillary electrophoresis system, QIAxcel (Qiagen, Hilden, Germany).

Statistical methods

Multidimentional scaling (MDS) analyses were done using BioNumerics v.6.01 software (Applied Maths, Sint-Martens-Latem, Belgium) and provided a three dimentional representation of *E. coli* phylogenetic groups based on similarities between HFERP DNA fingerprints. The correct assignment of *E. coli* strains to phylogenetic group was evaluated by using Jackknife analysis, with maximum similarities [32]. Self-organizing maps (SOMs) analyses using the SOM Toolbox implements (Laboratory of Computer and Information Science, Helsinki, Norway) of Matlab 2009b (MathWorks, Natick, MA) were used to determine if there were correlations between *E. coli* phylogenetic groups and environmental factors, including temperature, pH, BOD, and *E. coli* MPN values. Raw data for the SOMs analyses are organized as presented in Table S1 (supplementary file). The occurrence of four *E. coli* phylogenetic groups and the five water quality parameters described previously [7] were used as input variables for the SOMs analyses. The visualized map images, which were automatically optimized at 42 SOM cells, were generated with minimum quantization error (1.58) and the least topographic errors (0.034). Quantization error (QE) and topographic error (TE) were estimated to confirm reliable resolution and topology conservation of the SOMs analyses, and these are meanings of the statistical significance. QE is the average distance between each data vector and its best matching unit (BMU), measures map resolution. TE measures map quality, which represents the proportion of all data vectors for which 1st and 2nd BMUs are not adjacent, and is thus used for the measurement of topology preservation. The results of the SOMs analyses are based on statistical procedures, and map size and structure ($n \times m$) are automatically optimized and determined based on the least QE and TE values among numerous cases. Borders between clusters existing in the map were defined by using a hierarchical cluster analysis. Features and analytic procedures of SOMs analyses have been described previously [7,36–41].

Redundancy analysis (RDA), one of multivariate analyses included in software packages CANOCO v5.0 (Microcomputer Power, Ithaca, New York), was performed to test which environmental factors significantly correlated with the variation in the *E. coli* phylogenetic groups [42]. Five parameters including

Table 2. PCR primers used for phylogenetic grouping and virulence gene detection among *E. coli* strains.

Target genes or DNA fragment	Primer sequence	Amplicon size (bp)	Reference
chuA	5′-GAC GAA CCA ACG GTC AGG AT-3′	279	[26]
	5′-TGC CGC CAG TAC CAA AGA CA-3′		
yjaA	5′-TGA AGT GTC AGG AGA CGC TG-3′	211	
	5′-ATG GAG AAT GCG TTC CTC AAC-3′		
TspE4.C2	5′-ATG GAG AAT GCG TTC CTC AAC-3′	152	
	5′-GAG TAA TGT CGG GGC ATT CA-3′		
stx_1	5′-ATA AAT CGC CAT TCG TTG ACT AC-3′	180	[35]
	5′-AGA ACG CCC ACT GAG ATC ATC-3′		
stx_2	5′-GGC ACT GTC TGA AAC TGC TCC-3′	255	
	5′-TCG CCA GTT ATC TGA CAT TCT G-3′		
eaeA	5′-GAC CCG GCA CAA GCA TAA GC-3′	384	
	5′-CCA CCT GCA GCA ACA AGA GG-3′		
hlyA	5′-GCA TCA TCA AGC GTA CGT TCC-3′	534	
	5′-AAT GAG CCA AGC TGG TTA AGC T-3′		

temperature, pH, BOD, *E. coli* MPN values, and strain diversity were used for the explanatory variables constraining the occurrences of the *E. coli* phylogenetic groups A and B1, which occupied over 80% of total *E. coli* strains tested in this study. Response data in the analysis have a gradient 0.8 standard deviation (SD) units, suggesting that *E. coli* phylogenetic group response to the gradient of the environmental factors is linear. The forward selection method was used to determine significant environmental variables affecting the *E. coli* phylogenetic groups using the 999 Monte Carlo permutations at $P<0.05$ [43].

Results

Genotypic separation of *E. coli* strains in different phylogenetic groups

The 3,480 *E. coli* strains isolated from the Yeongsan River basin in Korea were comprised of 1,724 (49.5%), 1,192 (34.3%), 188 (5.4%), and 376 (10.8%) strains in phylogenetic groups A, B1, B2, and D, respectively. The MDS analyses were performed to investigate a correlation between HFERP DNA fingerprints and phylogenetic groups of *E. coli* strains (Fig. 1). As shown in Figure 1, each dot in the three dimensional space indicates HFERP DNA fingerprint of each *E. coli* strain, and it was given a color according to one of the phylogenetic groups the *E. coli* strain belongs to. Distances between dots simply reflect similarities between HFERP DNA fingerprints of *E. coli* strains. Interestingly, phylogenetic groups A and B1 showed a tendency to cluster along with genotype distributions (indicated by blue circles). Moreover, the both phylogenetic groups were separated even within the tightly clustered genotypic group (indicated by black circle) obtained from the cold months as described previously [7]. In contrast, phylogenetic groups B2 and D showed no significant clustering along with genotypic distributions in the MDS analyses. As shown in Table 3, the clustering trend of the MDS analyses was reconfirmed by using Jackknife analysis, in which strains were removed from the group one at a time and treated as unknowns for classification. Phylogenetic groups A and B1 showed significantly higher percentages of *E. coli* strains correctly assigned to their own group (90.1% and 89.4%) than phylogenetic groups B2 and D (65.4% and 77.7%) ($p<0.05$, two-sample *t*-test assuming equal variances using Microsoft Excel).

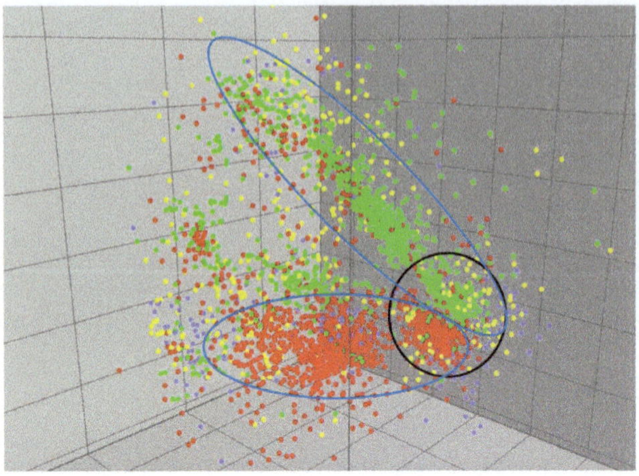

Figure 1. Multidimensional scaling (MDS) analyses of *E. coli* HFERP DNA fingerprints given colors according to their phylogenetic groups. Legend: A (red), B1 (green), B2 (sky-blue), and D (yellow).

SOMs analyses for trends of *E. coli* phylogenetic groups

Surface water samples were classified into self-organized map (SOM) cells based on their water quality parameters and prevalence of *E. coli* phylogenetic groups. These are used as input variables for the SOMs analyses, and water samples were assigned into SOM cells depending on the similarity among their variables (Fig. 2A). A hierarchical cluster analysis (Fig. 2B) revealed that clusters existed in the map. According to the similarities of the variables in the output cells, clusters were defined as combined cell units of the map (Fig. 2A). As shown in Figure 2, two main clusters (C1 and C2) were further subdivided into six sub-clusters (C1a, C1b1, C1b2, C2a, C2b1, and C2b2). Sub-clusters C2b1 and C2b2 in Cluster C2 were mainly comprised of *E. coli* obtained surface water samples collected from the cold months, October, November, and December. This suggested season-specificity of SOMs. Site-specificity was also observed in sub-cluster C2a since it includes only isolates from surface water samples collected from the urban-affected locations (sites GJ1, GJ2, GJ3, and YS3). Results in Figure 3A and 3B show the contribution of each input variable to the classification of SOMs cells by displaying the distribution of the variable intensities on the map. The intensities are expressed by colors between white and black, with the units of each variable. Dark and light colors indicate high and low values of each variable, respectively. The divisions of SOMs cells shown in Figure 3A can be explained by classification factors which are input variables showing intensity maps in Figure 3B. For instance, a sub-cluster C2a was mainly defined by *E. coli* MPN showing high values of the variable in the cluster. By the same token, a sub-cluster C2b2 was determined by phylogenetic group B2. Furthermore, these results indicated that the relationships between input variables can be observed by comparing the intensity maps to each other. Based on the analyses, the following interpretation could be deduced: (1) Elevated *E. coli* MPN counts determined cluster C2a, which showed specificity for urban-affected sites, and did not directly correlate with other variables; (2) Strain diversity and temperature were directly correlated, which was also described in a previous study[7], and these two variables are considered to be major contributors determining the two main clusters C1 and C2; (3) BOD was also correlated with strain diversity and temperature, and high BOD values appeared to influence of cluster C1b1; (4) No correlations were observed between pH and the other variables; (5) The high frequency of phylogenetic group A strains appeared to be correlated with low values of strain diversity and temperature. This correlation was weak and was not as significantly correlated as was the relationship between strain diversity and temperature. The high frequency of occurrence of phylogenetic group A strains with low temperature is consistent with the increasing ratio of phylogenetic group A in the cold months as described above. Strains in phylogenetic group A are also considered to be a determining factor for the main clusters C1 and C2; (6) Phylogenetic group B1 strains were mostly present in high temperatures and showed a frequency distribution contrary to phylogenetic group A strains; (7) Phylogenetic group B2 strains did not directly correlate with any other variables and appeared to affect cluster C2b2; and (8) The high frequency of phylogenetic group D strains correlated with high temperatures and contributed to assignment of cluster C1a. Overall, temperature affected strain diversity and the occurrence of phylogenetic groups A, B1, and D. In contrast, phylogenetic group B2 was not affected by any other variables. However, the lack of any patterns in B2 strains and to a lesser extent D strains probably reflects their rarity.

Table 3. The percentage of *E. coli* strains assigned to each phylogenetic group calculated by Jackknife analysis based on HFERP DNA fingerprints data.

Assigned phylgogenetic group	% *E. coli* strains assigned to each group			
	A	B1	B2	D
A	90.1	8.1	8.5	5.9
B1	6.8	89.4	18.1	11.7
B2	1.4	1.3	65.4	4.8
D	1.7	1.2	8.0	77.7

Correlation between distribution of *E. coli* phylogenetic groups and environmental factors

Redundancy analysis (RDA) was performed to determine correlations between prevalence of *E. coli* phylogenetic groups and environmental factors in surface water of the Yeongsan River basin. Five parameters such as BOD, temperature, pH, *E. coli* MPN, and strain diversity were used for the possible explanatory variables for constraining the occurrence of *E. coli* phylogenetic groups A and B1, which occupied more than 80% of total *E. coli* strains. The RDA explained 49.3% of the total variation in the phylogenetic groups A and B1.

The correlations between the two *E. coli* phylogenetic groups and the environmental data were shown in an ordination diagram (Fig. 4) showing the results which are similar to ones observed from the SOMs analyses. Temperature, strain diversity, and BOD were positively correlated to each other. The pH values were not likely to correlate with the other environmental factors and phylogenetic groups. Temperature, strain diversity, BOD and *E. coli* MPN were determined to be significant explanatory variables affecting the occurrences of the two *E. coli* phylogenetic groups ($P<0.05$) by using a forward selection method. High values of temperature, strain diversity, and BOD were correlated with high number of phylogenetic group B1, but with low number of phylogenetic group A. *E. coli* MPN, on the other hand, were positively correlated with phylogenetic group A, but negatively with B1. Furthermore, negative correlation was shown between phylogenetic group A and B1.

The correlation between *E. coli* MPN and phylogenetic group A may be affected by the fact that group A was most frequently found among total *E. coli* strains, suggesting that larger number of *E. coli* isolates provides more chance to obtain phylogenetic group A strains.

Distribution of potential pathogenic *E. coli* strains among phylogenetic groups

Based on previously accepted definitions of *E. coli* pathotypes [24,44], *E. coli* strains carrying virulence genes were assigned to the pathotypes: *stx1*, *stx2 or stx1/stx2*, shiga toxigenic *E. coli* (STEC); *eaeA* with *stx1* or/and *stx2*, enterohemorrhagic *E. coli* (EHEC); *eaeA* without *stx1* or *stx2*, enteropathogenic *E. coli* (EPEC). Among the

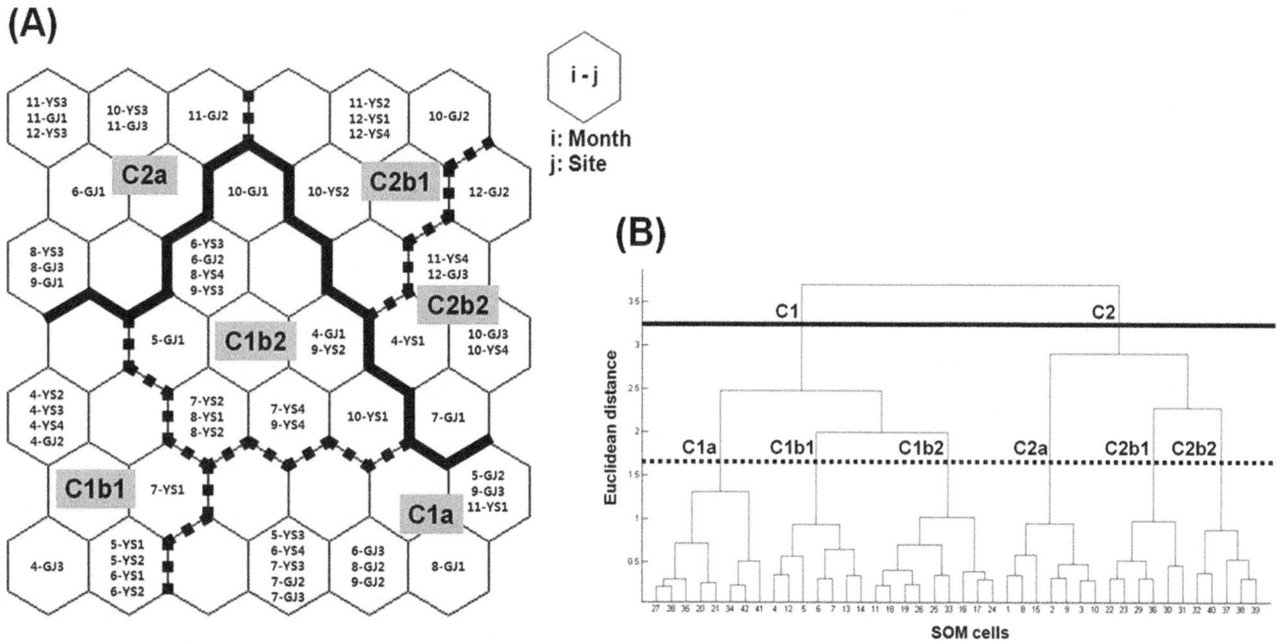

Figure 2. SOMs analyses classification and dendrogram. (A) Six clusters were determined by the SOMs analyses: two main clusters (C1 and C2) including six sub-clusters (C1a, C1b1, C1b2, C2a, C2b1 and C2b2), and (B) dendrogram created by clustering input variables based on the similarity among SOM cells.

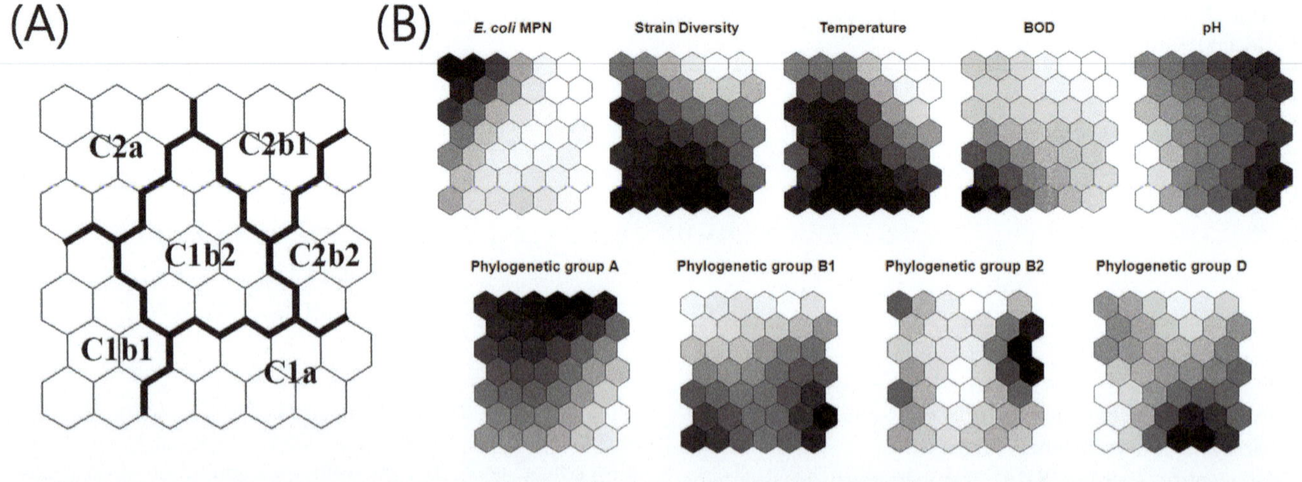

Figure 3. Color maps displaying the intensity of each input variables on the classified SOM cells. Dark and light colors indicate high and low values of each variable, respectively. The intensity of strain diversity was defined as percentage of number of *E. coli* genotypes for total isolates.

3,480 *E. coli* strains examined, 217 (6.2%), 64(1.8%), and 9 (0.3%) were assigned to STEC, EPEC, and EHEC groups, respectively. Results in Figure 5 show that the greatest percentage of STEC strains was observed among phylogenetic group D (9.3%, 35 of 376). In contrast, phylogenetic group B2 included the greatest percentage of EPEC strains (6.4%, 12 of 188).

Discussion

The separation of strains into phylogenetic groups A and B1 was also supported by genotypes defined by HFERP DNA fingerprint analyses as observed from MDS and jackknife analysis. MDS analyses indicated that although environmental factors such as temperature did affect the pattern of HFERP genotypes of *E. coli* strains [7], their phylogenetic groups that was correspondent with genotypic separation did not show any seasonal pattern of grouping (Fig. 1). This result is likely due to the fact that while phylogenetic grouping is based on indirect gross changes in enzyme mobility or stable genetic markers, rep-PCR technique directly detects changes in nucleotide base composition to identify possible rearrangements in bacterial genome structure. This suggests that the HFERP DNA fingerprinting method is more sensitive to genomic variations for adaption and plasticity in the environment than is phylogenetic grouping [31]. Furthermore, the seasonal specificity of *E. coli* HFERP genotypes, which was seen as a clustering of strains obtained from cold months, did not directly coincide with the distribution of *E. coli* phylogenetic groups as seen by the MDS analyses (Fig. 1). This also suggested that HFERP DNA fingerprint-defined genotypes of *E. coli* are more seasonally influenced by environmental factors, such as temperature than *E. coli* phylogenetic groups.

It is interesting that the SOMs analyses revealed that there was an inverse relationship between the occurrence of *E. coli* strains

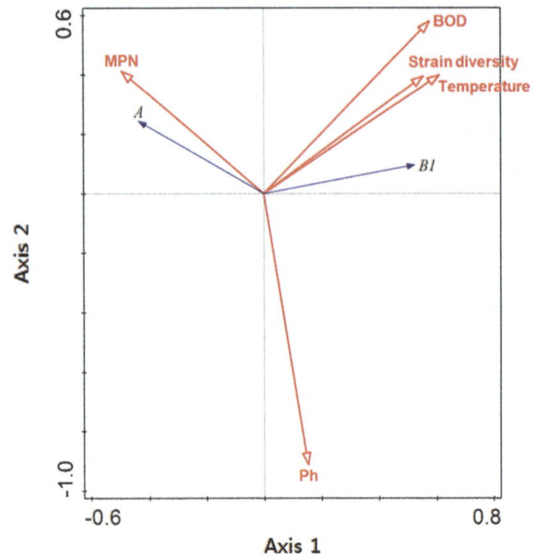

Figure 4. Ordination diagram from the redundancy analysis. The correlations of the environmental variables (explanatory variables) with the two phylognetic groups (response variables) can be predicted from angles between the arrows of the variables. The biplot projection of the phylogenetic group arrow tips onto the arrow of an environmental variable provides a more precise approximation.

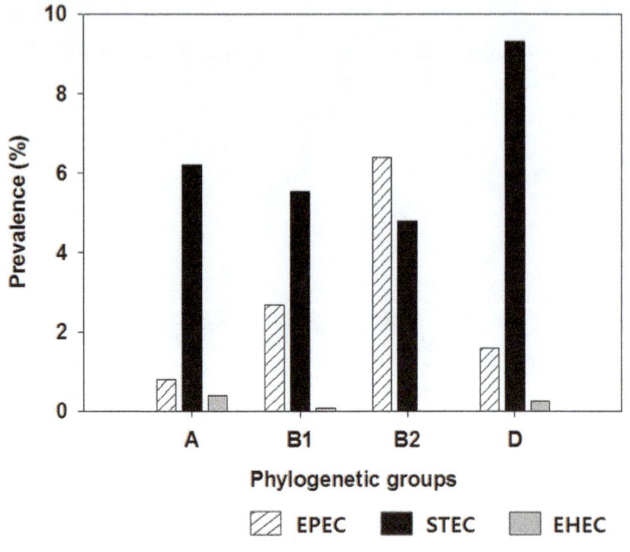

Figure 5. Prevalence of potential *E. coli* pathotypes EPEC, STEC, and EHEC among phylogenetic groups.

belonging to phylogenetic groups A and B1. Although the fact that phylogenetic groups A and B1 are inversely related to each other may simply reflect that they account for about 85% of the isolates, it is still remarkable since the inverse relation was correlated with the environmental factors. While *E. coli* strains belonging to phylogenetic group A was found to increase in the conditions such as low temperature and genotypical diversity, phylogenetic group B1 strains increased in the contrary conditions in the Yeongsan River basin, revealing that water temperature tends to be associated with the occurrence of specific *E. coli* phylogenetic groups. These correlations between *E. coli* phylogenetic groups and environmental factors were more clearly observed from the RDA. However, since phylogenetic groups were not strictly correlated with temperature as much as genotypic strain diversity did from the SOMs analyses, this suggests that other environmental factors may also cause differences between *E. coli* phylogenetic groups. In addition, the MDS analyses (Fig. 1) showed that phylogenetic group A was not comprised of strains obtained from only cold months, suggesting temperature is not only the factor affecting *E. coli* phylogeny.

Previously, Tenaillon et al. (2010) reported that the population structure of *E. coli* is predominantly clonal in spite of the occurrence of recombination events, allowing the delineation of major phylogenetic groups [14]. Based on the results reported here, our data suggest that *E. coli* phylogenetic groups are adaptable and genotypically-influenced by changes in environmental conditions. This conclusion is in large part supported by our observation that phylogenetic group A strains appear to be more adaptable to low temperature than the other groups examined in this study. While several previous studies already have reported different *E. coli* ecological structures depending on their origins, few studies described the specific environmental factors affecting *E. coli* phylogenetic groups in river water.

As described previously [7], no site-specificity of *E. coli* genotypes and phylogenetic groups was observed in this study as well. However, the representativeness of the *E. coli* strains for each surface water sample is quite variable because of the same number of the strains obtained from each sample, irrespective of how much total population of *E. coli* is in there. Particularly the representativeness of the *E. coli* isolates in urban sites showing generally high MPN values is likely to be lower than the other sites. Thus, it would be improper to discuss on site-specificity of the results in this study.

Although the total number of *E. coli* strains belonging to phylogenetic groups B2 and D occurred less frequently than ones belonging to the other phylogenetic groups, the percentage of potential *E. coli* pathotypes comprising EPEC and STEC strains were highest among members of groups B2 and D. This result is consistent with previous studies that phylogenetic groups A and B1 members were often comprised of commensal stains and that phylogenetic groups B2 and D were observed to contain more virulent strains than other groups [18,45–47]. Thus, the previously reported relationship between phylogeny and virulence of *E. coli* strains can similarly be applied to the *E. coli* population existing in surface waters of the Yeongsan River basin. However, it should be noted that typical strains of EHEC O157:H7 possessing no β-d-glucuronidase were excluded from the results since *E. coli* strains examined in this study were initially obtained by using mTEC agar plates which has selectivity for the bacterial species based on β-d-glucuronidase activity [48].

In conclusion, we show that there is a relationship between genotypic groupings of *E. coli* based on their HFERP DNA fingerprints and strains in phylogenetic groups A and B1. Moreover, we report here that the SOMs and RDA analyses indicate that the distributional changes of these groups are affected by the environmental factors. The compositional differences of phylogenetic groups among *E. coli* populations in different environments may be caused by differences in adaptability and plasticity of *E. coli* strains belonging to each phylogenetic group. Environmentally-adapted genotypes and changes in phylogenetic groups according to specific environmental conditions probably are not related to temporal deposition of fecal pollutant since they were observed regardless of sampling locations. The genetic bases for these changes are unknown, and further investigations of genomic DNA sequence variation among the *E. coli* strains examined in the current study will likely provide in-depth information about the genotypic changes imparted by environmental conditions of the river water. Moreover, it would be worthwhile to perform a further repetitive monitoring of *E. coli* populations in the environment to confirm the reproducibility of the study results during a longer sampling period since a river environment is not a static ecosystem.

Acknowledgments

We thank the members of the Environmental Systems Engineering Laboratory in the Gwangju Institute of Science and Technology who assisted with surface water sample collections from the Yeongsan River basin.

Author Contributions

Conceived and designed the experiments: JJ HH. Performed the experiments: JJ DYWD AL. Analyzed the data: JJ. Contributed reagents/materials/analysis tools: TU HH. Wrote the paper: JJ MJS HH.

References

1. USEPA (2006) Method 1603: *Escherichia coli* (*E. coli*) in Water by Membrane Filtration Using Modified membrane-Thermotolerant *Escherichia coli* Agar. U.S. Environmental Protection Agency, Washington, DC.
2. Byappanahalli MN, Fujioka RS (1998) Evidence that tropical soil environment can support the growth of *Escherichia coli*. Water Sci Technol 38:171–174.
3. Sadowsky MJ, Whitman RL (2010) Fecal Bacteria. ASM Press: Washington, DC.
4. Byappanahalli MN, Whitman RL, Shively DA, Sadowsky MJ, Ishii S (2006) Population structure, persistence, and seasonality of autochthonous *Escherichia coli* in temperate, coastal forest soil from a Great Lakes watershed. Environ Microbiol 8:504–513.
5. Ishii S, Ksoll WB, Hicks RE, Sadowsky MJ (2006) Presence and growth of naturalized *Escherichia coli* in temperate soils from lake superior watersheds. Appl Environ Microbiol 72: 612–621.
6. Dombek PE, Johnson LK, Zimmerley ST, Sadowsky MJ (2000) Use of repetitive DNA sequences and the PCR to differentiate *Escherichia coli* isolates from human and animal sources. Appl Environ Microbiol 66:2572–2577.
7. Jang J, Unno T, Lee SW, Cho KH, Sadowsky MJ, et al. (2011) Prevalence of season-specific *Escherichia coli* strains in the Yeongsan River Basin of South Korea. Environ Microbiol 13:3103–3113.
8. Herzer PJ, Inouye S, Inouye M, Whittam TS (1990) Phylogenetic distribution of branched RNA-linked multicopy single-stranded DNA among natural isolates of *Escherichia coli*. J Bacteriol 172:6175–6181.
9. Clermont O, Olier M, Hoede C, Diancourt L, Brisse S, et al. (2011) Animal and human pathogenic *Escherichia coli* strains share common genetic backgrounds. Infect Genet Evol 11:654–662.
10. Walk ST, Alm EW, Gordon DM, Ram JL, Toranzos GA, et al. (2009) Cryptic lineages of the genus *Escherichia*. Appl Environ Microbiol 75:6534–6544.

11. Luo C, Walk ST, Gordon DM, Feldgarden M, Tiedje JM, et al. (2011) Genome sequencing of environmental *Escherichia coli* expands understanding of the ecology and speciation of the model bacterial species. Proc Natl Acad Sci USA 108:7200–7205.
12. Moissenet D, Salauze B, Clermont O, Bingen E, Arlet G, et al. (2010) Meningitis caused by *Escherichia coli* producing TEM-52 extended-spectrum β-lactamase within an extensive outbreak in a neonatal ward: Epidemiological investigation and characterization of the strain. J Clin Microbiol 48:2459–2463.
13. Jaureguy F, Landraud L, Passet V, Diancourt L, Frapy E, et al. (2008) Phylogenetic and genomic diversity of human bacteremic *Escherichia coli* strains. BMC Genomics 9:560.
14. Tenaillon O, Skurnik D, Picard B, Denamur E (2010) The population genetics of commensal *Escherichia coli*. Nat Rev Microbiol 8:207-217.
15. Bergthorsson U, Ochman H (1998) Distribution of chromosome length variation in natural isolates of *Escherichia coli*. Mol Biol Evol 15:6–16.
16. Gordon DM (2004) The influence of ecological factors on the distribution and genetic structure of *Escherichia coli*. In *Escherichia coli* and *Salmonella typhimurium*: Cellular and Molecular Biology. Neidhardt, F., et al. (eds) ASM press: Washington, DC.
17. Walk ST, Alm EW, Calhoun LM, Mladonicky JM, Whittam TS (2007) Genetic diversity and population structure of *Escherichia coli* isolated from freshwater beaches. Environ Microbiol 9:2274–2288.
18. Nowrouzian FL, Adlerberth I, Wold AE (2006) Enhanced persistence in the colonic microbiota of *Escherichia coli* strains belonging to phylogenetic group B2: role of virulence factors and adherence to colonic cells. Microbes Infect 8:834–840.
19. Ratajczak M, Laroche E, Berthe T, Clermont O, Pawlak B, et al. (2010) Influence of hydrological conditions on the *Escherichia coli* population structure in the water of a creek on a rural watershed. BMC Microbiol 10:222.
20. Johnson JR, Stell AL (2000) Extended virulence genotypes of *Escherichia coli* strains from patients with urosepsis in relation to phylogeny and host compromise. J Infect Dis 181:261–272.
21. Le Gall T, Clermont O, Gouriou S, Picard B, Nassif X, et al. (2007) Extraintestinal virulence is a coincidental by-product of commensalism in B2 phylogenetic group *Escherichia coli* strains. Mol Biol Evol 24:2373–2384.
22. Escobar-Páramo P, Grenet K, Le Menac'h A, Rode L, Salgado E, et al. (2004) Large-scale population structure of human commensal *Escherichia coli* isolates. Appl Environ Microbiol 70:5698–5700.
23. Picard B, Garcia JS, Gouriou S, Duriez P, Brahimi N, et al. (1999) The Link between phylogeny and virulence in *Escherichia coli* extraintestinal infection. Infect Immun 67:546–553.
24. Unno T, Han D, Jang J, Lee S-N, Ko G, et al. (2009) Absence of *Escherichia coli* phylogenetic group B2 strains in humans and domesticated animals from Jeonnam province, Republic of Korea. Appl Environ Microbiol 75:5659–5666.
25. Goullet P, Picard B (1989) Comparative electrophoretic polymorphism of esterases and other enzymes in *Escherichia coli*. J Gen Microbiol 135:135–143.
26. Clermont O, Bonacorsi S, Bingen E (2000) Rapid and simple determination of the *Escherichia coli* phylogenetic group. Appl Environ Microbiol 66:4555–4558.
27. Gordon DM, Clermont O, Tolley H, Denamur E (2008) Assigning *Escherichia coli* strains to phylogenetic groups: multi-locus sequence typing versus the PCR triplex method. Environ Microbiol 10:2484–2496.
28. Duriez P, Clermont O, Bonacorsi S, Bingen E, Chaventré A, et al. (2001) Commensal *Escherichia coli* isolates are phylogenetically distributed among geographically distinct human populations. Microbiol-SGM 147:1671–1676.
29. Johnson JR, Kuskowski MA, Owens K, Gajewski A, Winokur PL (2003) Phylogenetic origin and oirulence genotype in relation to resistance to fluoroquinolones and/or extended-spectrum cephalosporins and cephamycins among *Escherichia coli* isolates from animals and humans. J Infect Dis 188:759–768.
30. Clermont O, Christenson JK, Denamur E, Gordon DM (2013) The clermont *Escherichia coli* phylo-typing method revisited: improvement of specificity and detection of new phylo-groups. Environ Microbiol Rep 5:58–65.
31. Ishii S, Sadowsky MJ (2009) Applications of the rep-PCR DNA fingerprinting technique to study microbial diversity, ecology and evolution. Environ Microbiol 11:733–740.
32. Johnson LK, Brown MB, Carruthers EA, Ferguson JA, Dombek PE, et al. (2004) Sample size, library composition, and genotypic diversity among natural populations of *Escherichia coli* from different animals influence accuracy of determining sources of fecal pollution. Appl. Environ Microbiol 70:4478–4485.
33. Ciebin BW, Brodsky MH, Eddington R, Horsnell G, Choney A, et al. (1995) Comparative evaluation of modified m-FC and m-TEC media for membrane filter enumeration of *Escherichia coli* in water. Appl Environ Microbiol 61:3940–3942.
34. Sambrook J, Russell WD (2001) Molecular cloning: a laboratory manual, 3rd ed. Cold Spring Harbor Laboratory Press: Cold Spring Harbor, NY.
35. Paton AW, Paton JC (1998) Detection and characterization of shiga Toxigenic *Escherichia coli* by using multiplex PCR assays for *stx1*, *stx2*,eaeA, Enterohemorrhagic *E. coli hlyA*, rfb O111, and rfb O157. J Clin Microbiol 36:598–602.
36. Park Y-S, Céréghino R, Compin A, Lek S (2003) Applications of artificial neural networks for patterning and predicting aquatic insect species richness in running waters. Ecol Model 160:265–280.
37. Park Y-S, Chang J, Lek S, Cao W, Brosse S (2003) Conservation strategies for endemic fish species threatened by the three gorges dam. Conserv Biol 17:1748–1758.
38. Tuang FN, Rademaker JLW, Alocilja EC, Louws FJ, de Bruijn FJ (1999) Identification of bacterial rep-PCR genomic fingerprints using a backpropagation neural network. FEMS Microbiol Lett 177:249–256.
39. Tudesque L, Gevrey M, Grenouillet G, Lek S (2008) Long-term changes in water physicochemistry in the Adour-Garonne hydrographic network during the last three decades. Water Res 42:732–742.
40. Weber M, Teeling H, Huang S, Waldmann J, Kassabgy M, et al. (2010) Practical application of self-organizing maps to interrelate biodiversity and functional data in NGS-based metagenomics. ISME J 5:918–928.
41. Gevrey M, Rimet F, Park YS, Giraudel J-L, Ector L, et al. (2004) Water quality assessment using diatom assemblages and advanced modelling techniques. Freshwater Biol 49:208–220.
42. Smilauer JLaP Multivariate Analysis of Ecological Data using CANOCO. Cambridge University Press: Cambridge, UK.
43. Boer SI, Hedtkamp SIC, van Beusekom JEE, Fuhrman JA, Boetius A, et al. (2009) Time- and sediment depth-related variations in bacterial diversity and community structure in subtidal sands. ISME J 3:780–791.
44. Tarr PI, Gordon CA, Chandler WL (2005) Shiga-toxin-producing *Escherichia coli* and haemolytic uraemic syndrome. The Lancet 365:1073–1086.
45. Nowrouzian F, Hesselmar B, Saalman R, Strannegard I-L, Aberg N, et al. (2003) *Escherichia coli* in infants' intestinal microflora: Colonization rate, strain turnover, and virulence gene carriage. Pediatr Res 54:8–14.
46. Gordon DM, Stern SE, Collignon PJ (2005) Influence of the age and sex of human hosts on the distribution of *Escherichia coli* ECOR groups and virulence traits. Microbiology-SGM 151:15–23.
47. Johnson JR, O'Bryan TT, Kuskowski M, Maslow JN (2001) Ongoing horizontal and vertical transmission of virulence genes and *papA* alleles among *Escherichia coli* blood isolates from patients with diverse-source bacteremia. Infect Immun 69:5363–5374.
48. Feng P, Lampel KA, Karch H, Whittam TS (1998) Genotypic and phenotypic changes in the emergence of *Escherichia coli* O157:H7. J Infect Dis 177:1750–1753.

14

Wetlands in Changed Landscapes: The Influence of Habitat Transformation on the Physico-Chemistry of Temporary Depression Wetlands

Matthew S. Bird[1]*, Jenny A. Day[2]

1 DST/NRF Research Chair in Shallow Water Ecosystems, Nelson Mandela Metropolitan University, Port Elizabeth, South Africa, 2 Freshwater Research, Department of Biological Sciences, University of Cape Town, Private Bag, Rondebosch, South Africa

Abstract

Temporary wetlands dominate the wet season landscape of temperate, semi-arid and arid regions, yet, other than their direct loss to development and agriculture, little information exists on how remaining wetlands have been altered by anthropogenic conversion of surrounding landscapes. This study investigates relationships between the extent and type of habitat transformation around temporary wetlands and their water column physico-chemical characteristics. A set of 90 isolated depression wetlands (seasonally inundated) occurring on coastal plains of the south-western Cape mediterranean-climate region of South Africa was sampled during the winter/spring wet season of 2007. Wetlands were sampled across habitat transformation gradients according to the areal cover of agriculture, urban development and alien invasive vegetation within 100 and 500 m radii of each wetland edge. We hypothesized that the principal drivers of physico-chemical conditions in these wetlands (e.g. soil properties, basin morphology) are altered by habitat transformation. Multivariate multiple regression analyses (distance-based Redundancy Analysis) indicated significant associations between wetland physico-chemistry and habitat transformation (overall transformation within 100 and 500 m, alien vegetation cover within 100 and 500 m, urban cover within 100 m); although for significant regressions the amount of variation explained was very low (range: ~2 to ~5.5%), relative to that explained by purely spatio-temporal factors (range: ~35.5 to ~43%). The nature of the relationships between each type of transformation in the landscape and individual physico-chemical variables in wetlands were further explored with univariate multiple regressions. Results suggest that conservation of relatively narrow (~100 m) buffer strips around temporary wetlands is likely to be effective in the maintenance of natural conditions in terms of physico-chemical water quality.

Editor: Dafeng Hui, Tennessee State University, United States of America

Funding: Funding for this project was provided by the National Research Foundation (NRF, Pretoria www.nrf.ac.za) and the Water Research Commission (WRC, Pretoria www.wrc.org.za). The funders had no role in study design, data collection and analysis, decision to publish, or preparation of the manuscript.

Competing Interests: The authors have declared that no competing interests exist.

* E-mail: matthew.bird@nmmu.ac.za

Introduction

Landscapes in human-populated regions have become intensively altered by anthropogenic activities and this landscape transformation has become a key driver of ecological systems worldwide (e.g. [1–4]). Information on the effects of terrestrial habitat transformation on wetland ecosystems is scarce, particularly so for small temporary wetlands, often a numerically dominant wetland type in seasonally dry areas [5,6].

Physico-chemical constituents of the water column (e.g. pH, nutrients, conductivity) are regarded as potentially important determinants of biotic assemblage composition in wetlands and other freshwater ecosystems (e.g. [7–14]). More specifically, in the south-western Cape (South Africa) De Roeck [15] established that physico-chemical factors exert a significant structuring effect on invertebrate assemblage composition in temporary depression wetlands. Alteration of these factors through anthropogenic disturbance has potential to mediate ecosystem changes in these wetlands, through bottom-up effects on biota such as aquatic invertebrates and amphibians. Previous studies have focussed on permanent wetland types, from which various authors have reported significant effects of habitat transformation on an array of individual physico-chemical variables including turbidity, pH, nutrients, conductivity and dissolved oxygen [16–22]. Very few studies have specifically addressed relationships between terrestrial habitat transformation and physico-chemical conditions within temporary wetlands. Carrino-Kyker and Swanson [23] found a significant positive relationship between agricultural land use and conductivity levels in a study of thirty temporary pools in northern Ohio, USA. Brooks et al. [24] studied four ephemeral forest pools in Massachusetts, USA, and reported higher pH and conductivity, and lower concentrations of dissolved oxygen, for two of the pools occurring in urban areas compared with the two pools situated in undisturbed areas. Rhazi et al. [25] found higher levels of nutrients (nitrogen and phosphorus) in wetlands surrounded by agricultural fields than for those in natural areas for a set of ten temporary wetlands in Morocco. It appears that no universally consistent impacts of habitat transformation on physico-chemical conditions within temporary wetlands have been established thus far.

This study assesses the extent and nature of alterations to wetland physico-chemistry associated with human landscape transformation around temporary depression wetlands of the south-western Cape mediterranean-climate region of South Africa. Wetland physico-chemical characteristics are presented in relation to gradients of surrounding terrestrial habitat transformation induced by human activities. The term 'habitat transformation' is used hereafter with reference to the loss of natural terrestrial habitat around wetlands due directly (e.g. agriculture) or indirectly (e.g. alien invasive vegetation) to human land-use practices. A broad approach was taken by sampling a large number of wetlands across the south-western Cape region (n = 90). The survey design aimed to sample wetland sites across gradients of habitat transformation, defined in terms of the amount of surrounding habitat converted to agriculture, urban area, or invaded by alien vegetation. These are the three major agents of habitat transformation in the south-western Cape [26–28] and are all particularly prevalent on the low-lying coastal plains of the region [29], thus threatening temporary depression wetlands. The assessment of alien invasive vegetation as a type of habitat transformation around wetlands is a key element of the approach to this investigation. The negative effects of invasive vegetation on the quantity of groundwater available to aquatic systems in the region have been well documented [30–33], yet empirical studies that have addressed the influence of alien vegetation on surface water quality (e.g. physico-chemistry) of aquatic systems (lentic or lotic) are lacking. A single exception is the study of Bird et al. [34], who studied a set of 12 temporary depression wetlands within a Sand fynbos ecosystem in Cape Town, where the landscape has become differentially invaded by kikuyu grass, *Pennisetum clandestinum* Höchst. ex Chiov, and Port Jackson willow *Acacia saligna* (Labill) Wendl. They found that replacement of indigenous Sand fynbos habitat with alien vegetation results in lowered humic input to wetlands, with knock-on effects on other wetland physico-chemical constituents such as pH. The results of Bird and co-workers indicate the potential for broader effects on wetlands in the region, as a result of large-scale replacement of sclerophyllous fynbos vegetation with invasive plant species. This study investigates such broad-scale patterns, assessing whether the small-scale patterns observed by Bird and co-workers are also salient across the region.

Despite a paucity of information on the principal drivers of physico-chemical conditions in temporary depression wetlands, certain key factors have emerged from the literature, which include local geological substrate (soil properties), morphology of the wetland basin, surrounding landscape topography, surrounding terrestrial vegetation type and local climate (for reviews see [35–37]). Given that one or more of these driving factors were expected to be significantly altered by human habitat transformation (e.g. soil physico-chemical properties may be affected by the type of land use), we hypothesized that physico-chemical conditions in the studied temporary wetlands would in turn show significant association with changes in these driving variables and thus we expected physico-chemistry to be affected by surrounding habitat transformation. Specific relationships between each type of habitat transformation and each measured physico-chemical variable were explored to generate further hypotheses regarding effects of habitat transformation on wetland physico-chemistry in the region.

Methods

Ethics statement

Permission for fieldwork in the Agulhus National Park was granted by South African National Parks. A scientific collection permit was granted by Cape Nature, which allowed access to conservation areas under control of the provincial administration of the Western Cape and privately owned land in the province of the Western Cape. This research did not involve capture or handling of animals and therefore did not require approval of animal care and use procedures. The field study did not create effects on endangered or protected species.

Study area

Ninety isolated depression wetlands (*sensu* Ewart-Smith et al. [38]) were sampled once during the winter-spring wet season (late-July to early-October) of 2007. The south-western Cape is unique in sub-Saharan Africa for having a mediterranean climate, typically encompassing cool, wet winters and warm, dry summers. The study area and selection of sites are described by Bird et al. [39] in a concurrent study on macroinvertebrate assemblages that utilised the same set of wetlands. Briefly, sampling covered an area from Cape Agulhus in the south to St Helena Bay in the north and incorporated three broadly distinguishable coastal plains (Figure 1). Wetlands were grouped *a priori* according to their natural (reference) state in terms of comparable soils and climate. Clusters of comparable wetlands were established using the vegetation groups of Rebelo et al. [40] as an indication of naturally comparable wetland groups, given the intimate link between vegetation type and local abiotic factors in the study region. In this regard, five wetland clusters were sampled in the study region (Appendix S1 in File S1), namely Sand fynbos (n = 44), Western strandveld (n = 28), Shale renosterveld (n = 6), Ferricrete fynbos (n = 6) and Sandstone fynbos (n = 6).

Habitat transformation

The assessment of transformed habitat around each wetland was based on the quantification of the cover of natural vegetation (untransformed land), alien vegetation (predominantly *A. saligna* and *P. clandestinum*), agriculture and urban land within 100 and 500 m of each wetland [39]. An estimate of the areal cover of each habitat category was obtained from circular areas corresponding to 100 and 500 m radii from the edge of each wetland (i.e. approximately 0.03 and 0.8 km^2). These scales were chosen because they could be accurately assessed on the ground without using GIS data. For both scales, the cover of each habitat type was estimated and assigned to one of four ordinal cover categories: 0 - none; 1 – sparse cover (<33%); 2 – moderate cover (33–66%); 3 – extensive cover (>66%). For those wetlands that were difficult to survey for a 500 m radius on foot, satellite imagery (Google Earth, accessed 2007) was combined with ground survey information, in order to score the ordinal categories of habitat cover. All estimates were made by the same person, in order to avoid inter-personal judgment biases. Appendix S2 in File S1 reports the habitat transformation scores for each wetland as raw data.

Environmental variables

In order to record the range of physico-chemical conditions in each wetland, various biotopes were sampled. Biotopes were differentiated based on the structural complexity of habitats, of which four major types were encountered: complex vegetation (generally submerged, inter-woven, rooted or non-rooted with fine dissected leaves, including species such as *Isolepis rubicunda*, *Stuckenia pectinata*, *Chara glomerata* and *Paspalum vaginatum*); simple vegetation (typically rooted and emerging from the water surface, reed- or sedge-like vegetation, including species such as *Typha capensis*, *Phragmites australis*, *Bolboschoenus maritimus* and *Juncus kraussii*); open water (no vegetation, deeper than 30 cm); and benthic un-vegetated habitat (no vegetation, shallower than 30 cm). The percentage surface area covered by each of these four different biotopes in each wetland was recorded visually in the field. During field sampling it was noted that a maximum of three biotopes

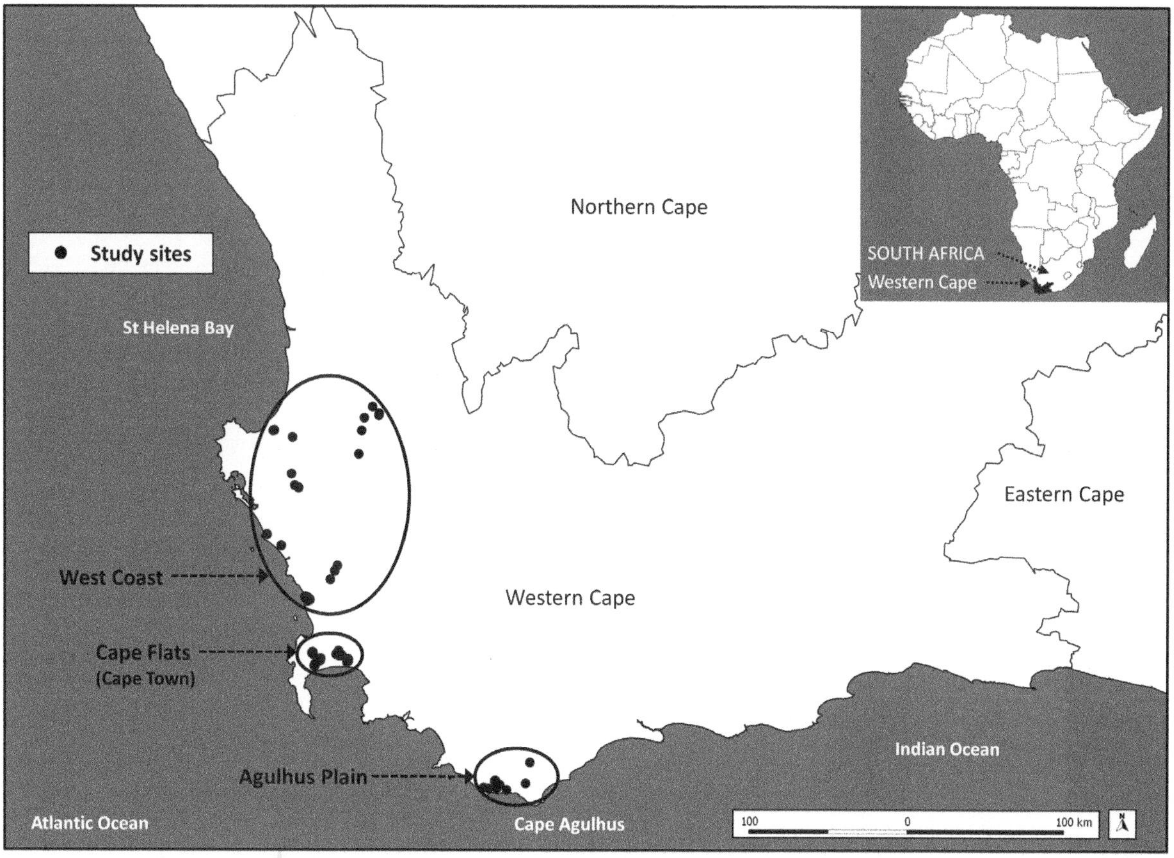

Figure 1. The south-western Cape study region showing sites sampled during the 2007 wet season (n = 90). Study sites were concentrated on three broadly distinguishable coastal plains (indicated by the bold circles). The region is bounded approximately by Cape Agulhus in the south and St Helena Bay in the north (modified from Bird et al. [39], for illustrative purposes only).

existed in any one wetland simultaneously. Thus, although all four biotope types were encountered among wetlands during field sampling, only three or fewer were represented within each wetland.

A number of *in situ* physico-chemical variables were measured in each of the biotopes within each wetland, producing three sets of *in situ* physico-chemical measures per wetland. For sites with only two biotopes, a double and a single set of physico-chemical readings were taken in the more and less abundant biotopes respectively. For sites where only one biotope covered the entire wetland, three replicate sets of physico-chemical readings were taken, with the aim of covering as much of the spatial extent of the wetland as possible among each set. All physico-chemical readings were taken at a standardized depth of 30 cm across all biotopes, with the exception of readings taken from habitats <30 cm deep. Measurements were taken as follows: pH using a Crison pH25 meter; dissolved oxygen with a Crison OXI45 oxygen meter; electrical conductivity (hereafter 'conductivity') with a Crison CM35 conductivity meter; and turbidity using a Hach 2100P turbidimeter. Temperature was recorded on the pH, oxygen and conductivity meters, although for analytical purposes an average of the readings across all three instruments was used.

Water column nutrient concentrations were measured at each wetland. Five 1L surface water samples were collected from each wetland, with the aim of covering the full spatial extent of each site, and pooled to form a bulk 5L sample. This was then thoroughly mixed and a 200 ml sub-sample was taken for analysis of nutrients levels in the laboratory. Samples were stored immediately in the dark at 4°C and upon return to the laboratory were frozen at −18°C. All samples were analysed within 30 days of collection from the field. NO_3^-+NO_2^-–N, PO_4^{3+}–P and NH_4^+–N concentrations were estimated using a Lachat Flow Injection Analyser. Approximate detection limits are: for PO_4^{3+} 15 μg.L^{-1} P; for NO_3^- and NO_2^- 2.5 μg.L^{-1} N; and for NH_4^+ 5 μg.L^{-1} N. These variables are hereafter referred to in the text as 'phosphates', 'nitrates + nitrites' and 'ammonium' respectively. The geographical position and altitude at the centre point of each wetland were recorded using a Garmin eTrex Vista handheld GPS device (point accuracy of 3 m). In order to make sure no permanently inundated wetlands were included in the dataset, only sites with maximum depth <2 m were sampled. Most of the deeper sites were re-visited in summer to confirm that they had dried up. Appendix S3 in File S1 reports the raw environmental data collected at each wetland.

Data analysis

Separate subsets of the dataset were used to analyse relationships between each type of habitat transformation and physico-chemical conditions in wetlands. These subsets were composed of sites affected by only one type of habitat transformation (e.g. agriculture). This was done to exclude sites that were affected by habitat transformations other than the type of interest. Each separate dataset was composed of least impaired sites (surrounded by extensive indigenous vegetation) and those sites that were impacted by varying degrees of habitat conversion for the given transformation type. Only least impaired sites occurring within the same wetland cluster as impacted sites were selected, to ensure comparison with naturally similar wetlands in the area. Two different but largely overlapping datasets were created for each habitat transformation type, corresponding to the 100 and 500 m scales of analysis, because in certain cases sites used in analysis of impact at one scale were not applicable at the other. An exception to this was for analyses relating to the amount of natural (indigenous) vegetation cover around wetlands, as this criterion was applicable to all cases.

Two broad types of response data were analysed in relation to surrounding gradients of habitat transformation, namely physico-chemical variables as a set (multivariate response) and individual physico-chemical variables (univariate response). Table 1 provides a list of the variables analysed in this study. The names used for these variables in Table 1 are hereafter used in the text. For multivariate analyses, response variables were first normalized and then converted into a Euclidean distance matrix, which was subsequently related to surrounding levels of habitat transformation. For the physico-chemical variables that were measured *in situ* for each biotope, an average of the three readings per wetland was used in subsequent analyses. Multivariate linear regressions were used to relate the physico-chemical response matrices to the habitat transformation predictor variables. Following the recommendations of Somerfield et al. [41], the gradient analyses in this study are best performed using regression rather than ANOVA techniques, given that the predictor variables representing gradients of impact are ordinal. Detrended Correspondence Analysis (DCA) indicated that gradient lengths in the physico-chemical dataset were best suited to linear rather than unimodal analyses [42]. Multivariate regressions were performed using distance-based Redundancy Analysis (dbRDA, [43,44]), a non-parametric multivariate regression procedure based on any given dissimilarity measure, in this case Euclidean distance. Patterns in the multivariate physico-chemical data were visually explored using Principal Components Analysis (PCA) ordination. Sites were coded on each PCA plot according to three factors of interest, namely surrounding overall levels of habitat transformation ('Natural 100 m' and 'Natural 500 m', Table 1), the wetland cluster into which they were classified (defined by vegetation type, Table 1), and at a broader level, the coastal plain on which they were situated (West Coast, Cape Flats and Agulhus Plain, Figure 1). These factors were incorporated in order to assess the variation in physico-chemical conditions among wetlands in relation to habitat transformation gradients, as well as natural spatial factors.

Univariate multiple linear regression models were used to test for relationships between individual physico-chemical variables and gradients of habitat transformation. The coefficient of partial determination (partial r^2) was incorporated into the univariate regression results by squaring the partial correlation coefficient (r) for the predictor variable of interest [45]. Univariate relationships were visualized using partial residual plots (*sensu* Larsen & McCleary [46]), which allow one to hold the covariables constant in each model and also allow visual examination for heterogeneity in the spread of residuals, deviations from linearity and outliers [45,47]. Potential outliers were quantitatively assessed using Cook's distances (Cook's D_i, *sensu* Cook and Weisberg [48]), where D_i values >1 or D_i values considerably larger than the rest of the values would warrant an outlier [45]. All univariate and multivariate regression models were conditioned upon covariables, so as to partial out the effects of potentially confounding factors. Covariables included the following measures for each wetland: longitude and latitude (decimal degrees); time (days since first sampling event); altitude (m); and vegetation type (five dummy variables defined the wetland clusters). Where there was urban land surrounding wetlands, there was often invasive vegetation and *vice versa*. To help address this overlap, when assessing relationships between invasive vegetation cover and wetland physico-chemical conditions, the amount of urban land cover was specified as a covariable, and *vice versa* when assessing the effects of urban land cover as the primary variable. To maximise parsimony, covariable subsets were pre-selected for each model using step-wise regression of each response variable or matrix on the full list of possible covariables (Table 1, Appendix S1 in File S1).

Environmental variables were $\log_{10}$ transformed where appropriate, in order to achieve normality. A standard α level of 0.05 was used to assess the statistical significance of regression relationships, except for tests related to agriculture, as the smaller sample size of the two agricultural datasets (100 m: n = 24; 500 m: n = 21) indicated that the possible lack of power to detect effects could be countered by interpreting P values <0.10 as offering some evidence against the null hypothesis. Due to the possibility of inflated Type 1 error for multiple testing the significance of regressions was also assessed at a more conservative α level of 0.01, as well as after sequential Bonferroni correction (*sensu* Holm [49]). The Bonferroni procedure is likely to yield inappropriately conservative results (inflated Type 2 error) for univariate testing in this study, given the large number of tests conducted. Significant results (for both α<0.05 and α<0.01) were interpreted with caution if there were only one or two significant variables out of a large group of tested variables [50,51]. DCA ordinations were performed using CANOCO for Windows v4.5 [52]. All dbRDA models (including the step-wise models) were implemented using the DISTLM routine of the PERMANOVA+ package [53]. P values for dbRDA models were tested by 9999 permutations of residuals under the reduced model. PCA ordinations were performed using PRIMER v6 software [54,55]. Univariate multiple regressions (including step-wise models and partial residual plots) were performed using STATISTICA v10 software (Statsoft Inc. 2010, Tulsa, Oklahoma, USA).

Results

Physico-chemical characteristics of wetlands

Wetlands generally had a neutral pH or were slightly alkaline, although several highly acidic sites were encountered in the Sand fynbos cluster and several highly alkaline sites were spread among the clusters (for pH ranges per wetland cluster see Appendix S4 in File S1). Conductivity levels (as a proxy for salinity) were generally low, with mean and median values per wetland cluster all below 5 $mS.cm^{-1}$. Turbidity levels were low on the whole and had mean and median values across all wetland clusters <10 NTU, except for the Shale renosterveld cluster, which stood out for having higher values (reflecting the naturally high quantity of clay particles in these shale-derived soils). Dissolved oxygen concentrations varied between moderate and high levels (range: 4.05–9.85 $mg.L^{-1}$) in terms of mean and median values among the wetland clusters. Mean and median nutrient concentrations were low, except for the high values reported for phosphates and ammonium in Shale renosterveld wetlands (phosphates range: 12.03–999.41 $\mu g.L^{-1}$; ammonium range: 61.72–2803.87 $\mu g.L^{-1}$). Several extremely nutrient-enriched sites were found in the Sand fynbos cluster, as reflected by the very high maximum values for all three nutrient variables in this cluster (nitrates + nitrites maximum: 8241.59 $\mu g.L^{-1}$; phosphates maximum:

Table 1. List of the physico-chemical response variables, habitat transformation predictor variables and spatio-temporal covariables incorporated into the analyses of this study.

Variable type	Variable scale	Category/set	Variable name	Description
Response variables	Quantitative (continuous)	Physico-chemistry		
			pH	Measured *in situ* for each biotope
			Conductivity	Measured *in situ* for each biotope
			Temperature	Measured *in situ* for each biotope
			Turbidity	Measured *in situ* for each biotope
			Dissolved oxygen	Measured *in situ* for each biotope
			Nitrates + nitrites	Integrated sample from across the wetland
			Phosphates	Integrated sample from across the wetland
			Ammonium	Integrated sample from across the wetland
Predictor variables	Semi-quantitative (ordinal)	Habitat transformation		
			Natural 100 m	Areal cover of indigenous vegetation within 100 m radius of wetland edge
			Natural 500 m	Areal cover of indigenous vegetation within 500 m radius of wetland edge
			Invaded 100 m	Areal cover of alien invasive vegetation within 100 m radius of wetland edge
			Invaded 500 m	Areal cover of alien invasive vegetation within 500 m radius of wetland edge
			Agriculture 100 m	Areal cover of agriculture within 100 m radius of wetland edge
			Agriculture 500 m	Areal cover of agriculture within 500 m radius of wetland edge
			Urban 100 m	Areal cover of urban surface within 100 m radius of wetland edge
			Urban 500 m	Areal cover of urban surface within 500 m radius of wetland edge
Covariables	Quantitative (continuous)	Spatio-temporal		
			Longitude	Taken at the wetland centre-point
			Latitude	Taken at the wetland centre-point
			Altitude	Taken at the wetland centre-point
			Time	Number of days since first sampling event
	Categorical	Spatio-temporal		
			Ferricrete fynbos	*Indigenous terrestrial vegetation type
			Sand fynbos	*Indigenous terrestrial vegetation type
			Sandstone fynbos	*Indigenous terrestrial vegetation type
			Shale renosterveld	*Indigenous terrestrial vegetation type
			Western strandveld	*Indigenous terrestrial vegetation type

* Either presently or historically surrounding each wetland (vegetation types *sensu* Rebelo et al. [40]).

2827.36 $\mu g.L^{-1}$; ammonium maximum: 4231.53 $\mu g.L^{-1}$, Appendix S4 in File S1).

Multivariate patterns

Physico-chemical conditions in wetlands were significantly related ($\alpha<0.01$ and also after Bonferroni correction) to the cover of natural (indigenous) vegetation within 100 m (P = 0.002) and 500 m (P = 0.009), and the cover of alien invasive vegetation within 100 m (P = 0.005) and 500 m (P = 0.005) (Table 2). Wetland physico-chemistry and urban cover within 100 m of wetlands were significantly related at $\alpha<0.05$ only (P = 0.022). Despite some of the results being significant, very little of the variation in physico-chemical conditions was explained by these land cover variables (range: 2.08–5.57%) in comparison to that explained by the spatio-temporal covariables (range: 35.57–43.09%). No significant relationships were found between physico-chemical conditions in wetlands and urban cover within 500 m or agricultural cover within 100 and 500 m (Table 2).

Although the variables 'Natural 100 m' and 'Natural 500 m' were significantly related to physico-chemical conditions (Table 2), the patterns are not obvious in the PCA plots (Figure 2a–b) with no clear grouping according to the different levels of natural

Table 2. Non-parametric multivariate regression tests (dbRDA) for relationships between habitat transformation gradients and physico-chemical conditions in wetlands.

Predictor variable	Res. df	F	P	% Var	Covariables	% Var (covariables)
Natural 100 m	82	3.962	0.002***	2.62	Time, longitude, latitude, altitude, SF, SR	43.09
Natural 500 m	82	3.106	0.009***	2.08	Time, longitude, latitude, altitude, SF, SR	43.09
Invaded 100 m	65	3.529	0.005***	3.26	Time, longitude, latitude, altitude	36.71
Invaded 500 m	66	3.441	0.005***	3.18	Time, longitude, latitude, altitude	35.79
Agriculture 100 m	18	1.333	0.243	2.50	Time, longitude, latitude, SR	63.71
Agriculture 500 m	16	1.186	0.299	3.45	Time, FF, SR	50.04
Urban 100 m	31	2.927	0.022*	5.57	Time, longitude, latitude, altitude	35.47
Urban 500 m	49	1.879	0.090	2.24	Time, latitude, altitude, Invaded 500 m	39.37

Natural - indigenous vegetation; Invaded - alien invasive vegetation; Agriculture - agricultural land; Urban - urban area. The areal cover of these variables is represented within 100 and 500 m radii of each wetland edge. To maximise parsimony, covariable subsets were pre-selected for each model using step-wise regression of each response matrix on the full list of possible covariables (see Table 1). % Var - the percentage of variation in each Euclidean distance matrix (normalized physico-chemical variables) that is explained by each respective predictor variable or covariable set in each model; Time – number of days since the first sampling event; SF – Sand fynbos; SR – Shale renosterveld; FF – Ferricrete fynbos; Res. df – residual degrees of freedom for each model. Significant P values at $\alpha<0.05$ (*), $\alpha<0.01$ (**) and after sequential Bonferroni correction (***) are indicated.

vegetation within 100 and 500 m. The PCA does not allow visualization of patterns with the effects of covariables partialled out (e.g. time, latitude), and thus may limit its usefulness in visualizing the effects of habitat transformation when covariables are involved in the analysis. In terms of physico-chemical constituents, sites showed better grouping according to the vegetation types in which they would naturally occur (i.e. the wetland clusters), but none of the vegetation types formed clusters that were clearly separated from the rest of the groups (Figure 2c). Variation within certain groups was large, particularly for the Sand fynbos cluster, which covered the broadest area of sampling and showed by far the most variation in physico-chemical conditions, as evidenced by scatter in the PCA plot among sites for this cluster (Figure 2c). The Western strandveld was the second largest cluster, showing the second-highest levels of physico-chemical variation. With the exception of Sandstone fynbos, which showed considerable variation among few sites, the remaining small wetland clusters displayed correspondingly lower levels of variation than for the bigger clusters. Physico-chemical conditions among the three broad coastal plain areas appear to be more distinguishable, although there is some overlap among areas towards the centre of the ordination (Figure 2d).

Univariate patterns

pH, phosphates, dissolved oxygen and turbidity were negatively related to indigenous vegetation cover within both 100 and 500 m radii of wetlands (Table 3). With the exception of turbidity, the same variables were positively related to invasive vegetation cover within 100 m of wetlands. Only dissolved oxygen concentrations were significantly related (positive slope) to invasive vegetation cover within 500 m. Phosphate concentrations in wetlands were positively related to agricultural cover within 100 m, whilst ammonium concentrations were negatively related to agricultural cover within 500 m. pH was positively related to urban cover within 100 m of wetlands, but none of the variables were significantly related to urban cover within 500 m. The relationships presented in Table 3, although significant, are generally weak, as inferred from the low amounts of explained variation in the response variables due to the habitat transformation predictor variables (partial r^2 values mostly <0.20 i.e. 20%, and none >0.30). As observed for the multivariate regressions (Table 2), the percentages of explained variation due to the spatio-temporal covariables (see 'r^2 - Covariables') in Table 3 were for the most part considerably higher than that explained by the habitat transformation predictor variables (see 'Partial r^2 – predictor'). Six of the univariate relationships were significant at $\alpha<0.01$ (Table 3a, 3e, 3i, 3l, 3o, 3p) and only two relationships (Table 3a, 3i) were significant after sequential Bonferroni correction.

The partial residual plots of Figure 3 offer visual representation of the regression relationships reported in Table 3, holding the covariables constant. Apparent in most of the plots is the considerable amount of vertical (Y axis) scatter in the residual points, which accounts for the low partial r^2 values observed in Table 3 and shows that relationships were generally weak. The plots also allow identification of outliers or groups of high leverage points. The pattern for pH appears to be highly leveraged by five very low pH sites occurring in one particular area on the Cape Flats, visible at the bottom of the plots in Figure 3a, 3e, 3i and 3p. To test their influence, a *post hoc* analysis was run without these sites and revealed that the partial relationships between pH and natural vegetation cover within 100 and 500 m remained significant at $\alpha = 0.05$, but were substantially weaker ('Natural 100 m': $t_{80} = -2.124$, $P = 0.037$, partial $r^2 = 0.053$; 'Natural 500 m': $t_{80} = -1.994$, $P = 0.049$, partial $r^2 = 0.047$). Partial relationships between pH and invasive vegetation cover within 100 m, and between pH and urban cover within 100 m, were rendered non-significant by exclusion of these sites from the models ('Invaded 100 m': $t_{61} = 1.303$, $P = 0.198$, partial $r^2 = 0.027$; 'Urban 100 m': $t_{29} = -0.187$, $P = 0.853$, partial $r^2 = 0.001$), indicating a strong influence of these sites on the regressions.

Relationships between phosphate concentrations and habitat transformation (Figure 3b, 3g, 3j and 3 m) showed high phosphate values associated with several of the extensively transformed wetlands, which may have influenced the reliability of these trends. Examination of Cook's distances for these models did not indicate that any of these high phosphate values had undue leverage on the trends (the maximum Cook's D_i value was 0.228). Dissolved oxygen concentrations showed similar patterns as for phosphates, although only one outlier was clearly apparent in these plots (Figure 3c and 3f, Figure 3k and 3l). Cook's distances once again indicated that no points had particularly undue leverage in these

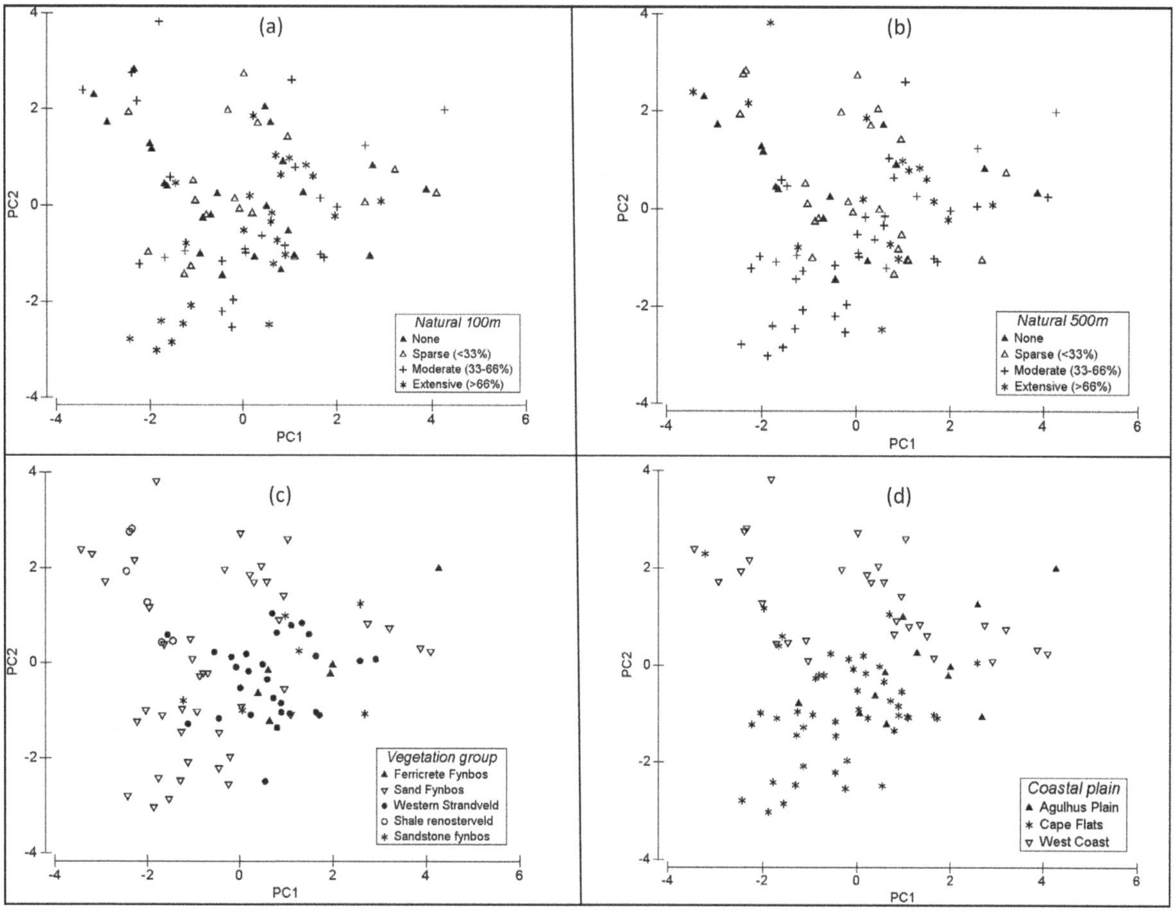

Figure 2. Principal Components Analysis ordinations of the physico-chemical variables (normalized) for all study sites (n = 90). The first two principal component axes are displayed. Sites are coded according to: (a) the areal cover of natural (indigenous) vegetation within a 100 m radius of each wetland edge; (b) the areal cover of natural (indigenous) vegetation within a 500 m radius of each wetland edge; (c) the natural vegetation type either presently or historically surrounding each wetland; and (d) the three broad coastal plains covered in this study.

models (maximum Cook's D_i value: 0.373). Turbidity displayed weak linear trends with the 'Natural 100 m' and 'Natural 500 m' variables (Figure 3d and 3h respectively), with low gradient slopes and considerable spread in the residual points on either side of the regression line. The positive relationship between turbidity levels and urban cover within 100 m (Figure 3o) was clearer and showed less scatter among points.

Discussion

Indigenous vegetation cover

The multivariate regression relationships presented in this study indicate that human transformation of the landscape surrounding temporary depression wetlands in the south-western Cape is associated with physico-chemical conditions in these wetlands. This statement refers to overall transformation of adjacent habitats as represented by the remaining indigenous vegetation cover. The pattern appears to be slightly stronger for habitat transformation taking place within 100 m of wetlands than for 500 m, although significant trends were seen at both scales. The contribution to the percentage variation in physico-chemical conditions explained by variables representing the different types of habitat transformation was very low (2.08–5.57%, Table 2), despite being significant in some cases. This explained variation was generally in the region of one order of magnitude lower than that explained by the spatio-temporal covariables in the multivariate models (35.47–63.71%, Table 2). At the broad scale of this study, the primary influence on physico-chemical conditions in wetlands thus appeared to come from spatio-temporal factors, although a significant signal was still detected for certain habitat transformation factors over and above the spatio-temporal influence. This indicates firstly, that spatio-temporal variation in physico-chemical conditions is high for these wetlands, and secondly, that habitat transformation has played a meaningful role (albeit relatively weak in comparison to that of spatio-temporal factors) in altering the physico-chemistry of these wetlands, as was hypothesized at the outset of this study.

Our results are in line with those of a similar study by Declerck et al. [20] on the water quality of 99 small permanent ponds (natural and artificial) affected by agriculture in Belgium. These authors recorded land use at multiple spatial scales up to 3.2 km around ponds and found that the maximum amount of variation in a set of physico-chemical variables explained by crop land was 2.3% and by the amount of indigenous forest cover was 4%, both measured at a scale of 100 m around ponds and both were statistically significant. Their study also corroborates our finding that habitat transformation influences on the physico-chemistry of small, isolated wetlands appear to be strongest within 100 m of wetlands, although only a slight difference was found between the 100 and 500 m scales in this study. Given that small isolated wetlands have been shown elsewhere to drain localised catch-

Table 3. Multiple linear regression models (a - p) of environmental response variables regressed against the habitat transformation variables (predictors), given the spatio-temporal covariables.

	Predictor variables	Response variables	β	SE	Partial r^2 (Predictor)	t	Res. df	P	Covariables	r^2
a)	(Covariables) Natural 100 m	pH	−0.354	0.081	0.187	−4.374	83	<0.001***	Longitude, latitude, time, altitude, Western strandveld	0.329
b)	Natural 100 m	Phosphates	−0.214	0.083	0.073	−2.591	86	0.011	Longitude, latitude, Shale renosterveld	0.371
c)	Natural 100 m	Dissolved oxygen	−0.216	0.092	0.061	−2.357	86	0.021	Altitude, Ferricrete fynbos	0.256
d)	Natural 100 m	Turbidity	−0.163	0.070	0.060	−2.326	85	0.022	Longitude, latitude, Shale renosterveld	0.543
e)	Natural 500 m	pH	−0.261	0.087	0.097	−2.993	83	0.004**	Longitude, latitude, time, altitude, Western strandveld	0.365
f)	Natural 500 m	Dissolved oxygen	−0.229	0.093	0.066	−2.467	86	0.016	Altitude, Ferricrete fynbos	0.255
g)	Natural 500 m	Phosphates	−0.195	0.084	0.059	−2.317	86	0.023	Longitude, latitude, Shale renosterveld	0.377
h)	Natural 500 m	Turbidity	−0.152	0.071	0.051	−2.138	85	0.035	Longitude, latitude, Shale renosterveld	0.548
i)	Invaded 100 m	pH	0.391	0.088	0.229	4.423	66	<0.001***	Time, altitude, Western strandveld	0.295
j)	Invaded 100 m	Phosphates	0.251	0.098	0.090	2.570	67	0.012	Longitude, latitude	0.274
k)	Invaded 100 m	Dissolved oxygen	0.248	0.102	0.079	2.418	68	0.018	Altitude	0.209
l)	Invaded 500 m	Dissolved oxygen	0.318	0.102	0.123	3.108	69	0.003**	Altitude	0.191
m)	Agriculture 100 m	Phosphates	0.371	0.161	0.210	2.307	20	0.032	Latitude, Sand fynbos	0.410
n)	Agriculture 500 m	Ammonium	−0.399	0.149	0.285	−2.676	18	0.015	Latitude	0.332
o)	Urban 100 m	Turbidity	0.457	0.135	0.253	3.395	34	0.002**	Latitude	0.158
p)	Urban 100 m	pH	0.341	0.131	0.167	2.610	34	0.013	Sand fynbos	0.252

Only significant relationships are presented here ($\alpha = 0.05$, with the exception of agriculture, where $\alpha = 0.10$). To maximise parsimony, covariable subsets were pre-selected for each model using step-wise regression of each response variable on the full list of possible covariables (see Table 1). For each predictor variable, results are listed in decreasing order of relationship strength based on P values. Only partial relationships between the response and predictor variables are reported here, not the full model results. Significant relationships at $\alpha < 0.01$ (**) and after sequential Bonferroni correction (***) are also indicated.

Natural - indigenous vegetation; Invaded - alien invasive vegetation; Agriculture - agricultural land; Urban - urban area. The areal cover of these variables is represented within 100 and 500 m radii of each wetland edge, measured on an ordinal scale. Time - Number of days since the first sampling event; β – standardized regression coefficient; SE – standard error of regression coefficient; Partial r^2 – coefficient of partial determination for each respective predictor variable; Res. df – residual degrees of freedom; r^2 (Covariables) = Full model r^2 - Partial r^2 (predictor).

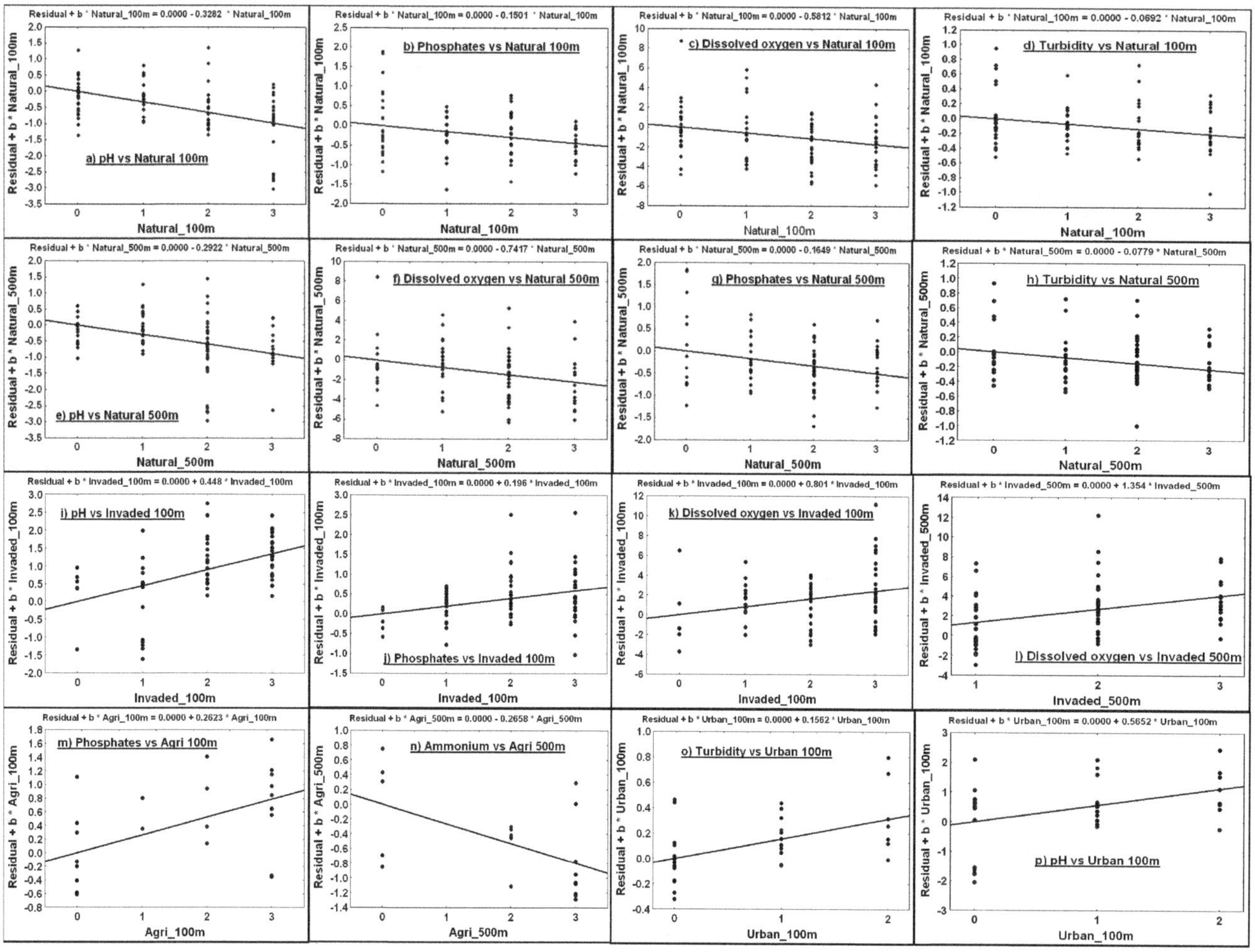

Figure 3. Partial residual plots displaying the relationships a – p presented in Table 3. Environmental response variables are depicted in relation to the habitat transformation variables (predictors, x axes), holding the spatio-temporal covariables constant. Natural - indigenous vegetation; Invaded - alien invasive vegetation; Agri - agricultural land; Urban – urban area. The areal cover of these variables is represented within 100 and 500 m radii of each wetland edge, measured on an ordinal scale: 0 – none; 1 – sparse; 2 – moderate; 3 – extensive. For more detailed information regarding each model, refer to Table 3. Note: The residuals on the vertical axis of each plot come from the regression of the response variable against all the predictors except the one of interest. The residuals for the horizontal axis of each plot come from the regression of the predictor variable of interest against all other predictors. Each residual scatterplot shows the relationship between a given univariate response variable and a predictor variable of interest, holding the other predictor variables constant. The regression equation for each relationship has been indicated, with each slope being equal to the non-standardized regression coefficient (b) in the full multiple regression model in which the parameter was included. '0.0000' indicates that the intercept value is <0.0001.

ments, they would not be expected to be affected by broader catchment-scale processes, as might be the case for rivers or lakes [56]. This may well explain why the strongest relationships between wetland physico-chemical conditions and surrounding land cover are reported at the 100 m scale in this study and that of Declerck et al. [20].

Negative linear relationships were reported between surrounding indigenous vegetation cover within 100 and 500 m of wetlands and pH, phosphates, dissolved oxygen and turbidity levels in these wetlands (Table 3). Turbidity has been shown to increase with transformation of the surrounding landscape for other wetland ecosystems, particularly as a result of sedimentation from agricultural or urban runoff [16,20]. Replacement of natural vegetation often leads to de-stabilization of soils [21] and thus various forms of habitat transformation could be responsible for increased sediment input to wetlands through increased surface water flows during rain events. The negative relationship between wetland pH and natural vegetation cover within 100 m is most likely an effect of removing fynbos, which is known to release acidic leachates into the soil [34,57–59]. The resultant physico-chemical effect would be an increase in the pH of wetlands as surrounding fynbos is lost. Bird et al. [34] found that the replacement of indigenous Sand fynbos with alien vegetation caused a reduction of humic input to wetlands, which in turn affected other physico-chemical constituents such as pH. They hypothesized that the alteration of pH with transformation of the landscape is only likely to occur in areas where soils are naturally acidic and the vegetation type is sclerophyllous fynbos. With reference to the current study, the Western strandveld cluster occurs on naturally alkaline soils due to the intrusion of calcareous sediments of marine origin [40], and the vegetation is not sclerophyllous, but dominated by succulents. Replacing this vegetation type with alien vegetation would not lead to effects on soil or surface water pH. However, Sand fynbos is a vegetation type that occurs on well-leached, naturally acidic soils and the vegetation itself is sclerophyllous, containing high levels of acidic tannins as a defence against herbivory [40]. Therefore, the loss of fynbos in this area can be hypothesized to raise soil and surface water pH. Closer inspection of Figure 3a and 3e reveals that five of the sites appeared to have a high leverage on the strength of the trend between indigenous vegetation cover and the pH of wetlands. These sites occurred inside the Kenilworth racetrack on the Cape Flats and were among the most pristine wetlands in the Sand fynbos cluster, due to an historical lack of disturbance inside the racetrack [60–64]. The low pH values are thus most likely a real reflection of the vast amount of undisturbed fynbos vegetation surrounding these sites. Evidence presented in this study provides further confirmation of the trend reported by Bird et al. [34], although the effect appears to be patchy at the broader scale and driven by sites occurring within Sand fynbos. One cannot expect to observe this relationship in areas where the natural vegetation type does not contain acidic tannins.

The negative relationships between surrounding indigenous vegetation cover and dissolved oxygen levels were surprising given that previous studies have found human disturbance of the landscape to be generally associated with decreased levels of oxygen and increased levels of nutrients in aquatic ecosystems (e.g. [65–67]). Further investigation is required to establish any underlying causes in this regard. The negative association between phosphate concentrations and natural vegetation cover within 100 and 500 m could be due to the effects of alien vegetation, agriculture or urban development.

Alien invasive vegetation

Considering the three types of habitat transformation separately, only invasive vegetation cover was significantly related to physico-chemical conditions at both 100 and 500 m spatial scales and thus appears to be an influential form of habitat transformation. Using various modelling approaches, Rouget et al. [28] predicted that between 27.2 and 30% of remaining untransformed habitat in the Cape Floristic Region is likely to be invaded by alien plants over the next 20 years (i.e. from the time of their study). Our results suggest that this predicted spread of alien invasive plants into untransformed areas in the near future could impact significantly on temporary wetland environments occurring in those areas without 'polluting' or physically altering them. Our results corroborate those of Bird et al. [34] in suggesting that terrestrial alien plants are indeed affecting the water quality of aquatic ecosystems in the region, despite numerous research efforts that have focussed only on water quantity effects of alien plants. Furthermore this has importance in the light of changes in biotic assemblages that could potentially be induced by these physico-chemical effects, given the potential importance of physico-chemistry in structuring biotic assemblages such as invertebrates.

The positive association between dissolved oxygen concentrations in wetlands and surrounding invasive vegetation cover within 100 and 500 m was difficult to explain and no literature appears to report similar findings. Further investigation is required to explore possible mechanisms governing this trend, although it should be noted that these relationships were not convincing as reflected by the low partial r^2 values (0.079 and 0.123 at 100 and 500 m respectively, Table 3). Phosphate concentrations (Appendix S3, S4 in File S1) were positively related to invasive vegetation cover within 100 m, but not 500 m, suggesting a localised nutrient input from invasive vegetation into groundwater. Once again the relationship was weak as judged by the small extent of explained variation in phosphates due to the predictor variable 'Invaded 100 m' (partial $r^2 = 0.090$). Bird et al. [34] also reported a significant positive relationship between alien vegetation cover within 100 m of wetlands and water column phosphate concentrations in temporary wetlands. A possible mechanism governing this trend is the elevation of soil phosphorus in adjacent terrestrial soils due to infestation by alien shrubs, which may then leach into wetlands. This is postulated based on the findings of Witkowski and Mitchell [68], who reported a significant increase in soil phosphorus in stands of *Acacia saligna* (also the dominant invader in the current study) compared to surrounding natural lowland fynbos vegetation. It was established that this was due to higher litterfall from acacias, which released leaves into the soil with significantly higher phosphorus content than those of lowland fynbos vegetation [68].

The relatively strong ($P = 0.001$, partial $r^2 = 0.229$, Table 3) positive relationship between pH and alien vegetation cover within 100 m is most likely a consequence of the loss of natural vegetation, which accompanies the transformation of habitats by invasive alien vegetation. As discussed earlier, we hypothesize that the loss of natural vegetation in the Sand fynbos area would cause an increase in soil and surface water pH through the loss of acidic tannins that characterise natural fynbos ecosystems. The predominant disturbance type in this area was alien vegetation and thus it was positively associated with levels of pH, even though it is not expected that alien vegetation itself raises the soil pH, but rather that it is associated with higher levels of pH as a consequence of the loss of naturally acidic vegetation to the system.

Agriculture

Previous studies, mostly on permanent wetlands, have indicated that agriculture has significant impacts on the water chemistry of wetlands (e.g. [20,21,25,69]), whilst no significant effects were detected in this study. This may, to some extent, be an artefact of the relatively small sample size for the agricultural datasets (100 m scale: $n = 24$; 500 m scale: $n = 21$), which reduces the statistical power to detect an effect [70]. The primary agricultural areas of the study region occur mostly on relatively fertile shale soils [40], where wheat agriculture has transformed the landscape so intensively that least impaired wetlands were difficult to find and it was necessary to search for small fragments of remaining natural vegetation that also happened to house temporary wetlands. In the Sand fynbos cluster, least impaired sites were not too difficult to find, and these were compared with sites occurring within pasture areas (the predominant form of agriculture in this area). However, the difficulty in this case was in finding enough sites within moderately and extensively transformed pasture areas. Our data on agriculture is thus limited and although no effect on wetland physico-chemistry was found, this should be interpreted with caution until a larger set of data is available. The lack of un-impacted depression wetlands that could be found within the extensively transformed wheat farming areas highlights the plight of these wetlands in lowland agricultural areas.

Urban development

The association between urban cover within 100 m and physico-chemical conditions in the studied wetlands is in line with the few previous studies which have addressed the topic for temporary wetlands [18,24], however certain affected variables appear to be different. The significant positive relationship between amount of urban area (within 100 m in this study) and wetland turbidity has been reported elsewhere [24,71] and could be attributed to sedimentation from increased surface runoff, amongst other factors. The positive relationship between pH and urban cover within 100 m is once again likely to be due to the rise in pH associated with the loss of fynbos vegetation, as habitat is converted to urban surfaces. It was surprising that nutrient concentrations were not associated with surrounding urban cover, as previous literature has reported this for other temporary wetland systems [18,24] and it was expected that the major form of disturbance for urban-exposed wetlands would be in the form of increased nutrient levels. Furthermore, it was expected that effects of urban development would extend beyond 100 m, given the intensity of this land use, and that a significant association would have been found between physico-chemical conditions and urban cover within 500 m. Follow-up work focussing on urban-encroached temporary wetlands in the area may help to establish why nutrient levels are not elevated in urban-impacted wetlands relative to unimpacted wetlands.

Spatial patterns of wetland physico-chemistry

The PCA ordinations indicated that the spatial scale of sampling was positively associated with the amount of variation in physico-chemical conditions in the temporary wetlands. Although wetland clusters did not clearly separate out on the basis of their physico-chemical constituents, the amount of variation within each cluster was linked to the spatial area covered. The ordinations further indicated that the spatial scale with the clearest pattern of influence on the physico-chemical variables was at the level of broad latitudinal regions (Agulhus Plain, Cape Flats and West coast, Figure 2d). This appears to be consistent with the above-mentioned pattern of increased variation with increased spatial scale and reinforces the pattern of a link between spatial extent of sampling and increasing variation of physico-chemical conditions in the region. This is perhaps not surprising given that one expects more variation in physico-chemistry as the area sampled broadens, due to an associated increased variation in natural environmental factors such as geology and local climate. However, very little information exists on these basic aspects of spatial variation of environmental conditions in temporary wetlands of the region (but see [15,34,72]) and thus it is important to document such patterns. There is a certain degree of confounding from temporal differences between clusters (they were sampled sequentially), which cannot be accounted for in the PCA ordinations, but which were partialled out of the multiple regression models.

Conclusions and recommendations

Significant physico-chemical signals (both multivariate and univariate) for effects of habitat transformation on temporary wetlands were detected over and above strong spatio-temporal influences in this study, confirming the hypothesis stated at the outset. We acknowledge that the dataset upon which our conclusions have been drawn does contain a degree of spatial and temporal confounding due to the sampling of wetlands over a period of several months and over a fairly large geographical area. However, this was largely unavoidable given the scale of the study and we have best accounted for spatio-temporal confounding through explicit incorporation of covariables into statistical models. Further studies at smaller spatial scales and better accounting for temporal variation are suggested to elucidate more specific information on the nature of the impacts of habitat transformation on temporary wetland ecosystems (e.g. [34]). The potential knock-on effects on wetland biota of the region due to physico-chemical changes associated with habitat transformation should also be investigated to help further understand the extent of ecosystem impacts (e.g. macroinvertebrate assemblages, [39,73]). Our data indicate that the physico-chemical environment of these temporary wetlands is significantly influenced by human transformation of natural habitat within adjacent landscapes (<500 m). Relationships were generally stronger at the scale of 100 m around wetlands than for 500 m, indicating that preservation of narrow buffer strips of indigenous vegetation around these wetlands may afford significant protection of water quality. Restoration of even small fragments of terrestrial vegetation surrounding temporary wetlands is likely to yield significant improvements in water quality towards the original least impaired state.

Supporting Information

File S1 This file contains Appendix S1–S4. Appendix S1. Data for the candidate covariables incorporated into the analyses of this study. Time was incorporated as a quantitative covariable measured as days since first sampling event ('date sampled' is provided for general reference). dd - decimal degrees. Appendix S2. Ordinal scores for each type of land cover around wetlands used to proxy habitat transformation gradients. Natural – indigenous vegetation; Invaded – alien invasive vegetation; Agriculture – land converted to agriculture; Urban – land converted to urban surfaces. The areal cover of these categories was scored within 100 and 500 m radii of each wetland edge using an ordinal scale: 0 - none; 1 – sparse cover (<33%); 2 – moderate cover (33–66%); 3 – extensive cover (>66%). Appendix S3. Environmental variables measured in this study. CV – complex vegetation biotope; SV – simple vegetation biotope; OW – open water biotope; BU – benthic un-vegetated biotope; TSA – total surface area; Max. depth – maximum depth. Appendix S4. Summary statistics of the physico-chemical variables (untransformed data) collected in this study, reported per wetland cluster (defined by terrestrial vegetation group). EC – electrical conductivity.

Acknowledgments

Disclaimer: Any opinion, finding and conclusion or recommendation expressed in this material is that of the author(s) and the National Research Foundation (NRF) does not accept any liability in this regard.

For fieldwork assistance we wish to thank Musa Mlambo. Our thanks to Renzo Perissinotto for reviewing the manuscript and providing valuable feedback.

Author Contributions

Conceived and designed the experiments: MSB JAD. Performed the experiments: MSB. Analyzed the data: MSB JAD. Contributed reagents/materials/analysis tools: MSB JAD. Wrote the paper: MSB JAD.

References

1. Meffe CK, Caroll CR (1997) Principles of Conservation Biology. Sunderland, MA: 2nd Edition. Sinauer. 779 p.
2. Vitousek PM, Aber JD, Howarth RW, Likens GE, Matson PA, et al. (1997) Human alteration of the global nitrogen cycle: sources and consequences. Ecological Applications 7: 737–750.
3. Löfman S, Kouki J (2001) Fifty years of landscape transformation in managed forests of southern Finland. Scandinavian Journal of Forest Research 16: 44–53.
4. Naveh Z (2007) Landscape ecology and sustainability. Landscape Ecology 22: 1437–1440.
5. Semlitsch RD, Bodie JR (1998) Are small, isolated wetlands expendable? Conservation Biology 12: 1129–1133.
6. Angeler DG, Moreno JM (2007) Zooplankton community resilience after press-type anthropogenic stress in temporary ponds. Ecological Applications 17: 1105–1115.
7. Harrison AD (1962) Hydrobiological studies on alkaline and acid stillwaters in the Western Cape Province. Transactions of the Royal Society of South Africa 36: 213–235.

8. Pedersen ER, Perkins MA (1986) The use of benthic invertebrate data for evaluating impacts of urban runoff. Hydrobiologia 139: 13–22.
9. Richards C, Host GE, Arthur JW (1993) Identification of predominant environmental factors structuring stream macroinvertebrate communities within a large agricultural catchment. Freshwater Biology 29: 285–294.
10. Clenaghan C, Giller PS, O'Halloran J, Hernan R (1998) Stream macroinvertebrate communities in a conifer-afforested catchment in Ireland: relationships to physico-chemical and biotic factors. Freshwater Biology 40: 175–193.
11. De Szalay FA, Resh VH (2000) Factors influencing macroinvertebrate colonization of seasonal wetlands: responses to emergent plant cover. Freshwater Biology 45: 295–308.
12. Schell JM, Santos-Flores CJ, Hunker BM, Kloehn S, Michelson A, et al. (2001) Physical-chemical influences on vernal zooplankton community structure in small lakes and wetlands of Wisconsin, USA. Hydrobiologia 445: 37–50.
13. Blinn DW, Halse SA, Pinder AM, Shiel RJ, McRae JM (2004) Diatom and micro-invertebrate communities and environmental determinants in the Western Australian wheatbelt: a response to salinisation. Hydrobiologia 528: 229–248.
14. Batzer DP, Cooper R, Wissinger SA (2006) Wetland animal ecology. In: Batzer DP, Sharitz RR, editors. Ecology of freshwater and estuarine wetlands. Berkeley, USA: University of California Press. pp. 242–284.
15. De Roeck ER (2008) Status and ecology of temporary wetlands in the Western Cape, South Africa: PhD Thesis, Laboratory of Aquatic Ecology and Evolutionary Biology, Catholic University of Leuven, Belgium. 143 p.
16. Crosbie B, Chow-Fraser P (1999) Percentage land use in the watershed determines the water and sediment quality of 22 marshes in the Great Lakes basin. Canadian Journal of Fisheries and Aquatic Sciences 56: 1781–1791.
17. Bruland GL, Hanchey MF, Richardson CJ (2003) Effects of agriculture and wetland restoration on hydrology, soils, and water quality of a Carolina bay complex. Wetlands Ecology and Management 11: 141–156.
18. Faulkner S (2004) Urbanization impacts on the structure and function of forested wetlands. Urban Ecosystems 7: 89–106.
19. Houlahan JE, Findlay CS (2004) Estimating the 'critical' distance at which adjacent land use degrades wetland water and sediment quality. Landscape Ecology 19: 677–690.
20. Declerck S, De Bie T, Ercken D, Hampel H, Schrijvers S, et al. (2006) Ecological characteristics of small farmland ponds: Associations with landuse practices at multiple spatial scales. Biological Conservation 131: 523–532.
21. Skagen SK, Melcher CP, Haukos DA (2008) Reducing sedimentation of depressional wetlands in agricultural landscapes. Wetlands 28: 594–604.
22. Akasaka M, Takamura N, Mitsuhashi H, Kadono Y (2010) Effects of land use on aquatic macrophyte diversity and water quality of ponds. Freshwater Biology 55: 909–922.
23. Carrino-Kyker SR, Swanson AK (2007) Seasonal physicochemical characteristics of thirty Northern Ohio temporary pools along gradients of GIS-delineated human land-use. Wetlands 27: 749–760.
24. Brooks RT, Miller SD, Newsted J (2002) The impact of urbanization on water and sediment chemistry of ephemeral forest pools. Journal of Freshwater Ecology 17: 485–488.
25. Rhazi L, Grillas P, Mounirou Toure A, Tan Ham L (2001) Impact of land use and activities on water, sediment and vegetation of temporary pools in Morocco. Life Sciences 324: 165–177.
26. Rebelo AG (1992) Red data book species in the Cape Floristic Region: threats, priorities and target species. Transactions of the Royal Society of South Africa 48: 55–86.
27. Lombard AT, Cowling RM, Pressey RL, Mustart PJ (1997) Reserve selection in a species-rich and fragmented landscape on the Agulhas Plain, South Africa. Conservation Biology 11: 1101–1116.
28. Rouget M, Richardson DM, Cowling RM, Lloyd JW, Lombard AT (2003) Current patterns of habitat transformation and future threats to biodiversity in terrestrial ecosystems of the Cape Floristic Region, South Africa. Biological Conservation 112: 63–85.
29. Heijnis C, Lombard AT, Cowling RM, Desmet PG (1999) Picking up pieces: a biosphere reserve for a fragmented landscape - the Coastal Lowlands of the Western Cape, South Africa. Biodiversity and Conservation 8: 471–496.
30. Enright WD (2000) The effect of terrestrial invasive alien plants on water scarcity in South Africa. Physics and Chemistry of the Earth, Part B: Hydrology, Oceans and Atmosphere 25: 237–242.
31. Le Maitre DC, Versfeld DB, Chapman RA (2000) The impact of invading alien plants on surface water resources in South Africa: a preliminary assessment. Water SA 26: 397–408.
32. Görgens AHM, van Wilgen BW (2004) Invasive alien plants and water resources: an assessment of current understanding, predictive ability and research challenges. South African Journal of Science 100: 27–34.
33. Richardson DM, van Wilgen BW (2004) Invasive alien plants in South Africa: how well do we understand the ecological impacts? South African Journal of Science 100: 45–52.
34. Bird M, Day J, Rebelo A (2013) Physico-chemical impacts of terrestrial alien vegetation on temporary wetlands in a sclerophyllous Sand fynbos ecosystem. Hydrobiologia 711: 115–128.
35. Colburn EA (2004) Vernal Pools: Natural History and Conservation. Blacksburg VA: The McDonald and Woodward Publishing Company. 426 p.
36. Magnusson AK, Williams DD (2006) The roles of natural temporal and spatial variation versus biotic influences in shaping the physicochemical environment of intermittent ponds: a case study. Archiv für Hydrobiologie 165: 537–556.
37. Williams DD (2006) The biology of temporary waters. Oxford: Oxford University Press. 337 p.
38. Ewart-Smith J, Ollis D, Day JA, Malan H (2006) National Wetland Inventory: Development of a Wetland Classification System for South Africa. Pretoria: WRC Report No. K8/652. Water Research Commission (WRC) and the South African National Biodiversity Institute (SANBI). 103 p.
39. Bird M, Mlambo M, Day J (2013) Macroinvertebrates as unreliable indicators of human disturbance in temporary depression wetlands of the south-western Cape, South Africa. Hydrobiologia 720: 19–37. DOI: 10.1007/s10750-10013-11618-10752.
40. Rebelo AG, Coucher C, Helme N, Mucina L, Rutherford MC (2006) Fynbos Biome. In: Mucina L, Rutherford MC, editors. The Vegetation of South Africa, Lesotho and Swaziland. Pretoria: South African National Biodiversity Institute. pp. 53–219.
41. Somerfield PJ, Clarke KR, Olsgard F (2002) A comparison of the power of categorical and correlational tests applied to community ecology data from gradient studies. Journal of Animal Ecology 71: 581–593.
42. Lepš J, Šmilauer P (2003) Multivariate analysis of ecological data using CANOCO. Cambridge: Cambridge University Press. 269 p.
43. Legendre P, Anderson MJ (1999) Distance-based redundancy analysis: testing multi-species responses in multi-factorial ecological experiments. Ecological Monographs 69: 1–24.
44. McArdle BH, Anderson MJ (2001) Fitting multivariate models to community data: a comment on distance-based redundancy analysis. Ecology 82: 290–297.
45. Quinn GP, Keough MJ (2002) Experimental Design and Data Analysis for Biologists. Cambridge, United Kingdom: Cambridge University Press. 537 p.
46. Larsen WA, McCleary SJ (1972) The Use of Partial Residual Plots in Regression Analysis. Technometrics 14: 781–790.
47. Mansfield ED, Conerly MD (1987) Diagnostic value of residual and partial residual plots. The American Statistician 41: 107–116.
48. Cook RD, Weisberg S (1982) Residuals and influence in regression. London: Chapman and Hall. 230 p.
49. Holm S (1979) A simple sequentially rejective multiple test procedure. Scandinavian Journal of Statistics 6: 65–70.
50. Bland JM, Altman DG (1995) Multiple significance tests: the Bonferroni method. British Medical Journal 310: 170.
51. Cabin RJ, Mitchell RJ (2000) To Bonferroni or not to Bonferroni: when and how are the questions? Bulletin of the Ecological Society of America 81: 246–248.
52. Ter Braak CJF, Šmilauer P (2002) CANOCO for Windows Version 4.5. Wageningen: Biometrics - Plant Research International. 272 p.
53. Anderson MJ, Gorley RN, Clarke KR (2008) PERMANOVA+ for PRIMER: Guide to software and statistical methods. Plymouth, UK: PRIMER-E. 214 p.
54. Clarke KR, Warwick RM (2001) Change in Marine Communities: An Approach to Statistical Analysis and Interpretation, 2nd edition. Plymouth, UK: PRIMER-E: Plymouth Marine Laboratory. 190 p.
55. Clarke KR, Gorley RN (2006) PRIMER v6: User Manual/Tutorial. Plymouth, UK: PRIMER-E: Plymouth Marine Laboratory. 190 p.
56. Davies BR, Biggs J, Williams PJ, Lee JT, Thompson S (2008) A comparison of the catchment sizes of rivers, streams, ponds, ditches and lakes: implications for protecting aquatic biodiversity in an agricultural landscape. Hydrobiologia 597: 7–17.
57. Gardiner AJC (1988) A study on the water chemistry and plankton in blackwater lakelets of the south-western Cape: PhD Thesis, Zoology Department, University of Cape Town. 240 p.
58. Raubenheimer CM, Day JA (1991) In vitro leaching of two sclerophyllous fynbos plants. Hydrobiologia 224: 167–174.
59. Midgley J, Schafer G (1992) Correlates of water colour of streams rising from Southern Cape catchments vegetated by fynbos and/or forest. Water SA 18: 93–100.
60. McDowell C (1989) Conservation and horse racing: the unseen connection, part 1. Veld & Flora 75: 36–39.
61. Hitchcock A (2006) Restoration conservation at Kirstenbosch. Veld & Flora 92: 40–44.
62. Hopkins S (2006) The ecology of tadpoles in a temporary pond in the Western Cape with comparisons to other habitats: PhD Thesis, University of the Western Cape, South Africa. 244 p.
63. Turner R (2006) The Kenilworth Atlas: a brief summary of pre- and post-burn field study at the Kenilworth Racecourse, Cape Town. Veld & Flora 92: 140–145.
64. Hitchcock A, Cowell C, Stauch M (2008) Weaving the golden circle. Veld & Flora 94: 31–35.
65. Carpenter SR, Caraco NF, Correll DL, Howarth RW, Sharpley AN, et al. (1998) Nonpoint pollution of surface waters with phosphorus and nitrogen. Ecological Applications 8: 559–568.
66. Leavitt PR, Brock CS, Ebel C, Patoine A (2006) Landscape-scale effects of urban nitrogen on a chain of freshwater lakes in central North America. Limnology and Oceanography 51: 2262–2277.
67. Pham SV, Leavitt PR, McGowan S, Peres-Neto P (2008) Spatial variability of climate and land-use effects on lakes of the northern Great Plains. Limnology and Oceanography 53: 728–742.

68. Witkowski ETF, Mitchell DT (1987) Variations in soil phosphorus in the fynbos biome, South Africa. Journal of Ecology 75: 1159–1171.
69. Lougheed VL, Crosbie B, Chow-Fraser P (2001) Primary determinants of macrophyte community structure in 62 marshes across the Great Lakes basin: latitude, land use, and water quality effects. Canadian Journal of Fisheries and Aquatic Sciences 58: 1603–1612.
70. Nakagawa S (2004) A farewell to Bonferroni: the problems of low statistical power and publication bias. Behavioural Ecology 15: 1044–1045.
71. Azous AL, Horner RR (1997) Wetlands and urbanization: implications for the future: final report of the Puget Sound Wetlands and Stormwater Management Research Program. Seattle, WA, USA: Washington State Department of Ecology, King County Water and Land Resources Division, and the University of Washington. 227 p.
72. Silberbauer MJ, King JM (1991) Geographical trends in the water chemistry of wetlands in the south-western Cape province, South Africa. South African Journal of Aquatic Science 17: 82–88.
73. Mlambo MC, Bird MS, Reed CC, Day JA (2011) Diversity patterns of temporary wetland macroinvertebrate assemblages in the south-western Cape, South Africa. African Journal of Aquatic Science 36: 299–308.

High Adherence Is Necessary to Realize Health Gains from Water Quality Interventions

Joe Brown*, Thomas Clasen

Department of Disease Control, Faculty of Infectious and Tropical Diseases, London School of Hygiene and Tropical Medicine, London, United Kingdom

Abstract

Background: Safe drinking water is critical for health. Household water treatment (HWT) has been recommended for improving access to potable water where existing sources are unsafe. Reports of low adherence to HWT may limit the usefulness of this approach, however.

Methods and Findings: We constructed a quantitative microbial risk model to predict gains in health attributable to water quality interventions based on a range of assumptions about pre-treatment water quality; treatment effectiveness in reducing bacteria, viruses, and protozoan parasites; adherence to treatment interventions; volume of water consumed per person per day; and other variables. According to mean estimates, greater than 500 DALYs may be averted per 100,000 person-years with increased access to safe water, assuming moderately poor pre-treatment water quality that is a source of risk and high treatment adherence (>90% of water consumed is treated). A decline in adherence from 100% to 90% reduces predicted health gains by up to 96%, with sharpest declines when pre-treatment water quality is of higher risk.

Conclusions: Results suggest that high adherence is essential in order to realize potential health gains from HWT.

Editor: Steven J. Drews, University of Calgary & ProvLab Alberta, Canada

Funding: This research was supported in part by Unilever, Ltd., whose subsidiary, Hindustan Unilever, Ltd., (www.hul.co.in) manufactures and sells water treatment products. The funders had no role in study design, data collection and analysis, decision to publish, or preparation of the manuscript. No additional external funding received for this study.

Competing Interests: This study was supported in part by a grant to the London School of Hygiene & Tropical Medicine by Unilever, Ltd., which produces and sells point-of-use water treatment products including the Pureit® water filter. Unilever had no role in the design of the study, analysis of the data, or preparation and submission of the manuscript for publication. As an employee of LSHTM and otherwise, TC and JB provide research and consulting services to UN organizations, government agencies, NGOs and private companies that promote water, sanitation and hygiene interventions including point-of-use water treatment products. These include, without limitation, the World Health Organization, UNICEF, the United States Agency for International Development, the Department for International Development, Medentech, Unilever and Vestergaard-Frandsen. This does not alter the authors' adherence to all the PLoS ONE policies on sharing data and materials.

* E-mail: joe.brown@lshtm.ac.uk

Introduction

Over 780 million people now lack access to an "improved" water source [1], and one study has estimated the number of people who rely on microbiologically or chemically unsafe water to be 1.8 billion, or about 28% of the global population [2]. Unsafe drinking water is a leading cause of preventable disease, with the burden borne primarily by children in low and middle-income countries. Pathogens transmitted in drinking water account for an unknown but presumed significant percentage of the estimated 4 billion cases and 1.9 million deaths from diarrhoeal diseases each year [3,4]. According to WHO and UNICEF monitoring data from 2004, diarrhoeal diseases are the second most common cause of deaths among children under 5 years of age worldwide, or 16% of the total [5].

Providing safe, reliable, piped-in water to every household is an essential goal, yielding optimal health gains while contributing to the Millennium Development Goal targets for poverty reduction, nutrition, childhood survival, school attendance, gender equity and environmental sustainability. Recent research has shown, however, that even such improved water supplies are often subject to faecal contamination [2] and that even occasional interruptions in the quality of water from piped water supplies can undermine the health benefits from safe drinking water [6]. While careful not to encourage diversion of resources away from expansion of safe piped water, public health officials have called for alternative approaches that may provide some of the health benefits of safe drinking-water as progress is made in improving infrastructure services [7].

One such alternative is household water treatment and safe storage (HWT) [8]. In many settings, both rural and urban, populations may have access to sufficient quantities of water, but that water may be unsafe for consumption as a result of microbial or chemical contamination. Effective HWT, such as boiling, filtration and chlorination, has been shown significantly improve microbial water quality [9,10,11,12]. Moreover, by focusing on the point of use rather than the point of delivery, treating water at the household level minimizes the risk of recontamination [13] and the risk of contamination associated with poorly functioning or intermittent water supply systems. There is evidence that HWT can be effective in preventing self-reported diarrhoeal disease [14,15,16,17], although systematic reviews have noted that the limited number of placebo-controlled trials have not reported a statistically significant reduction in risk and estimates of effect from unblinded trials may be subject to a number of well-known biases [15,18]. Like other environmental health interventions such as sanitation or hygiene, blinding intervention studies of HWT has proved to be challenging, but additional blinded trials may be useful in evaluating the evidence for this strategy of safe water provision.

One of the other challenges that promoters of HWT have reported is lack of adherence—*correct, consistent and sustained use*, sometimes referred to as *compliance.* Unlike centrally treated, piped-in water supplies, HWT is normally a batch process that must be undertaken by the end users on a daily or other frequent basis in order to provide consistent protection against waterborne pathogens. A systematic review of chlorine-based HWT, where adherence can be objectively assessed by residual free chlorine in treated water, found several studies with adherence under 70% [14]. Even when uptake of an HWT filter intervention was high when measured by current use, 83% of adults and 95% of children were reported to also consume untreated water [19]. A number of other studies of HWT reported reduced use of interventions over time, raising questions about whether low adherence may limit the usefulness of HWT as a strategy for securing access to safe water [19,20,21,22,23,24].

To date, however, there is limited evidence characterizing the impact of adherence on health gains attributable to HWT. A systematic review of water quality interventions to prevent diarrhoea reported a pooled risk ratio of 0.46 (95% CI: 0.25–0.84) among 16 trials reporting adherence over 50%, and 0.75 (95% CI: 0.63–0.90) among 5 trials reporting adherence of less than 50% [14]. In a review of chlorine interventions, Arnold and Colford (2007) found that the reduction of diarrhoeal disease was greater in studies that reported a higher percentage of samples with detectable levels of chlorine at the point of use [14]. Despite the limited evidence from reviews, the relationship between adherence and health impact has not been clearly characterised in field trials [16]. This is partly due to the difficulty of measuring adherence in a field setting for many technologies, especially at the level of the individual. Adherence has not been measured or reported in the majority of field trials of HWT to date.

In this paper, we present a basic quantitative microbial risk model examining the influence of pre-treatment water quality, treatment effectiveness, and adherence on predicted health gains. We seek to estimate the magnitude of the effect of variable adherence on health impacts achievable through safe water interventions, particularly HWT.

Methods

We constructed a quantitative microbial risk assessment (QMRA) model using assumptions for ranges of untreated and treated water quality, published dose-response relationships for reference pathogens, per-case severity weighting, population susceptibility, adherence, and per-capita consumption of drinking water. In most cases, we used the default assumptions and methods articulated in and recommended by the World Health Organization (WHO)'s *Guidelines for Drinking-water Quality (GDWQ)*, 4th Edition [25], where the rationale, caveats, and methodological issues that apply to this approach to estimating health risks are described in greater detail. We briefly describe the method here and provide a summary of calculations.

We report model output in disability-adjusted life years (DALYs) averted per 100,000 people per year attributable to HWT under the conditions we describe. DALYs can be used to compare health outcomes using severity and duration estimates of disease(s), combining quality and quantity of life factors [25,26]. This measure has been extensively described previously [27] and is used here for ease of comparison with other health impact estimates.

Reference pathogens

We used *Campylobacter jejuni*, rotavirus, and *Cryptosporidium* as reference pathogens in this risk assessment because of their relatively low infectious doses, high per-case DALY severity weighting, and the fact that we have published dose-response models for them [25]. The DALY estimates we calculated were intended to estimate the DALY contributions from bacteria, viruses, and protozoa (as classes of pathogens) that may be found in untreated water, following the approach of the WHO recommended model [25]. It is important to note that the use of these specific microbes is not a statement about their relative importance as aetiologic agents of diarrhoeal diseases globally, or their importance as specifically waterborne pathogens. Rotavirus, for example, is thought to be more frequently transmitted via pathways other than drinking water. They are all known and relatively common causes of gastrointestinal illness. We list assumptions for pathogen-specific per-case DALY weighting, population susceptibility, and risk of illness following infection in Table 1 and present an overview of dose-response calculations below.

Water quality assumptions

The use of QMRA models to predict reductions in waterborne disease risk requires estimation of pre- and post-treatment concentrations of microbes for which we have dose-response models. Faecal-oral pathogen occurrence and distribution are variable in drinking water sources and depend on many factors, including infection (and shedding) rates of specific pathogens in a given population, sanitation and hygiene conditions that may affect excreta containment, zoonotic reservoirs, seasonality and weather events, and other context-specific variables. No sufficiently detailed, country-level or international data exist for waterborne pathogens that would permit the meaningful estimation of means to be used in a generalizable and scientifically credible risk assessment model for drinking water risk. We have used assumptions for microbial counts in pre-treatment waters over a range from 0.0001 per litre ("low risk") to 1 per litre ("high risk"). For the purposes of this modelling exercise, we have assigned a log-normal probability density function to water quality assumptions, which is one distribution that has been observed to characterize microbial density in water [28,29].

Treatment effectiveness

We modelled DALYs averted for improved drinking water quality under several different assumptions about microbiological effectiveness, including: (1), a hypothetical "best case" technology that would be able to reduce bacteria, viruses, and protozoa by 3.01–6 $\log_{10}$ (99.9%–99.9999%), typical of the most advanced single or multiple-barrier technologies for household water treatment [30,31,32] and consistent with well operated conventional or advanced water treatment; (2), a mid-range treatment option that could reduce all classes of microbes by 2.01–3 $\log_{10}$ (99%–99.9%), in the range of basic drinking water treatment and consistent with effective HWT options; and (3), a basic level of protection corresponding to poorly performing water treatment systems, systems without adequate disinfection, or some single-barrier HWT options (1–2 $\log_{10}$ reduction, or 90%–99%) [25]. In the probabilistic model, we assumed a uniform probability density function between the upper and lower $\log_{10}$ bounds we specify.

Table 1. Overview of assumptions used in calculating DALYs averted, adapted from Table 7.6, *Guidelines for Drinking-water Quality*, 4th Edition [24].

	Units	Probability density function (pdf) for independent variables	Values (independent variables) or formulae (dependent variables)
Untreated water quality (C_R)	Mean organisms per litre	Log-normal (mean = standard deviation)	Means of 0.0001, , 0.001, 0.01, 0.1, or 1
Log_{10} reduction	Unitless	Uniform between bound specified	1–2, 2.01–3, or 3.01–6
Drinking-water quality (C_D)	Organisms per litre		$C_R \times (1-\%$ reduction)
Consumption of drinking-water (V)	Litres per person per day	Mean (point estimate)	1, 2, or 5
Adherence or consistency of drinking water treatment (A)	Percent of water consumed that is treated (%)	Uniform between bound specified	50–70%, 71–90%, or 91–100%
Exposure by drinking-water (D)	Organisms per day ingested	-	Equation (1)
Risk of infection ($P_{inf,d}$)	Per day	-	Equations (2, 3)
Risk of infection ($P_{inf,y}$)	Per year	-	Equation (4)
Risk of diarrhoeal illness given infection ($P_{ill\|inf}$)		Mean (point estimate)	*Campylobacter*: 0.3, rotavirus: 0.5, *Cryptosporidium*: 0.7
Risk of diarrhoeal illness (P_{ill})	Per year	-	Equation (5)
Disease burden (DB)	DALYs per case	Uniform between bound specified (for rotavirus only)	*Campylobacter*: 4.6×10^{-3}, rotavirus: 0.014–0.48, *Cryptosporidium*: 1.5×10^{-3}
Susceptible fraction (S)	Percentage of population	Mean (point estimate)	*Campylobacter*: 100%, rotavirus: 6%, *Cryptosporidium*: 100%
Disease burden per 100,000 persons	DALYs per year per 100,000	-	Equation (6)

Adherence

For safer water to translate into health gains, the percentage of treated water that is consumed is an important factor. For piped-in, domestic water service, this might be called *consistency of treatment*. In household water treatment, this may be called *adherence* and sometimes *compliance*. For the purposes of this modelling exercise, we use adherence and define it as the percentage of water an individual uses that is treated (0%–100%). We have divided adherence into *high* (91%–100%), *medium* (71%–90%), and *low* (50%–70%), with an assumed uniform probability density function describing the likelihood of values within this range in the probabilistic model. We assume here that the balance of the drinking water consumed by the user is untreated water. In reality, individuals and families may rely on a number of drinking water sources and treatment options, and these options may be a function of location, season, preferences, beliefs, knowledge, convenience, infrastructure, cost, previous household investment (e.g., a well or rainwater storage), or other factors. Our simplified assumption is that users have two options: treated water and untreated water that result in microbial risks that vary according to the assumptions we have specified.

Model structure

We used Oracle Crystal Ball, Fusion Edition (release 11.1.2.1.000, www.oracle.com) to build a stochastic (Monte Carlo) model based on probability density functions (pdfs) specified for key high-leverage variables of interest, while conserving point estimates for other variables in the base model [25] and using a single-hit version for sensitivity analysis. Model output was collected following 10,000 simulations of each model configuration and descriptive statistics were computed in Crystal Ball. Model predictive and structural validity were tested using both the single-hit and probabilistic models using a structured comparison of model output across a range of default assumptions.

Table 1 describes the basic model structure. Pre-treatment water quality is assumed with means of 0.0001–1 per litre, distributed log-normally (C_R). Based on effectiveness of treatment in reducing reference pathogens, treated drinking-water quality is calculated (C_R). We then used assumed consumption of water per day (V) and adherence or consistency of water treatment (A) to calculate the predicted exposure by drinking-water (d) as the number of microbes ingested per day.

$$d = C_D VA + C_R(1-A) \tag{1}$$

We used the exponential (*Cryptosporidium*) and the Beta-Poisson (rotavirus, *Campylobacter*) models to estimate the risk of infection from exposure to microbial pathogens [26]. We calculated the risk of infection of a single microbe (d) and then calculated the annual risk from ingestion of multiple microbes over the period of a year.

The exponential model is:

$$P_{inf}(d) = 1 - e^{-rd} \tag{2}$$

Where *P(d)* is the probability of an individual becoming infected after ingesting a dose (d) and r is an infectivity constant. The value of r is 0.2 for *Cryptosporidium* (Table 1) [25].

The equation for the Beta-Poisson model is:

$$P_{inf}(d) = 1 - \left(1 + \frac{d}{N_{50}}\left(2^{\frac{1}{\alpha}} - 1\right)\right)^{-\alpha} \tag{3}$$

N_{50} is the median infective dose, which has been experimentally derived as the dose estimated to cause infection in 50% of individuals exposed, and α is a dimensionless infectivity constant. The values for N_{50} and α are 896 and 0.145 for *Campylobacter*, respectively, and 6.17 and 0.253 for rotavirus [25,26].

Yearly infection risk is then calculated as:

$$P_{inf,y} = 1 - \left(1 - P_{inf,d}\right)^{365} \quad (4)$$

The risk of illness is calculated as:

$$P_{ill,y} = P_{inf,d} \times P_{ill|inf} \quad (5)$$

where the probability of illness given infection ($P_{ill|inf}$) is 0.3 for *Campylobacter*, 0.5 for rotavirus, and 0.7 for *Cryptosporidium* [24].

To calculate the disease burden in DALYs associated with yearly risk of illness, we used the following formula for each reference microbe:

$$DALYs = P_{ill} \times DB \times S \quad (6)$$

Where DB is the estimated per-case disease burden in DALYs and S is the susceptible fraction of the population for the outcome of interest. Per-case DALY weights are 4.6×10^{-3} (campylobacteriosis), 0.014–0.48 (rotavirus), and 1.5×10^{-3} (cryptosporidiosis). The model further assumes that 100% of the population is susceptible to cryptosporidiosis and campylobacteriosis, but that only 6% is susceptible to illness associated with rotavirus infection (Table 1). These DALYs weights and estimates for susceptible fraction are those used in the base model as described in the WHO *Guidelines for Drinking-water Quality* [25]. To calculate the estimated disease burden among 100,000 person-years across all three reference pathogens, we multiplied the microbe-specific disease burden by 100,000 and summed estimates for *Campylobacter*, *Cryptosporidium*, and rotavirus. Figures for averted DALYs were then calculated as the difference between DALYs before and after treatment, assuming that the only alternative to treated water was untreated source water.

We conducted a four-way sensitivity analysis to determine the relative sensitivity of mean DALYs averted to the key independent variables of background water quality, litres consumed per day, treatment effectiveness, and consistency/adherence. For the sensitivity analysis, we used the single-hit model using means for all inputs to examine variability in output resulting from changes to independent variables of interest.

Results

Predicted health impacts associated with water quality interventions under pre-treatment water quality, technology effectiveness, and adherence assumptions are presented in Table 2 as DALYs averted per 100,000 people per year. For some combinations of parameters resulting in high impact (moderate pre-treatment water quality, medium to high treatment effectiveness, and high adherence), point estimates for mean DALYs averted per 100,000 person-years are over 500. With lower levels of adherence, DALYs averted decrease dramatically when untreated water is of high risk, since adherence is of greater importance when untreated water is a source of disease. The range of estimates of effect when pre-treatment water is of moderate to high risk indicate that the potential health gains are very high as adherence approaches 100% (up to 1840 DALYs per 100,000 per year).

Results suggest that when pre-treatment water is of moderate to high risk, adherence is more important than treatment effectiveness. To take one example from Table 2, when pre-treatment water is of moderate high risk, the mean estimated DALYs averted per 100,000 person-years for a technology with low effectiveness (1–2 $\log_{10}$ reduction of bacteria, viruses, and protozoa) but that is used very consistently (91–100% adherence) is 363, compared with 68.1 for a technology that is more effective (3.01–6 $\log_{10}$ reduction of bacteria, viruses, and protozoa) but that is used less consistently (50–70% adherence).

Assuming lower risk untreated water, both the potential health gains from water treatment and the sensitivity to adherence are reduced, suggesting that the importance of water treatment and high adherence are elevated in high-risk waters and reduced in low-risk waters. We used a uniform probability density function in the model to indicate "high" adherence between 91%–100% (individual simulation estimates being equally likely to be assigned a value between these in individual simulations), a range which probably represents highest likely achievable adherence *in situ*.

Figure 1 presents a graphical illustration of how the predicted health gains vary as a function of both adherence and pre-treatment water quality. Calculation of DALYs averted in Figure 1 assumed a 2 $\log_{10}$ reduction across all reference pathogens and 1 litre per day consumption. When pre-treatment water was of high risk [A] and [B], there is a dramatic decline in predicted health gains with decreasing adherence. Water that is of moderate risk [C] is less affected by reduced adherence, although sharp declines are observed. Waters that are already low risk [D] and [E] do not show a strong association between DALYs averted and adherence because the risk of drinking untreated water is already low.

Results of the sensitivity analysis (Table 3) indicate that model output is highly sensitive to pre-treatment water quality and adherence, moderately sensitive over the range of assumptions to treatment effectiveness against waterborne pathogens, and relatively insensitive to volume consumed per person per day over realistic values (1–5 litres per day per person). A straightforward comparison of point estimates across changes in water quality, adherence, effectiveness and consumption suggest that, within pre-treatment water risk categories, DALYs averted estimates are most sensitive to adherence. Reducing adherence from 100% to 90% reduces the number of DALYs averted by over 90% under some model assumptions and up to 96%, with sharper declines when pre-treatment water quality is higher risk (Table 3 and Figure 1). The model was also very sensitive to the per-case DALY disease burden estimates (data not shown), which have been derived elsewhere and are key assumptions included in the base model we used [25,26] (Table 1).

Discussion

When and where water is an important source of pathogen exposure, water quality interventions can reduce exposure to pathogens and result in improved health. Our results suggest, however, that the potential health gains are reduced sharply with even occasional consumption of untreated drinking water.

Our results are consistent with the findings of a similar QMRA analysis by Hunter et al. (2009) [6]. This study concluded that the health benefits of improved quality drinking water (as delivered by centralised treatment and distribution) were limited if even a small percentage of overall water consumed was of lower quality, for example during interruptions of service in piped water supply of when the alternative drinking-water source was surface water. Hunter *et al.* concluded as we have that the overall risk attributable to drinking water is controlled by those periods of higher exposure risk when no quality protection is in place, reducing overall impacts of water quality improvements significantly if the intervention is not present a high percentage of the time. We

Table 2. Probabilistic model output with estimated mean DALYs averted per 100,000 persons pear year as a function of pre-treatment water quality; treatment effectiveness against bacteria, viruses, and protozoa; and adherence.

Pre-treatment water quality (log-normal distribution)	Treatment effectiveness: bacteria, viruses, and protozoa	Adherence, percentage of total water consumed that is treated	Mean DALYs averted per 100,000 persons (standard deviation)	Range	Mean standard error
High risk:	High	High (91–100%)	196 (178)	3.54–1,570	1.78
reference	(3.01–6 $\log_{10}$)	Medium (71–90%)	50.4 (28.7)	0–523	0.29
pathogens		Low (50–70%)	21.5 (16.8)	0–129	0.17
present,	Medium	High (91–100%)	173 (140)	0–1,740	1.40
mean:	(2.01–3 $\log_{10}$)	Medium (71–90%)	49.2 (28.2)	0–494	0.28
1 per litre		Low (50–70%)	20.8 (16.7)	0–86.0	0.17
	Low	High (91–100%)	104 (62.0)	0–1,010	0.62
	(1–2 $\log_{10}$)	Medium (71–90%)	43.3 (26.1)	0–347	0.26
		Low (50–70%)	19.4 (15.9)	0–85.8	0.16
Moderate	High	High (91–100%)	520 (326)	29.5–1,840	3.26
high risk:	(3.01–6 $\log_{10}$)	Medium (71–90%)	174 (149)	4.91–1,130	1.49
reference		Low (50–70%)	68.1 (62.6)	3.95–659	0.63
pathogens	Medium	High (91–100%)	498 (316)	36.0–1,890	3.16
present,	(2.01–3 $\log_{10}$)	Medium (71–90%)	170 (146)	6.82 (1,110)	1.46
mean:		Low (50–70%)	69.0 (63.5)	1.14–676)	0.63
0.1 per litre	Low	High (91–100%)	363 (257)	20.0–1,460	2.57
	(1–2 $\log_{10}$)	Medium (71–90%)	147 (127)	5.41–998)	1.27
		Low (50–70%)	63.7 (57.6)	2.93–604	0.58
Moderate	High	High (91–100%)	533 (313)	26.9–1,630	3.13
risk:	(3.01–6 $\log_{10}$)	Medium (71–90%)	369 (213)	19.3–1,180	2.13
reference		Low (50–70%)	216 (131)	6.56–682	1.31
pathogens	Medium	High (91–100%)	527 (304)	31.4–1,650	3.04
present,	(2.01–3 $\log_{10}$)	Medium (71–90%)	365 (211)	18.6–1,140	2.11
mean:		Low (50–70%)	215 (129)	6.69–724	1.29
0.01 per litre	Low	High (91–100%)	483 (277)	24.5–1,520	2.77
	(1–2 $\log_{10}$)	Medium (71–90%)	338 (196)	17.9–1,140	1.96
		Low (50–70%)	206 (124)	8.23–676	1.24
Moderate	High	High (91–100%)	133 (127)	2.15–1,100	1.27
low risk:	(3.01–6 $\log_{10}$)	Medium (71–90%)	109 (102)	3.14–966	1.02
reference		Low (50–70%)	78.5 (71.3)	2.29–557	0.71
pathogens	Medium	High (91–100%)	133 (125)	3.49–1,180	1.25
present,	(2.01–3 $\log_{10}$)	Medium (71–90%)	109 (101)	2.80–865	1.01
mean:		Low (50–70%)	79.3 (70.5)	2.20–615	0.71
0.001 per	Low	High (91–100%)	129 (121)	2.06–1,070	1.21
litre	(1–2 $\log_{10}$)	Medium (71–90%)	105 (98.6)	2.55–885	0.99
		Low (50–70%)	74.9 (67.0)	1.82–511	0.67
Low risk:	High	High (91–100%)	18.3 (15.9)	0–231	0.18
reference	(3.01–6 $\log_{10}$)	Medium (71–90%)	15.5 (13.3)	0–231	0.15
pathogens		Low (50–70%)	11.5 (9.92)	0–212	0.12
present,	Medium	High (91–100%)	19.3 (15.8)	0–546	0.19
mean:	(2.01–3 $\log_{10}$)	Medium (71–90%)	16.1 (13.3)	0–311	0.16
0.0001 per		Low (50–70%)	11.9 (9.94)	0–221	0.12
litre	Low	High (91–100%)	18.5 (15.5)	0–319	0.18
	(1–2 $\log_{10}$)	Medium (71–90%)	15.2 (12.9)	0–377	0.15
		Low (50–70%)	11.3 (9.60)	0–157	0.11

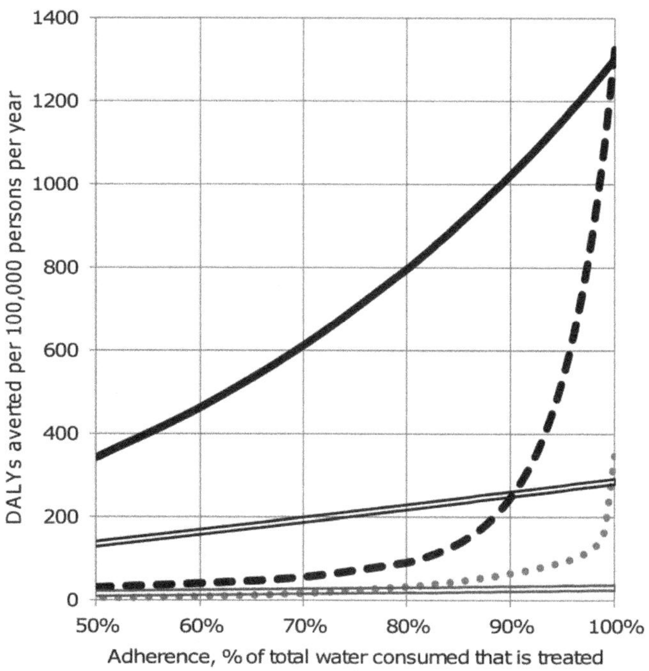

Figure 1. DALYs averted per 100,000 population, per year, based on assumptions about technology effectiveness (2 $\log_{10}$ reduction in each pathogen class), adherence, and background water quality assumptions from Table 2.

propose that our results are applicable to water quality interventions generally, including centralized water treatment.

There is reason to believe, however, that there is even greater risk among the estimated 1.2 billion who report treating water in the home. First, most of those that practice HWT reside in low-income settings where the risk of highly contaminated water is greatest. Second, unlike residents of higher-income settings who may receive warning when their water treatment systems fail, these individuals are unable to judge the quality of their drinking water and must follow the practice daily. Third, those who consistently treat their water at home are likely to be exposed to untreated drinking water when they are away from the home—a fundamental shortcoming of HWT compared to effective community-wide improvements in water quality.

Reconciling these results with the existing evidence base for water quality interventions raises certain questions. Our findings suggest that the dramatic protective effects against diarrhoeal disease reported in recent systematic reviews of household water treatment [14,15,16,17] and other water quality interventions [33,34] would seem to be unlikely in the absence of high adherence. Unfortunately, even studies that report on coverage and overall use rarely report on the extent to which participants consume untreated water. Nevertheless, our findings are consistent with two recent studies of the same water filter among young children in low-income settings. While investigators found no protective effect from an intervention in which most participants acknowledged regularly drinking untreated water [19], the intervention was highly protective against diarrhoea when few participants consumed untreated water (Rachel Peletz, personal communication).

Another potential explanation is that the large protective effects reported by open trials of water quality interventions are exaggerated due to reporting bias [15,18]. Correct, consistent use has been an ongoing challenge for HWT, and few studies appear to reach the levels of uptake that our results suggest would be necessary to achieve the reported effects on diarrhoea. In this respect, our results lend support to the assertion that published health impact estimates of water quality interventions may have been overestimated. An analysis of reported DALYs averted by HWT interventions and a comparison with this and other modelling efforts may be a useful next step in contextualizing the evidence base for water quality interventions.

Another possibility that may explain higher than expected estimates of health effects from trials with low apparent adherence is that water quality interventions may have been more likely to have been used consistently when untreated water was perceived to be of higher risk due to aesthetic or other indicators, and those perceptions may have correlated with actual levels of pathogens (i.e., that adherence is higher when it is more important, when water is of higher risk). More research is needed on whether and how HWT behaviours are driven by perceptions of water safety, and whether and how perceptions of water safety are related to actual water safety.

Our results may also help explain the diminished health impact that has been observed from longer-term evaluations of HWT interventions [14,16]. A number of studies of HWT have shown reduced use of HWT interventions over time, raising questions about the potential for sustained use [19,20,21,22,23,24]. Technologies with a high user burden, recurrent costs, or those that involve substantial behaviour change may be especially challenging to achieve the continued high adherence that our results indicate is necessary to sustain health impact [35]. Boiling has overcome many of these same obstacles and become the leading HWT method worldwide [36]. However, studies of the effectiveness of boiling demonstrate the challenge of achieving consistently safe drinking water even from this highly normative practice.

Comparing predicted health gains resulting from two levels of microbiological treatment effectiveness (2 $\log_{10}$ and 6 $\log_{10}$ reduction of reference pathogens) suggests decreasing marginal health gains with increasing efficiency (Table 3). A focus on

Table 3. Sensitivity analysis across a range of pre-treatment microbial counts of reference pathogens and adherence, at "high" and "low" effectiveness (6 $\log_{10}$ and 2 $\log_{10}$, respectively) and at 2 levels of water consumption (1 litre per person per day and 5 litres per person per day).

Consumption, litres per person per day	Treatment effectiveness: bacteria, viruses, and protozoa	Adherence, percentage of total water consumed that is treated	DALYs averted per 100,000 people per year over pre-treatment water risk categories				
			High: 1 per litre	Moderate high: 0.1 per litre	Moderate: 0.01 per litre	Moderate low: 0.001 per litre	Low: 0.0001 per litre
1	6 $\log_{10}$ (99.9999%)	100%	1680	1618	1339	290	32
		90%	69	278	1049	258	29
		80%	35	96	815	227	25
		70%	18	57	626	196	22
		60%	9	41	473	166	19
		50%	5	31	349	137	16
1	2 $\log_{10}$ (99%)	100%	347	1324	1302	285	31
		90%	65	245	1022	254	28
		80%	33	91	796	223	25
		70%	17	56	612	193	22
		60%	9	41	463	164	19
		50%	5	31	342	135	16
5	6 $\log_{10}$ (99.9999%)	100%	1679	1612	1333	288	32
		90%	69	278	1045	257	28
		80%	35	96	812	226	25
		70%	18	57	624	195	22
		60%	9	41	471	166	19
		50%	5	31	348	136	16
5	2 $\log_{10}$ (99%)	100%	100	626	1181	272	30
		90%	50	157	934	243	27
		80%	26	75	732	213	24
		70%	14	50	567	185	21
		60%	7	38	432	157	18
		50%	4	29	321	129	15

Reported means are from single-hit models.

increasing microbiological effectiveness of water quality interventions, for example from 99% to 99.9999%, may not result in proportionately lower risk, especially if adherence will be less than 100%; the swamping effect of lower than 100% adherence essentially makes the real differences between microbial reduction efficiencies meaningless with respect to outcomes as predicted in this modelling exercise. We acknowledge that the value of requiring higher efficiencies for water treatment technologies is due to the higher efficiencies that may be required under outbreak scenarios, when untreated water may be of higher risk, or if there are no alternatives to high risk source waters. Given the fact that some technologies may be adopted more readily and used more consistently than other technologies (even those that may be more effective in reducing microbes), these results would seem to suggest that a focus on promoting adherence may be more critical in delivering health gains than focusing exclusively on increasing microbiological effectiveness [31].

The health impact estimates we report here should be interpreted in light of the uncertainty of the assumptions and necessary simplifications used to produce them. Pre-treatment water quality, for example, is likely to be highly variable in any given setting. Dose-response models for reference pathogens have been derived from few studies, mostly using data from healthy adults in wealthier countries. As an approach to estimating health risk, QMRA incorporates a number of assumptions whose values and ranges are uncertain. Where possible, we have attempted to use assumptions that would tend to result in conservative estimates of health impacts, and realistic ranges of pre-treatment water quality, adherence, treatment effectiveness in reducing microbes, and other key variables. QMRA is a quickly evolving approach and models like the one we have used will benefit from further refinement of methods and assumptions.

The greatest risks of waterborne disease globally are from microbial pathogens, although chemical contaminants are locally significant risks to public health. This risk assessment does not address the potential problem of anthropogenic or naturally occurring radiological or chemical contaminants that may be present in drinking-water, including but not limited to pesticides, arsenic, fluoride, heavy metals, nitrate, excess salts, disinfection by-products, pharmaceuticals, or others. Our study also omits helminthic and other diseases that may be transmitted by drinking water. Including these additional factors would require additional data and different modelling approaches.

Current guidelines supporting HWT emphasize the need for correct, consistent and sustained use [31] and our results underscore the need to make high adherence an essential priority in HWT design and program implementation. In order to maximise health gains from interventions, it is necessary to provide HWT solutions that people will use exclusively or nearly so. This may require not only intensified behaviour change interventions, but also process improvements in HWT products and technologies that minimize the likelihood of non-adherence. In order to monitor and evaluate these and other water quality interventions, it is also important to develop and improve tools for monitoring and to ensure that field studies report adherence. Given the challenges of achieving this high level of adherence through point-of-use water treatment, these results also confirm the need to continue efforts to provide safe, reliable, piped-in water to household taps in order to realize the full promise of health gains from water quality interventions.

Acknowledgments

The authors gratefully acknowledge input by colleagues at the Hygiene Centre, LSHTM, and thoughtful critique from anonymous reviewers of this manuscript.

Author Contributions

Conceived and designed the experiments: JB TC. Performed the experiments: JB. Analyzed the data: JB. Contributed reagents/materials/analysis tools: JB. Wrote the paper: JB TC.

References

1. WHO and UNICEF (2012) Progress on Sanitation and Drinking Water: 2012 Update. Geneva: WHO Press, Available: www.wssinfo.org. Accessed 2012 March 15.
2. Onda K, LoBuglio J, Bartram J (2012) Global Access to Safe Water: Accounting for Water Quality and the Resulting Impact on MDG Progress. Int J Environ Res Public Health 9(3): 880–894.
3. WHO (2005) World Health Report 2005. Geneva: World Health Organization.
4. Blakely T, Hales S, Kieft C, Wilson N, Woodward A (2005) The global distribution of risk factors by poverty level. Bulletin of the World Health Organization 83: 118–126.
5. UNICEF (2009) Diarrhea: why children are still dying and what can be done. New York: United Nations Children's Fund, Geneva: World Health Organization.
6. Hunter PR, Zmirou-Navier D, Hartemann P (2009) Estimating the impact on health of poor reliability of drinking water interventions in developing countries. Science of the Total Environment 407(8): 2621–2624.
7. Mintz ED, Reiff FM, Tauxe RV (1995) Safe water treatment and storage in the home: a practical new strategy to prevent waterborne disease. JAMA 273(12): 948–953.
8. WHO (2007) Combating waterborne disease at the household level. Geneva: World Health Organization.
9. Luby S, Agboatwalla M, Raza A, Sobel J, Mintz E, et al. (2001) A low-cost intervention for cleaner drinking water in Karachi, Pakistan. International Journal of Infectious Diseases 5: 144–150.
10. Clasen TF, Thao do H, Boisson S, Shipin O (2008) Microbiological effectiveness and cost of boiling to disinfect drinking water in rural Vietnam. Environmental Science and Technology 42(12): 4255–60.
11. Sobsey MD (2002) Managing Water in the Home: Accelerated Health Gains from Improved Water Supply. Geneva: World Health Organization.
12. Brown J, Sobsey MD, Loomis D (2008) Drinking water filters reduce diarrheal disease in Cambodia: a randomized, controlled trial of locally made ceramic filters. American Journal of Tropical Medicine and Hygiene 79(3): 394–400.
13. Wright J, Gundry S, Conroy R (2004) Household drinking water in developing countries: a systematic review of microbiological contamination between source and point-of-use. Tropical Medicine and International Health 9(1): 106–117.
14. Arnold B, Colford J (2007) Treating water with chlorine at point-of-use to improve water quality and reduce child diarrhea in developing countries: A systematic review and meta-analysis. American Journal of Tropical Medicine and Hygiene 76(2): 354–364.
15. Clasen T, Roberts I, Rabie T, Schmidt W, Cairncross S (2006) *Interventions to improve water quality for preventing diarrhoea (A Cochrane Review)*. In: The Cochrane Library, Issue 3, 2006.
16. Waddington H, Snilstveit, White H, Fewtrell L (2009) Water, sanitation and hygiene interventions to combat childhood diarrhoea in developing countries. Delhi: International Initiative for Impact Evaluation.
17. Fewtrell L, Kaufmann RB, Kay D, Enanoria W, Haller L, et al. (2005) Water, sanitation, and hygiene interventions to reduce diarrhea in less developed countries: a systematic review and meta-analysis. Lancet Infectious Diseases 5: 42–52.
18. Schmidt WP, Cairncross S (2009) Household water treatment in poor populations: is there enough evidence for scaling up now? Environmental Science and Technology 43(4): 986–992.
19. Boisson S, Kiyombo M, Sthreshley L, Tumba S, Makambo J, et al. (2010) Field assessment of a novel household-based water filtration device: a randomised, placebo-controlled trial in the Democratic Republic of Congo. PLoS One 10;5(9): e12613.
20. Mausezahl D, Christen A, Duran Pacheco G, Tellez FA, Iriarte M, et al. (2009) Solar Drinking Water Disinfection (SODIS) to Reduce Childhood Diarrhoea in Rural Bolivia: A Cluster-Randomized, Controlled Trial. PLoS Med 6(8): e1000125. doi:10.1371/journal.pmed.1000125.
21. Mong Y, Kaiser R, Ibrahim D, Razafimbololona L, Quick RE (2001) Impact of the safe water system on water quality in cyclone-affected communities in Madagascar. American Journal of Public Health 91(10): 1577–1579.
22. Brown J, Sobsey M, Proum S (2007) Use of Ceramic Water Filters in Cambodia. Washington, DC: WSP-World Bank Field Note.

23. Colindres RE, Jain S, Bowen A, Mintz E, Domond P (2007) After the flood: an evaluation of in-home drinking water treatment with combined flocculent-disinfectant following Tropical Storm Jeanne – Gonaives, Haiti, 2004. Journal of Water & Health 5: 367–374.
24. Luby S, Agboatwalla M, Raza A, Sobel J, Mintz E, et al. (2001) A low-cost intervention for cleaner drinking water in Karachi, Pakistan. International Journal of Infectious Diseasess 5: 144–150.
25. WHO (2011) Guidelines for Drinking-water Quality, 4th Edition. Geneva: World Health Organization.
26. Havelaar A, Melse JM (2003) *Quantifying Public Health Risk in the WHO GDWQ*. RIVM report 734301022/2003.
27. Gold MR, Stevenson D, Fryback DG (2002) HALYS and QALYS and DALYS, Oh My: similarities and differences in summary measures of population Health. Annual Review of Public Health 23: 115–134.
28. Haas CN (1996) How to average microbial densities to characterize risk. Water Research 30(4): 1036–1038.
29. Christian RR, Pipes WO (1983) Frequency distribution of coliforms in water distribution systems 45(2): 603–609.
30. EPA (United States Environmental Protection Agency) (1987) Guide Standard and Protocol for Testing Microbiological Water Purifiers. Office of Drinking Water.
31. NSF (National Sanitation Foundation, now NSF-International) (2003) NSF P231- Microbiological Water Purifiers. Ann Arbor, USA: NSF International.
32. WHO (2011) Evaluating Household Water Treatment Options: Health-Based Targets and Performance Specifications. Geneva: World Health Organization.
33. Esrey SA, Feachem RG, Hughes JM (1985) Interventions for the control of diarrheal diseases among young children: improving water supplies and excreta disposal facilities. Bulletin of the World Health Organization 63: 757–72.
34. Esrey SA, Potash JB, Roberts L, Shiff C (1991) Effects of improved water supply and sanitation on ascariasis, diarrhoea, dracunculiasis, hookworm infection, schistosomiasis, and trachoma. Bulletin of the World Health Organization 69(5): 609–621.
35. Brown J, Outlaw T, Clasen T, Wu J, Sobsey M (2009) Safe Water for All: Harnessing the Private Sector to Reach the Underserved. Washington, D.C.: World Bank/International Finance Corporation.
36. Rosa G, Clasen T (2010) Estimating the scope of household water treatment in low- and medium-income countries. American Journal of Tropical Medicine and Hygiene 2010 Feb;82(2): 289–300.

16

Coupled Effects of Natural and Anthropogenic Controls on Seasonal and Spatial Variations of River Water Quality during Baseflow in a Coastal Watershed of Southeast China

Jinliang Huang[1,2]*, Yaling Huang[2], Zhenyu Zhang[2]

1 Coastal and Ocean Management Institute, Xiamen University, Xiamen, People's Republic of China, 2 Fujian Provincial Key Laboratory for Coastal Ecology and Environmental Studies, Xiamen University, Xiamen, People's Republic of China

Abstract

Surface water samples of baseflow were collected from 20 headwater sub-watersheds which were classified into three types of watersheds (natural, urban and agricultural) in the flood, dry and transition seasons during three consecutive years (2010–2012) within a coastal watershed of Southeast China. Integrating spatial statistics with multivariate statistical techniques, river water quality variations and their interactions with natural and anthropogenic controls were examined to identify the causal factors and underlying mechanisms governing spatiotemporal patterns of water quality. Anthropogenic input related to industrial effluents and domestic wastewater, agricultural activities associated with the precipitation-induced surface runoff, and natural weathering process were identified as the potential important factors to drive the seasonal variations in stream water quality for the transition, flood and dry seasons, respectively. All water quality indicators except SRP had the highest mean concentrations in the dry and transition seasons. Anthropogenic activities and watershed characteristics led to the spatial variations in stream water quality in three types of watersheds. Concentrations of NH_4^+-N, SRP, K^+, COD_{Mn}, and Cl^- were generally highest in urban watersheds. NO_3^-N Concentration was generally highest in agricultural watersheds. Mg^{2+} concentration in natural watersheds was significantly higher than that in agricultural watersheds. Spatial autocorrelations analysis showed similar levels of water pollution between the neighboring sub-watersheds exhibited in the dry and transition seasons while non-point source pollution contributed to the significant variations in water quality between neighboring sub-watersheds. Spatial regression analysis showed anthropogenic controls played critical roles in variations of water quality in the JRW. Management implications were further discussed for water resource management. This research demonstrates that the coupled effects of natural and anthropogenic controls involved in watershed processes, contribute to the seasonal and spatial variation of headwater stream water quality in a coastal watershed with high spatial variability and intensive anthropogenic activities.

Editor: Maura (Gee) Geraldine Chapman, University of Sydney, Australia

Funding: This study was supported by the Natural National Science Foundation of China (Grant No. 40810069004, Grant No. 40901100), and the Science Foundation of Fujian Province (2010Y0064). The funders had no role in study design, data collection and analysis, decision to publish, or preparation of the manuscript.

Competing Interests: The authors have declared that no competing interests exist.

* E-mail: jlhuang@xmu.edu.cn

Introduction

River water quality has become one of important concern worldwide. On the one hand, rivers constitute the main water resource for drinking, irrigation, and industrial purposes in inlands [1–2]. On the other hand, as receiving water bodies, rivers assimilate or carry industrial and domestic wastewater, and runoff from agricultural fields, roadways and streets, thereafter discharging them into downstream estuarine and coastal water [3–4]. It is reported that nearly 80% (4.8 billion) of the world's population (for 2000) lives in areas where either incident human water security or biodiversity threats exceed 75th percentile [5]. In addition, water quality all over the world presents a trend of increasingly severe deterioration [6]. As such, it is important to have reliable information on river water quality for water resource management from the local to global scale. This necessity is even more pronounced in coastal watersheds due to escalating environmental pressure and their special role in regional ecosystem services.

Water quality is affected by a combination of natural factors (e.g. precipitation, temperature, bedrock, soil, terrain) and anthropogenic factors (e.g. agricultural practices, domestic wastewater/industrial influent) [7–9]. Understanding how anthropogenic and natural factors control water quality and how the relationships changes over time and space will help water resource managers to target appropriate scales and factors for the improvement of their water quality management efforts.

The factors and processes involved in natural and anthropogenic controls govern the seasonal and spatial variability in stream water quality in watersheds. It is important to determine the watershed processes that regulate stream water quality under increasing pressure from natural and anthropogenic disturbance [5], [10–13]. Hydrological and biogeochemical processes are two important mechanisms to explain the seasonal variations of water quality in the watersheds [3], [14–15]. The nutrients are temporarily stored in the ground surface, vadose zone or in the groundwater and then transported into stream via subsurface water or precipitation-induced surface runoff [15–16]. The biogeochemical processes including in-stream immobilization, denitrification, mineralization, and bedrock weathering, associated with the hydro-meteorological regime, also determine the temporal variability of stream water quality [15], [17–18].

On the other hand, anthropogenic activities and watershed characteristics drive spatial variability in stream water quality in the watersheds. Watershed land use impacts water quality through nonpoint sources, which are major contributors of pollution to the catchment-coast continuum [19–20]. Linkage of land use pattern and water quality has been well documented for developing watershed management practices [7], [21–23]. In addition, the point sources of nutrients, which are usually synonymous with domestic wastewaters and industrial effluents [24–26], contribute greatly to stream water quality degradation in developing countries with relatively low wastewater treatment capacity [27]. The watershed characteristics including topography and bedrock geology constitutes the important natural factors that drives the spatial variation in stream water quality [21], [28]. The higher slope variability leads to higher rates of erosion, which subsequently increase the rates of particulate matter entering the watershed [29]. The chemical composition of streams is influenced by distinct bedrock type and the distribution of Ca^{2+} and Mg^{2+} concentrations in rivers depends strongly upon weathering of soil and bedrock geology [30–31].

The coupled effects of natural and anthropogenic controls, together with underlying hydrologic and biogeochemical processes, contribute to the seasonal and spatial variation of stream water quality. Magnitude of in-stream N immobilization is controlled by stream order and headwater streams are important sites for N processing and retention [32–33]. Martin et al. (2004) reported that high subsurface flow with high nitrate concentration during high water periods and active denitrification during low water period explained the higher streamwater nitrate concentration in winter than in summer in a watershed in France [15]. Duan and Kaushal (2013) found that Peak inorganic N concentrations occur throughout the winter and decline considerably during the growing season due to the effect of N immobilization in the Chesapeake Bay watershed [18]. Meantime, they pointed out that there is potential to increase winter N delivery to streams because of human activity, e.g. periodic agricultural activities. Bowes et al. (2005) found that streams receiving wastewater effluent typically show a characteristic pattern of high P concentration during summer low flow and more diluted concentration during winter storm events in an English catchment [34]. Rothwell et al. (2010) found that the elevated Mg^{2+} concentrations in the lowlands are due to underlying geology, rather than urban or arable land use [12].

The relative influences of the natural and anthropogenic factors change over the range of temporal and spatial scales investigated, and this results in a pressing challenge for studies on the mechanism determining the variability in river water quality, especially in the situation that water quality in most Chinese rivers and groundwater sources is poor and declining under pressures from industrial and municipal wastewater discharges and NPS pollution from agricultural and aquacultural runoffs of fertilizers, pesticides and manure and freshwater quality is a prime concern in China, especially in the relatively developed regions such as the eastern coastal areas of China [27], [35].

Analyzing the spatiotemporal variations in stream water quality generally relies on multivariate statistical techniques combined with GIS and remote sensing. Multivariate statistical techniques including principal component analysis (PCA) and redundancy analysis are useful for data reduction and interpretation of apportionment of the pollution sources [3], [36]. The increasing availability of remotely-sensed data enables landscape-water quality studies to be more easily performed on both local and regional scales [22], [37]. The conventional ordinary least square (OLS) method combined with GIS is the main method for estimating the empirical relationships between ambient water quality parameters and watershed characteristics [2], [38]. Recently, a powerful spatially statistical method, the geographically weighted regression technique, was developed and applied to examine the spatially varying relationships between land use and water quality [23]. [39–41]. However, few studies consider the spatial dependence of water quality explicitly. Moreover, many previous studies do not examine both spatial and temporal change simultaneously [42–43], which make it hard to fully uncover the spatiotemporal variation of river water quality.

The Jiulong River watershed (JRW), a typical medium-sized subtropical coastal watershed, has experienced continuing degradation in water quality over the last 20 years. This plays an important role in the region's economic and ecological health. A better understanding of the spatiotemporal variations in water quality and the underlying mechanisms are critical for regional water quality management. The primary objective of the present study is to examine the coupled effects of anthropogenic and natural controls on the seasonal and spatial variations in water quality in 20 headwater watersheds of the JRW. We test two hypotheses: (1) anthropogenic activities, combined with the hydrological and biogeochemical processes, contribute to the seasonal variability in stream water quality; (2) coupled effects of anthropogenic activities and watershed characteristic, result in the spatial variability in stream water quality.

Materials and Methods

2.1. Ethics Statement

No specific permits were required for the described field studies and our field studies did not involve endangered or protected species.

2.2. Study Area

The JRW covers about 14,700 km^2 in the eastern coastal area of China (Fig. 1). The watershed includes the North and West Rivers, which meet in Zhangzhou, and produce an annual flow of twelve billion cubic meters into the Jiulong River estuary and the Xiamen-Kinmen coastal waters. It is situated in a subtropical zone with a monsoon climate: the annual average temperature is 19–21°C, and annual precipitation averages 1400–1800 mm, of which 70% occurs between April and September. Red earth and lateritic red earth are the main soil types in the JRW, with pH values ranging from 4.0 to 4.8 and a mean value of 4.5. The upstream region is mountainous and 68% of the watershed has a topographic slope in excess of 18% [44]. The major geology type of the JRW is granite, volcanic tuff, and sandstone [45]. The North River is dominated by granite and sandstone while the West River is mainly comprised of granite and volcanic tuff.

Zhangzhou plain, located at the downstream end of the JRW, constitutes one of China's most developed regions in terms of agricultural production due to its subtropical monsoon climate and agricultural policies, which are influenced by the closeness to Taiwan [2]. The plain is intensively agricultural with orchards of

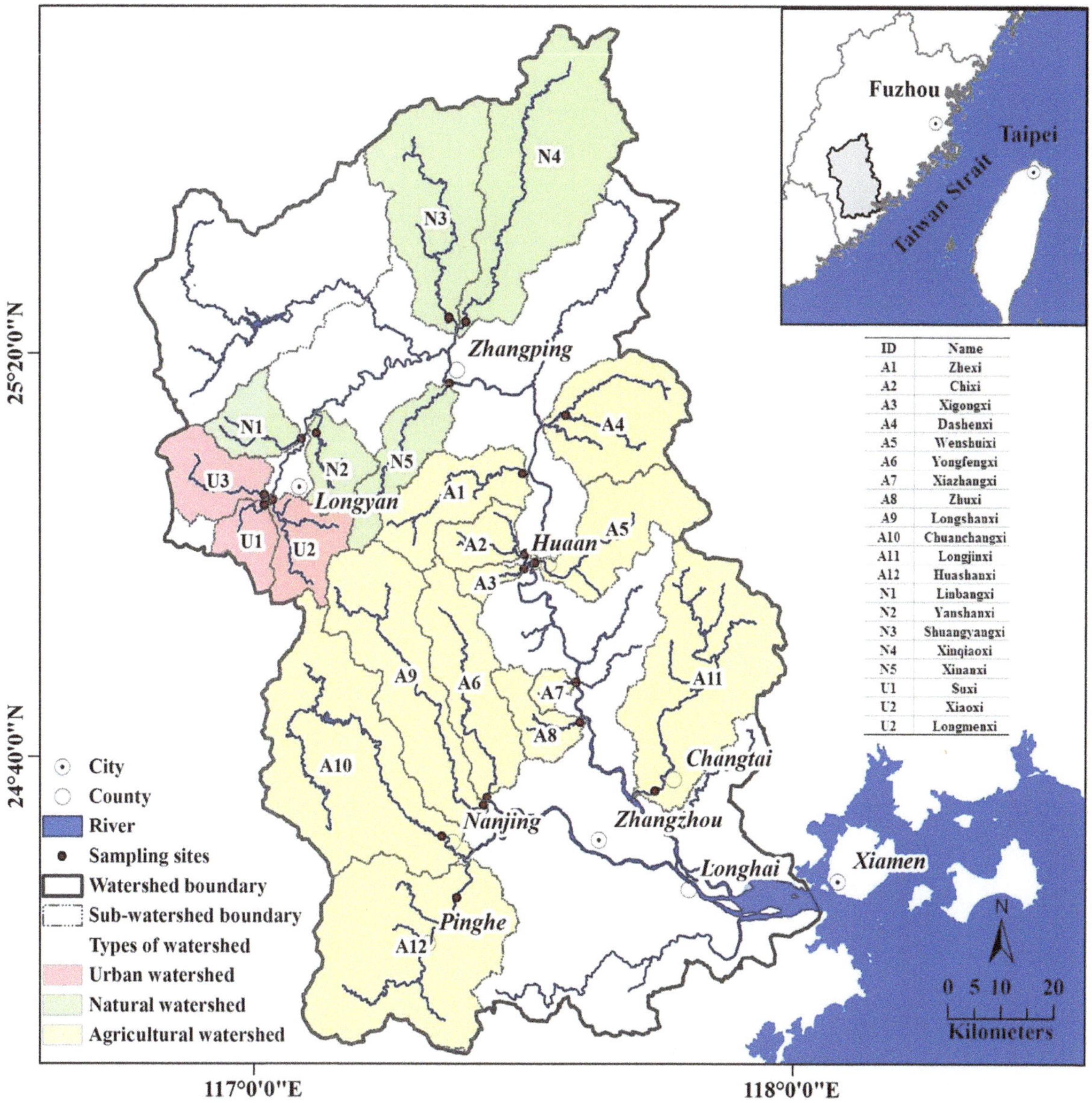

Figure 1. Sampling sites in the watershed studied.

banana, longan, litchi, pomelo, citrus, and flowers. The N budget for the JRW indicates that fertilizer and animal feedstuff contribute 83.6% of the total N input [46]. For most crop land, surface application of fertilizers is used with high rates in spring (over 200 kg N/hm^2) and the predominant N fertilizers used in the JRW including urea, ammonium hydrocarbonate and NPK compound fertilizers [45].

The JRW consists mainly of eight counties/districts: Zhangzhou, Xinlou, Zhangping, Hua'an, Changtai, Pinghe, Longhai and Nangjing. More than ten million residents use the Jiulong River as their source of water for residential, industrial and agricultural activities. The watershed's gross domestic product (GDP) accounts for a quarter of Fujian Province's economic output while it is approximately one tenth of Fujian Province. Population and economic growth while relatively low municipal wastewater treatment rate make the point source pollution from industrial wastewater and sewage discharged into the river be an innegligible pollution source in the JRW [47]. Additionally, Longyan municipality, including Xinluo and Zhangping, located at the upstream JRW, has the major mining areas of Fujian province. More than 64 kinds of mineral resources can be found there.

2.3. Sampling and Experimental Design

Based on the typical land use patterns in the JRW, 20 headwater sub-watersheds classified into three types of groups, namely, natural, urban and agricultural were chosen for sampling. Urban watersheds (U1–U3), agricultural watersheds (A1–A12) and natural watersheds (N1–N5) are the sub-watersheds where the proportion of developed land, cropland and forest land is over 6%, 10% and 80%, respectively. The surface water was sampled during baseflow (i.e. 7-day minimum streamflow) period with nine sampling campaigns from 20 headwater sub-watersheds in three sampling seasons during three consecutive years (Fig. 1). Three investigations were carried out in each of the flood seasons (on August 28th, 2010, August 15th, 2011, and August 16th, 2012), dry seasons (on November 28th, 2010, November 29th, 2011, and

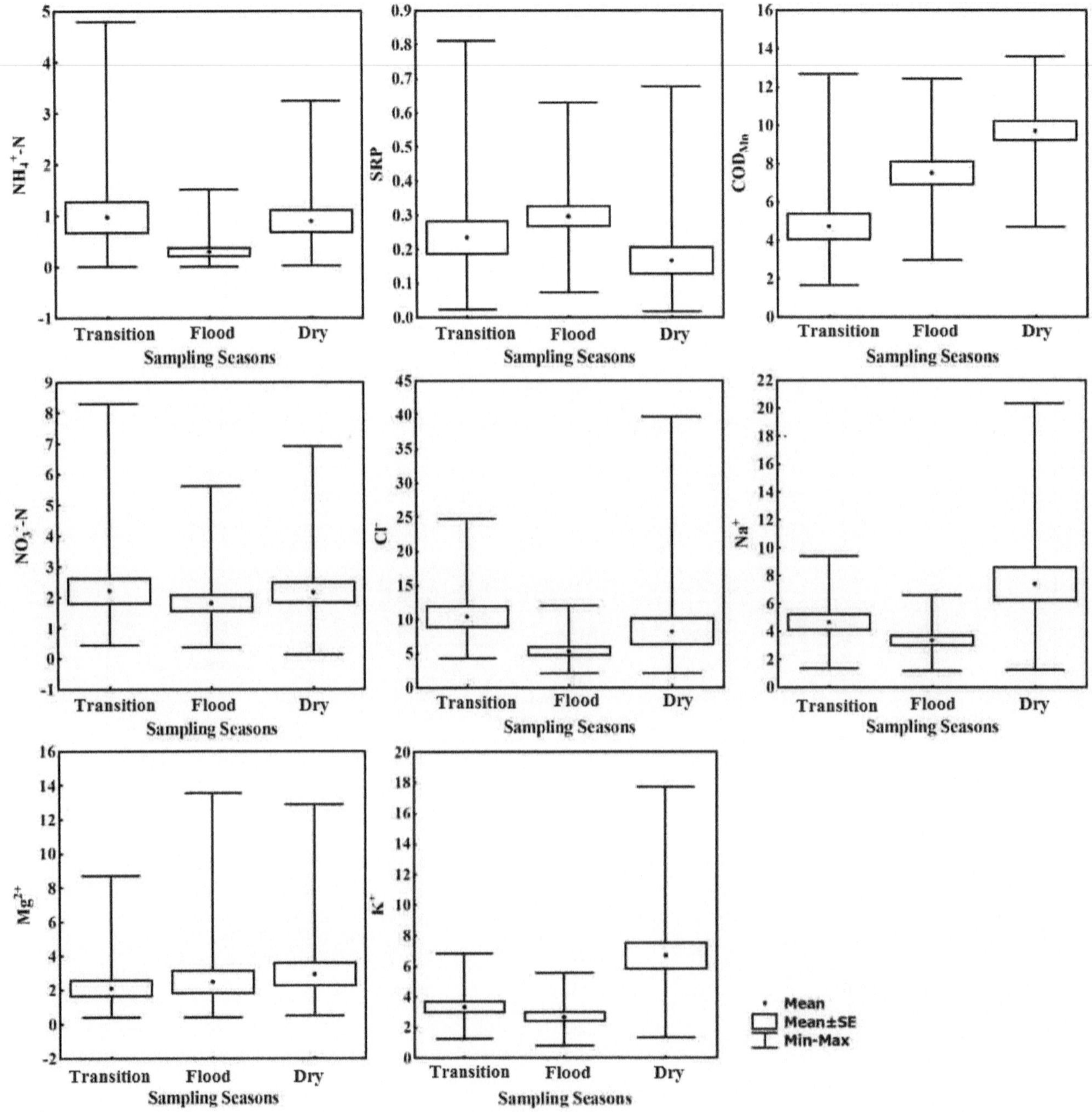

Figure 2. Comparison between concentrations of water quality parameters among the three sampling seasons.

November 20th), and transition seasons (on February 28th, 2010, March 1st, 2011, and March 25th, 2012). Note that the 25th and 75th percentile values of the monthly streamflow are usually calculated to define these three seasons, namely, dry season ($P < 25\%$), transition season ($P = 25\%–75\%$) and flood season ($P > 75\%$).

Water quality was characterized using the mean values of these nine-time sampling data sets. To minimum the influence of the sediment and to reduce the influence of hydrologic regime on organic and polyphosphate molecules dissolved in the sample, we sampled the surface water of stream and filtered the water immediately. The samples were kept at 4°C and transported to the laboratory for advanced analysis. Eight chemical parameters were analyzed following standard methods [48] and completed within 24 h after sampling. These parameters were ammonium N (NH_4^+-N), the potassium permanganate index (COD_{Mn}), soluble reactive phosphate (SRP), nitrate N (NO_3^-N), chloride (Cl^-), sodium (Na^+), magnesium (Mg^{2+}) and potassium (K^+). These parameters were chosen because they can reflect the influence on water quality in terms of natural control (including bedrock geology) and anthropogenic controls (including domestic wastewater/industrial effluents, agricultural activities).

2.4. Data Sources

Landscape patterns play an important role in water quality variation at the watershed scale. Land use/land cover (LULC) and

Table 1. K independent samples of water quality among the different sampling seasons.

	NH_4^+-N	SRP	COD_{Mn}	NO_3^-N	Cl^-	Na^+	Mg^{2+}	K^+
Mean Rank(Flood)	23.05	40.40	31.50	28.40	24.20	22.20	28.20	20.70
Mean Rank(Transition)	32.73	29.23	16.95	30.85	38.95	30.45	28.10	26.55
Mean Rank(Dry)	35.73	21.88	43.05	32.25	28.35	38.85	35.20	44.25
Chi-Square	5.755	11.413	22.433	0.498	7.588	9.090	2.173	19.718
Asymp. Sig.	0.056	0.003	0.000	0.780	0.023	0.011	0.337	0.000

Sample No. = 60; Asymp. Sig. <0.05 indicates significant variation.

landscape pattern metrics (LPMs) were used in this study to delineate the spatial patterns. The spatial pattern of land use and landscape can reflect the underlying human activities [49]. Landsat Thematic Mapper satellite imagery of 2010 with 25 m resolution was used to create LULC data. The land categories were generated using a combination of unsupervised classification and spatial reclassification based on manual on-screen digitizing (for details, see [50]). Land cover was aggregated for six major categories: forest (natural forest, without fertilizing), cropland (including economic forests, with fertilizing), developed land, orchard, water and bare land, and the four most predominant categories (forest, cropland, developed land, and orchard) were retained for analysis in this study. Three LPMs: Patch Density (PD), Largest Patch Index (LPI), and Shannon's Diversity Index (SHDI), were chosen in this study, to explore the linkage between landscape pattern and water quality.

The natural factors considered in this study were topography and geology. Standard deviation of slope was derived from 25 m-resolution Digital Elevation Model in the JRW. A geologic map in the JRW was generated based on the geology map in Fujian province (scale 1:1600000), thus two main types of bedrock geology were extracted, namely, sandstones and siltstones (Geology 1), and granites, lavas, and volcanic tuff (Geology 2). Anthropogenic factors also includes socioeconomic development indices represented by population density (Pop_density), GDP, primary industry output value (GDP1), secondary industry output value (GDP2), and tertiary industry output value (GDP3) were collected from the Statistical Yearbook.

2.5. Conventional Statistical Analysis

The K independent samples nonparametric test was used to determine the significance of variations of water quality during different sampling seasons. Kruskal-Waillis test was used to calculate mean rank values of each water quality indicator with the following equation:

$$K=(N-1)\frac{\sum_{i=1}^{g} n_i(\bar{r}_i-\bar{r})^2}{\sum_{i=1}^{g}\sum_{j=1}^{n_1}(r_{ij}-\bar{r})^2} \quad (1)$$

Where n_i is the number of observations in group; r_{ij} is the rank of observation j from group i; N is the total number of observations across all groups.

The Post Hoc multiple comparisons were used to determine the significance of variations in water quality among headwater watersheds with different dominant land use types. Least Square Difference (LSD) method was used in this study to identify significance of variation in difference sample seasons. The test statistic is calculated as follows:

$$t_{LSD}=\frac{\bar{x}_i-\bar{x}_j}{\sqrt{s_w^2\left(\frac{1}{n_i}+\frac{1}{n_j}\right)}} \quad (2)$$

where s_w^2 represents the "variance within groups" and is equal to the mean square within in the ANOVA table. This test statistic has N-k degrees of freedom.

Pearson analysis was used in this study to examine the strength and significance of the relationships between selected influencing factors and water quality parameters. Based on a sample of paired data (x_i, y_i), the sample Pearson correlation coefficient r is defined as the following formula:

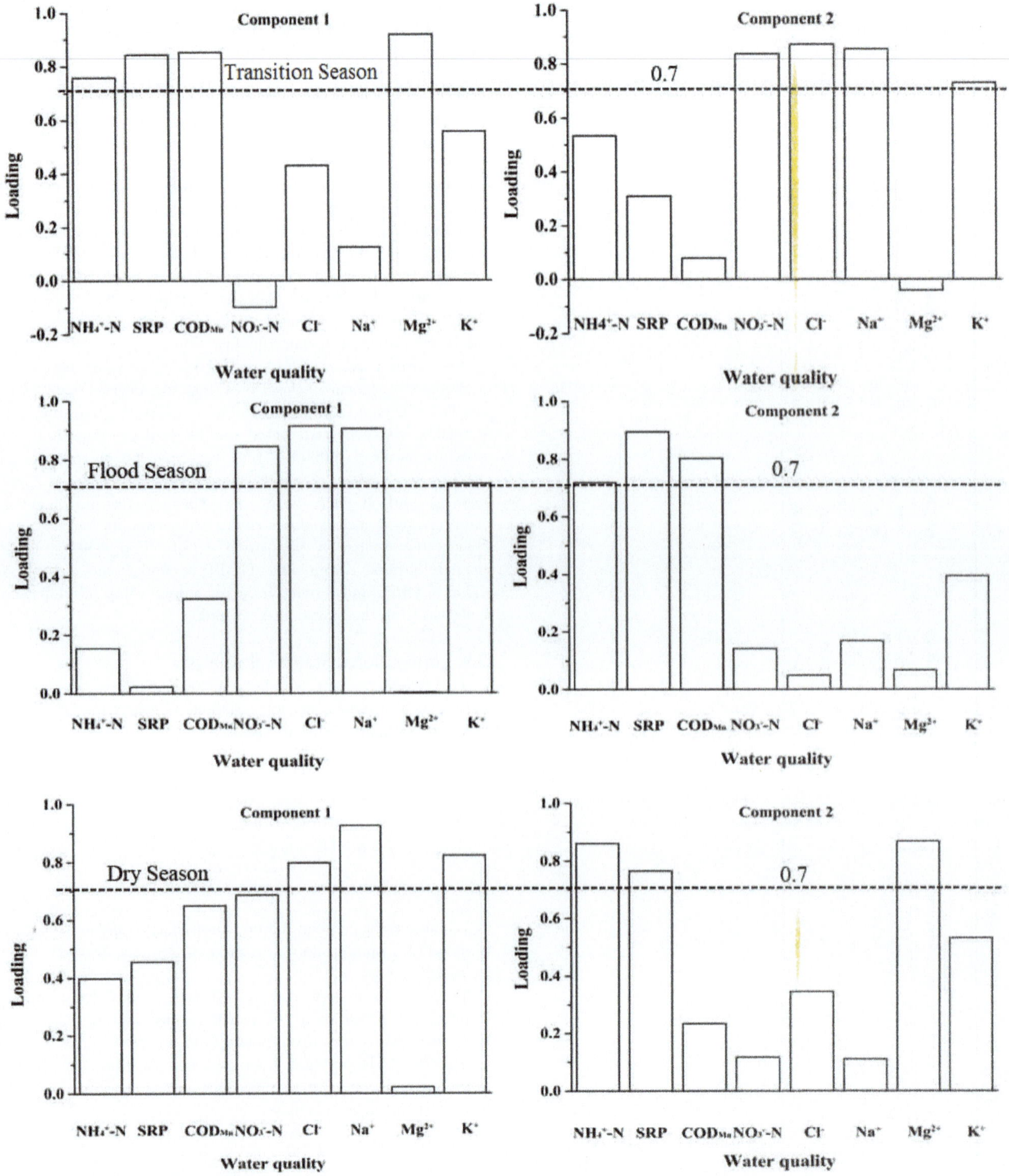

Figure 3. Rotated Component Matrix for water quality parameters among different sampling seasons.

$$r = \frac{\sum_{i=1}^{n}(x_i-\bar{x})(y_i-\bar{y})}{\sqrt{\sum_{i=1}^{n}(x_i-\bar{x})^2(y_i-\bar{y})^2}} \quad (3)$$

PCA was used to identify important components that explained most of the variance of water quality in different sampling seasons. This is designed to reduce the number of variables to a small number of indices while attempting to preserve the relationships present in the original data [3]. Kaiser-Meyer-Olkin (KMO) and Bartlett's test was often used to examine the sensitivity of the data for PCA [9] [51], KMO is a measure of sampling adequacy and a

Table 2. Total variance explained in the three sampling seasons.

Sampling seasons	Component	Initial Eigen-values			Extraction Sums of Squared Loadings			Rotation Sums of Squared Loadings		
		Total	% of Variance	Cumulative%	Total	% of Variance	Cumulative%	Total	% of Variance	Cumulative%
Transition season	1	4.753	59.411	59.411	4.753	59.411	59.411	3.373	42.158	42.158
	2	1.736	21.696	81.107	1.736	21.696	81.107	3.116	38.948	81.107
Flood season	1	3.894	48.677	48.677	3.894	48.677	48.677	3.010	37.630	37.630
	2	1.735	21.690	70.367	1.735	21.690	70.367	2.170	27.123	64.753
Dry season	1	4.800	60.003	60.003	4.800	60.003	60.003	3.427	42.832	42.832
	2	1.185	14.816	74.818	1.185	14.816	74.818	2.559	31.986	74.818

high KMO value (close to 1) generally indicates that PCA may be useful.

$$Z_{ij} = P_{i1}x_{1j} + P_{i2}x_{2j} + P_{i3}x_{3j} + \cdots P_{im}x_{mj} \quad (4)$$

where Z is the *PC* value, P is the *PC* loading, x is the measured value of variable, i is the PC number, j the sample number, and m is the total number of measured variables.

2.7. Spatial Statistics

Moran's I, a global measure of spatial autocorrelation, was used to identify the degree of spatial dependence on water quality parameters over time. Moran's I is defined as the following formula:

$$I = \frac{n}{\sum_{i=1}^{n}(x_i - \bar{x})^2} \frac{\sum_{i=1}^{n}\sum_{j=1}^{n} w_{ij}(x_i - \bar{x})(x_j - \bar{x})}{\sum_{i=1}^{n}\sum_{j=1}^{n} w_{ij}} \quad (5)$$

where, x_i and x_j refer to water quality in station i and station j, respectively $\bar{x}$ is the overall mean water quality, and w_{ij} is the weight matrix. Because not all-sub watersheds are adjacent to each other, four nearest neighbors were chosen as cutoff points when creating the weight matrix: that is, if station i and j are within the neighbor threshold, $w_{ij} = 1$, otherwise $w_{ij} = 0$ [42]. Global Moran's I evaluates whether the pattern expressed is clustered, dispersed, or random. A Moran's I value near +1.0 indicates clustering while a value near −1.0 indicates dispersion, and a value of 0 indicates perfect spatial randomness [52].

To identify landscape factors explaining water quality variations, we used OLS stepwise multiple linear regressions in SPSS. OLS is a type of global statistic, which assumes the relationship under study in constant over space, and so the parameters are estimated to be the same for all the study area. The model can be stated as follows:

$$y = \beta_0 + \sum_{i=1}^{p} \beta_i X_i + \varepsilon \quad (6)$$

where y is the dependent variable, β_0 is the intercept, β_i is the parameter estimate (coefficient) for independent variable x_i, p is the number of independent variables, and ε is the error term.

Compared to the OLS regression models, spatial regression models incorporate spatial dependence in the form of lag or error dependence. The spatial error regression (Eq.7) and spatial lag regression (Eq. 8) are defined as follows:

$$y_i = X_i\beta_i + \varepsilon$$
$$\varepsilon = \lambda w_\varepsilon + \zeta \quad (7)$$

$$y_i = X_i\beta_i + \rho w y_j + \varepsilon \quad (8)$$

where y_i and y_j represents the dependent variable at sampling site i and j, x_i is the independent variable at i, β_i is the regression coefficient; ε is the random error term, λ is the spatial autoregressive coefficient of spatial regression, w_ε is the spatially lagged error term, ζ is the homoskedastic and independent error term. ρ is the spatial autoregressive coefficient, and wy_j is the spatially lagged dependent variable.

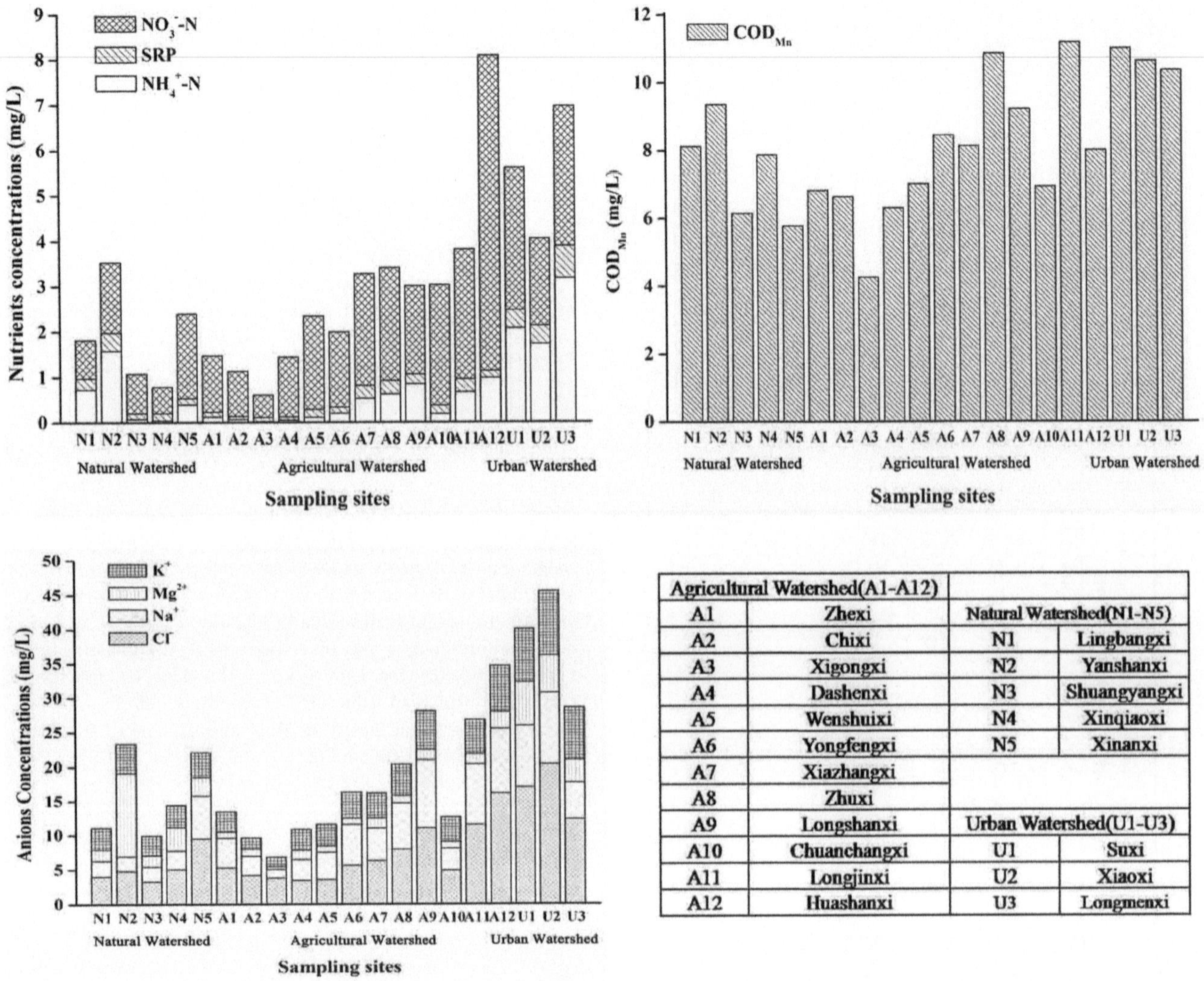

Agricultural Watershed(A1-A12)			
A1	Zhexi	Natural Watershed(N1-N5)	
A2	Chixi	N1	Lingbangxi
A3	Xigongxi	N2	Yanshanxi
A4	Dashenxi	N3	Shuangyangxi
A5	Wenshuixi	N4	Xinqiaoxi
A6	Yongfengxi	N5	Xinanxi
A7	Xiazhangxi		
A8	Zhuxi		
A9	Longshanxi	Urban Watershed(U1-U3)	
A10	Chuanchangxi	U1	Suxi
A11	Longjinxi	U2	Xiaoxi
A12	Huashanxi	U3	Longmenxi

Figure 4. Comparison between concentrations of water quality parameters among the three types of watersheds.

To identify the influencing factors from the perspective of spatial dependence on river water quality, we compared the OLS models (stepwise multiple linear regression) and spatial regression models by comparison of R^2, AIC and Moran's I values in order to

Table 3. LSD Post Hoc multiple comparisons of water quality variables among the three types of watersheds.

Water quality	Urban watersheds-Natural watersheds		Urban watersheds-Agricultural watersheds		Agricultural watersheds-Natural watersheds	
	Mean difference	Sig.	Mean difference	Sig.	Mean difference	Sig.
NH_4^+-N	1.757*	0.000	1.958*	0.000	−0.201	0.450
SRP	0.291	0.001	0.329*	0.000	−0.037	0.510
COD_{Mn}	3.204*	0.021	2.842*	0.021	0.362	0.699
NO_3^-N	1.566	0.136	0.450	0.617	1.116	0.144
Cl^-	11.245*	0.001	9.593*	0.001	1.652	0.410
Na^+	5.082*	0.016	2.879	0.104	2.203	0.129
Mg^{2+}	0.719	0.668	3.777*	0.019	−3.058*	0.021
K^+	4.827*	0.000	4.532*	0.000	0.295	0.678

*indicates significant at $p<0.05$.

Table 4. Moran's I values for water quality indicators among the three sampling seasons.

	NH_4^+-N	SRP	COD_{Mn}	NO_3^-N	Cl^-	Na^+	Mg^{2+}	K^+
Flood season	0.81**	−0.21	0.80**	0.03	0.21	0.48*	0.24*	0.66**
Transition season	1.42**	1.39**	0.29	0.18	0.76**	0.33	0.49*	0.99**
Dry season	1.29**	1.28**	0.28	0.31*	0.51**	0.41*	0.47*	1.27**
Mean value	1.35**	1.12**	0.86**	0.19	0.89**	0.47*	0.39	1.32**

**indicates significant at $p<0.01$;
*indicates significant at $p<0.05$.

select the suitable models. Higher R^2 in this study meant that influencing factors could explain more variance in the water quality. A lower AIC value suggested that a closer approximation of model to reality and had better model performance. Moran's I values were used to calculate the residuals from each regression in order to determine whether spatial autocorrelations existed. The dependent variables are mean values of each water quality indicator for all sampling times. Independent variables included the factors derived from the PCA. The same variables were used in both the OLS and spatial regression.

The landscape metrics were calculated using Fragstat 3.3. The calculation of Moran's I values and GIS analyses were performed using ArcGIS 9.3. OLS and spatial regression were performed in the GeoDa 9.5i. Data pre-processing (including Kaiser-Meyer-Olkin (KMO) and Bartlett's test) and PCA were conducted using SPSS16.0. Box-plots were produced using Statistical 7.

Results

3.1. Temporal Variation Analyses

3.1.1. Seasonal variation of stream water quality. The concentrations of water quality parameters among the three sampling seasons are shown in Fig. 2. All water quality parameters except SRP had their highest values of mean concentration in the dry and transition seasons, whereas the concentrations were the lowest in the flood season. The highest mean concentrations of COD_{Mn}, Mg^{2+}, Na^+ and K^+ were found in the dry season. The mean concentrations of NH_4^+-N, NO_3^-N, Cl^- in the transition season were higher than those in the dry and flood seasons. Interestingly, the highest concentration of SRP was detected in the flood season and the mean concentrations of SRP were in the order of flood season >transition season >dry season.

The K independent samples test to examine the significance of seasonal variations of river water quality (Table 1) shows that the mean rank values for most of the water quality indicators except for SRP and Cl^- were higher in the dry season than those in the transition and flood seasons, indicating that the water quality in the dry season was remarkably worse. The mean rank value of Cl^- was highest in the transition season while the mean rank value of SRP was highest in the flood season, implying that the most seriously polluted season for Cl^- and SRP was the transition and flood seasons, respectively.

3.1.2. Identifying potential pollution sources among three sampling seasons. KMO and Barlett's test was performed before PCA analysis in this study. The KMO value is 0.699 and the p value of Barlet's test is less than 0.0001, which means there are no significant relationships among water quality variables and it is suitable for PCA analysis. The PCA examining the differences of water pollution characteristics among the three sampling seasons identified the principal components (PCs) with initial Eignen-values >1 (Table 1) and the important water quality parameters with absolute values of component loading >0.7 (Fig. 3). The two PCs identified for the three sampling seasons exhibited distinct differences in terms of the water quality parameters involved (Table 2 and Fig. 3).

For the transition season, two PCs could explain the majority of total variation (81.107%). PC1, accounting for 42.158% of the total variance, had strongly positive loadings on the organic-related parameters (COD_{Mn}), inorganic nutrient-related water quality parameters (NH_4^+-N and SRP) and Mg^{2+}. Thus, this group might be interpreted as the latent factors from anthropogenic input related to industrial effluents and domestic wastewater, and the natural weathering process. PC2, accounting for 38.948% of the total variance, had strongly positive loadings on NO_3^-N, Cl^-,

Table 5. Correlations between selected water quality parameters and environmental variables using Pearson analysis.

	NH_4^+-N	SRP	COD_{Mn}	NO_3^-N	Cl^-	Na^+	Mg^{2+}	K^+
forest	−0.138	−0.117	−0.387*	−0.716**	−0.384*	−0.495*	0.132	−0.314
cropland	−0.254	−0.209	0.024	0.444*	−0.112	0.132	−0.554**	−0.176
developed land	0.769**	0.777**	0.807**	0.441*	0.681**	0.593**	0.376	0.830**
orchard	0.095	0.155	0.383*	0.630**	0.210	0.395*	0.002	0.254
PD	0.140	0.168	0.403*	0.687**	0.328	0.475*	−0.241	0.277
SHDI	0.274	0.242	0.525**	0.741**	0.477*	0.593**	−0.124	0.439*
LPI	−0.139	−0.108	−0.312	−0.651**	−0.328	−0.435*	0.114	−0.294
Pop_density	0.777**	0.701**	0.633**	0.457*	0.879**	0.709**	0.285	0.898**
GDP	0.754**	0.799**	0.706**	0.102	0.406*	0.261	0.552**	0.574**
GDP1	0.246	0.147	0.286	0.782**	0.440*	0.482*	0.036	0.459**
GDP2	0.763**	0.791**	0.637**	0.010	0.374	0.188	0.589**	0.543**
GDP3	0.645**	0.740**	0.760**	0.168	0.377	0.326	0.429*	0.527**
Slope_std	0.501*	0.368	0.381	0.369	0.468*	0.340	0.131	0.472*
Geology 1	0.490**	0.461*	0.116	−0.301	0.108	−0.264	0.684**	0.178
Geology 2	−0.278	−0.144	0.022	−0.243	−0.198	0.045	−0.468*	−0.220

*indicates significant at $p<0.05$;
**indicates significant at $p<0.01$.

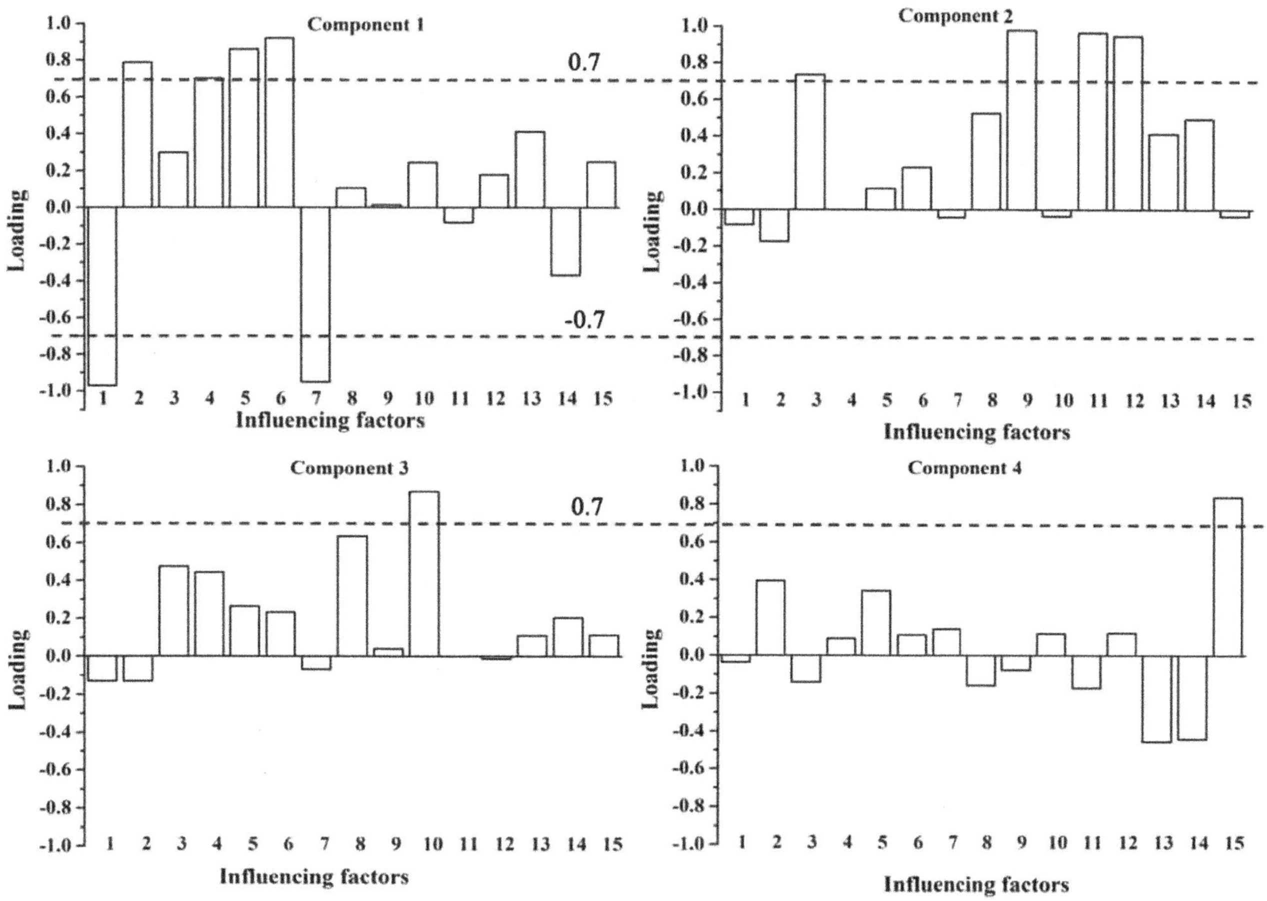

Figure 5. Rotated Component Matrix for environmental variables identified for variation in water quality. (1, 2, 3,4,5,6,7,8,9,10,11,12,13,14,15 stands for forest, cropland, developed land, orchard, PD, SHDI, LPI, Pop_density, GDP, GDP1, GDP2, GDP3, Slope_std, Geology 1, Geology 2).

Na^+ and K^+, which might reflect that the impact was associated with agricultural activities.

In the flood season, PC1 explained 37.63% of the total variance and had strongly positive loadings on NO_3^-N, Cl^-, Na^+ and K^+, which might have represented the influences of agricultural activities, associated with the precipitation-induced surface runoff. PC2 accounted for 27.132% of the total variance and had strongly positive loadings on NH_4^+-N, SRP and COD_{Mn}. Thus, PC2 represented industrial effluents and domestic wastewater.

In the dry season, PC1 accounted for 42.832% of the total variance, where Cl^-, Na^+ and K^+ had the greatest loadings. This factor might be interpreted as representing the natural weathering process. PC2 explained 31.986% of the total variance and was largely contributed by nutrients and metal ions related with industrial effluents and domestic wastewater.

3.2. Spatial Variation Analyses

The water quality among agricultural, natural and urban watersheds showed great spatial variations (Fig. 4).

The concentrations of NH_4^+-N, SRP, and K^+ were higher in urban watersheds than those in natural and agricultural watersheds. NO_3^-N concentration in agricultural watersheds especially in sub-watershed A12 was generally higher than that in natural and urban watersheds. The concentrations of Cl^-, and COD_{Mn} were generally higher in urban watersheds than those in natural and agricultural watersheds. It should be noted that the concentrations of Mg^{2+}, NH_4^+-N, NO_3^-N, and COD_{Mn} in sub-watershed N2 were the highest among the natural sub-watersheds, which were even higher than those in some of the sub-watersheds of the urban and agricultural watersheds.

The LSD Post Hoc multiple comparison method used to identify how land use patterns had a clear impact on spatial variations of water quality (Table 3). Table 3 revealed that the urban watersheds exhibited more significance in terms of NH_4^+-N, COD_{Mn}, Cl^-, and K^+ than natural and agricultural watersheds, indicating that urbanized areas were easily exposed to water quality pollution associated with industrial and domestic effluents. The Mg^{2+} concentration in natural watersheds was significantly higher than that in agricultural watersheds, suggesting that the natural weathering process might have played a more important role on Mg^{2+} concentration than agricultural activities. Although most of the water quality parameters collected in agricultural watersheds had higher concentrations than those in natural sub-watersheds, no significant variations were detected.

3.3. Spatial and Temporal Variation in Water Quality

Moran's I values (used to identify the degree of spatial dependence on water quality parameters over time) for the water quality indicators in the three different sampling seasons are shown in Table 4.

Significant positive spatial autocorrelations were found for NH_4^+-N, Mg^{2+} and K^+ among the three sampling seasons,

Table 6. Total variance explained for environmental factors.

Component	Initial Eigen-values			Extraction Sums of Squared Loadings			Rotation Sums of Squared Loadings		
	Total	% of Variance	Cumulative %	Total	% of Variance	Cumulative %	Total	% of Variance	Cumulative %
PC1	6.027	40.179	40.179	6.027	40.179	40.179	5.105	34.033	34.033
PC2	4.258	28.385	68.564	4.258	28.385	68.564	4.123	27.489	61.522
PC3	1.226	8.175	76.740	1.226	8.175	76.740	1.807	12.047	73.569
PC4	1.053	7.017	83.757	1.053	7.017	83.757	1.528	10.188	83.757

indicating that the water quality problem in terms of these three water quality parameters shared regional anthropogenic or natural factors with regards to river pollution over time. The Moran's I values for most of the water quality indicators (except for COD_{Mn} and Na^+) in the flood season were the lowest, indicating that significant water quality variations were exhibited between neighboring sub-watersheds and river water quality in the flood season, which might have been contributed by non-point source pollution related to heterogeneous land use patterns in watersheds. The Moran's I values were highest for NH_4^+-N, SRP, COD_{Mn}, Cl^- and Mg^{2+} in the transition season, and highest for NO_3^-N and K^+ in the dry season, suggesting that most of the water quality parameters in the transition and dry seasons had significant spatial autocorrelations, and exhibited similar levels of water pollution between the neighboring sub-watersheds in these two seasons. Generally, most of the water quality parameters in the dry and transition seasons had more significant spatial autocorrelations than those in the flood season.

3.4. Coupled Effects of Natural and Anthropogenic Controls on Water Quality Variations

River water quality is influenced by a combination of natural and anthropogenic factors. In this study, natural factors involved the standard deviation of slope and geology factors, anthropogenic factors included LULC (i.e. four land use types), LPMs (i.e. three landscape indicators), socioeconomic development indices (i.e. Pop_density, GDP, GDP1, GDP2, and GDP3).

3.4.1. General correlation between water quality parameters and environmental variables. Table 5 shows the general correlation between selected influencing factors and water quality parameters using Pearson analysis.

a) Relationship between water quality and land use/cover: Developed land had positive correlations with most water quality parameters (except for Mg^{2+}), suggesting that developed land was an important factor associated with degraded water quality. Forest showed significant negative correlation with COD_{Mn}, NO_3^-N, Cl^- and Na^+. Cropland was significantly positively correlated with NO_3^-N, whereas it had significant negative correlation with Mg^{2+}. Orchard had significant positive correlation with COD_{Mn}, NO_3^-N and Na^+.

b) Relationship between water quality and LPMs: PD was significantly positively correlated with COD_{Mn}, NO_3^-N and Na^+; And SHDI had significant positive linkages with COD_{Mn}, NO_3^-N, Cl^-, Na^+ and K^+, whereas LPI had negative associations with most of the water quality parameters.

c) Relationship between water quality and social-economic factors: Pop_density had significantly positive correlation with most of the water quality indicators (except for Mg^{2+}), similar to build-up. GDP was significant positively related with NH_4^+-N, SRP, COD_{Mn}, Cl^-, Mg^{2+} and K^+. Significantly positive relationships between GDP1 and NH_4^+-N, SRP, COD_{Mn}, Cl^-, Mg^{2+} and K^+ were obtained, whereas GDP2 and GDP3 were significantly positively correlated with NH_4^+-N, SRP, COD_{Mn}, Mg^{2+} and K^+.

d) Relationship between water quality and natural factors: Slope_std had significant negative associations with NH_4^+-N, NO_3^-N and K^+, indicating that increased deviation in slope would force the flow to carry pollutants on impermeable surfaces and discharge into river in shorter time. Geology 1 was significantly positively correlated with NH_4^+-N, SRP and

Table 7. Comparison of R^2, AIC and Moran's I values between OLS models and spatial regressions.

Water quality parameters		R^2	AIC	Moran's I
NH_4^+-N	OLS	0.739	31.813	−0.29
	Spatial lag	0.745	33.546	−0.28
	Spatial error	0.793	29.568	−0.24
SRP	OLS	0.716	−34.402	−0.31
	Spatial lag	0.719	−32.54	−0.28
	Spatial error	0.747	−35.500	−0.24
COD_{Mn}	OLS	0.711	67.929	−0.25
	Spatial lag	0.713	69.856	−0.25
	Spatial error	0.763	65.436	−0.12
NO_3^-N	OLS	0.767	50.399	0.06
	Spatial lag	0.833	47.305	−0.16
	Spatial error	0.846	45.967	0.12
Cl^-	OLS	0.587	113.39	−0.25
	Spatial lag	0.797	106.346	−0.25
	Spatial error	0.707	110.047	0.05
Na^+	OLS	0.530	93.660	−0.17
	Spatial lag	0.664	91.319	−0.15
	Spatial error	0.647	90.483	−0.08
Mg^{2+}	OLS	0.360	96.857	−0.41
	Spatial lag	0.365	98.766	−0.39
	Spatial error	0.568	92.963	−0.39
K^+	OLS	0.774	65.727	0.09
	Spatial lag	0.869	60.085	−0.24
	Spatial error	0.833	62.930	0.16

Mg^{2+}. Geology 2 was significantly negatively correlated with Mg^{2+}.

3.4.2. Significant explanatory variables identified for variations in water quality. PCA was used to obtain appreciable data reduction and to identify the important potential factors explaining spatiotemporal variations in water quality. Follow this, OLS and spatial regression were performed to uncover the mechanism of water quality variations from the perspective of spatial dependence of river water quality. The results obtained from PCA are presented in Table 6 and Fig. 5. Table 6 shows the R^2, AIC and Moran's I values for water quality

Table 8. Spatial regression models established in the JRW.

Water quality parameters	Spatial Regression models	R^2	Sig.
NH_4^+-N^a	y = 0.710+0.650*factor2+0.264*factor3−0.236*factor4 (LAMBDA = −0.561)	0.793	**
SRP^a	y = 0.233+0.128*factor2+0.033*factor3(LAMBDA = −0.395)	0.747	*
$COD_{Mn}{}^a$	y = 8.194+0.654*factor1+1.504*factor2+0.597*factor3 (LAMBDA = −0.452)	0.763	**
$NO_3^-N^a$	y = 2.043+0.774*factor1+0.828*factor3 (LAMBDA = 0.709)	0.846	**
Cl^{-b}	y = 14.456+3.878*factor2+4.058*factor3−0.799*WY	0.797	**
Na^{+a}	y = 8.195+1.229*factor1+1.182*factor2+2.155*factor3−0.567*WY	0.664	**
Mg^{2+a}	y = 2.511−0.883*factor1+1.561*factor2−0.804*factor4(LAMBDA = −0.716)	0.568	*
K^{+b}	y = 7.207+2.032*factor2+1.772*factor3−0.380*factor4−0.656*WY	0.869	*

Note: Factor1, 2, 3, and 4 corresponds to the four components identified and presented in Fig. 6.
a denotes the results of spatial error models, *b* denotes the results of spatial lag models.
WY: weighted mean of the dependent variable for adjacent sub-basins.
*indicates significant at $p<0.05$.
**indicates significant at $p<0.01$.

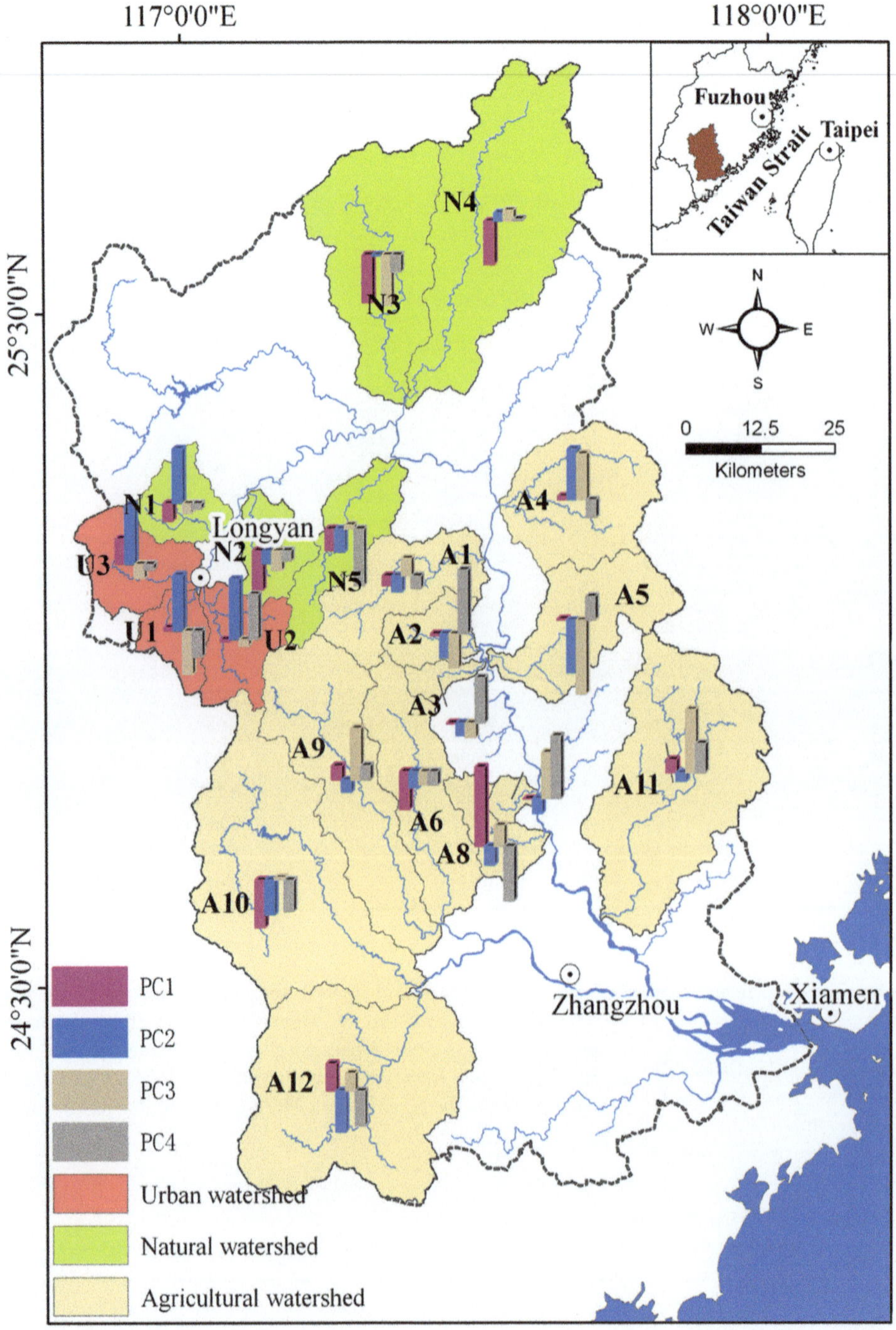

Figure 6. The four potential pollution sources identified to explain spatiotemporal variations in water quality for 20 headwater watersheds in the JRW. (PC1, PC2, PC3, and PC4 represents landscape patterns, urbanization and socioeconomic development, agricultural activity, and natural control, respectively).

residuals for both spatial regression (including spatial lag models and spatial error models) and OLS models.

Four components with initial Eigen-values greater than 1 could explain 83.757% of the total variance (Table 6). PC1, PC2, PC3 and PC4 explained 34.033, 27.489, 12.047 and 10.188% of the total variance. As shown in Fig. 5, PC1 was positively correlated to and largely contributed by cropland, orchard, PD and LPI and negatively affected by forest and LPI. All these important parameters included in PC1 were related to the LULC and LPMs of the watershed, indicating that landscape patterns played an important role in water quality variations in the JRW. PC2 which was positively and largely correlated with developed land, GDP, GDP2 and GDP3, represented the coupled effects of urbanization and socioeconomic development. PC3, which was positively correlated with GDP1, represented the contribution from agricultural activities. PC4 was positively contributed by Geology 2, and this component represented the influence from the natural control. Overall, three potential factors (i.e. PC1, PC2, PC3) identified were related to anthropogenic controls.

Non-significant spatial autocorrelations were found in all models, indicating that both OLS models and spatial regressions could be chosen to determine the relationship between environmental factors and water quality parameters (Table 7). An improvement in R^2 using the spatial regression model over the OLS model was observed, especially for Cl^-, Na^+ and Mg^{2+} (Table 7). In addition, most of the spatial error models had higher R^2 values than spatial lag models, which suggests that the lower error in the spatial error models. Comparing the AIC values from spatial regressions and those from OLS models, all the spatial error and most of the spatial lag models had lower AIC values than the corresponding AIC values from the OLS models, indicating that spatial regressions, especially spatial error models had better model performance than OLS models.

Given the higher R^2 values and lower AIC values from spatial regression model, spatial regression model were chosen for further study to identify significant explanatory variables for each parameters. The models explain approximately up to 87% of the variation in water quality (Table 8).

As shown in Table 8, factor2 and factor3 were the two most important factors associated with water quality variations, implying that anthropogenic controls played critical roles in variations of water quality in the JRW. Factor2 was related to all the water quality parameters except for NO_3^-N while factor3 was included in all the regression models except for Mg^{2+}. In particular, factor2 had significant impact on NH_4^+-N, SRP, COD_{Mn}, Mg^{2+} and K^+, implying environmental factors related to urbanization and socioeconomic development were the important explanatory variables for these parameters. It is understandable that NO_3^-N was largely and positively contributed by factor3 associated with agricultural activities. Factor1 representing landscape pattern was also an important factor, which had significant positive effects on COD_{Mn}, NO_3^-N, and Na^+. Factor4 appeared in three of the regression models related to NH_4^+-N, Mg^{2+} and K^+, and the relationships were significantly negative.

Generally, agricultural activities and landscape patterns had coupled effects on variations related to NO_3^-N and Na^+. Urbanization and socioeconomic development were the dominant explanatory variables for COD_{Mn} and NH_4^+-N were also the primary predictor for SRP, Mg^{2+}, and K^+. This suggested that urban development was the driving source of oxygen demand and nutrient concentration. Agricultural activities and urbanization and socioeconomic development were the two important predictors for Cl^-.

Fig. 6 shows the four potential factors identified to explain spatiotemporal variations in water quality for 20 headwater watersheds in the JRW. Compared to other two types of watershed, PC2 (i.e. urbanization and socioeconomic development) is the largest contributor for variations among four factors identified in three urban watersheds. In terms of natural watersheds, environmental factors mostly related to anthropogenic controls had a negative influence on water quality.

Discussion

4.1. Causal Factors of Spatiotemporal Variations in Water Quality

Water quality is affected by the combination of natural and anthropogenic factors, the relative influences of which change with temporal and spatial scale [1]. Our study effectively differentiated the impacts from anthropogenic factors (including landscape pattern, agricultural activities, urban and socioeconomic development) and natural factors on stream water quality. Our results showed that urban and socioeconomic development acted as the primary predictor for COD_{Mn} and NH_4^+-N, and the relationships between developed land and water quality parameters were consistently positive. Many previous studies had similar findings concerning the contribution of urbanization to degraded water quality [2], [43], [53–54]. Our findings also supported prior observations concerning more pollutants in urbanized areas with higher population density [54–55] and the potential occurrence of water quality pollution under rapid economic development [56–57]. Moreover, our study also found that the highest concentration of Cl^- in urban watershed, which indicated the potential risk of increasing salinity in urban areas [1], [57–58].

The percentage of cropland was significantly positively correlated with NO_3^-N and agricultural activities were identified in this study as an important predictor for NO_3^-N through the spatial regression model, which verified that cropland and agricultural activities were an important "source" for NO_3^-N..Kannel et al. (2007) also note that the increasing concentration of NO_3^-N with intensified use of chemical fertilizers together with urbanization [59]. In our study, the percentage of orchard was significantly positively correlated with COD_{Mn}, NO_3^-N, and Na^+. As one of China's most developed areas in terms of agricultural production, orchards are intensive especially in the downstream of the JRW, therefore, a large amount of animal waste and organic fertilizer used in the orchards, together with the rotten fruits, would make a contribution to the organic pollutants and NO_3^-N. Other studies in the JRW also found that large amount of nitrogen fertilizers are applied to crop which increase the source of nitrogen to stream [45–46]. The result from the spatial regression analysis in this study also suggested that agricultural activities were the important predictor for water quality variation especially Na^+ and Cl^- (Table 8). In this study, the percentage of forest had negative correlations with most of the water quality parameters, which was consistent with the previous findings that forest possesses the ecological function of filtering pollutants [29], [60].

In terms of natural factors, significantly positive relationships between Slope_std and NH_4^+-N, Cl^- and K^+ were found in this study. Slope represents an aggregation of rainfall, soil, geology, vegetation and temperature [12]. Similar observations were obtained by other researchers that slope variable act as a sink for some water quality parameters [29], [41–42]. In this study, significantly positive correlation was found between Geology 1 (sandstones and siltstones) and NH_4^+-N, SRP, Mg^{2+} as the result of weathering and dissolution of minerals from the local agricultural soil and bedrock [61]. On the other hand, Geology 2 (granites, lavas, and volcanic tuff) was negatively correlated with Mg^{2+} since the granitic bedrock is too hard to be weathered and the negative effect of pore solution by the volcanic tuff on Mg^{2+} release [62–63]. In this study, we found that Mg^{2+} concentration in natural watersheds was significantly higher than that in agricultural watersheds. Pearson analysis also showed that the correlation between concentration of Mg^{2+} and Geology 1 were stronger ($r = 0.684$, $p<0.01$) than with other factors (Table 5). Rothewell et al. (2010) had a similar observation that the Mg^{2+} concentration is related to underlying geology rather than the effect of human activities [12].

4.2. Spatial Dependence of Water Quality in Different Seasons

River water quality exhibits spatial autocorrelation, since adjacent sites are dominated by similar natural ecosystems and experience comparable human disturbances [42–43]. Our results showed that global Moran's I values for most of the water quality parameters (except for COD_{Mn} and Na^+) were lowest in the flood season, whereas most water quality parameters showed significant spatial autocorrelation in the dry season. In the flood season, nonpoint source pollution constituted the primary pollution source (suggested in subsection 4.1.2) and we inferred hydrological process, particularly agricultural and urban stormwater runoff containing different amounts of nutrients from different land cover types such as cropland and street, might make different contribution to water quality degradation in sub-watersheds with great spatial heterogeneity. Thus, it might result in the water quality problems becoming more localized. In contrast, river water quality in the dry season was mostly affected by point source pollution such as industrial and domestic effluents, suggesting that water quality is a reflection of regional anthropogenic activities or natural factors [42]. In addition, NH_4^+-N had the highest Moran's I value among all the water quality indicators, which was further verified the results from the spatial regression model, which indicated that NH_4^+-N was more associated with point source pollution, such as industrial effluents and domestic wastewater. In other words, watersheds with similar urbanization and socioeconomic development might show a similar NH_4^+-N pollution pattern. NH_4^+-N and COD_{Mn} are two important indicators which can to some extent reflect the local wastewater treatment capacity [2], [64], and improving the capacity for industrial and domestic wastewater treatment is important for regional water management in the context of accelerating urbanization. Comparatively, NO_3^-N had the lowest Moran's I value in our study, indicating that NO_3^-N pollution was localized. NO_3^-N was largely contributed by agricultural activities and natural factors including soil, geology and slope can also affect the pollutant export through hydrologic pathways such as runoff to a water body [29].

4.3. Seasonal Pattern and the Underlying Mechanisms

Most seasonal variations in river water chemistry are driven by climatic and biotic factors and are therefore largely governed by the watershed processes related to natural or human induced disturbances [29], [65]. The water soluble nutrients such as inorganic N may be leaching into the groundwater and transport into stream via subsurface water [66]. However, insoluble nutrient species such as P often as the form of particulates, thus transporting with sediments through the surface water pathways especially in the rainy day [67–68]. As a result, bioreactive elements can cycle differently and undergo release or retention from sediments in response to shifting redox conditions and warming. Thus, the source of nutrients can be variable, for example during flood season, high nitrate concentration in streamwater results from higher contribution of nitrate-rich hillslope shallow groundwater, whereas during dry season streamwater mainly comes from denitrified bottomland groundwater [69–70] and deep fractured aquifers [71].

In our study, the pronounced seasonal changes in NH_4^+-N, NO_3^-N, Cl^-, COD_{Mn}, Mg^{2+}, Na^+ and K^+ concentrations in the 20 headwater streams (Fig. 2 and Fig. 4), with highest concentrations in the dry and transition seasons and lowest concentrations in the flood season, suggests that river water quality dynamics in the JRW may be determined by the anthropogenic controls such as domestic wastewater & industrial effluent, and agricultural N input associated with vadose zone leaching and groundwater seepage, as well as lowest in-stream N immobilization rate due to lowest temperature in the dry and transition seasons. Besides, increased streamflow in the flood season may reduce the concentration if the source of nutrients is invariant, thus effective dilution effect associated hydrologic regime might lead to the lowest concentrations in the flood season in the JRW. Potential pollution sources identified by PCA analysis in subsection 4.1.2 could also interpret for the seasonal variations in river water quality. Point source pollution associated with industrial and domestic wastewater was identified as the important potential factor to make a great contribution to the variations in NH_4^+-N and COD_{Mn}, and agricultural activities were the primary predictor for NO_3^-N (Table 8).

In this study, SRP concentration is highest in the flood season (in Aug.), which might be related to more sediment-bound P input associated with hydrological processes such as stormwater runoff in this season. The SRP load exported from catchment non-point sources was associated with stormwater runoff and released from in-stream internal sources with the increasing river flow [2], [72]. As the result of that, the net effect of the amount of P entering and leaving an ecosystem as quantified by a mass balance provides a measure of P retention (e.g. during low stream flow) or P export (e.g. during high stream flow) over a given spatial and temporal scale [73]. It should be noted that other studies found SRP can be controlled by the biogeochemical process and anthropogenic input. Duan et al. (2012) notes the effect of temperature on seasonal change SRP based on the observation that high SRP concentrations exhibit in the low flow during summer as result of biochemical reactive [13]. Bowes et al. (2005) found that streams receiving wastewater effluent typically show a characteristic pattern of high P concentration during summer low flow and more diluted concentration during winter storm events in an English catchment [34]. Hydrological process rather than biogeochemical process drives the seasonal variations of P export in the JRW, which might be due to greater variability of inter-annual precipitation than of inter-annual temperature in the watershed studied.

4.4. Spatial Variations of Water Quality among three Types of Watershed

Land use and land cover play a central role in fate and transport of water quality [48], [74]. Runoff from agricultural land and effluents from urban and industrial area are major sources of nutrients and fine sediment in river systems [26], [75–76]. Urbanization and agricultural activities are common source of elevated water pollution, which is expected to continue increasing in the coming decades due to growing populations, further development and greater demand for food production [1], [13], [77]. Our study showed that the water in urban watersheds was more seriously polluted than in agricultural and natural watersheds, and the natural watersheds often had better water quality. This was not surprising due to the fact that the selected urban sub-watersheds had the highest urbanization level and there are many mining lots and steel plants located at the upstream of these watersheds. Acid mine drainage from active and abandoned strongly influences stream water chemistry by acidifying stream water which in turn increase the dissolution of mineral and high concentrations of several elements [78]. In addition, livestock farming was also intensive there. As a result, the water quality in the urban watersheds was influenced by a combination of point source and non-point source pollution associated with industrial, domestic and livestock waste. Additionally, most of the water quality parameters in agricultural watersheds had higher concentrations than those in natural watersheds, but no significant variance was obtained. This could be explained by the fact that the amount of natural watersheds is smaller than that of agricultural watersheds and natural sub-watershed, N2 was a special sub-watershed that had a high percentage of forest, but was seriously polluted by mining activities.

4.5. Implications for Water Quality Management

The findings of our study are helpful for water resource management within the watershed. Water quality was worst in the dry season and point source pollution associated with industrial and domestic wastewater made a great contribution to the variations of NH_4^+-N and COD_{Mn} in this season. Cl^- was relatively higher in urban watersheds than those in natural and agricultural watersheds, indicating the potential increasing salinization in urbanized areas. Moreover, the most significant spatial autocorrelation for most water quality parameters found in the transition and dry seasons, especially in the dry season. Thus, regional water management alternatives such as improving the capacity of wastewater treatment are more suitable than localized measures in the dry season. Comparatively, nonpoint source pollution associated with agricultural activities made the greatest contribution to the spatiotemporal variations in water quality in the flood and transition seasons. Water quality problems become more localized in the flood season due to the fact that agricultural nonpoint source pollution in different watersheds might make a different contribution to water quality degradation in sub-watersheds with great spatial heterogeneity. Thus, more attention should be paid to agricultural non-point source pollution and more localized water management measures should be taken in the flood season. Considering that urbanization and socioeconomic development and landscape pattern were the two dominant influencing factors for water quality variation, spatial planning for urbanization and socioeconomic development at watershed scale should be recognized by the local government when balancing economic growth and environmental conservation.

Conclusion

Anthropogenic input related to industrial effluents/domestic wastewater, agricultural activities associated with the precipitation-induced surface runoff, and natural weathering process were identified as the potential important factors to drive the seasonal variations in stream water quality in 20 headwater watershed of the JRW for the transition, flood and dry seasons, respectively. The NH_4^+-N, NO_3^-N, Cl^-, COD_{Mn}, Mg^{2+}, Na^+ and K^+ in the 20 headwater streams had the highest concentration concentrations in the dry and transition seasons and lowest concentrations in the flood season. SRP had the highest concentrations in the flood season as a result of hydrological control.

Anthropogenic activities and watershed characteristic led to the spatial variations in stream water quality in three types of watersheds. Anthropogenic input associated with industrial effluents/domestic wastewater led to the concentrations of NH_4^+-N, SRP, K^+, COD_{Mn}, and Cl^- were generally higher in urban watersheds than those in natural and agricultural watersheds. NO_3^-N were generally higher in agricultural watersheds than those in urban and natural watersheds as a result of anthropogenic input related to agricultural activities. Mg^{2+} concentration in natural watersheds was significantly higher than that in agricultural watersheds, which is largely due to the watershed characteristic (bedrock geology). Spatial autocorrelations analysis showed similar levels of water pollution between the neighboring sub-watersheds exhibited in the dry and transition seasons while non-point source pollution contributed to the significant water quality variations between neighboring sub-watersheds.

Spatial regression analysis showed anthropogenic controls played critical roles in variations in stream water quality in the JRW. Urbanization and socioeconomic development were the dominant explanatory variables for NH_4^+-N, SRP, COD_{Mn}, Mg^{2+}, Cl^- and K^+, and agricultural activities and landscape patterns had coupled effects on variations related to NO_3^-N and Na^+. Management implications were further discussed for water resource management. This research demonstrates that the coupled effects of natural and anthropogenic controls, together with underlying hydrologic and biogeochemical processes, contribute to the seasonal and spatial variation of headwater stream water quality in a coastal watershed with high spatial variability and intensive anthropogenic activities.

Acknowledgments

We thank Professor John Hodgkiss of The University of Hong Kong for assistance with English and anonymous reviewers supplied constructive feedback that helped to improve this paper.

Author Contributions

Conceived and designed the experiments: JlH YlH ZyZ. Performed the experiments: YlH ZyZ. Analyzed the data: JlH YlH. Contributed reagents/materials/analysis tools: JlH YlH. Wrote the paper: JlH YlH.

References

1. Mouri M, Takizawa S, Oki T (2011) Spatial and temporal variation in nutrient parameters in stream water in a rural-urban catchment, Shikoku, Japan: effects of land cover and human impact. Journal of Environmental Management 92: 1837–1848.
2. Huang JL, Li QS, Pontius RG, Klemas V, Hong HS (2013) Detecting the Dynamic Linkage between Landscape Characteristics and Water Quality in a Subtropical Coastal Watershed, Southeast China. Environmental Management 51: 32–44.
3. Ouyang Y, Nkedi-Kizza P, Wu QT, Shinde D, Huang CH (2006) Assessment of seasonal variations in surface water quality. Water Research 40(20): 3800–3810.
4. Somura H, Takeda I, Arnold JG, Mon Y, Jeong J, et al. (2012) Impact of suspended sediment and nutrient loading from land uses against water quality in the Hii River basin, Japan. Journal of Hydrology 450: 25–35.
5. Vorosmarty CJ, McIntyre PB, Gessner MO, Dudgenon D, Prusevich A, et al. (2010) Global threats to human water security and river biodiversity. Nature 467(7351): 555–561.
6. UN WWAP (World Water Assessment Programme) (2009) Water in a changing world. The United Nations World Water Development Report 3.
7. Baker A (2003) Land use and water quality. Hydrological process 17(12): 2499–2501.
8. Smith AJ, Thomas RL, Nolan JK, Velinsky DJ, Klein S, et al. (2013) Regional nutrient thresholds in wadeable streams of New York State protective of aquatic life. Ecological Indicators 29: 455–467.
9. Li SY, Xia XL, Tan X, Zhang QF (2013) Effects of catchment and riparian landscape setting on water quality chemistry and seasonal evolution of water quality in the upper Han River Basin, China. PLoS ONE e53163.
10. Pearl HW (2006) Assessing and managing nutrient-enhanced eutrophication in estuarine and coastal water: interactive effects of human and climate perturbations. Ecological Engineering 26: 40–54.
11. Kaushal SS, Groffman PM, Band LE, Shieds CA, Morgan RP, et al. (2008) Interaction between urbanization and climate variability amplifies watershed nitrate export in Maryland. Environmental Science and Technology 42(16): 5872–5878.
12. Rothwell JJ, Dise NB, Taylor KG, Allott TEH, Scholefield P, et al. (2010) A spatial and seasonal assessment of river water chemistry across North West England. Science of the Total Environment 408(4): 841–855.
13. Duan SW, Kaushal SS, Groffman PM, Band LE, Belt KT (2012) Phosphorus export across an urban to rural gradient in the Chesapeake Bay watershed. Journal of Geophysical Research 117: G01025.
14. Mitchell (2001) Linkages of nitrate losses in watersheds to hydrological process. Hydrological Processes 15(17): 3305–3307.
15. Martin C, Aquilina L, Gascuel-Odoux C, Molenat J, Faucheux M, et al. (2004) Seasonal and interannual variations of nitrate and chloride in stream waters related to spatial and temporal patterns of groundwater concentrations in agricultural catchment. Hydrological Process 18: 1237–1254.

16. Laurent J, Mazumder A (2014) Influence of seasonal and inter-annual hydrometeorological variability under varying land-use composition. Water Research 48: 170–178.
17. Worrall F, Burt TP (2001) Inter-annual controls on nitrate export from an agricultural catchment-how much land-use change is safe. Journal of Hydrology 243(3–4): 228–241.
18. Duan SW, Kaushal SS (2013) Warming increases carbon and nutrient fluxes from sediments in streams across land use. Biogeosciences 10: 1193–1207.
19. Randhir TO, Hawes AG (2009) Watershed land use and aquatic ecosystem response: Ecohydrologic approach to conservation policy. Journal of Hydrology 364(1–2): 182–199.
20. Swaney DP, Hong BT, Ti CP, Howarth RW, Humborg C (2012) Net anthropogenic nitrogen inputs to watersheds and riverine N export to coastal waters: a brief overview. Current Opinion in Environmental Sustainability 4(2): 203–211.
21. Gardner KK, McGlynn BL (2009) Seasonality in spatial variability and influence of land use/cover and watershed characteristics on stream water nitrate concentrations in a developing watershed in the Rockey Mountain West. Water Resource Research 45: W08411.
22. Huang JL, Klemas V (2012) Using Remote Sensing of Land Cover Changes in Coastal Watersheds to Predict Downstream Water Quality Trend. Journal of Coastal Research 28: 930–944.
23. Tu J (2013) Spatial variations in the relationships between land use and water quality across an urbanization gradient in the watersheds of Northern Georgia, USA. Environmental Management 51: 1–17.
24. Haygarth P, Turner BL, Fraser A, Jarvis S, Harrod T, et al. (2004) Temporal variability in phosphorus transfers: classifying concentration-discharge event dynamics. Hydrology and Earth System Sciences 8(1): 88–97.
25. Jordan P, Amscheidt A, McGrogan H, McCormick S (2007) Characterising phosphorus transfers in rural catchments using a continuous bank-side analyser. Hydrology and Earth System Sciences 11(1): 372–381.
26. Edwards AC, Whithers PJA (2008) Transport and delivery of suspended solid, nitrogen and phosphorus from various sources to freshwaters in the UK. Journal of Hydrology 350(3–4): 144–153.
27. Liu JG, Diamond (2005) China's environment in a globalizing world. Nature 435(7046): 1179–1186.
28. Weller DE, Jordan TE, Correl DL, Liu ZJ (2003) Effects of land-use change on nutrient discharge from the Patuxent River watershed. Estuaries 26: 244–266.
29. Sliva L, Williams DD (2001) Buffer zone versus whole catchment approaches to studying land use impact on river water quality. Water Research 35(14): 3462–3472.
30. Jarvie HP, Oguchi T, Neal C (2002) Exploring the linkages between river water chemistry and watershed characteristics use GIS-based catchment and locality analyses. Regional Environmental Change 3(1–3): 36–50.
31. Ko KS, Lee JS, Klim JG, Lee (2009) Assessments of natural and anthropogenic contròls on the spatial distribution of stream quality in Southeastern Korea. Geosciences Journal 13(2): 191–200.
32. Alexander RB, Smith RA, Schwarz GE (2000) Effect of stream channel size on the delivery of nitrogen to the Gulf of Mexico. Nature 403(6771): 756–761.
33. Peterson BJ, Wollheim WM, Mulbolland PJ, Webster JR, Meyer JL, et al. (2001) Control of nitrogen export from watershed by headwater streams. Science 292(5514): 86–90.
34. Bowes MJ, Hilton J, Irons GP, Hornby DD (2005) The relative contribution of sewage and diffuse phosphorus sources in the River Avon catchment, southern England: implications for nutrient management. Science of the Total Environment 344(1–3): 67–81.
35. Ministry of Environmental Protection (MEP) (2011) Report on the state of the environment In China 2009. Available: http://english.mep.gov.cn/standards_reports/soe/soe2009/201104/t20110411_208976.htm.
36. Vanlandeghem MM, Meyer MD, Cox SB, Sharma B, Patiño R (2012) Spatial and temporal patterns of surface water quality and ichthyotoxicity in urban and rural river basins in Texas. Water Research 46: 6638–6651.
37. Zhou F, Huang GH, Guo H, Zhang W, Hao Z (2007) Spatiotemporal patterns and source apportionment of coastal water pollution in eastern Hong Kong. Water Research 41(15): 3429–3439.
38. Yang X, Jin W (2010) GIS-based spatial regression and prediction of water quality in river networks: A case study in Iowa. Journal of Environmental Management 91: 1943–1951.
39. Fotheringham AS, Brunsdon C, Charlton M (2002) Geographically weighted regression: The analysis of spatially varying relationships. Wiley, Chichester.
40. Tu J, Xia ZG (2008) Examining spatially varying relationships between land use and water quality using geographically weighted regression I: Model design and evaluation. Science of Total Environment 407(1): 358–378.
41. Pratt B, Chang H (2012) Effects of land cover, topography, and built structure on seasonal water quality at multiple spatial scales. Journal of Hazardous Materials 209–210: 48–58.
42. Chang H (2008) Spatial analysis of water quality trends in the Han River basin, South Korea. Water Research 42: 3285–3304.
43. Su SL, Xiao R, Mi XY, Xu XY, Zhang ZH, et al. (2013) Spatial determinants of hazardous chemicals in surface water of Qiantang River, China. Ecological Indicators 24: 375–381.
44. Huang JL, Hong HS, Zhang LP (2004) Study on predicting soil erosion in Jiulong River watershed based on GIS and USLE. Journal of Soil and Water Conservation 18(5): 75–79. (in Chinese)
45. Cao WZ, Hong HS, Zhang YZ, Chen NW, Zeng Y, et al. (2006) Anthropogenic nitrogen sources and export in a village-scale catchment in Southeast China. Environmental Geochemistry and Health 28(1–2): 45–51.
46. Chen NW, Hong HS, Zhang LP, Cao WZ (2008) Nitrogen sources and exports in an agricultural watershed in Southeast China. Biogeochemistry 87(2): 169–179.
47. Huang JL, Lin J, Zhang YZ, Hong HS (2013) Analysis of Phosphorus Concentration in a Subtropical River Basin in China: Implications for River Basin Management. Ocean and Coastal Management 81: 29–37.
48. State Environmental Protection Administration of China (SEPAC) (2002) Methods of Monitoring and Analyzing for Water and Wastewater (4th Edition). Beijing: China Environmental Science Press. (In Chinese)
49. Redman CL (1999) Human dimensions of ecosystem studies. Ecosystems 2(4): 296–298.
50. Huang JL, Pontius RG Jr, Li QS, Zhang YJ (2012) Use of Intensity Analysis to Link Patterns with Processes of Land Change from 1986 to 2007 in a Coastal Watershed of Southeast China. Applied Geography 34: 371–384.
51. Sun R, Wang Z, Chen L, Wang W (2013) Assessment of surface water quality at large watershed scale: land-use, anthropogenic, and administrative impacts. Journal of the American Water Resources Association (JAWRA) 49(4): 741–752.
52. Ishizawa H, Stevens G (2007) Non-English language neighborhoods in Chicago, Illinois: 2000. Social Science Research 36: 1042–1064.
53. Lee SW, Hwang SJ, Hwang HS (2009) Landscape ecological approach to the relationships of land use patterns in watersheds to water quality characteristics. Landscape and Urban Planning 92: 80–89.
54. Carey RO, Migliaccio KW, Li Y, Schaffer B, Kiker GA, et al. (2011) Land use disturbance indicators and water quality variability in the Bicayne Bay Watershed, Floria. Ecological Indicator 11(5): 1093–1104.
55. Line DE (2013) Effect of development on water quality for seven streams in North Carolina Environmental Monitoring and Assessment 185: 6277–6289.
56. Su SL, Xiao R, Xu XY, Zhang ZH, Mi XY, et al. (2013) Multi-scale spatial determinants of dissolved oxygen and nutrients in Qiantang River, China. Regional Environmental Change 13(1): 77–89.
57. Brando PM, Coe MT, DeFries R, Azevedo AA (2013) Ecology, economy and management of an agroindustrial frontier landscape in the southeast Amazan. Philosophical Transactions of the Royal Society B-Biological Sciences DOI: 10.1098/rstb.2012.0152.
58. Porter-Goff ER, Frost PC, Xenopoulos MA (2013) Changes in riverine benthic diatom community structure along a chloride gradient. Ecological Indicators 32: 97–106.
59. Kannel PR, Lee S, Kanel SR, Khan SP, Lee YS (2007) Spatial-temporal variation and comparative assessment of water quality of urban river system: a case study of the River Bagmati (Nepal). Environmental Monitoring and Assessment 129(1–3): 433–459.
60. de Souza ALT, Fonseca DG, Liborio RA (2013) Influence of riparian vegetation and forest structure on the water chemistry with the ecological status of Irish rivers. Water Research 298: 12–18.
61. Kney AD, Brandes D (2007) A graphical screening method for assessing stream water quality using specific conductive and alkalinity data. Journal of Environmental Management 82: 519–528.
62. Jones DP, Graham RC (1993) Water-holding characteristics of weathered granitic rock in chaparral and forest ecosystems. Soil Science Society of America Journal 57: 256–261.
63. Dultz S, Simonyan AV, Pastrana J, Behrens H, Plotze M, et al. (2013) Implications of pore space characteristics on diffusive transport in basalts and granites. Environmental Earth Science 69: 969–985.
64. Ahearn DS, Sheibley RW, Dahlgren RA, Anderson M, Johnson J, et al. (2005) Land use and land cover influence on water quality in the last free-flowing river draining the western sierra Nevada, California. Journal of Hydrology 313(3–4): 234–247.
65. Duan S, Kaushal SS, Groffman PM, Band LE, Belt KT (2012) Phosphorus export across an urban to rural gradient in the Chesapeake Bay watershed. Journal of Geophysical Research: Biogeosciences 117(G01025).
66. Petry J, Malcolm IA, Youngson AF (2002) Hydrological controls on nutrient concentrations and fluxes in agricultural catchments. Science of the Total Environment 294(1–3): 95–110.
67. Willett VB, Reynolds BA, Stevens PA, Ormerod SJ, Jones DL (2004) Dissolved organic nitrogen regulation in freshwaters. Journal of Environmental Quality 33(1): 201–209.
68. Chang H, Kwon WT (2007) Spatial variations of summer precipitation trends in South Korea, 1973–2005. Environmental Research Letters 2(4): 045012.
69. Altman SJ, Parizek RR (1995) Dilution of nonpoint-source nitrate in groundwater. Journal of Environmental Quality 24(4): 707–718.
70. Hill AR (1996) Nitrate removal in stream riparian zones. Journal of Environmental Quality 25(4): 743–755.
71. Pauwels H, Lachassagne P, Bordenave P, Foucher JC, Martelat A (2001) Temporal variability of nitrate concentration in a schist aquifer and transfer to surface waters. Applied Geochemistry 16(6): 583–596.
72. May L, House WA, Bowes M, McEvoy J (2001) Seasonal export of phosphorus from a lowland catchment: upper River Cherwell in Oxfordshire, England. Science of the Total Environment 269(1–3): 117–130.
73. House WA (2003) Geochemical cycling of phosphorus in rivers. Applied Geochemistry 18(5): 739–748.

74. Yin XM, Lu XG, Xue ZS, Liu ZM (2012) Influence of land use change on water quality in Naoli River watershed, northeast China. Journal of Food Agriculture and Environment 3–4: 1214–1218.
75. Jarvie HP, Neal C, Withers PJA (2006) Sewage-effluent phosphorus: A greater risk to river eutrophication than agricultural phosphorus? Science of the Total Environment 360(1–3): 246–253.
76. Neal EG, Hood E, Smikrud K (2010) Contribution of glacier runoff to freshwater discharge into the Gulf of Alaska. Geophysical Research Letters 37: L06404.
77. Carpenter SR, Caraco NF, Correll DL, Howarth RW, Sharpley AN, et al. (1998) Nonpoint pollution of surface waters with phosphorus and nitrogen. Ecological applications 8(3): 559–568.
78. Pasieczna A, Lis J (2008) Environmental geochemical mapping of the Olkusz 1:25000 scale map sheet, Silesia-Cracow region, southern Poland. Geochemistry-Exploration Environment Analysis 8: 323–331.

Spatial Variation as a Tool for Inferring Temporal Variation and Diagnosing Types of Mechanisms in Ecosystems

Matthew P. Hammond*, Jurek Kolasa

Department of Biology, McMaster University, Hamilton, Ontario, Canada

Abstract

Ecological processes, like the rise and fall of populations, leave an imprint of their dynamics as a pattern in space. Mining this spatial record for insight into temporal change underlies many applications, including using spatial snapshots to infer trends in communities, rates of species spread across boundaries, likelihood of chaotic dynamics, and proximity to regime shifts. However, these approaches rely on an inherent but undefined link between spatial and temporal variation. We present a quantitative link between a variable's spatial and temporal variation based on established variance-partitioning techniques, and test it for predictive and diagnostic applications. A strong link existed between spatial and regional temporal variation (estimated as Coefficients of Variation or CV's) in 136 variables from three aquatic ecosystems. This association suggests a basis for substituting one for the other, either quantitatively or qualitatively, when long time series are lacking. We further show that weak substitution of temporal for spatial CV results from distortion by specific spatiotemporal patterns (e.g., inter-patch synchrony). Where spatial and temporal CV's do not match, we pinpoint the spatiotemporal causes of deviation in the dynamics of variables and suggest ways that may control for them. In turn, we demonstrate the use of this framework for describing spatiotemporal patterns in multiple ecosystem variables and attributing them to types of mechanisms. Linking spatial and temporal variability makes quantitative the hitherto inexact practice of space-for-time substitution and may thus point to new opportunities for navigating the complex variation of ecosystems.

Editor: Bazartseren Boldgiv, National University of Mongolia, Mongolia

Funding: This study was supported by Natural Sciences and Engineering Research Council of Canada (NSERC - http://www.nserc-crsng.gc.ca) grant 5-31314-6100. The funders had no role in study design, data collection and analysis, decision to publish or preparation of the manuscript.

Competing Interests: The authors have declared that no competing interests exist.

* E-mail: mhammond@gmx.com

Introduction

The spatial texture of a landscape is a fundamental reflection of the ecological processes underpinning it. Thus, spatial distributions of population [1], geological [2] and climatological variables [3] can impart key details about the forces, operating over time, that forged them. Spatial patterns are *diagnostic* when they are used to uncover hidden mechanisms in the landscape, and *predictive* when they indicate the likely future behavior of a process. Ecology is full of examples of the former, diagnostic approach where spatial patterns are mined for evidence of mechanisms like dispersal, competition or environmental structuring [4–7]. But the latter, predictive approach is also commonplace. Because obtaining long time series is difficult, inferring temporal patterns from spatial data is used in such varied contexts as: (i) chronosequences, where gradients of different-aged sites are used to track how a process (e.g., succession) changes from one state to another over time [8–11], (ii) boundary dynamics, where spatial snapshots can reveal the rate of species spread [12], (iii) complex dynamics, where spatial data helps detect chaos [13], and (iv) regime or phase shifts, where changes in spatial variation can expose the incipient reorganization of an ecosystem [14–16].

Using spatial patterns to infer temporal patterns ("space-for-time substitution"; *sensu* [8]) or mechanisms quickly encounters the hard problem of interpreting dynamics [17,18]. We, as others [18,19], note that progress will require a deep understanding of what spatial patterns reveal about temporal patterns, and how these come together in the spatiotemporal patterns of landscapes. We further suggest that such insight will help both *predictive* and *diagnostic* efforts.

Here we focus on the variability of values over time (i.e., the inverse of stability [20] or constancy [21]) as opposed to properties like resistance or resilience [21]. We focus, therefore, on the dissimilarity of values and whether this variability (e.g., fluctuations in density) can be predicted from a snapshot of spatial variability. This application, in particular, would be useful given the rarity of long time series in ecology [9], the widespread nature of anthropogenic impacts [22], the increasing attention to ecological stability [20,23–26], and the need for clearer links between spatial and temporal concepts [27].

To our knowledge, no links explicitly tie the temporal and spatial variation of a variable. However, a theoretical foundation for doing so is available through ANOVA variance partitioning [28–30] because overall spatiotemporal variation can be broken down into its spatial and temporal components. Crucially, these components can be re-expressed in terms of temporal variance at the regional scale (i.e., spatially-aggregated at time *k*; given as

Var(Y) where $Y_k = \sum_{i=1}^{n} X_{ik}$). This scale reflects the net sum of what occurs in all patches and thus reflects variation in resources and ecological functions at the landscape level. Our derivation makes regional temporal variance itself decomposable (Fig. 1A). Traditional variance partitioning methods, in contrast, only decompose total spatiotemporal variance. We show in Fig. 1A that regional temporal variance (Var(Y)) obeys a simple relationship with a spatial quantity - the sum of spatial variances measured at time k ($\sum \mathrm{var}(X_k)$). This relationship, in turn, is modified by two spatiotemporal patterns [28–31], *inter-patch synchrony* and *persistence of spatial variation* (Figs. 1 & 2; see File S1 for derivation and details).

Inter-patch synchrony (summed inter-patch covariances; $\sum \mathrm{cov}(X_i,X_j)$) refers to temporal changes that happen simultaneously in patches i and j. It is well known to boost temporal variation at the regional scale [32] (e.g., widespread population decline during drought). On the other hand, *persistence of spatial variation* (summed inter-time covariances; $\sum \mathrm{cov}(X_k,X_l)$) - *persistence* for short - describes differences or gradients between patches i and j that are retained from time k to l (e.g., fixed or permanent differences between locations). Opposite in sign to synchrony, persistence is associated with lower temporal variance. This is because a pattern of spatial variation is retained over time only if most patches are relatively stable. Temporal variance thus depends critically on the balance of synchrony and persistence.

Implications of Analytical Framework

The relationship captured in Fig. 1A points out the basis for *predictive* applications like space-for-time substitution. It does so by showing that spatial and temporal variance will scale exactly (by a factor of n_i/n_k) for stochastic processes. Stochastic processes enable

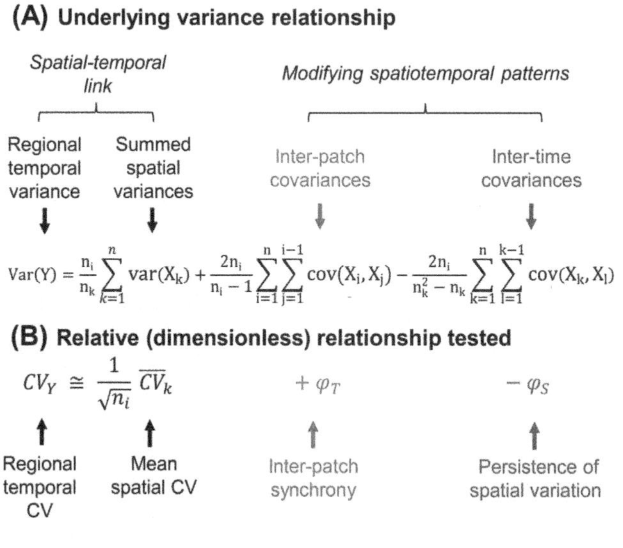

Figure 1. The spatial-temporal variability link. (A) We derived an analytical relationship linking regional temporal variance of a process (Var(Y)) to summed spatial variances at time k ($\sum \mathrm{var}(X_k)$). Inter-patch synchrony ($\sum \mathrm{cov}(X_i,X_j)$) and persistence ($\sum \mathrm{cov}(X_k,X_l)$) modify this link and lead temporal and spatial variance to scale as a function of number of patches (n_i) and time points (n_k) when these terms are zero. (**B**) We evaluate the usefulness, for prediction and description, of the corresponding (relative) relationship that uses dimensionless coefficients: Regional temporal CV (CV_Y), mean spatial CV ($\overline{CV_k}$), and indices of synchrony (φ_T) and persistence (φ_S). While an exact solution exists (Eq. S28, File S1), we use a more useful approximation, $CV_Y \cong (1/n_i^{1/2})\overline{CV_k}$, that gives an expected temporal CV for a given spatial CV when synchrony and persistence are negligible (Eq. S31, File S1).

this because their values are uncorrelated between patches i and j, as well as between times k and l, and this sets synchrony and persistence terms to zero ($\sum \mathrm{cov}(X_i,X_j) = 0$, $\sum \mathrm{cov}(X_k,X_l) = 0$). This is a form of ergodicity [33] that can be illustrated by an analogy. Imagine a seascape in which wave peaks are independent of each other: In this null case, wave amplitudes from trough to peak would be equally large or small whether waves were measured from a fixed point (i.e., waves passing over time) or from a transect across the seascape (i.e., a snapshot of waves across space). Our formulation merely adds that this match between temporal and spatial variability applies at the regional (seascape) scale as well as at the patch (wave) scale. Fig. 2 summarizes this mechanism, showing how temporal fluctuations are recorded as spatial variability.

Diagnosis, where inferences are made about how patterns came about, may also be made possible by the analytical solution. This is because components of temporal variance from Fig. 1A also describe and summarize spatiotemporal patterns that are the net result of ecological mechanisms. Moreover, because these terms are linked to temporal variability, they may provide a new view of dynamics and their consequences for stability.

Because they are commonly used in ecology, we extended our analytical framework to include common indices (Fig. 1B) like the Coefficient of Variation (CV), and indices of synchrony (φ_T) and persistence (φ_S). We test the validity of these formulations and turn them to answering three questions: (i) Is the spatial CV of a landscape variable a meaningful proxy for its (regional) temporal CV? (ii) Under what conditions is it predictive? And (iii) what do departures from an exact match between spatial and temporal CV tell us about the forces shaping dynamics of variables? We apply our approach to 136 biotic and physicochemical variables from three landscape types: Laboratory arrays of connected aquatic microcosms (measured for 20 weeks), a natural array of Jamaican coastal rock pool ecosystems (13 years), and a set of seven lakes from the North Temperate Lakes LTER site (30 years). Results shed light on what real world inferences can be drawn when the relationship between spatial and temporal variation is known.

Materials and Methods

Ethics Statement

Invertebrate species were sampled with permission on land owned by University of West Indies (Discovery Bay Marine Lab) and are not protected by law. Laboratory experiments used invertebrate species that do not require permits or procedural approvals.

Analytical Relationship: Linking Spatial and Regional Temporal Variance

Values in a landscape vary over time ($k\ldots n$), and across patches ($i\ldots n$). These dimensions of variation both contribute to regional temporal variance, which is the variance of the spatially-aggregated time series (i.e., Var(Y) where $Y_k = \sum_{i=1}^{n} X_{ik}$). Spatial and temporal variation can be precisely linked through two mathematical truisms: (i) Spatial and temporal variances, estimated from the same site × time data matrix, are related by rules that underlie ANOVA variance partitioning and (ii) these variances, which capture variation at the aggregate scale for both time (i.e., Var(Y)) and for space (i.e., temporally-aggregated; Var(Z) where $Z_i = \sum_{k=1}^{n} X_{ik}$), can be further decomposed into variances and covariances of patches i and j or time points k and l [34]. See File S1 for derivation. Var(Y) can thus be re-expressed as in Fig. 1A where; $\mathrm{var}(X_k)$ is the spatial variance at time k, $\mathrm{cov}(X_i,X_j)$ is the covariance of patch i with j (synchrony), and $\mathrm{cov}(X_k,X_l)$ is the

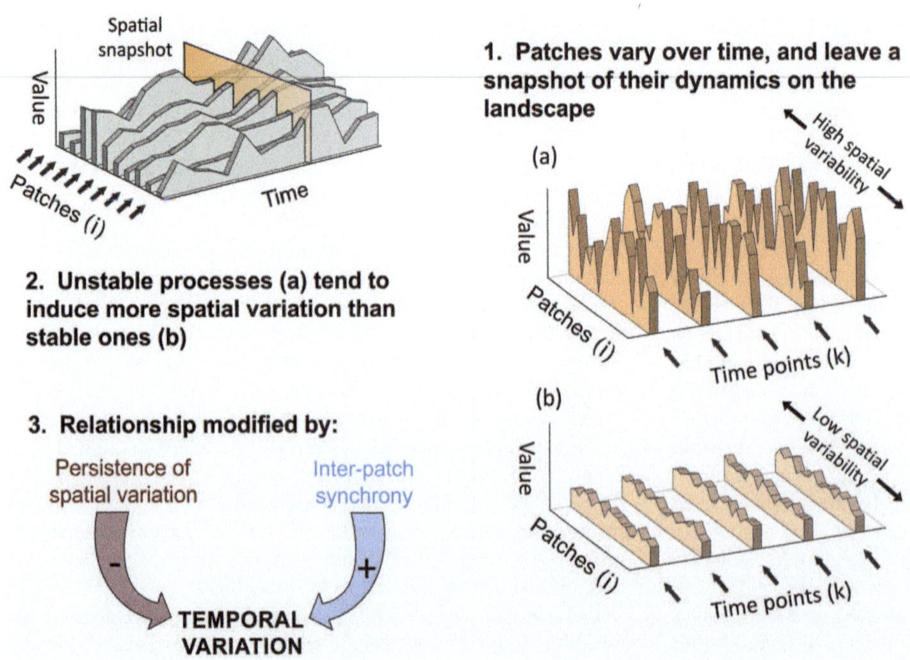

Figure 2. Spatial imprinting of ecosystem processes. Theorized mechanism by which temporal fluctuations of patches create spatial variability in the landscape, which may in turn be a proxy for temporal variability. Spatiotemporal patterns (inter-patch synchrony and persistence) modify the correspondence of spatial and temporal variability (Fig. 1), so it is unknown whether the link is strong enough for predictive (e.g., space-for-time substitution) applications and whether modifying terms have diagnostic/descriptive value.

covariance of time *k* with *l* (persistence of spatial variation). These three components of spatiotemporal pattern are consistent with prior theory and statistical concepts [28–31].

We converted the above analytical relationship into dimensionless quantities (Eq. S28; File S1) – regional temporal CV (CV_Y), spatial CV (CV_k), and indices of synchrony (φ_T) and persistence (φ_S). While exact, this relationship does not yield a clear null relationship between CV_Y and spatial CV. We therefore used an approximation of it (Eq. S31; File S1) that gives the expected value of temporal CV from spatial CV in the absence of synchrony or persistence (Fig. 1B). Temporal CV values calculated using this approximation were 94–98% correlated (1:1) with values from random number simulations where synchrony and persistence were close to zero. These null values, in turn, were used to plot the lines of "independent dynamics" shown in spatial-temporal CV plots (Figs. 3–5).

Data Analysis

Biotic variables included population densities of invertebrate and fish species and ecosystem-level quantities like NPP, while physicochemical variables ranged from temperature and pH to ion concentrations (Table S1, S2). For each variable, we estimated all indices in Fig. 1B. Regional temporal CV was estimated as the quotient of the time series standard deviation and mean. Mean spatial CV of a variable was defined as the average of spatial CV's measured at time *k*. These were calculated either across the three microcosms of each experimental replicate, across 49 Jamaican rock pools, or across the seven LTER lakes. If a species was absent from all water bodies spatial CV could not be calculated for that time point. In these cases, mean spatial CV was calculated as an average of the time points in which it was present (mean frequency of occurrence = 65% of years).

We estimated inter-patch synchrony using a variance ratio φ_T [35] which is, roughly speaking, a ratio of aggregate (regional) to component variances:

$$\varphi_T = \frac{\mathrm{Var(Y)}}{(\sum \mathrm{SD}(X_i))^2} \tag{1}$$

where $\mathrm{SD}(X_i)$ is the standard deviation of a patch. As patches synchronize, the value of φ_T grows from zero to one. The counterpart of inter-patch synchrony is persistence where, instead of temporal changes being similar from patch *i* to *j*, spatial differences are similar from time *k* to time *l*. We therefore estimated persistence with the spatial counterpart of φ_T, which we call φ_S. φ_S was calculated by replacing Var(Y) and $\mathrm{SD}(X_i)$ in Eq. 1 with their spatial equivalents; the variance of the temporally-aggregated series (i.e., Var(Z)) and spatial standard deviation at time *k* ($\mathrm{SD}(X_k)$). Analogous to φ_T, φ_S values increase from zero to one as differences among patches persist more through time.

We used General Linear Models (GLM) and multiple regression in Statistica v. 8.0 (StatSoft Inc., 2007) to predict regional temporal CV from mean spatial CV, φ_T, and φ_S (i.e., ~Fig. 1B). Temporal and spatial CV values were log transformed for analysis because, when plotted, they tended to form fan-shaped data clouds that were best described by power functions. We tested residuals of all analyses for normal distribution using Kolmogorov-Smirnov tests. Surfaces (Fig. 5) were fitted by distance-weighted least squares.

Microcosm Connectivity Experiment

We assembled replicate arrays of three ×700 mL aquatic microcosms. Each array contained community types that were relatively stable under laboratory conditions: (i) impoverished, containing ubiquitous microbes initially surviving in distilled water, (ii) phytoplankton and microbes, and (iii) 10 invertebrate species (cladocerans, ostracods) and phytoplankton and microbes. We arranged microcosms such that each initially contained a

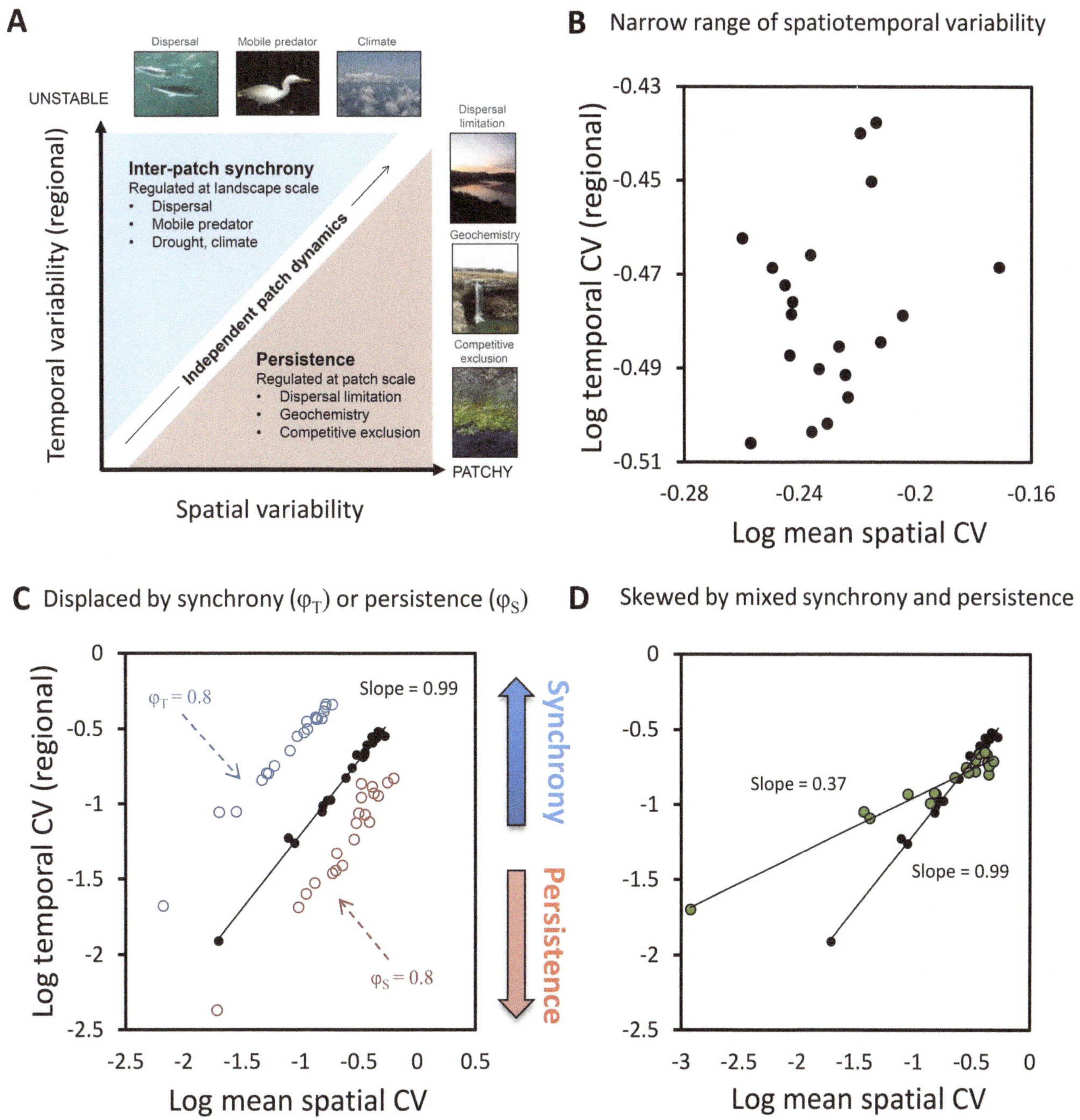

Figure 3. Anatomy of a plot between spatial and regional temporal variability. A stochastic null model of a three patch mosaic illustrated several features of a plot between log mean spatial CV and log regional temporal CV. (**A**) Three regions exist in which a variable (point) can fall - an "independent dynamics region" when values are independent between patches *i* and *j* and time points *k* and *l*, a "synchrony region" when inter-patch synchrony boosts temporal CV, and a "persistence region" when spatial gradients are retained over time; (**B**) Weak linear relationship when variables share similar spatiotemporal variability, leading to scatter from small variations in synchrony or persistence; (**C**) Strong linear relationships when variables differ in spatiotemporal variabililty and occupy the "independent dynamics region" (black circles), but also when all variables are equally dispaced by synchrony (blue circles) or by persistence (red circles); (**D**) Deviation of regression slope from ~1 (black circles) when variables change in synchrony or persistence as a function of variability. Here, a gradient exists from variables with low variability and synchrony to variables with high variability and persistence. Spatial CV values are means of spatial CV measured at time point *k*. Each point represents a variable and is a mean of ten replicates.

distinct community type, with all three types represented in an array.

Spatial exchange was manipulated by connecting component microcosms with clear Tygon® tubes. Treatments were: No connection among microcosms and bi-directional connection among all three microcosms. Connector tube diameters were increased by ~70% at week 10 of the 20-week experiment. Seven ecosystem-level variables were measured weekly in each micro-

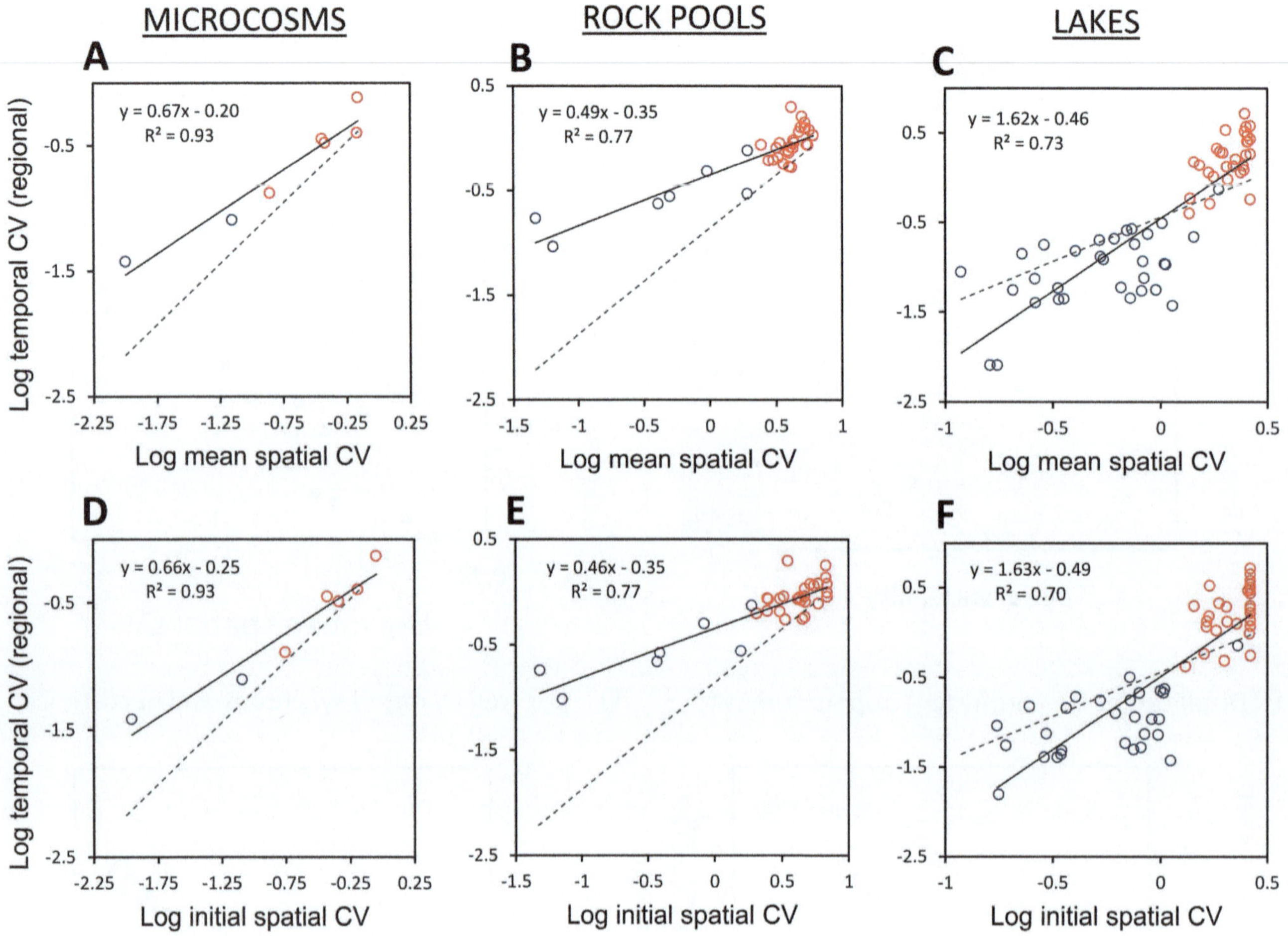

Figure 4. Empirical CV plots illustrating an underlying spatial-temporal link (Fig. 1B). The regional temporal CV of an ecosystem variable (data point) was predictable from its spatial CV in microcosm (n = 7) (**A, D**), rock pool (n = 33) (**B, E**), and lake systems (n = 60) (**C, F**). The predictive value of spatial variability was consistent in that linear associations emerged whether spatial variability was estimated as the mean of spatial CV's at time *k* (**A–C**), or whether a spatial CV from an initial time point (*k*) was used to predict temporal CV of the remaining (*k+1…n*) time series (**D–F**). Dashed lines denote the relationship expected for stochastic processes i.e., when values are independent across space and time. These were obtained by simulating random numbers with the same data structure as empirical data sets. Abiotic variables (blue circles) were consistently more stable and less spatially patchy than biotic variables (red circles).

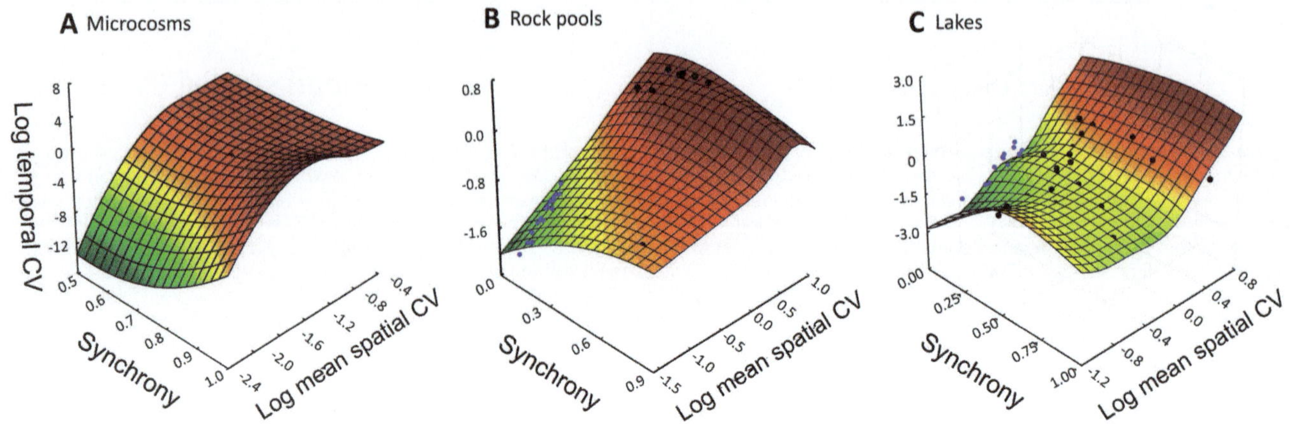

Figure 5. The modifying role of inter-patch synchrony. Relationship between spatial CV and regional temporal CV as modified by the degree of inter-patch synchrony in (**A**) microcosm, (**B**) rock pools and (**C**) lakes. Synchrony increased regional temporal CV relatively little over that explained by spatial CV. Black points = empirical variables, blue points = simulated, randomly-generated variables (n = 20; see File S1) to represent "independent dynamics region."

cosm of the array for 20 weeks using light-dark bottle methods, chlorophyll extractions and environmental sensor probes (Table S1). Microcosm NPP data were rescaled, bringing the lowest value to zero to correct spatial and temporal CV's for negative values.

Natural Rock Pool Ecosystem

We collected data over thirteen annual surveys (1989–2002) in a Jamaican rock-pool system of 78 invertebrate species, dispersing among 49 rock pools. The system lies near Discovery Bay Marine Laboratory, University of the West Indies, on the northern coast of Jamaica (18°28′ N, 77°25′ W). Pools create a mosaic 25 m in radius on a fossil reef no further than 10 m from the ocean and have volumes ranging from 0.5 to 78.4 L. Pools are, on average, within 1 m of the nearest neighbor and never more than 5 m away. Ocean tides occasionally flood a few of the most seaward pools. But most are refilled only by precipitation or, on some occasions, ocean spray. We treated the 49 pools as a single system linked by material fluxes and organism dispersal.

The 70+ invertebrate species in rock pools disperse predominantly by propagules transported via wind, ocean spray, animal vectors and, very occasionally, by overflow from neighboring pools after heavy rainfall [36]. Invertebrate species include: Ostracods (20 species), copepods (five species), cladocerans (five species), worms (15 species), aquatic insects (18 species) and other crustaceans (six species). Most species occurred rarely, some only once (for more details, see [37]). We therefore confined all analyses, except for contributions of variance components (Fig. 6) to 26 common species and temperature, pH, salinity, dissolved oxygen, oxygen saturation, and chlorophyll-a.

Invertebrate densities were estimated for each pool as the number of animals in a 0.5 L sample of water, which was withdrawn after stirring the pool to dislodge organisms from rock walls and to homogenize contents. Each sample was filtered through 63 μm mesh to isolate invertebrates, which were immediately preserved in 50% ethanol. Community samples were sorted, identified to highest possible taxonomic resolution and counted by microscope.

Environmental variables like salinity and pH (Table S2) were measured in each pool using multiprobe sondes (DataSonde, Yellow Springs Instruments, Yellow Springs, Ohio, USA or Hydrolab, Austin, Texas, USA) during biotic surveys for 6–11 of the survey years, depending on the variable.

Small rock pools occasionally dried up, preventing community sampling. These events were recorded as blank data entries, and were <10% of total observations. For our main analyses (Figs. 4–5), we replaced blank entries with zeroes, assuming that a desiccated pool harbored no living, adult invertebrates. To check if this assumption introduced bias, analyses were repeated using two alternative procedures; (i) leaving blank cells unchanged or (ii) interpolating by replacing blanks with the pool mean. All procedures produced similarly significant results indicating no major effect of our assumption. For abiotic variables, blank entries had no logical association with zero (e.g., desiccation does not suggest a 0°C temperature), so these cells were left blank.

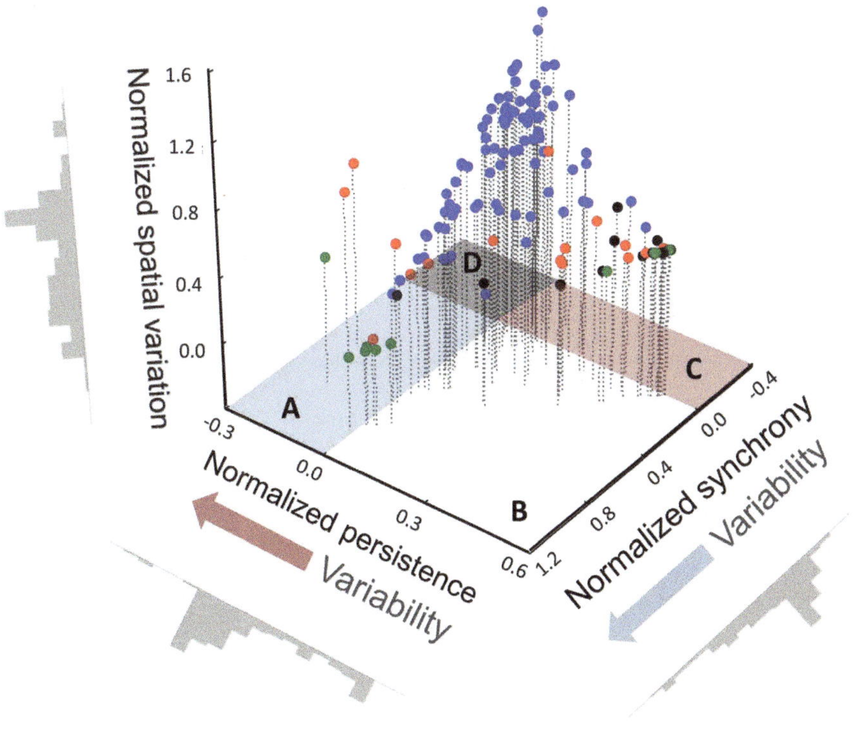

Figure 6. Patterns of spatiotemporal variation underlying the temporal variability of ecosystem variables. The interplay of the three components of temporal variance - spatial variation, synchrony and persistence - was captured by plotting the normalized values of each term in Fig. 1A against each other. Values of each term were standardized to the sum of all three terms such that the resulting proportions summed to one. Variables were assigned to *a priori* groupings based on their likely genesis and mode of regulation, where blue points = species populations, green = atmospheric, red = non-population biotic, black = watershed. n = 136, and includes an additional 36 rare rock pool species that were excluded from earlier analyses due to sparseness of data. Points scatter across theorized modes of dynamics described in Table 1: A = destabilized by synchrony, B = stabilized by persistence, destabilized by synchrony, C = stabilized by persistence, D = stabilized by compensatory dynamics, Intersection of A-D = stabilized by asynchrony. Gray histograms show frequency distributions for each component of temporal variance.

Table 1. Theorized modes of dynamics in landscape variables, their effect on regional temporal variation, and ecological examples.

Graph region	Synchrony/persistence	Dominant effect on temporal variation	Ecological scenario
A	High/Low	Destabilized by synchrony	Local factors less important, landscape-scale factors synchronize dynamics. E.g., synchrony of isolated mammal populations via weather [63]
B	High/High	Stabilized by persistence and destabilized by synchrony	Local factors establish permanent spatial gradients, landscape-scale factors synchronize dynamics. E.g., synchrony of source-sink fish populations via dispersal [64]
C	Low/High	Stabilized by persistence	Local factors establish permanent spatial gradients, dynamics differ among sites. E.g., stable spatial distributions of organisms across habitats [65]
D	Low/Low	Stabilized by compensatory dynamics	Dynamics negatively correlated from time to time, site to site. E.g., spatiotemporal refugia of competing soil nematodes [60]
Intersection of A,B,C,D	Zero/Zero	Stabilized by asynchrony	Dynamics appear stochastic, independent from time to time, site to site. E.g., settlement of broadcast oceanic larvae [66]

Modes reflect different mixtures of inter-patch synchrony and persistence, and correspond to regions of Fig. 6, where variables from three natural ecosystems are plotted by their spatiotemporal patterns.

North Temperate Lakes Long-Term Ecosystem Research Program

We used data from seven Wisconsin lakes (Allequash, Big Muskellunge, Crystal, Sparkling, Trout, Crystal Bog and Trout Bog), collected by the North Temperate Lakes Long-Term Ecological Research program. Lake data were obtained from a public database hosted by the North Temperate Lakes LTER, NSF, Center for Limnology, University of Wisconsin-Madison, available at http://lter.limnology.wisc.edu. We included up to 30 years of data from 60 biotic and abiotic variables across five datasets (Table S2). The following datasets, collected and maintained by LTER associates, were used:

- Chemical limnology of primary study lakes: Major ions
- Chemical limnology of primary study lakes: Nutrients, pH and carbon
- Physical limnology of primary study lakes
- Pelagic macroinvertebrate abundance
- Fish abundance.

Collection methods corresponding to datasets can be found as metadata on the online database (http://lter.limnology.wisc.edu). For fish data, only fyke net catches were used and were standardized by effort (i.e., catch per unit effort) to facilitate comparison. Values for a given lake were annual, obtained by averaging organism densities or physicochemical values across depths, across sampling dates and across stations. Density data were used to equalize the contribution from each lake because Trout Lake is up to 3200 times larger than other lakes, and therefore dominates the landscape spatiotemporal pattern. Results therefore emphasize patterns owing to ecological differences among lakes, rather than to lake size.

Data are available from the LTER database or upon request from the authors.

Results

Three quantities jointly accounted for 87–100% of a variable's regional temporal CV (Table S3) - mean spatial CV, inter-patch synchrony φ_T, and persistence φ_S. This result verifies that spatial and temporal CV's are related and substitutable to the degree that synchrony or persistence does not interfere. Perfect substitution occurs when values are uncorrelated between patches i and j and between times k and l. In this case, regional temporal CV is roughly $1/n_i^{1/2}$ times the mean spatial CV (Fig. 1B; Eq. S31). A null, stochastic model (see File S1) showed that such a variable (data point) lies within an "independent dynamics region" on a plot of spatial and temporal CV (Fig. 3A). It falls on a line of slope $\sim 1/n_i^{1/2}$ or ~ 1 on a log-scale (Fig. 3C). Inter-patch synchrony (e.g., from climatic forcing) increases regional temporal variability, shifting a variable to an upper "synchrony region" of the plot (Fig. 3A). But the data point shifts to a lower, "persistence region" (Fig. 3A) when spatial gradients are retained over time (e.g., from patch-specific factors).

A range of plot patterns can emerge depending on the variables sampled and their spatiotemporal dynamics. First, weak or no linear association exists when variables' spatial and temporal CV's span only a narrow range and stochastic differences in synchrony and persistence create scatter (Fig. 3B). Second, strong linear association emerges if all landscape variables have stochastic behavior (black circles, Fig. 3C), similar degrees of inter-patch synchrony (blue circles, Fig. 3C) or similar degrees of persistence (red circles, Fig. 3C). The latter two cases occur because all points are equally displaced up or down from the independent dynamics region. Finally, regression slopes deviate from the expected slope of 1 when some variables display synchrony and others persistence to produce skew (Fig. 3D).

Significant linear regressions existed between spatial CV and regional temporal CV for real ecosystem variables (Fig. 4). In all three data sets, the most stable variables (e.g., hydrological and environmental variables) had low spatial variability, while unstable variables (e.g., species populations) were spatially patchy in the landscape (Fig. 4A–C). Some statistical dependence exists in these plots because mean spatial CV and regional temporal CV are calculated from the same data matrix. However, results were almost perfectly conserved when a single sampling event (k) served to estimate spatial CV and predict subsequent (k+$1\ldots n$) temporal CV (Fig. 4D–F). Results are also unlikely to have arisen from biased estimators, since trends were confirmed using several alternative indices (Table S3).

Not all variables lay close to the independent dynamics region. This was reflected in regression slopes and intercepts (solid lines, Fig. 4) which departed from the independent dynamics case (dashed lines, Fig. 4). For instance, regression slopes for microcosms and rock pools were reduced because physicochemical variables (blue circles) at one end of the data cloud exhibited more inter-patch synchrony (Fig. 4A,B,D,E). Meanwhile, lake physicochemical variables showed considerable persistence, lowering temporal variability and steepening the slope (Fig. 4C,F). Though slopes and intercepts deviated from the independent dynamics region, r^2 values of 0.70–0.93 suggest that the rank order of variable's CV's was preserved. We do note some outliers, however, such as pH, Mg and Ca in the persistence region of Fig. 4C, F. Also, some variable types (e.g., species populations; red circles) had greater scatter. Overall, however, synchrony and persistence interfered relatively little with the scaling of spatial and temporal CV. Multiple-regression beta coefficients, for instance, revealed that temporal CV increased 2.9–6.2 times more with a variable's spatial CV than with its synchrony (Fig. 5).

Variables differed in the relative importance of spatial variability, synchrony or persistence to their landscape dynamics. We generated a fingerprint of these dynamics by normalizing the right hand terms of the variance equation (Fig. 1A) for each variable and then plotting them (Fig. 6). A division existed between biotic and physicochemical dynamics, and variables spread across several regions of the plot (quadrants A-D) corresponding to different spatiotemporal behaviors leading to temporal variability. Table 1 synthesizes results by describing these modes of behavior and how they may arise in nature.

Species populations (blue points) clumped together, their dynamics dominated by spatial variability with little synchrony or persistence (Fig. 6; intersection of quadrants A-D). Meanwhile, most atmospherically-driven processes (green points e.g., temperature, dissolved oxygen) were set apart by little persistence but were destabilized by synchrony (quadrant A). In contrast, many watershed-associated variables (black points e.g., ion concentrations) were characterized by persistence but little synchrony (quadrant B), a combination leading to lower temporal variability at the regional scale.

Discussion

Spatial Signatures of Temporal Variability

Variables from three aquatic ecosystems showed a striking and tight correspondence between their regional temporal CV and mean spatial CV. This trend may be considered predictive because it held even when the spatial CV was known from only one time point. Moreover, trends emerged in ecosystems ranging from large to small, and from tropical to temperate, suggesting a potentially general and widespread phenomenon. Applying a space-time correspondence follows more than a century of studies involving substitution [8–10,38], but our formulation extends usefulness in two ways; (i) it is quantitative in the form of equations in Fig. 1 rather than qualitative (e.g., chronosequence studies; see [33] for critical review) and (ii) the logic applies equally when substituting CV's of a single variable or when plotting many ecosystem variables for a multivariate view of landscape variation.

Tight linear dependence between spatial and temporal CV's likely owes to two reasons: First, when dynamics are stochastic or independent, variation in space roughly matches that in time as in ergodicity. Thus small fluctuations in time render equally small fluctuations across space. Second is the empirical observation that the factors which theoretically interfere with this correspondence - synchrony and persistence - do so little, at least when using the CV. While variables can lie anywhere on the plot (shifted up the y-axis by synchrony, down by persistence; Fig. 3C), they adhered more to the "independent dynamics" region than being shifted (Fig. 5). This makes sense in that a variable (e.g., a population) with low temporal CV in each patch will still be relatively stable regionally even when patches partially synchronize. This, in turn, registers as a low spatial CV because of how small temporal fluctuations beget small spatial variation (Fig. 2). There are exciting hints that this type of correspondence also applies to other temporal properties, such as recovery time or deterministic chaos, that leave a telling trace of their temporal dynamics in space [13,16].

Analytical solutions and simulations show that spatial CV has value as a signature of temporal variability under certain conditions. When values of a variable are relatively uncorrelated in time and space (e.g., Fig. 4A), the temporal CV can be recovered with an analytical approximation (Fig. 1B; Eq. S31). Accuracy wanes when ecological forces synchronize patches and shift the variable into the synchrony region of the plot (Fig. 3) where spatial CV underestimates temporal CV. Here, synchrony simultaneously boosts temporal variability and lowers spatial variability by aligning the peaks and troughs of fluctuations. Accuracy is also lower when ecological forces cause spatial gradients to persist through time and shift a variable into the persistence region of the plot (Fig. 3). Here, spatial variability exists, but it is created by patches that are stable over time.

Dynamics must be reasonably assumed to be stochastic to use spatial CV as a quantitative proxy for temporal CV. This assumption will often not hold in nature (e.g., when climate swings induce synchrony). And whether it does hold will likely depend on the types of variables chosen (fast or slow, broad-scale or fine-scale) as well as the spatial and temporal sampling scales (Table 2). Even when the assumption does not hold, however, regressions suggest that the rank order of temporal CV's (e.g., highest to lowest) might still be recovered from spatial CV's for qualitative substitution. This should be possible when: (i) All variables are thought to be synchronized or persistent to the same degree (Fig. 3C), (ii) they smoothly intergrade from synchrony to persistence (Figs. 3D; 4C,F), and/or (iii) the distorting effect of synchrony and persistence can be estimated and corrected for (Table 2). Thus, the link between spatial and temporal variation may be valuable for understanding when CV's are interchangeable, and how to interpret them when they are not.

Some variable types may be better suited to substitution than others. Interestingly, species populations (Fig. 4; red circles) illustrated a tradeoff between precision and accuracy. Precision to distinguish species with high versus low temporal CV's using spatial CV's was weak. This was for the sampling reasons that species had a limited range of variability (i.e., range of CV's< scatter) or possibly greater measurement error; or for the ecological reason that they differed in synchrony or persistence which created scatter. Such differences can occur when species respond even slightly differently to the spatial over the temporal environment [39]. But on the other hand, accuracy was probably higher for species because these variables lay closer to the independent dynamics line than physicochemical variables.

Results captured strong associations between spatial and temporal CV when data were separated by one time point (Fig. 4C–F) or were overlapping (Fig. 4A–C). Future work must address how far into the future a spatial CV can be a proxy for temporal CV. Additional error will certainly accrue for substitution if unexpected events alter a variable's temporal variability in ways not indicated by its initial spatial variability. The exact impact of this non-stationarity, however, is a matter of scale and of research question asked. For instance, high resolution prediction of single-species dynamics following a stochastic disturbance (e.g., forest fire) may be untenable if the disturbance drastically alters variability, synchrony or spatial persistence. But spatial CV should

Table 2. Conditions under which the spatial variability of an ecological process is a precise or accurate substitute for its regional temporal variability.

Condition	Description	Substitution	Requirements	Ecological scenario
Independent dynamics	Values must be reasonably assumed independent from place to place, time to time	Spatial and temporal CV values roughly interchangeable (see Fig. 1B)	Scale of process ≤ scale of measurement in both space and time	Isolated communities with fast dynamics; intercontinental, long-term comparisons
Constant levels of synchrony or persistence across variables	Spatiotemporal behavior (e.g., synchrony) that is shared by all variables shifts all points equally on plot (Fig. 3C). Values shifted by a constant	Order (rank) of temporal CV's is conserved among variables	Approximately equivalent spatiotemporal behavior among variables	Variables similarly shaped by spatial constraints (e.g., microclimates) or by landscape-scale temporal shocks (e.g., weather)
Mixed synchrony and persistence of known magnitude	Variables differ in their synchrony and persistence (Fig. 3D), but degree of divergence is known and corrected for	Temporal CV value or order (rank) can be back-calculated	Estimates of synchrony and/or persistence; Fairly stationary dynamics	Variables responding differentially to spatiotemporal variation in the landscape with synchrony or persistence

be a rough proxy – either quantitative or qualitative – for regional temporal CV under the ecological and sampling conditions in Table 2.

Diagnostic Signatures of Complex Dynamics

The terms of Fig. 1 also allowed a window into diagnosing landscape dynamics. Ecosystems conceal ecological information in an eclectic range of spatiotemporal patterns [40,41]. Populations of exploited species [42], biodiversity hotspots [43], harmful algal blooms [44], pest outbreaks [45] and wildfires [46] all display complex patterning in space and time. A range of analyses explore these patterns by examining underlying frequencies (e.g., spectral [47] and wavelet analysis [48]) and patterns of correlation (e.g., correlograms [49]), or by fitting predictive models (e.g., auto-regressive [50]). But these approaches are not designed for linking spatiotemporal pattern to a variable's temporal variation.

We used plots of spatial versus temporal CV's as convenient and unique summaries of landscape dynamics (Fig. 4). Some variables occupied the synchrony region (above dashed line), others the persistence region (below dashed line). Such a mixture may be typical when a wide array of ecosystem processes is sampled. These mixtures left their mark on regression slopes. Slopes <1 in Fig. 4A–B show that abiotic variables in microcosms and rock pools were more stable, but also more prone to synchrony, than populations. Y-intercepts also changed to the degree that multiple variables displayed synchrony or persistence. These therefore offer multivariate indices of the degree of synchrony (β_0> expected) or persistence (β_0< expected) experienced by a landscape (Fig. 3C). Such plots may be fruitful ground for streamlined comparisons of landscapes containing diverse variables and dynamics, like pre-and post-disturbance ecosystems.

The components of temporal variance themselves – spatial variance, synchrony and persistence (Fig. 1A) – may also prove useful for describing patterns and mechanisms driving temporal variation. As we have seen, spatial variance can signal instability at the local, patch scale (e.g., from demographic [51], community [52], natural enemies [53], spatial [26] or local environmental causes; Table S4). Synchrony, in turn, indicates landscape-scale causes of variation like dispersal [53,54] or weather [55,56]. Finally, persistence points to the existence of long-term differences in mean value or state among patches, such as regulation by local communities or physical conditions. Combined, these components of variation gave an alternative view of dynamics.

Fine distinctions emerged when all elements of Fig. 1A were normalized to create a fingerprint or signature of dynamics. The grouping of variables controlled by different parts of the biosphere, like the atmosphere (e.g., temperature, dissolved oxygen) and the watershed (e.g., pH, ion concentrations), suggests that unique signatures may exist for types of ecological processes. Potential may thus exist for predicting the likely dynamics of a variable based on its type (e.g., atmospheric). Meanwhile, the breadth of spatiotemporal behaviors seen suggests a range of spatial variation, synchrony and persistence combinations leading to temporal variation in nature. Our framework may be useful for cataloguing these types of spatiotemporal dynamics in ecosystems (e.g., Table 1), and for making broad-stroke inferences about spatio-temporal mechanisms (e.g., population rescue effects [26,57], predator-prey cycles [58], species coexistence [59,60], and invasive species spread [61]).

Conclusions

Unexplained variation is common in nature, both across heterogeneous landscapes and over timespans of interest. By illustrating the link between spatial and temporal variation, we bring more clarity to the problems of *prediction* and *diagnosis* from spatial or spatiotemporal patterns. More work is needed to test the limits of prediction across scales, variables and ecosystem types. Yet, indications here suggest usefulness in; substituting spatial for temporal variability (either quantitatively or qualitatively), judging when substitution will not work, and interpreting the manifold changes of multivariate landscapes. Such efforts are hoped to add momentum towards the Rosetta Stone of landscape and ecosystem ecology, in which process and mechanism may be deeply and easily discerned from landscape pattern [62].

Supporting Information

File S1.

Table S1 Community and ecosystem variables measured over the 20-week microcosm connectivity experiment.

Table S2 Variables from lake, rock pool and microcosm data sets used in analyses. An additional 36 rock pool species (not shown), known from fewer occurrences, were included for calculating spatiotemporal signatures (Fig. 6).

Table S3 Comparison of CV-based results with alternative indices of variability. General Linear Models were fit between indices of aggregate temporal variability and three spatiotemporal descriptors: Spatial variability, inter-patch synchrony and persistence (see Fig. 1B). Indices of variability included CV and four others. Asterisks denote statistical significance and p-value. *$p<0.05$, **$p<0.01$, ***$p<0.001$. R^2 values in parentheses.

Table S4 Mechanisms that may dampen regional variability and reduce spatial variability by stabilizing local patches.

Acknowledgments

K. Abbott, P. Abrams, M. Anand, S. Carpenter, L. Dai, C. Jones, K. McCann, T. Parker, J. Quinn, D. Rollo, D. Vasseur, R. Vogt, D. Waring, J. Wu and several anonymous reviewers provided comments on the manuscript. We also thank the North Temperate Lakes LTER and Discovery Bay Marine Lab as data sources.

Author Contributions

Conceived and designed the experiments: MPH JK. Performed the experiments: MPH JK. Analyzed the data: MPH. Contributed reagents/ materials/analysis tools: MPH JK. Wrote the paper: MH JK.

References

1. Reed DC, Raimondi PT, Carr MH, Goldwasser L (2000) The role of dispersal and disturbance in determining spatial heterogeneity in sedentary organisms. Ecology 81: 2011–2026.
2. Barnes JB, Ehlers TA, Insel N, Mcquarrie N, Poulsen CJ (2012) Linking orography, climate, and exhumation across the central Andes. Geology 40: 1135–1138.
3. Leavitt PR, Fritz SC, Anderson NJ, Baker PA, Blenckner T, et al. (2009) Paleolimnological evidence of the effects on lakes of energy and mass transfer from climate and humans. Limnology and Oceanography 54: 2330–2348.
4. Seabloom EW, Bjørnstad ON, Bolker BM, Reichman OJ (2005) Spatial signature of environmental heterogeneity, dispersal, and competition in successional grasslands. Ecological Monographs 75: 199–214.
5. Fangliang H, Legendre P, Lafrankie J V (2008) Distribution patterns of tree species in a Malaysian tropical rain forest. Journal of Vegetation Science 8: 105–114.
6. Burrows MT, Harvey R, Robb L, Poloczanska ES, Mieszkowska N, et al. (2009) Spatial scales of variance in abundance of intertidal species: Effects of region, dispersal mode, and trophic level. Ecology 90: 1242–1254.
7. Anderson KE, Hilker FM, Nisbet RM (2012) Directional biases and resource-dependence in dispersal generate spatial patterning in a consumer-producer model. Ecology Letters 15: 209–217.
8. Pickett STA (1989) Space-for-time substitution as an alternative to long-term studies. Long-term studies in ecology: Approaches and alternatives. 110–135.
9. Fukami T, Wardle DA (2005) Long-term ecological dynamics: Reciprocal insights from natural and anthropogenic gradients. Proceedings of the Royal Society B 272: 2105–2115.
10. Walker LR, Wardle DA, Bardgett RD, Clarkson BD (2010) The use of chronosequences in studies of ecological succession and soil development. Journal of Ecology 98: 725–736.
11. Blois JL, Williams JW, Fitzpatrick MC, Jackson ST, Ferrier S (2013) Space can substitute for time in predicting climate-change effects on biodiversity. Proceedings of the National Academy of Science 110: 9374–9379.
12. Eppinga MB, Pucko CA., Baudena M, Beckage B, Molofsky J (2013) A new method to infer vegetation boundary movement from "snapshot" data. Ecography 36: 622–635.
13. Sole R V, Bascompte J (1995) Measuring chaos from spatial information. Journal of Theoretical Biology 175: 139–147.
14. Litzow MA, Urban JD, Laurel BJ (2008) Increased spatial variance accompanies reorganization. Ecological Applications 18: 1331–1337.
15. Guttal V, Jayaprakash C (2009) Spatial variance and spatial skewness: Leading indicators of regime shifts in spatial ecological systems. Theoretical Ecology 2: 3–12.
16. Dai L, Korolev KS, Gore J (2012) Slower recovery in space before collapse of connected populations. Nature 496: 355–358.
17. Marcos-Nikolaus P, Martin-Gonzalez JM (2002) Spatial forecasting: Detecting determinism from single snapshots. International Journal of Bifurcation and Chaos 12: 369–376.
18. McIntire EJB, Fajardo A (2009) Beyond description: The active and effective way to infer processes from spatial patterns. Ecology 90: 46–56.
19. Levin SA (1992) The problem of pattern and scale in ecology: The Robert H. MacArthur Award lecture. Ecology 73: 1943–1967.
20. Lehman CL, Tilman D (2000) Biodiversity, stability, and productivity in competitive communities. The American Naturalist 156: 534–552.
21. Grimm V, Wissel C (1997) Babel, or the ecological stability discussions: An inventory and analysis of terminology and a guide for avoiding confusion. Oecologia 109: 323–334.
22. Turner MG (2010) Disturbance and landscape dynamics in a changing world. Ecology 91: 2833–2849.
23. Pimm SL (1984) The complexity and stability of ecosystems. Nature 307: 321–326.
24. Micheli F, Cottingham KL, Bascompte J, Bjørnstad ON, Eckert GL, et al. (1999) The dual nature of community variability. Oikos 85: 161–169.
25. Cottingham KL, Brown BL, Lennon JT (2001) Biodiversity may regulate the temporal variability of ecological systems. Ecology Letters 4: 72–85.
26. Abbott K (2011) A dispersal-induced paradox: Synchrony and stability in stochastic metapopulations. Ecology Letters 14: 1158–1169.
27. Bolliger J, Wagner HH, Turner MG (2007) Identifying and quantifying landscape patterns in space and time. In: Kienast F, Wildi O, Ghosh S, editors. A changing world: Challenges for landscape research. Springer. 177–194.
28. Lewis WM (1978) Comparison of temporal and spatial variation in the zooplankton of a lake by means of variance components. Ecology 59: 666–671.
29. Larsen DP, Kincaid TM, Jacobs SE, Urquhart NS (2001) Designs for evaluating local and regional scale trends. BioScience 51: 1069–1078.
30. Legendre P, De Caceres M, Borcard D (2010) Community surveys through space and time: Testing the space-time interaction in the absence of replication. Ecology 91: 262–272.
31. Chesson PL (1985) Coexistence of competitors in spatially and temporally varying environments: A look at the combined effects of different sorts of variability. Theoretical Population Biology 287: 263–287.
32. Earn DJD, Levin SA, Rohani P (2000) Coherence and conservation. Science 290: 1360–1364.
33. Paine ADM (1985) "Ergodic" reasoning in geomorphology: Time for a review of the term? Progress in Physical Geography 9: 1–15.
34. Feller W (1950) An introduction to probability theory and its applications. New York: Wiley.
35. Loreau M, De Mazancourt C (2008) Species synchrony and its drivers: Neutral and nonneutral community dynamics in fluctuating environments. The American Naturalist 172: E48–E66.
36. Sciullo L, Kolasa J (2012) Linking local community structure to the dispersal of aquatic invertebrate species in a rock pool metacommunity. Community Ecology 13: 203–212.
37. Pandit S, Kolasa J, Cottenie K (2009) Contrasts between habitat generalists and specialists: An empirical extension to the basic metacommunity framework. Ecology 90: 2253–2262.
38. Cowles HC (1899) The ecological relations of vegetation on the sand dunes of Lake Michigan. Botanical Gazette 27: 167–202.
39. Jongejans E, De Kroon H (2005) Space versus time variation in the population dynamics of three co-occurring perennial herbs. Journal of Ecology 93: 681–692.
40. Kareiva P, Wennergren U (1995) Connecting landscape patterns to ecosystem and population processes. Nature 373: 299–302.
41. Bascompte J (1998) Spatiotemporal patterns in nature. Trends in Ecology and Evolution 13: 173–174.
42. Erisman B, Aburto-oropeza O, Gonzalez-abraham C, Mascarenas-Osorio I, Moreno-Baez M, et al. (2012) Spatio-temporal dynamics of a fish spawning aggregation and its fishery in the Gulf of California. Nature Scientific Reports: 1–11.
43. Rutledge DT, Lepczyk CA, Xie J, Liu J (2001) Spatiotemporal dynamics of endangered species hotspots in the United States. Conservation Biology 15: 475–487.
44. Edwards M, Johns DG, Leterme SC, Svendsen E, Richardson AJ (2012) Regional climate change and harmful algal blooms in the northeast Atlantic. Limnology and Oceanography 51: 820–829.
45. Aukema BH, Carroll AL, Zheng Y, Zhu J, Raffa KF, et al. (2008) Movement of outbreak populations of mountain pine beetle: Influences of spatiotemporal patterns and climate. Ecography 31: 348–358.
46. Senici D, Chen HYH, Bergeron Y, Cyr D (2010) Spatiotemporal variations of fire frequency in central boreal forest. Ecosystems 13: 1227–1238.
47. Michele C De, Bernardara P (2005) Spectral analysis and modeling of space-time rainfall fields. Atmospheric Research 77: 124–136.
48. Grenfell BT, Bjornstad ON, Kappey J (2001) Travelling waves and spatial hierarchies in measles epidemics. Nature 414: 716–723.
49. Ranta E, Kaitala V, Lundberg P (1997) The spatial dimension in population fluctuations. Science 278: 1621–1623.
50. Crabtree R, Potter C, Mullen R, Sheldon J, Huang S (2009) A modeling and spatio-temporal analysis framework for monitoring environmental change using NPP as an ecosystem indicator. Remote Sensing of the Environment 113: 1486–1496.

51. Watson A, Moss R, Rothery P, Parr R (1984) Demographic causes and predictive models of population fluctuations in Red Grouse. Journal of Animal Ecology 53: 639–662.
52. Jiang L, Pu Z (2009) Different effects of species diversity on temporal stability in single-trophic and multitrophic communities. The American Naturalist 174: 651–659.
53. Vogwill T, Fenton A, Brockhurst MA (2009) Dispersal and natural enemies interact to drive spatial synchrony and decrease stability in patchy populations. Ecology Letters 12: 1194–1200.
54. Dey S, Joshi A (2006) Stability via asynchrony in Drosophila metapopulations with low migration rates. Science 312: 434–436.
55. Liebhold A, Koenig WD, Bjørnstad ON (2004) Spatial synchrony in population dynamics. Annual Review of Ecology, Evolution and Systematics 35: 467–490.
56. Paradis E, Baillie R, Sutherland WJ, Gregory RD (1999) Dispersal and spatial scale affect synchrony in spatial population dynamics. Ecology Letters 2: 114–120.
57. Blasius B, Huppert A, Stone L (1999) Complex dynamics and phase synchronization in spatially extended ecological systems. Nature 399: 354–359.
58. Tobin PC, Bjørnstad ON (2003) Spatial dynamics and cross-correlation in a transient predator–prey system. Journal of Animal Ecology 72: 460–467.
59. Hassell MP, Comins HN, May RM (1994) Species coexistence and self-organizing spatial dynamics. Nature 370: 290–292.
60. Ettema CH, Rathbun SL, Coleman DC (2000) On spatiotemporal patchiness and the coexistence of five species of Chronogaster (Nematoda: Chronogasteridae) in a riparian wetland. Oecologia 125: 444–452.
61. Pysek P, Hulme PE (2005) Spatio-temporal dynamics of plant invasions: Linking pattern to process 1. Ecoscience 12: 302–315.
62. Keitt TH, Urban DL (2005) Scale-specific inference using wavelets. Ecology 86: 2497–2504.
63. Post E, Forchhammer MC (2002) Synchronization of animal population dynamics by large-scale climate. Nature 420: 168–171.
64. Isaak DJ, Thurow RF (2006) Network-scale spatial and temporal variation in Chinook salmon (Oncorhynchus tshawytscha) redd distributions: Patterns inferred from spatially continuous replicate surveys. Canadian Journal of Fisheries and Aquatic Sciences 296: 285–.
65. Morfin M, Fromentin J-M, Jadaud A, Bez N (2012) Spatio-temporal patterns of key exploited marine species in the Northwestern Mediterranean Sea. PLoS One 7: 1–14.
66. Siegel DA, Mitarai S, Costello CJ, Gaines SD, Kendall BE, et al. (2008) The stochastic nature of larval connectivity among nearshore marine populations. Proceedings of the National Academy of Sciences 105: 8974–8979.

18

Sediment and Turbidity Associated with Offshore Dredging Increase Coral Disease Prevalence on Nearby Reefs

F. Joseph Pollock[1,2,3]*, Joleah B. Lamb[1,2], Stuart N. Field[4,5], Scott F. Heron[6,7], Britta Schaffelke[3], George Shedrawi[4], David G. Bourne[3], Bette L. Willis[1,2]

1 ARC Centre of Excellence for Coral Reef Studies, School of Marine and Tropical Biology, James Cook University, Townsville, Queensland, Australia, **2** AIMS@JCU, Australian Institute of Marine Science and James Cook University, Townsville, Queensland, Australia, **3** Australian Institute of Marine Science, PMB 3, Townsville, Queensland, Australia, **4** Department of Parks and Wildlife, Marine Science Program, Kensington, Western Australia, Australia, **5** Oceans Institute, University of Western Australia, Crawley, Western Australia, Australia, **6** NOAA Coral Reef Watch, Townsville, Queensland, Australia, **7** Marine Geophysical Laboratory, School of Engineering and Physical Sciences, James Cook University, Townsville, Queensland, Australia

Abstract

In recent decades, coral reef ecosystems have declined to the extent that reefs are now threatened globally. While many water quality parameters have been proposed to contribute to reef declines, little evidence exists conclusively linking specific water quality parameters with increased disease prevalence *in situ*. Here we report evidence from *in situ* coral health surveys confirming that chronic exposure to dredging-associated sediment plumes significantly increase the prevalence of white syndromes, a devastating group of globally important coral diseases. Coral health surveys were conducted along a dredging-associated sediment plume gradient to assess the relationship between sedimentation, turbidity and coral health. Reefs exposed to the highest number of days under the sediment plume (296 to 347 days) had two-fold higher levels of disease, largely driven by a 2.5-fold increase in white syndromes, and a six-fold increase in other signs of compromised coral health relative to reefs with little or no plume exposure (0 to 9 days). Multivariate modeling and ordination incorporating sediment exposure level, coral community composition and cover, predation and multiple thermal stress indices provided further confirmation that sediment plume exposure level was the main driver of elevated disease and other compromised coral health indicators. This study provides the first evidence linking dredging-associated sedimentation and turbidity with elevated coral disease prevalence *in situ*. Our results may help to explain observed increases in global coral disease prevalence in recent decades and suggest that minimizing sedimentation and turbidity associated with coastal development will provide an important management tool for controlling coral disease epizootics.

Editor: Christopher J. Fulton, The Australian National University, Australia

Funding: This work was partially funded by an Australian National Network in Marine Science (ANNiMS) PhD mobility scholarship awarded to FJP and an ANNiMS Industry Research grant awarded to JBL. The Gorgon Joint Venture (http://www.chevronaustralia.com/our-businesses/gorgon) also largely funded the project through DPaW's Dredging Audit and Surveillance Program as part of environmental offsets. The funders had no role in study design, data collection and analysis, decision to publish, or preparation of the manuscript.

Competing Interests: The authors have declared that no competing interests exist.

* Email: Frederic.Pollock1@ jcu.edu.au

Introduction

Over the last 30 years, hard coral cover has decreased by an average of 50% on Indo-Pacific reefs and 80% on Caribbean reefs [1–3]. While these declines have been attributed to a number of factors, including water pollution, habitat destruction, overfishing, invasive species, and global climate change [1,2], coral diseases have recently emerged as a significant driver of global coral reef decline [4,5]. The destructive potential of coral disease is most clearly exemplified in the Caribbean where successive disease outbreaks have decreased acroporid cover by up to 95% and contributed substantially to observed ecological phase shifts from coral to algal-dominated reefs [6–8]. Furthermore, the incidence of coral epizootics has increased globally over the last 20 years [3,6,9,10], highlighting the need to understand and manage the factors underlying coral disease outbreaks.

Reduced water quality caused by explosive human population growth is often cited as an important factor driving coral disease epizootics [11–13]. Land clearing exposes 1% of the Earth's surface to eroding processes annually and urbanization of coastal areas is expanding disproportionally to population growth [14,15]. Consequently, coastal coral reefs, like many other marine ecosystems, are increasingly subjected to elevated levels of eutrophication, sedimentation and turbidity, factors proposed to compromise disease resistance of corals and/or increase pathogen virulence [13,16]. Coastal dredging for land reclamation, beach nourishment and port construction further exacerbates terrestrial nutrient and sediment influx by resuspending benthic sediments [17]. Additionally, more frequent and intense storms associated with climate change amplify water quality declines by promoting coastal runoff and sediment resuspension [18]. Eutrophication, and more specifically nutrient enrichment, has been shown to exacerbate existing coral disease infections and shift coral-associated microbial communities towards communities typical of diseased corals, but little is known about the role of nutrients in disease initiation [13,19,20].

Sedimentation and turbidity, associated with both weather events and anthropogenic activities, are also frequently proposed to contribute to increased coral disease prevalence [16], although empirical evidence is lacking. Hodgson [21] suggested sedimentation as a potential mechanism for the transmission of coral pathogens from marine or terrestrial substrates onto nearby corals. Silt-associated bacteria were identified as a possible cause of necrosis in sediment-damaged corals, since antibiotic-treated water reduced tissue damage in experimentally silted corals. In field-based observations, Haapklyä et al. [22] noted a correlation between seasonal coastal runoff, including increased sedimentation and turbidity, and the prevalence of coral disease on inshore reefs. Elevated turbidity reduces the amount and quality of ambient light available for photosynthesis by the corals' endosymbiotic algae (*Symbiodinium*) and excess sedimentation inhibits the heterotrophic feeding efficiency of corals, reducing the energy intake of both symbiotic and asymbiotic corals [23,24]. While corals are able to shed some sediment through mucus production and ciliary action, these mechanisms are energetically expensive and further burden the corals' already reduced energy budgets [25,26]. Despite a wealth of circumstantial evidence indicating sedimentation and turbidity as potential coral disease drivers [15,22,27], no studies have directly linked sedimentation, turbidity and coral disease in the field. Given that nearly 40% of coral reefs are located adjacent to large population centers and coastlines under rapid development to accommodate expanding urban activities [28], effective coastal management will increasingly depend upon a comprehensive understanding of the impacts of sediment, turbidity and associated water quality decline, on all aspects of coral reef health.

Here, we describe the first study to examine the influence of elevated sedimentation and turbidity on coral disease levels *in situ*. We performed detailed coral health assessments along a gradient of exposure to a sediment-laden dredge plume within the Montebello and Barrow Islands off the northern coast of Western Australia. The otherwise relatively pristine conditions of these offshore reefs enabled an empirical examination of the relationship between sedimentation, turbidity and coral disease prevalence in the absence of other confounding influences. Our results indicate that elevated sedimentation and turbidity can significantly increase coral disease prevalence and highlight the urgent need to manage coastal development near coral reef ecosystems.

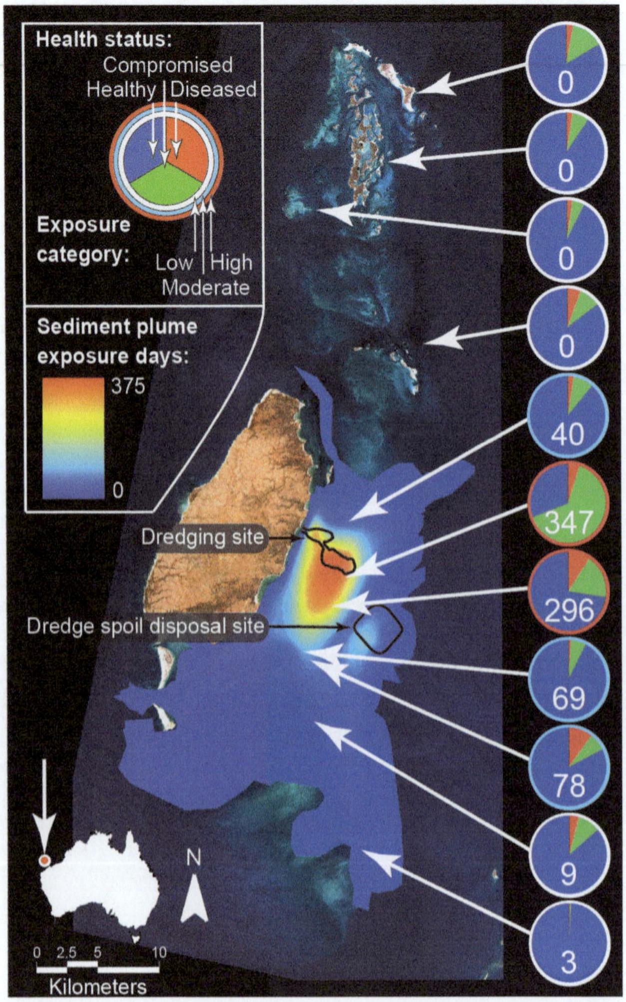

Figure 1. Map showing study site and coral health survey locations at Montebello and Barrow Islands, Western Australia. Colored overlays (gradient from red to blue) indicate satellite-derived sediment plume exposure days determined by hot spot analysis of MODIS satellite imagery. Pie charts indicate the proportion of colonies at each site (n = 3 transects per site) recorded as apparently healthy (blue), diseased (red) or displaying other signs of compromised coral health (green). Numbers inset on pie charts indicate satellite-derived sediment plume exposure days at each site. Colors ringing pie charts indicate plume exposure categories, i.e. white: low (0 to 9 exposure days); blue: moderate (40 to 78 exposure day); and red: high (296 to 347 exposure days).

Methods

Ethics statement

This research was conducted under the following permits: Western Australia Department of Environment and Conservation Collection Permit number SF008340 and Western Australia Department of Environment and Conservation Export Permit number ES002169.

Study site

The Montebello and Barrow Islands are situated in the Pilbara region of Northwest Australia, approximately 1,600 km north of Perth (see Figure 1). The Montebello and Barrow Islands Marine Protected Areas (MBIMPA), incorporating the Montebello Islands Marine Park, Barrow Island Marine Park and the Barrow Island Marine Management Area, were gazetted in 2004. The environment within the MBIMPA is considered to be relatively pristine as a consequence of low human usage, minimal terrestrial influence and strict management controls on industrial developments in the area [29].

The Gorgon Project, based on Barrow Island (20.80°S, 115.40°E), is one of the world's largest natural gas projects and the largest single resource natural gas project in Australia's history. The Gorgon Project dredging program involved the removal and dumping of approximately 7.6 million tons of marine sediment over an 18-month period from 19 May 2010 to 7 November 2011.

Satellite-derived assessment of sediment plume extent

The area affected by the dredging-induced sediment plume area was quantified daily over the duration of the dredging program using Moderate Resolution Imaging Spectroradiometer (MODIS) satellite imagery, as described by Evans et al. [30]. Briefly, the sediment plume boundary was interpreted manually using one of up to two MODIS images captured daily. A single practitioner manually assessed plumes visible in MODIS images and distinguished these sediment plumes from benthic substrates. Quality control measures were utilized to ensure the collection of consistent, high-quality data (see Evans et al. [30] for details). A 'hotspot' analysis was performed on the cumulative daily plume boundaries, for the total number of days available, to provide a dataset describing the number of days the sediment plume was present at any position within the waters surrounding the Montebello and Barrow Islands. These data were used to determine sediment plume exposure days, which are defined here as the cumulative number of days a suspended sediment plume was visible in satellite images at a given location throughout the duration of dredging operations. One year of pre-dredging MODIS imagery was also analyzed to identify a baseline for naturally occurring turbidity events.

Coral health and community composition surveys

Coral health surveys were conducted in December 2011, one month after the completion of the 18-month Gorgon Project dredging program. Eleven sites were selected, extending both north and south from the dredging site, representing a gradient of sediment plume exposure (Figure 1). At each site, three 15 m × 2 m belt transects, placed haphazardly at least 5 m apart, were surveyed along depth contours at 2 to 6 m, consistent with standardized protocols developed by the Global Environment Facility and World Bank Coral Disease Working Group [31]. Within each 30 m^2 belt transect, every scleractinian coral colony over 5 cm in diameter was identified to genus-level and classified as either diseased [i.e., affected by one or more of the following diseases: white syndromes (WS), brown band disease, skeletal eroding band, black band disease, and/or growth anomalies]; showing other signs of compromised health (i.e., tissue necrosis associated with sediment accumulation, bleaching, pigmentation response, and/or sponge, red algae, or green algal overgrowth); or healthy (i.e., no visible signs of disease lesions or other indicators of compromised health) using indicators described by Willis et al. [32]. Additionally, signs of coral predation by *Drupella sp.* and/or *Acanthaster planci* (crown-of-thorns seastar; COTS) were recorded. Standard line-intercept surveys were used to determine coral cover and coral community composition to the genus-level by estimating the linear extent of each coral to the nearest centimeter along the central line of each 15 m belt transect. These protocols allow the data collected in this study to be directly compared to other similar standardized coral disease datasets worldwide.

Assessment of temperature-based risk of disease

Relationships between diseases and anomalously warm temperature have been determined for various marine and terrestrial organisms [15]. To evaluate the role that thermal stress might have played in shaping the spatial patterns of coral disease and other signs of compromised coral health observed, we analyzed temperature-based predictors of disease outbreak risk based on published empirical relationships between temperature metrics, coral cover and disease abundance (summarized in Heron et al. [33]). While these temperature-disease relationships were derived for only one disease type (WS) affecting one coral genus (*Acropora* spp.) on the Great Barrier Reef, the thermal stress predictors of disease risk provide useful indicators of host susceptibility and potentially pathogen loads [33]. Briefly, retrospective Pathfinder satellite sea-surface temperature (SST) time-series for the period 1985 to 2009 were concatenated with NOAA's near real-time 11 km SST time-series (February 2009 to December 2011); concatenation was performed by bias-adjusting the latter dataset to match the former based on the overlap period. The resulting weekly dataset provided a SST time-series for each survey location throughout the dredging period (May 2010 to November 2011) and an internally consistent climatological baseline for the calculation of thermal stress metrics. Five temperature-based stress metrics associated with disease likelihood were derived (see Table 1 for definitions): Hot Snap, Cold Snap and Winter Conditions (see [34] for full details); and mean positive summer anomaly (MPSA) and predicted abundance [35]. Predicted abundance of disease cases per 1,500 m^2 ($A_{disease}$) was calculated using MPSA and total hard coral cover for all species (TCC) from the field surveys, following the model of Maynard et al. [35];

$$A_{disease} = MPSA^a TCC^b \tag{1}$$

where $a = 1.07$ and $b = 1.59$ [37]. All temperature-based metrics were assessed at the site level (i.e., no replication at the transect level) due to the limited resolution of satellite-derived SST data.

Data analyses

Prevalence values for coral diseases and other signs of compromised health were calculated within each 30 m^2 belt transect by dividing the number of colonies with signs of any of 5 diseases recorded or of 6 other indicators of compromised health by the total number of colonies present. To assess the effect of dredging on disease prevalence and on other indicators of compromised health along the plume gradient, sites were assigned to one of three exposure categories based on the number of days a dredging-associated suspended sediment plume was visible in usable MODIS satellite images (Figure 1):

- low-exposure (0–9 sediment plume exposure days; 18 transects),
- moderate-exposure (40–78 sediment plume exposure days; 9 transects) and
- high-exposure (296–347 sediment plume exposure days; 6 transects).

To analyze patterns of coral disease and other signs of compromised health among broad coral growth forms within each sediment plume exposure category, coral genera were assigned to one of three growth form categories: massive, plating or branching (see Methods S1 in File S1). Associations between the prevalence of disease and other compromised coral health indicators and sediment plume exposure days were tested with Pearson product-moment correlations. Differences in mean prevalence levels among the three sediment plume exposure groups were analyzed using two-way (sediment plume exposure category, site) nested analyses of variance (ANOVA), with site treated as a random factor that was nested within the fixed factor, plume exposure. Plume exposure days, coral predation by COTS and *Drupella*, and total hard coral cover were compared among plume exposure groups using the two-way ANOVA design described above. Differences in mean prevalence of disease and compromised health indicators were compared among growth forms within each dredge exposure category using one-way ANOVAs. Similarly, all temperature-based measures of disease likelihood were compared using one-way ANOVAs. Associations between sediment-exposure days and prevalence of coral health and disease were tested with Pearson product-moment correlations. Prior to analyses, assumptions of normality (Shapiro-Wilks)

Table 1. Environmental predictor variables assessed in a multivariate multiple regression analysis (DISTLM) at sites within three sediment plume exposure categories determined by MODIS satellite imagery: low (0 to 9 plume exposure days, n = 6 sites), moderate (40 to 68 plume exposure days, n = 3 sites) and high (296 to 347 plume exposure days, n = 2 sites); and results of an analysis of variance (ANOVA) for each predictor variable among dredge plume exposure groups.

Factor		Sediment plume Exposure Category					
		Low (n = 6 sites)	Moderate (n = 3 sites)	High (n = 2 sites)	ANOVA	Assessed in	
		mean (SE)	mean (SE)	mean (SE)	$F_{(2,11)}$, p	DISTLM?	Source
Dredging:	Sediment plume exposure[a] days[a]	2.0 (1.5)	62.3 (11.5)	321.5 (25.5)	285.7, <0.001*	Yes	[30]
Predation:	COTS scars[b]	0.0 (0.0)	1.6 (1.1)	0.0 (0.0)	1.5, 0.29	Yes	This study
	Drupella scars[c]	3.6 (0.6)	11.7 (3.3)	1.4 (0.9)	2.9, 0.11	Yes	This study
Coral cover:	Total hard coral cover[d]	36.0 (15.0)	53.5 (21.1)	26.0 (20.4)	1.7, 0.24	Yes	This study
Thermal stress:	Peak SST[e]	30.7 (0.2)	30.8 (0.3)	30.9 (0.0)	0.2, 0.81	No[k]	[34]
	Peak SSTA[f]	2.8 (0.1)	2.7 (0.2)	2.7 (0.0)	0.07, 0.94	No[k]	[34]
	Hot Snap[g]	2.0 (1.7)	1.8 (1.4)	2.2 (0.4)	0.04, 0.96	Yes[k]	[34]
	Winter Conditions[h]	10.3 (0.7)	9.0 (0.1)	8.0 (0.2)	2.2, 0.18	Yes	[34]
	MPSA[i]	0.3 (0.1)	0.2 (0.03)	0.3 (0.01)	1.2, 0.35	Yes	[35]
Modeled disease:	Predicted disease abundance[j]	101 (94)	88 (39)	52 (57)	0.3, 0.75	No	[35]

[a]Number of days the sediment plume was recorded over a site for the duration of dredging operations (days).
[b]Prevalence of coral colonies with crown-of-thorns seastar lesions determined by *in situ* coral health surveys (%).
[c]Prevalence of coral colonies with *Drupella* lesions determined by *in situ* coral health surveys (%).
[d]Total hard coral cover on transects determined by line intercept method (%).
[e]Maximum sea surface temperature recorded during dredging operations (May 2010 to November 2011) (°C).
[f]Maximum excursion of sea surface temperature from the long-term climatological value during dredging operations (May 2010 to November 2011) (°C).
[g]Accumulation of thermal anomalies greater than the long-term summer mean temperature plus one standard deviation (°C-weeks).
[h]Accumulation of winter anomalies (+ and −) from the long-term winter mean temperature (°C-weeks).
[i]Average of summer temperature anomalies greater than zero calculated from the monthly mean temperature plus one monthly standard deviation.
[j]Modeled numeric prediction of disease abundance based upon MPSA and total coral cover (disease cases per 1,500 m^2).
[k]Peak SST and Peak SSTA were excluded from the DISTLM due to a strong correlation with Hot Snap ($r = 0.87$ and 0.79, respectively).

and homogeneity of variance (Levene's test of homogeneity) were tested. Post-hoc comparisons between groups were performed using Tukey's HSD tests. All univariate statistical analyses were performed using STATISTICA 10 (StatSoft, Tulsa, Oklahoma).

Coral community composition (at the genus-level) was compared among reefs within the three sediment plume exposure categories to ensure that community structure was homogenous across treatments. To test for differences in community composition, a nested permutational analysis of variance (PERMANOVA) was used, in which site was treated as a random factor nested within the fixed factor, plume exposure [36]. Similarities among coral communities were illustrated using a non-metric multidimensional scaling plot (nMDS) [36].

A non-parametric distance-based linear model (DISTLM) was used in combination with distance-based redundancy ordination analysis (dbRDA) to explore the hypotheses that variability in patterns of disease and other compromised health indicators could be explained by environmental variables known to impact coral condition and distribution (i.e., sediment plume exposure days, hard coral cover, predation, and calculated thermal stress; see Table 1). The DISTLM models the relationship between these environmental variables and the multivariate coral health prevalence dataset based on a multiple regression model. This routine finds the linear combination of variables that explains the greatest amount of variation in the coral health dataset and examines the amount of variance explained by each environmental variable, providing a pseudo-F statistic. The best-fit model, based on corrected Akaike's Information Criterion (AICc), was then visualized in multidimensional space using dbRDA ordination [36]. Preliminary diagnostics to assess multi-collinearity among predictor variables using draftsman plots reveal two thermal stress indicators, Peak SST and Peak SSTA, were highly correlated with Hot Snap ($r = 0.87$ and 0.79, respectively). To avoid redundancy, Peak SST and Peak SSTA were not included in the DISTLM or dbRDA. Predictors that best explained the data were overlaid as biplots representing the strength (vector length) and direction of influence [36]. All multivariate analyses were conducted in PRIMER v6 [37] and PERMANOVA+ [36] using Bray-Curtis similarity matrices based on fourth-root transformed data.

Results

Satellite-derived assessment of sediment plume extent

Satellite images of the Gorgon Project sediment plume, of a quality suitable for deriving plume extent, were available for 411 of the 538 dredging days (i.e. 76% of days). Poor quality images (e.g., due to cloud cover or the sensor not capturing the study region) during the remaining 127 days (24% of days) were omitted from the analysis. Therefore, the number of sediment plume exposure days reported here is conservative and likely underestimate the true number of days sites spent under the sediment plume. Hotspot analysis of satellite imagery revealed that the sediment plume was most commonly detected around the dredge channel and sediment spoil dumping sites (Figure 1). Cumulative sediment plume exposure declined away from these sites, with the plume typically dispersing to the south of the dredge and spoil sites in response to prevailing wind and current patterns [30].

Impact of dredging on coral disease prevalence

A significant, positive correlation was found between overall coral disease prevalence and sediment plume exposure days (Figure S1a in File S1, Pearson's $r_9 = 0.49$, $p<0.05$). Mean disease prevalence ($\pm$ SE) at high-exposure sites ($7.26\pm1.56\%$) was greater than 2-fold higher than at low-exposure sites ($3.1\pm0.6\%$) and 1.5-fold higher than at moderate-exposure sites ($4.7\pm1.5\%$) (Figure 2a, Table S1 in File S1, $F_{2,8} = 9.1$, $p<0.002$). When results from all sites were combined, WS (69%) and skeletal eroding band (17%) dominated the disease cases observed. At the high-exposure sites, elevated disease prevalence was largely the result of high WS levels, which were 2.5-fold greater than at low- and moderate-exposure sites (Figure 2a, Table S1 in File S1, $F_{2,8} = 17.5$, $p< 0.001$). In contrast, the highest prevalence of brown band disease was recorded at moderate-exposure sites, where it was more than 9 times greater than at high- or low-exposure sites ($F_{2,8} = 0.9$, $p< 0.001$). The prevalence of black band disease, growth anomalies and skeletal eroding band did not differ significantly between exposure categories (Figure 2a, Table S1 in File S1, all $p>0.05$).

Impact of dredging on other signs of compromised coral health

There was a significant, positive correlation between the prevalence of other compromised health indicators and sediment plume exposure days (Figure S1b in File S1, Pearson's $r_9 = 0.79$, $p<0.001$). Mean prevalence of these indicators was more than 6-fold greater at high-exposure sites ($47.9\%\pm11.2$) than at low- ($8.0\pm1.4\%$) or moderate-exposure sites ($7.9\pm0.9\%$) (Figure 2b, Table S1 in File S1, $F_{2,8} = 50.8$, $p<0.001$). Sediment-associated tissue necrosis was 57 times more prevalent at high-exposure sites compared to low- and moderate-exposure sites (Figure 2b, $F_{2,8} = 154.9$, $p<0.001$). Bleaching, sponge overgrowth and pigmentation responses were also significantly greater at high-exposure sites relative to low- or moderate-exposure sites (Figure 2b, Table S1 in File S1, all $p<0.001$). The prevalence of red and green algae did not differ significantly between exposure categories (all $p>0.05$).

Influence of coral community composition and morphology on disease and other signs of compromised health

There was no significant difference in coral community composition between sediment plume exposure categories, indicating that reefs within the three groupings were comparable in regards to coral structure (Figure S2 in File S1, pseudo-F = 1.38, $p>0.1$). However, coral community composition did vary significantly among sites within exposure categories (Figure S2 in File S1, pseudo-F = 7.54, $p<0.001$).

Disease levels did not differ significantly among growth forms (i.e., massive, plating and branching colonies) at high or low exposure sites (Figure S3a in File S1, all $p>0.05$). However, massive corals at moderate-exposure sites sustained significantly less disease than branching and plating colonies (Figure S3a in File S1, all $p>0.05$). The prevalence of other compromised coral health indicators did not differ between growth forms within any sediment plume exposure category (Figure S3b in File S1, all $p> 0.05$).

Environmental drivers of disease and compromised health

ANOVA and DISTLM (visualized through dbRDA) both identified sediment plume exposure level as the most significant environmental driver of elevated levels of disease and other indicators of compromised health. Among all environmental parameters assessed (see Table 1), sediment plume exposure days was the only metric that differed significantly among exposure categories ($F_{2,11} = 285.7$, $p<0.001$). Furthermore, the abundance of disease predicted by satellite-derived temperature-based stress

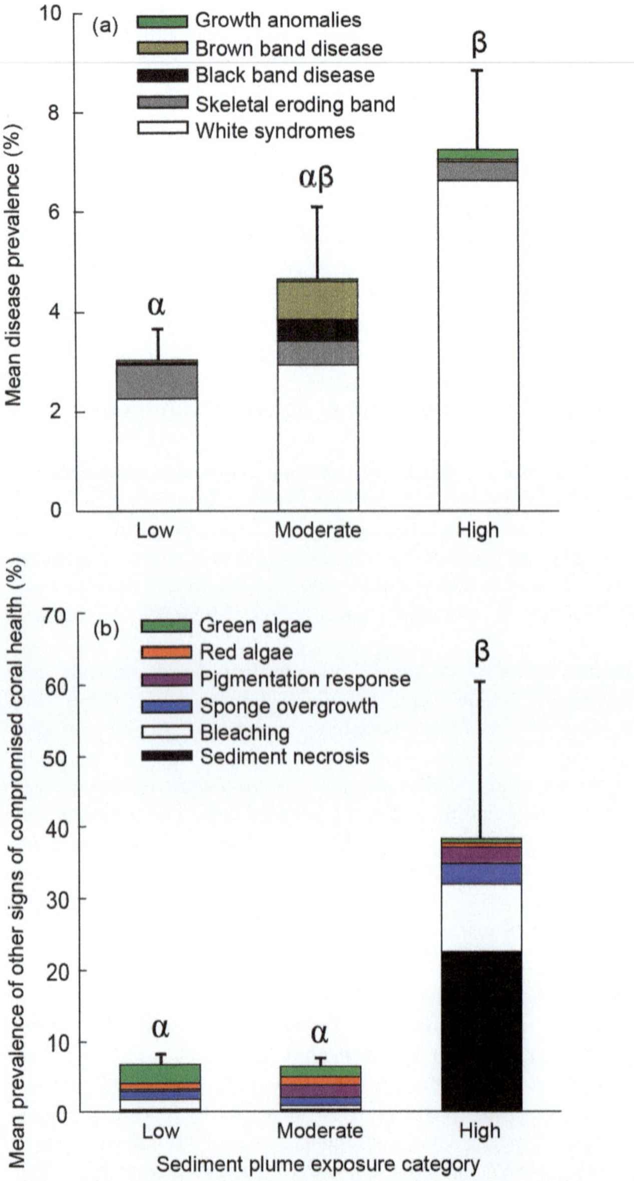

Figure 2. Mean prevalence of (a) coral disease and (b) other signs of compromised coral health at sites within three sediment plume exposure categories: low (0 to 9 plume exposure days; n = 18 transects), moderate (40 to 78 plume exposure days; n = 9 transects), and high (296 to 347 plume exposure days; n = 6 transects). Stacked bars indicate disease or other compromised coral health indicator prevalence by category and error bars indicate standard error among transects for total prevalence of disease or other compromised coral health indicators. Letters indicate post-hoc groupings (Tukey's HSD, $p<0.05$) between sediment plume exposure categories.

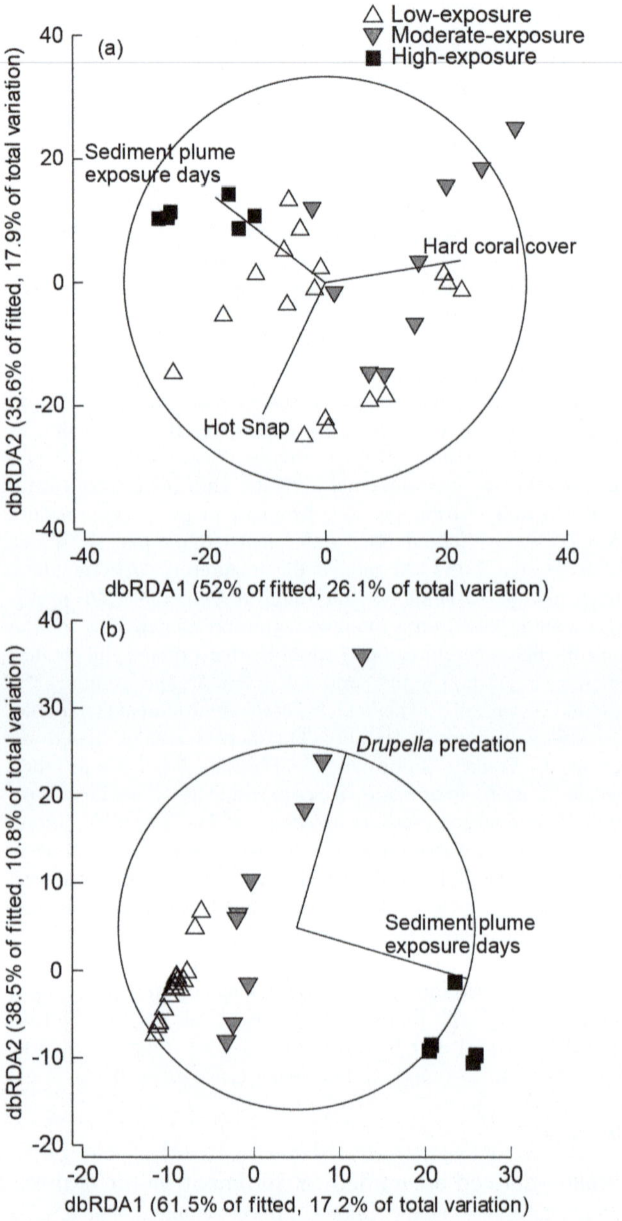

Figure 3. Distance-based redundancy analysis (dbRDA) ordination plots illustrating the relationship between environmental predictors that best explain the variation of (a) coral disease and (b) other compromised coral health indicators among sites. The dbRDA was constrained by the best-fit explanatory variables from a multivariate multiple regression analysis (DISTLM) and vectors overlays are shown for predictor variables explaining a significant proportion of the variation in the prevalence of (a) coral disease and (b) other compromised coral health indicators.

metrics did not differ significantly among sediment plume exposure groups (Table 1, $p>0.05$).

The dbRDA diagram depicting coral disease composition (based on the simplest best fit DISTLM, $AICc = 215.12$, $R^2 = 0.47$), showed a cluster of high-exposure sites away from moderate- and low-exposure sites (Figure 3a). The greatest amount of variation in the disease prevalence data was explained by sediment plume exposure days ($pseudo\text{-}F_{4,30} = 3.97$, $p<0.05$) and total hard coral cover ($pseudo\text{-}F_{4,30} = 4.28$, $p<0.05$) (Figure 3a). The temperature metric Hot Snap ($pseudo\text{-}F_{4,30} = 3.31$, $p<0.05$) explained a lesser, but still significant amount of variation in the disease dataset (Figure 3a). The overlay vector for sediment plume exposure days corresponded to the axis separating high-exposure sites from low- and moderate-exposure sites, while the vectors for hard coral cover and Hot Snap largely corresponded to the axis separating sites within low- and moderate-exposure categories (Figure 3a).

The dbRDA diagram depicting the composition of other compromised health indicators (based on the simplest best fit DISTLM, $AICc = 215.57$, $R^2 = 0.28$) also showed a clear separation between tightly clustered high-exposure transects and the more dispersed low- and moderate-exposure transects (Figure 3b). Sediment plume exposure days ($pseudo\text{-}F_{4,30} = 5.99$, $p < 0.005$) and *Drupella* ($pseudo\text{-}F_{4,30} = 3.65$, $p < 0.05$) explained the greatest amount of variation in the datacloud. The overlay vector for plume exposure days corresponded to the axis separating high-exposure sites from low- and moderate-exposure sites, while the vector for *Drupella* largely corresponded to the axis separating sites within low and moderate-exposure categories (Figure 3b).

Discussion

This study provides the first empirical evidence linking turbidity and sedimentation with elevated levels of coral disease and other indicators of compromised coral health *in situ*. We found two-fold higher disease prevalence, largely driven by increases in WS, and six-fold higher levels of other compromised health indicators at high sediment plume exposure sites. Since these *in situ* health assessments were conducted more than 18 months after commencement of dredging, it is likely that the most susceptible corals experienced complete mortality prior to surveys being undertaken. Therefore, these prevalence figures likely underestimate the true impact of dredging-associated sedimentation and turbidity on coral health.

While previous studies have suggested a myriad of environmental stressors as potential drivers of coral disease [21,22,38], the current study highlights a direct link to sedimentation and turbidity. On Australia's east coast, the UNESCO World Heritage Committee recently cited increasing coastal development and catchment runoff as serious threats to the "outstanding universal value" of Australia's Great Barrier Reef World Heritage Area [39], and on Australia's west coast, an estimated 200 million cubic meters of sediment will be dredged and dumped in projects currently passing through Western Australia's approvals and/or regulatory system alone [40]. Clearly, these findings will have direct implications for coastal managers charged with balancing economic development with the imperative to maintain healthy coral reefs. As the rate of coastal development near coral reef ecosystems continues to escalate in many parts of the world, a comprehensive understanding of the impacts of sediment and turbidity on coral health will become increasingly important.

Impact of dredging on coral disease prevalence

Elevated disease levels at high sediment plume exposure sites were primarily the result of the more than 2.5-fold higher prevalence of WS, an important group of diseases that have affected reefs throughout the Indo-Pacific and which are characterized by a distinct band of sloughing coral tissue revealing underlying coral skeleton [32,41,42]. WS prevalence is often correlated with environmental stress, and strong correlations between warm temperature anomalies and elevated WS levels have shown that thermal stress is an important driver of some types of WS [34,35]. However, we found no differences in multiple thermal stress metrics or predicted disease abundance among the three sediment plume exposure groups and while Hot Snap and total hard coral cover helped to explain the distribution of disease among sites, this metric largely correlated with differences among low and moderate-exposure sites. Both univariate analysis and multivariate modeling identified sediment plume exposure as the main driver of elevated WS levels in the current study, providing further evidence for the role of environmental stress, specifically increased sedimentation and turbidity, in WS pathogenesis.

The greater prevalence of brown band disease at moderate-exposure sites compared to high- and low-exposure sites may reflect differences in ciliate proliferation rates under specific turbidity/light levels. Brown band disease is characterized by a dense, brown mat of ciliates packed with *Symbiodinium* derived from consumed coral tissue [32,43]. Since *Symbiodinium* cells residing within ciliates are photosynthetically competent during brown band disease progression, it has been proposed that brown band disease-associated ciliates could derive nutrition from photosynthates produced by ingested *Symbiodinium*, while also benefiting from an additional oxygen source within the densely populated and presumably oxygen-limited brown band mat [32,43]. At highly turbid, high sediment plume exposure sites, *Symbiodinium* photosynthesis would be hindered, potentially removing the advantage of brown band ciliates over their presumably immune-compromised coral hosts. However, medium-exposure sites could provide the right balance of compromised host immunity and sufficient light availability to facilitate infection and proliferation of brown band ciliates. Further studies specifically investigating the influence of reduced light levels on the partitioning of photosynthates between *Symbiodinium* and ciliates are required to test this hypothesis.

Total mean disease levels at low-exposure sites (3.1±0.6%) were similar to levels reported from nearby Ningaloo Marine Park [44], indicating that these low-exposure sites provide a good approximation of background disease levels in the region. Prevalence levels of black band disease, skeletal eroding band disease and growth anomalies did not differ significantly among sediment plume exposure groups and all were consistent with levels reported from Ningaloo Reef [44].

Impact of dredging on other compromised coral health indicators

The greater prevalence of other indicators of compromised coral health at high sediment plume exposure sites was largely the result of elevated levels of sediment-associated necrosis and bleaching, which were 57-fold and 9-fold higher, respectively. Increased turbidity reduces the amount of light available for photosynthesis, while sediment deposition further shades corals and taxes energy budgets through the need to allocate energy to sediment removal. Although corals are able to actively remove sediment particles through ciliary and tentacular movement, combined with polyp distension and mucus production [45–47], these mechanisms can become overwhelmed during periods of intense and/or chronic sediment deposition. When sediment stress is chronic, even low-levels can dramatically alter coral energy budgets by reducing *Symbiodinium* densities (i.e., bleaching) and by decreasing the photochemical efficiencies (F_v/F_m) of the *Symbiodinium* that remain [45–47]. If resulting energy deficits are not relieved through either metabolic depression or heterotrophic feeding, bleaching can lead to mortality of the affected coral tissue (i.e., sediment necrosis).

While bleaching and sediment necrosis observed in this study were mostly confined to depressions on the coral surface, these patches of partial mortality could expose the coral to further mortality or subsequent infection by disease, even after the completion of dredging operations. In laboratory sedimentation experiments, Flores et al. [48] reported that corals with only 10% partial mortality at the end of a period of sediment exposure subsequently suffered complete mortality during a 4-week "recovery" period. Although bleaching and sediment necrosis are known sources of coral mortality during periods of prolonged sediment exposure [49,50], future studies should investigate the potential of sediment-induced lesions to develop into coral disease, which could progress even after the sediment stress is removed.

Elevated prevalence of pigmentation responses (PR) at high exposure sites provides further evidence of sediment plume-induced coral stress. PR has been observed in corals subjected to a suite of stressors and has been proposed to be a general immune response to a variety of physical and biological challenges, including competitors and pathogens [32,51,52]. Tissues associated with PR possess high levels of melanin, an important component of invertebrate innate immunity that can act as a defensive barrier against foreign bodies [52]. Elevated PR levels at high exposure sites may represent an inflammatory-like response by the coral to either sediment particles clogging tentacles and polyp surfaces or to invading pathogens.

Effect of coral growth form on sediment plume-induced disease and other compromised health indicators

The prevalence of disease and other compromised coral health indicators did not vary between coral growth forms (i.e., massive, plating and branching) at high sediment plume exposure sites. The influence of coral morphology (i.e., growth from) on sediment clearing rates has been well investigated [45,53,54], with branching corals generally considered to be more effective at passive sediment removal, while massive and plating forms retain more sediment due to their shapes, which inhibit passive rejection and removal [55]. However, sediment rejection rates and sediment tolerance are not directly related [54]. For example, Stafford-Smith [54] reported some plating species (e.g. *Montipora aequituberculata*) with high sediment tolerance despite poor sediment rejection capacity, whereas some massive species (e.g. *Favia stelligera* and *Leptoria phrygia*) are efficient sediment rejecters but exhibit low sediment tolerance. While previous investigations have focused on only a few coral species in relatively artificial conditions, this is the first field study to investigate the relationship between growth form, sedimentation, turbidity and coral health among all hard coral species present on a natural reef. Although this study indicates that growth form is not a strong predictor of corals' susceptibility to disease and other compromised health indicators associated with increased sedimentation and turbidity, some caution is required in this interpretation, as the most susceptible corals may have died before these surveys were undertaken.

Conclusions. This study provides empirical, field-based evidence linking sedimentation and turbidity to elevated levels of coral disease and other compromised coral health indicators on reefs. While poor water quality has been suggested as a driver of coral disease [13,56,57], little ecological evidence exists linking specific water quality parameters and coral disease data. WS responded strongly to elevated sedimentation and turbidity demonstrating a clear link between water quality and coral disease, though the mechanisms underlying this response remain unclear. Future studies that investigate the response of the coral host (e.g., immune function and energy reserves) and potential pathogens (e.g., shifts in bacterial and viral communities on corals and in surrounding seawater) to elevated sedimentation and turbidity are required. As coastal development intensifies in many parts of the world, the health of coral reefs will depend upon a thorough understanding of the impacts of water quality changes on coral reef health.

Supporting Information

File S1 This file contains supplementary Methods S1, Discussion, Table S1, and Figure S1–Figure S3. Methods S1, Coral genera classified by growth form. Table S1, Mean prevalence of individual coral diseases and other health indicators. Figure S1, Correlations between sediment plume exposure days and the prevalence of disease and other compromised coral health indicators. Figure S2, Non-metric multidimensional scaling (nMDS) plot of coral assemblages. Figure S3, Prevalence of disease and other compromised health by growth form.

Acknowledgments

Disclaimer: The manuscript contents are solely the opinions of the authors and do not constitute a statement of policy, decision, or position on behalf of NOAA or the U.S. Government.

The authors would like to acknowledge Allan Shields and Ben Gryta of the Fisheries Vessel PV Edwards and Rachael Middlebrook and Kevin Crane of the Western Australia Department of Parks and Wildlife and the Office of the Environmental Protection Authority for their assistance in the field. The authors also thank Jean-Baptiste Raina for his assistance with statistical analyses and Timothy Simmonds for providing graphical assistance. This project was funded as part of the Dredging Audit and Surveillance Program by the Gorgon Joint Venture as part of the environmental offsets. The Gorgon Project is a joint venture of the Australian subsidiaries of Chevron, Exxon Mobil, Shell, Osaka Gas, Tokyo Gas and Chubu Electric Power.

Author Contributions

Conceived and designed the experiments: FJP JBL SNF SFH BS GS DGB BLW. Performed the experiments: FJP JBL GS. Analyzed the data: FJP JBL SNF SFH GS. Contributed reagents/materials/analysis tools: SNF SFH DGB BLW. Contributed to the writing of the manuscript: FJP JBL SNF SFH BS GS DGB BLW.

References

1. Gardener T, Cote IM, Gill JA, Grant A, Watkinson AR (2003) Long-term region-wide declines in Caribbean corals. Science 301: 958–960. doi:10.1126/science.1086050.
2. Bruno JF, Selig ER (2007) Regional decline of coral cover in the Indo-Pacific: timing, extent, and subregional comparisons. PLoS One 2: e711. doi:10.1371/journal.pone.0000711.
3. Weil E, Rogers CS (2011) Coral Reef Diseases in the Atlantic-Caribbean. In: Dubinsky Z, Stambler N, editors. Coral Reefs: An Ecosystem in Transition. Dordrecht: Springer Netherlands. pp. 465–491. Available: http://www.springerlink.com/index/10.1007/978-94-007-0114-4_27. Accessed 2014 Feb 9.
4. Rosenberg E, Kelloff CA, Rohwer F (2007) Coral microbiology. Oceanography 20: 146–154.
5. Bourne DG, Garren M, Work TM, Rosenberg E, Smith GW, et al. (2009) Microbial disease and the coral holobiont. Trends Microbiol 17: 554–562. doi:S0966-842X(09)00207-8.
6. Aronson RB, Precht WF (2001) White-band disease and the changing face of Caribbean coral reefs. Hydrobiologia 460: 25–38. doi:10.1023/A:1013103928980.
7. Weil E, Smith G, Gil-Agudelo DL (2006) Status and progress in coral reef disease research. Aquat Organ 69: 1–7.
8. Vollmer SV, Kline DI (2008) Natural disease resistance in threatened staghorn corals. Biol Fac Publ. Available: http://works.bepress.com/svollmer/3/. Accessed 14 August 2013.
9. Harvell CD (1999) Emerging Marine Diseases-Climate Links and Anthropogenic Factors. Science 285: 1505–1510. doi:10.1126/science.285.5433.1505.
10. Porter J, Dustan P, Jaap W, Patterson K, Kosmynin V, et al. (2001) Patterns of spread of coral disease in the Florida Keys. Hydrobiologia 460: 1–24. doi:10.1007/978-94-017-3284-0_1.
11. Hayes ML, Bonaventura J, Mitchell TP, Prospero JM, Shinn EA, et al. (2001) How are climate and marine biological outbreaks functionally linked? Hydrobiologia 460: 213–220. doi:10.1007/978-94-017-3284-0_19.
12. Harvell CD, Mitchell CE, Ward JR, Altizer S, Dobson AP, et al. (2002) Climate warming and disease risks for terrestrial and marine biota. Science 296: 2158–2162. doi:10.1126/science.1063699.
13. Bruno JF, Petes LE, Drew Harvell C, Hettinger A (2003) Nutrient enrichment can increase the severity of coral diseases. Ecol Lett 6: 1056–1061. doi:10.1046/j.1461-0248.2003.00544.x.

14. UNEP/GPA (2006) Protecting coastal and marine environments from land-based activities: a guide for national action. UNEP/Earthprint. Available: http://books.google.com/books?hl = en&lr = &id = VcmpRaT9o44C&oi = fnd&pg = PR1&dq = %22designate+the+overall+lead+that+will+have+to+carry+out+initial+preparation%22+%22taking+national+action+to+protect+the+marine+and+coastal+environment+from%22+%22strategic+planning+tools+such+as+the+National+Programmes+of+Action+(npa+s)%22+&ots = qM2dVdUDgt&sig = RCMMV42GWDaW1bG1Rhx-IzW_SXs. Accessed 2013 Aug 14.
15. Fabricius KE (2005) Effects of terrestrial runoff on the ecology of corals and coral reefs: review and synthesis. Mar Pollut Bull 50: 125–146. doi:10.1016/j.marpolbul.2004.11.028.
16. Harvell D, Jordán-Dahlgren E, Merkel S, Rosenberg E, Raymundo L, et al. (2007) Coral disease, environmental drivers, and the balance between coral and microbial associates. Oceanography 20: 172–195. doi:10.5670/oceanog.2007.91.
17. The World Association for Waterborne Transport Infrastructure (2010) Dredging and port construction around coral reefs. PIANC, 75.
18. Nearing MA, Jetten V, Baffaut C, Cerdan O, Couturier A, et al. (2005) Modeling response of soil erosion and runoff to changes in precipitation and cover. CATENA 61: 131–154. doi:10.1016/j.catena.2005.03.007.
19. Voss JD, Richardson LL (2006) Nutrient enrichment enhances black band disease progression in corals. Coral Reefs 25: 569–576.
20. Vega Thurber R, Willner-Hall D, Rodriguez-Mueller B, Desnues C, Edwards RA, et al. (2009) Metagenomic analysis of stressed coral holobionts. Env Microbiol 11: 2148–2163. doi:10.1111/j.1462-2920.2009.01935.x.
21. Hodgson G (1990) Tetracycline reduces sedimentation damage to corals. Mar Biol 104: 493–496. doi:10.1007/BF01314355.
22. Haapkylä J, Unsworth RKF, Flavell M, Bourne DG, Schaffelke B, et al. (2011) Seasonal rainfall and runoff promote coral disease on an inshore reef. PLoS ONE 6: e16893. doi:10.1371/journal.pone.0016893.
23. Falkowski PG, Jokiel PL, Kinzie RA (1990) Irradiance and corals. In Ecosystems of the World: Coral Reefs (ed Z. . Dubinsky), pp. 89–107. Amsterdam: Elsevier.
24. Richmond RH (1993) Coral reefs: present problems and future concerns resulting from anthropogenic disturbance. Am Zool 33: 524–536. doi:10.1093/icb/33.6.524.
25. Peters EC, Pilson MEQ (1985) A comparative study of the effects of sedimentation on symbiotic and asymbiotic colonies of the coral Astrangia danae Milne Edwards and Haime 1849. J Exp Mar Biol Ecol 92: 215–230. doi:10.1016/0022-0981(85)90096-6.
26. Riegl B, Branch GM (1995) Effects of sediment on the energy budgets of four scleractinian (Bourne 1900) and five alcyonacean (Lamouroux 1816) corals. J Exp Mar Biol Ecol 186: 259–275. doi:10.1016/0022-0981(94)00164-9.
27. Sutherland KP, Porter JW, Torres C (2004) Disease and immunity in Caribbean and Indo-Pacific zooxanthellate corals. Mar Ecol Prog Ser 266: 273–302. doi:10.3354/meps266273.
28. Burke LM, Reytar K, Spalding M, Perry A (2011) Reefs at risk revisited. World Resources Institute Washington, DC. Available: http://www.ww.endoverfishing.org/uploadedFiles/PEG/Publications/Other_Resource/Pew-KeepingtheGreenEconomyBlue-Burke.pdf. Accessed 2013 Aug 21.
29. Western Australia Department of Environment and Conservation (2007) Management plan for the Montebello/Barrow Islands Marine Conservation Reserves 2007–2017. Management Plan.
30. Evans RD, Murray KL, Field SN, Moore JAY, Shedrawi G, et al. (2012) Digitise this! A quick and easy remote sensing method to monitor the daily extent of dredge plumes. PLoS ONE 7: e51668. doi:10.1371/journal.pone.0051668.
31. Beeden R, Willis B, Raymundo L, Page C, Weil E (2008) Underwater cards for assessing coral health on Indo-pacific reefs. Melbourne: Currie Communications.
32. Willis BE, Page CA, Dinsdale EA (2004) Coral Disease on the Great Barrier Reef. Coral Health and Disease. Berlin: Springer-Verlag Publishing. pp. 69–104.
33. Heron SF, Maynard J, Willis B, Christensen T, Harvell CD, et al. (2012) Developments in understanding relationships between environmental conditions and coral disease. Proceedings of the 12th International Coral Reef Symposium. Cairns, Australia. Available: http://eprints.jcu.edu.au/22400/. Accessed 2013 Jan 30.
34. Heron SF, Willis BL, Skirving WJ, Eakin CM, Page CA, et al. (2010) Summer hot snaps and winter conditions: modelling white syndrome outbreaks on Great Barrier Reef corals. PLoS ONE 5: e12210. doi:10.1371/journal.pone.0012210.
35. Maynard JA, Anthony KRN, Harvell CD, Burgman MA, Beeden R, et al. (2011) Predicting outbreaks of a climate-driven coral disease in the Great Barrier Reef. Coral Reefs 30: 485–495. doi:10.1007/s00338-010-0708-0.
36. Anderson MJ, Gorley RN, Clarke KR (2008) PERMANOVA+ for PRIMER: guide to software and statistical methods. Plymouth, UK: PRIMER-E Ltd.
37. Clarke KR, Gorley RN (2006) PRIMER v6: user manual/tutorial. Plymouth, UK: PRIMER-E Ltd.
38. Kline DI, Kuntz NM, Breitbart M, Knowlton N, Rohwer F (2006) Role of elevated organic carbon levels and microbial activity in coral mortality. Mar Ecol Prog Ser 314: 119–125. doi:10.3354/meps314119.
39. Douvere F, Badman T (2012) Mission report: reactive monitoring mission to Great Barrier Reef (Australia) 6th to 14th March 2012. UNESCO World Heritage Centre - IUCN.
40. Masini, R., Jones, R. & Sim, C. 2011 Node 1 - Dredging science: Science plan. Western Australia Marine Science Institute, 23.
41. Aeby GS (2005) Outbreak of coral disease in the Northwestern Hawaiian Islands. Coral Reefs 24: 481–481. doi:10.1007/s00338-005-0493-3.
42. Sussman M, Willis BL, Victor S, Bourne DG (2008) Coral pathogens identified for White Syndrome (WS) epizootics in the Indo-Pacific. PLoS One 3: e2393. doi:10.1371/journal.pone.0002393.
43. Bourne DG, Boyett HV, Henderson ME, Muirhead A, Willis BL (2008) Identification of a ciliate (Oligohymenophorea: Scuticociliatia) associated with brown band disease on corals of the Great Barrier Reef. Appl Environ Microbiol 74: 883–888. doi:10.1128/AEM.01124-07.
44. Onton K, Page CA, Wilson SK, Neale S, Armstrong S (2011) Distribution and drivers of coral disease at Ningaloo reef, Indian Ocean. Mar Ecol Prog Ser 433: 75–84. doi:10.3354/meps09156.
45. Hubbard JAEB, Pocock YP (1972) Sediment rejection by recent scleractinian corals: a key to palaeo-environmental reconstruction. Geol Rundsch 61: 598–626. doi:10.1007/BF01896337.
46. Rogers C (1990) Responses of coral reefs and reef organisms to sedimentation. Mar Ecol Prog Ser 62: 185–202. doi:10.3354/meps062185.
47. Philipp E, Fabricius K (2003) Photophysiological stress in scleractinian corals in response to short-term sedimentation. J Exp Mar Biol Ecol 287: 57–78. doi:10.1016/S0022-0981(02)00495-1.
48. Flores F, Hoogenboom MO, Smith LD, Cooper TF, Abrego D, et al. (2012) Chronic exposure of corals to fine sediments: lethal and sub-lethal impacts. PloS One 7: e37795. doi:10.1371/journal.pone.0037795.
49. Glynn PW (1996) Coral reef bleaching: facts, hypotheses and implications. Glob Change Biol 2: 495–509. doi:10.1111/j.1365-2486.1996.tb00063.x.
50. Brown BE (1997) Coral bleaching: causes and consequences. Coral Reefs 16: 129–138. doi:10.1007/s003380050249.
51. Ravindran J, Raghukumar C (2006) Pink-line syndrome, a physiological crisis in the scleractinian coral *Porites lutea*. Mar Biol 149: 347–356. doi:10.1007/s00227-005-0192-1.
52. Palmer CV, Mydlarz LD, Willis BL (2008) Evidence of an inflammatory-like response in non-normally pigmented tissues of two scleractinian corals. Proc R Soc B Biol Sci 275: 2687–2693. doi:10.1098/rspb.2008.0335.
53. Rogers CS (1983) Sublethal and lethal effects of sediments applied to common Caribbean reef corals in the field. Mar Pollut Bull 14: 378–382. doi:10.1016/0025-326X(83)90602-1.
54. Stafford-Smith MG (1993) Sediment-rejection efficiency of 22 species of Australian scleractinian corals. Mar Biol 115: 229–243. doi:10.1007/BF00346340.
55. Brown BE, Howard LS (1985) Assessing the effects of "stress" on reef corals. Adv Mar Biol 22: 1–63.
56. Wooldridge S, Brodie J, Furnas M (2006) Exposure of inner-shelf reefs to nutrient enriched runoff entering the Great Barrier Reef Lagoon: Post-European changes and the design of water quality targets. Mar Pollut Bull 52: 1467–1479.
57. Wooldridge SA, Done TJ (2009) Improved water quality can ameliorate effects of climate change on corals. Ecol Appl 19: 1492–1499.

19

Hypoxia and Acidification Have Additive and Synergistic Negative Effects on the Growth, Survival, and Metamorphosis of Early Life Stage Bivalves

Christopher J. Gobler*, Elizabeth L. DePasquale, Andrew W. Griffith, Hannes Baumann

School of Marine and Atmospheric Sciences, Stony Brook University, Southampton, New York, United States of America

Abstract

Low oxygen zones in coastal and open ocean ecosystems have expanded in recent decades, a trend that will accelerate with climatic warming. There is growing recognition that low oxygen regions of the ocean are also acidified, a condition that will intensify with rising levels of atmospheric CO_2. Presently, however, the concurrent effects of low oxygen and acidification on marine organisms are largely unknown, as most prior studies of marine hypoxia have not considered pH levels. We experimentally assessed the consequences of hypoxic and acidified water for early life stage bivalves (bay scallops, *Argopecten irradians*, and hard clams, *Mercenaria mercenaria*), marine organisms of significant economic and ecological value and sensitive to climate change. In larval scallops, experimental and naturally-occurring acidification (pH, total scale = 7.4–7.6) reduced survivorship (by >50%), low oxygen (30–50 μM) inhibited growth and metamorphosis (by >50%), and the two stressors combined produced additively negative outcomes. In early life stage clams, however, hypoxic waters led to 30% higher mortality, while acidified waters significantly reduced growth (by 60%). Later stage clams were resistant to hypoxia or acidification separately but experienced significantly (40%) reduced growth rates when exposed to both conditions simultaneously. Collectively, these findings demonstrate that the consequences of low oxygen and acidification for early life stage bivalves, and likely other marine organisms, are more severe than would be predicted by either individual stressor and thus must be considered together when assessing how ocean animals respond to these conditions both today and under future climate change scenarios.

Editor: Sam Dupont, University of Gothenburg, Sweden

Funding: This research was supported by the NOAA's Ocean Acidification Program through award #NA12NOS4780148 from the National Centers for Coastal Ocean Science, the New Tamarind Foundation, and NSF's Biological Oceanography Program (Award #1129622). The funders had no role in study design, data collection and analysis, decision to publish, or preparation of the manuscript.

Competing Interests: The authors have declared that no competing interests exist.

* E-mail: christopher.gobler@stonybrook.edu

Introduction

Low oxygen zones are ubiquitous features in coastal and open ocean environments. Within coastal zones, the excessive delivery of nutrients from agriculture and urban centers stimulates algal productivity, and the subsequent microbial degradation of this organic matter reduces oxygen levels, contributing towards hypoxia [1]. Hypoxic regions of coastal zones have expanded in recent decades [1,2] to the detriment of resident organisms and fisheries [2–6]. In the open ocean, oxygen minimum zones are persistent features of subsurface waters with rates of microbial oxygen consumption that exceed oxygen ventilation rates and are expanding due to ocean warming [7–9]. A less frequently considered consequence of microbial respiration is the production of CO_2 and the resultant reduction in pH through the formation and dissociation of carbonic acid. Through the process of respiration, levels of CO_2 and O_2 are stoichiometrically linked in marine ecosystems, being subsequently altered by differences in gas exchange and chemical equilibria. Recent studies have indeed shown a close correspondence between low oxygen and acidification in many coastal and open ocean systems [10–13], with pCO_2 levels in low oxygen zones often of a magnitude greater than predicted for surface oceans later this century (>1,000 μatm)[10,11,14]. The continued combustion of fossil fuels and subsequent warming of the planet are expected to expand low oxygen and acidified regions in the world's oceans this century [15–17].

Despite the common co-occurrence of hypoxia and acidification in marine systems, their concurrent effects on ocean life are poorly understood. Acidification negatively impacts an array of marine organisms [17–19]. Oxygen is required to sustain aerobic life, and the thresholds at which low dissolved oxygen causes mortality and/or harm have been established for many marine organisms [2–6]. However, the large majority of laboratory studies assessing the effects of low oxygen on marine organisms have manipulated oxygen levels by administering nitrogen gas [e.g. 21–27], a process that reduces both oxygen and pCO_2 levels and thus significantly increases, rather than decreases seawater pH (Figure 1). Hence, the majority of hypoxia studies performed to date have created a laboratory condition (low oxygen, low CO_2, high pH) that does not reflect the actual physiological challenges to marine life in low oxygen zones (Figure 1).

Here, we examine the concurrent impacts of acidification and low oxygen on the early life stages of two calcifying bivalves: bay scallops, *Argopecten irradians*, and hard clams, *Mercenaria mercenaria*. As foundational species within coastal ecosystems [28], calcifying

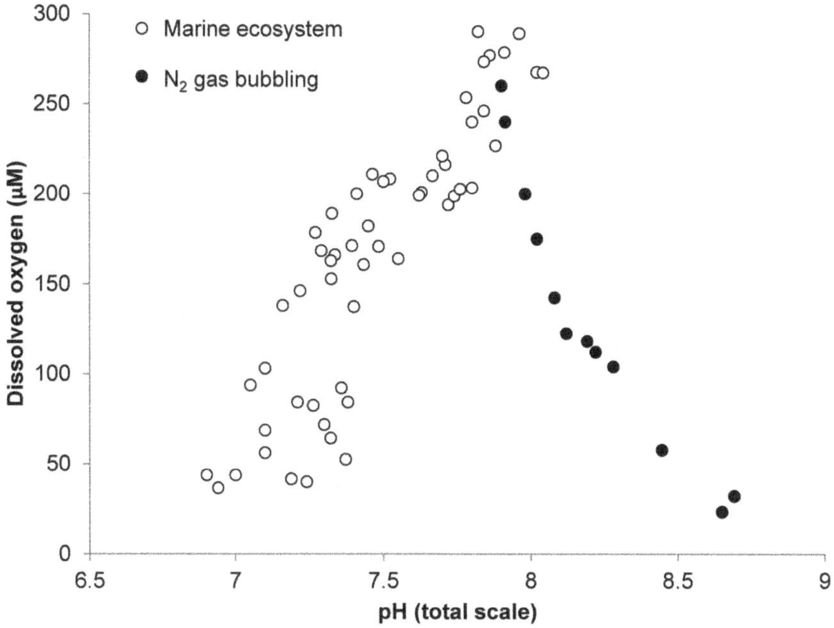

Figure 1. Changes in pH_T (total scale) and dissolved oxygen obtained via bubbling estuarine seawater (salinity = 30) with N_2 gas (black circles) and found naturally (white circles) in the Forge River, NY, USA, during summer 2012.

bivalves such as those studied here are of substantial value to man: mollusk harvests approach one billion USD annually in the US alone [29] with ecosystem services far exceeding that value [30]. The negative impacts of climate change on the early life history stages such as those studied here may promote population bottlenecks in bivalve populations [31]. While early life stage clams and scallops are sensitive to ocean acidification [19,20], the effects of low oxygen on these organisms under acidified conditions have yet to be examined. For this study, we specifically hypothesized that early life growth, development (metamorphosis), and survival of these bivalves would be differently affected by hypoxic vs. acidified conditions, and that concurrent hypoxia and acidification would have additive negative effects. During experiments, levels of dissolved oxygen, pCO_2, and pH_T (pH measured on the total scale) in seawater were achieved by mixing air, CO_2, and N_2 gases, as well as by amending naturally hypoxic and acidified coastal water from the Forge River Estuary, NY, USA, with aeration and sodium carbonate (Figure 2; Tables S1, S2, S3, S4).

Methods

Two types of experiments are presented in this manuscript. The first experiments involved the exposure of early life stage bivalves (larvae and first year juveniles) to normal and low pH/low dissolved oxygen conditions achieved via the delivery of varying mixtures of CO_2 gas, N_2 gas, and air. The second experiment involved the exposure of larval bivalves to water low in low pH and low dissolved oxygen obtained from a hypoxic and acidified ecosystem that was subsequently manipulated via aeration and the addition of sodium carbonate. The first experiment was performed on larval stage bay scallops (*A. irradians*) and two different stages of juvenile hard clams (*M. mercenaria*), whereas the second experiment was performed on larval stage *A. irradians*. For all experiments, replicate ($n = 4$) experimental vessels with early life stage bivalves (described below) were maintained at $24 \pm 1°C$ using commercially available aquarium heaters and chillers (Aquatic Eco-systems, Inc., Florida, USA). This temperature is optimal for growth and survival of early life stages of the two species used here [32,33].

Manipulating pH and dissolved oxygen levels with tanked gas

A series of gas proportionators (Cole Parmer Flowmeter system, multitube frame) were used to deliver CO_2 gas, N_2 gas, and air to seawater at different rates to maintain conditions that were normoxic with a normal pH (air only), hypoxic with a normal pH (tanked N_2 gas pre-mixed with ~390 µatm CO_2 gas), normoxic but acidified (tanked 5% CO_2 gas only), and hypoxic and acidified (tanked N_2 gas, tanked 5% CO_2 gas, and air; Figure 2A). The gas proportionators delivered these gases to experimental vessels filled with 0.2 µm filtered seawater from eastern Shinnecock Bay, New York, USA, to yield the levels of pH and dissolved oxygen desired for experiments using volumetric flow rates that turned over the volume of experimental vessels >100 times daily preventing equilibration with atmospheric CO_2 and oxygen. For experiments with larvae, the gas mixtures from the proportionator system were continuously delivered to the bottom of 1 L high density polyethylene beakers with polycarbonate lids whereas 10 L polyethylene vessels with polyethylene lids were used for experiments with juvenile bivalves. Vessels were bubbled for least 24 h prior to commencing experiments. Dissolved oxygen levels were quantified with a Clark-type electrode YSI 5100 oxygen meter, which provided an analytical precision of ±4% and was calibrated daily. Dissolved oxygen measurements made with this probe were nearly identical to, and never significantly different from, those obtained via Winkler titrations [34]. To quantify precise levels of total dissolved inorganic carbon attained in experimental treatments, water samples were analyzed during experiments with an EGM-4, Environmental Gas Analyzer® (PP Systems) system that quantified total dissolved inorganic carbon levels after separating the gas phase from seawater using a Liqui-Cel® Membrane (Membrana). This instrument provided a methodological precision ±3% for replicated measurements of total dissolved inorganic

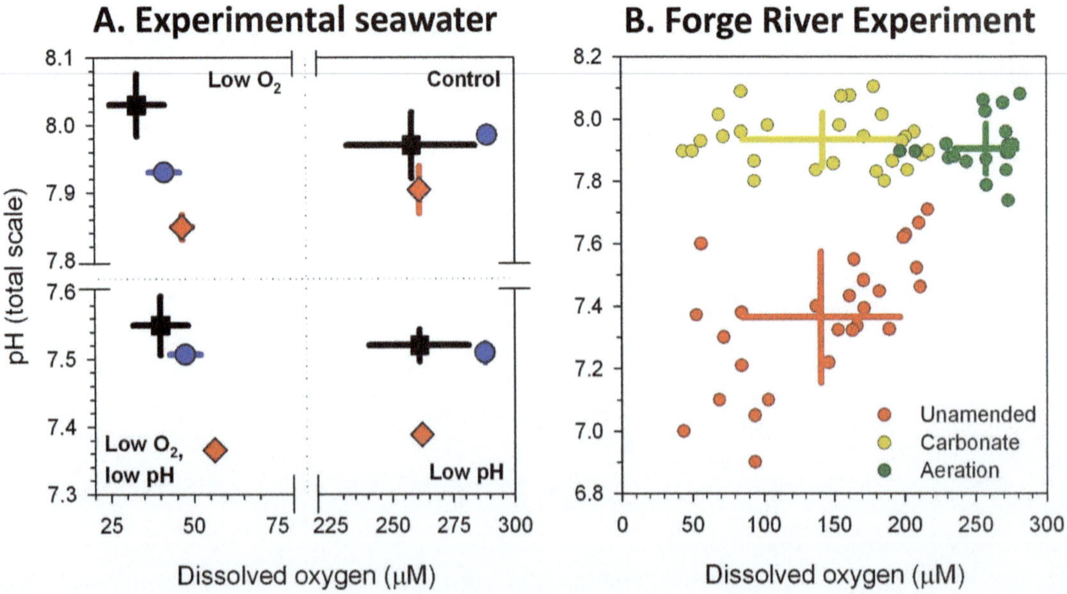

Figure 2. Levels of dissolved oxygen and pH_T during experiments. **A.** Experiments delivering N_2 gas, CO_2 gas, and air to seawater with *A. irradians* larvae (black squares), two month old *M. mercenaria* (red diamonds), or four month old *M. mercenaria* (blue circles) as presented in Figures 2, 4 and 5. Points are means ±1 S.D. **B.** Experiment with water from the Forge River, NY, USA, and *A. irradians* larvae left unamended (red), amended with sodium carbonate (yellow), or aerated (green). Points are means of measurements during individual days of the experiment. Crosses present the mean ±S.D. for the entire experiment.

carbon and provided full recovery (104±5%) of Dr. Andrew Dickson's (University of California San Diego, Scripps Institution of Oceanography) certified reference material for total inorganic carbon in seawater (Batch 117 = DIC 2009.99 μmol dissolved inorganic carbon kg seawater^{-1}). Levels of CO_2 were calculated based on measured levels of total inorganic carbon, pH_T (mol kg seawater^{-1}), temperature, salinity, phosphate, silicate, and first and second dissociation constants of carbonic acid in seawater [35] using the program CO2SYS (http://cdiac.ornl.gov/ftp/co2sys/). Daily measurements of pH used a Honeywell Durafet Ion Sensitive Field Effect Transistor (ISFET)-based pH sensor calibrated with Dr. Dickson's seawater pH standard [36], which indicated that experimental vessels maintained target pH levels throughout experiments (<0.1% relative standard deviation within treatments). Spectrophotometric measurements of pH made using *m*-cresol purple as described by Dickson *et al.* [37] were nearly identical to, and never significantly different from, those obtained with the Durafet pH sensor.

Experiments with naturally hypoxic and acidified water

The Forge River estuary (40°47′71″N, 72°49′85″W) is a small (3 km^2), shallow (2–4 m), microtidal (0.7 m) system exchanging with Moriches Bay, a part of Long Island's south shore lagoonal system in NY, USA [38]. The Forge River receives excessive nutrient loads from dense residential populations and duck farms and experiences hypoxia during summer months [38]. While the two bivalves studied here (*M. mercenaria, A. irradians*) can be found along the entire eastern shore of North America including estuaries that exchange with and surround the Forge River such as such as Moriches Bay and the Peconic Estuary (see below), benthic surveys reveal these organism do not persist in the Forge River (C. Gobler, pers. obs.). For experiments, water from the Forge River was collected ten times during July and August of 2012 from the Town of Brookhaven (NY, USA) public dock at dawn from a depth of ~2 m with a horizontal Van Dorn bottle. Water was transferred without bubbling to 20 L carboys and transported to the Stony Brook-Southampton Marine Sciences Center. Specific permission was not needed for this activity since this was a public location and seawater collection is not regulated by any municipal government. The pH and dissolved oxygen of the water was measured as described above and the salinity of the water was determined using a YSI© 556 sonde. Water from carboys was transferred by gravity, without bubbling to replicate (n = 4) 1 L high density polyethylene beakers with polycarbonate lids for experiments. Experimental water was manipulated with sodium carbonate or aeration to contrast changes in pH (carbonate addition only) to changes in dissolved oxygen and pH concurrently (aeration). Specifically, replicate vessels (n = 4) were left unamended as a control, aerated at 50 ml min^{-1} with ambient air, or administered 26 μM sodium carbonate per 0.1 pH unit difference between the experimental water and a pH value of 8.0. Prior to manipulation, water was always low in oxygen (15–90 μM) and pH (7.1–7.7). The addition of sodium carbonate brought pH, carbonate ion, and aragonite saturation states to values nearly identical to the aeration treatment but did not alter dissolved oxygen while bubbling typically brought pH to ~8 and dissolved oxygen to ~250 μM (Figure 2B). Dissolved oxygen, pH, and precise carbonate chemistry levels were monitored as described above and are reported in the results section and Table S2. Water was exchanged ~ every three days, and levels of dissolved oxygen and pH generally rose between water changes.

Larval experiments

Larval bay scallops (*A. irradians*) were obtained from broodstock collected from the eastern Peconic Estuary that experiences levels of pH, dissolved oxygen, and salinity common for a meso-trophic estuary (pH_{NBS} range: 7.8–8.2, mean ±S.D. = 8.0±0.16; dissolved oxygen: 170–380 μM, mean = 250±60 μM; salinity = 24.6–30.8, mean = 28.1±1.34; [39]). Broodstock were collected by the East Hampton Town Shellfish Hatchery via use of their New York

State Department of Environmental Conservation Collector's Permit that gives them permission for such activity. Neither bivalve species investigated here is endangered or protected. Scallops were spawned at the East Hampton Town Shellfish Hatchery located in Montauk, NY, which uses water from Gardiner's Bay, a meso- to oligotrophic, open water body that exchanges with the Atlantic Ocean (pH_{NBS} range: 7.8–8.2, mean $= 8.0 \pm 0.14$; dissolved oxygen: 210–380 µM, mean $= 280 \pm 60$ µM; salinity $= 25.9$–31.1, mean $= 28.4 \pm 1.42$; [39]. These pH and dissolved oxygen levels are significantly higher than those found in more eutrophic systems such as the Forge River (Figure 2B) and will be considered 'normal' for the purposes of this study. For larval experiments, experimental vessels (1 L) were filled with 0.2 µm filtered seawater from eastern Shinnecock Bay, New York, USA, and within hours of fertilization, larvae were distributed to each treatment beaker at a concentration of ~400 individuals L^{-1}, consistent with post-spawning densities in estuaries [40]. Larvae were fed an ideal food source, *Isochrysis galbana*, (Tahitian strain, T-Iso) at a rate known to be *ad libitum* (4×10^4 cells ml^{-1} d^{-1}) and to yield maximal bivalve larval growth and survivorship through metamorphosis [32,33]. Indeed, *I. galbana* densities declined daily demonstrating that larvae were feeding, but never declined by more than 50%. Cultures of *I. galbana* were maintained in exponential phase growth using standard algal culture conditions. To promote high survivorship, containers in contact with larvae were never exposed to chemicals [41]. To discourage the growth of bacteria during experiments, an antibiotic solution (Sigma-Aldrich No. 4083, penicillin, streptomycin, neomycin) was administered to each beaker at 1% its original concentration at the beginning of each experiment and during each water change. This antibiotic mixture at this concentration has been shown to have no negative effects on the growth and survivorship of shellfish larvae [42]. Every three days, larvae were gently poured onto a 64 µm mesh, and condition (live or dead) and developmental stage (veliger, pediveliger, or metamorphosed) of all larvae were recorded. The percent of the population that had metamorphosed at each stage was calculated for surviving individuals only. Dead larvae were identified by a lack of swimming and movement of the velum as well as a loss of pigmentation and/or fully open valves. Surviving larvae from each beaker were transferred into a new vessel with new filtered seawater, food, and antibiotics within a 30 minute period. Experiments lasted 30–40 days post-spawning, thereby covering the early, most critical period in bivalve ontogeny [31,43,44]. At the end of the experiments, the diameter of surviving individuals was determined using a dissecting microscope equipped with a Nikon DigiSight Color Digital Camera System (DSVil) and ImageJ software. Precise CO_2 levels, dissolved oxygen levels, and complete carbonate chemistry from larval experiments appear in Tables S1 and S2.

Juvenile experiments

For experiments with juvenile *M. mercenaria*, individuals spawned from eastern Peconic Estuary broodstock (estuarine chemistry as described above) were raised on an algal diet under normal pH and dissolved oxygen conditions (levels as described above) at the East Hampton Town Shellfish Hatchery until experiments were performed. Individuals were transferred to seawater matching the salinity and temperature of the experiment for at least 48 h prior to the start of experiments. *M. mercenaria* of two ages and size classes (two month old $= 6.09 \pm 0.65$ mm and four month old $= 8.51 \pm 0.21$ mm) were used for experiments with acidified and low oxygen water achieved via tanked gases. Fifteen individuals of each species were placed into replicate (n = 4), 10 L, high-density polyethylene vessels that were maintained at 20 ± 1°C as described above. Survival of individuals and water chemistry were assessed daily; any dead individuals were removed within <18 h of expiring. Precise CO_2 levels, dissolved oxygen levels, and complete carbonate chemistry from these experiments appear in Tables S3 and S4. *M. mercenaria* were fed *Isochrysis galbana*, (Tahitian strain, T-Iso) at a density of 4×10^4 cells ml^{-1} daily. Initial and final shell lengths (distance from the umbo to the ventral edge) of small individuals were determined by means of image capture and digital imaging software (ImageJ), whereas lengths of larger individuals were determined with calipers. Length-based growth rates were calculated by dividing the change in height by the days of the experiment.

Statistical analyses

Statistical analyses were performed with SigmaStat 4.0 Percent values were arc-sin square root transformed before analyses. Two-way ANOVAs were performed to assess differences among survival, metamorphosis, and growth rates at each CO_2 and dissolved oxygen level, and were followed by Tukey's multiple comparison tests. One-way ANOVAs were performed to assess the effects of aeration and sodium carbonate on the survival, metamorphosis, growth rates on scallop larvae and were followed by Tukey's multiple comparison tests. Standard or Kruskal-Wallis ANOVAs were performed to assess differences in dissolved oxygen and pH attained in each experiment and were followed by Tukey's multiple comparison tests. Linear regression models of larval scallop survival during experiments were developed to derive daily mortality rates. Values reported in the results sections are means ± one standard deviation.

Results

In experiments where air, CO_2, and N_2 gases were mixed to create hypoxic and acidified conditions for *A. irradians* larvae, levels of dissolved oxygen were significantly lower in the hypoxic and combined low oxygen and acidified treatment (36.5 ± 8.93 µM) relative to the other treatments (260 ± 23.5 µM), while levels of pH were significantly lower in the acidified and combined low oxygen and acidified treatments ($pH_T = 7.53 \pm 0.04$) relative to the other treatments (8.00 ± 0.03; $p<0.05$; Figure 2A). These lower pH levels were accompanied by aragonite saturation states (0.79 ± 0.06) that were significantly lower than the control and low oxygen treatments (1.94 ± 0.14; $p<0.05$; Table S1). Under these conditions, larval scallops (*A. irradians*) experienced significantly reduced survival (Figure 3A; $p<0.05$; Table S5) in response to acidification but not hypoxia (Figure 2A), and there was no interaction between the treatments (Table S5). Specifically, daily mortality rates of individuals exposed to normal pH and oxygen, low oxygen, low pH, and low oxygen combined with low pH treatments were 1.5, 1.5, 1.9, and 1.8% per day while total survivorship after 40 d was 42%, 41%, 24%, and 20%, respectively (Figure 3A). In contrast to survival, low oxygen levels significantly reduced the size and delayed metamorphosis of scallop larvae (Figures 3B, 4A; $p<0.001$ for each, respectively Tables S6, S7), while low pH alone did not. The treatments interacted to yield less than additive percentage of metamorphosed larvae while there was no interaction with regard to size (Tables S6, S7). After 25 d, 72% and 56% of surviving individuals had metamorphosed in the control and acidified treatments, while none had metamorphosed in the low oxygen treatments with or without low pH (Figure 3B). After 40 d, 97% and 91% of individuals had metamorphosed in the control and acidified treatments, while only 9% and 2% had done so in the low oxygen treatments with or without low pH (Figure 3B).

Metamorphosed individuals from normal oxygen treatments at normal and low pH conditions measured 1,130±16.8 μm and 996±36.7 μm in length, respectively, whereas metamorphosed individuals from low oxygen treatments at normal or low pH conditions were only 445±32.0 and 447±35.0 μm, respectively. Individuals that failed to metamorphose after 40 d in these treatments were even smaller (175±10.9 and 177+14.1 μm; Figure 4A).

Water from Forge River Estuary used for experiments had levels of dissolved oxygen (140±56 μM), pH_T (7.37±0.21), carbonate ion (15±9.3 μM), and aragonite saturation (0.23±0.15) that were all significantly lower than the levels produced via the aeration of this water (260±22 μM, 7.91±0.08, 120±24 μM, 1.89±0.15, respectively; $p<0.05$; Figure 2B; Table S2). Aeration did not alter the alkalinity of the water (1848±119 μM for aerated and unamended treatments). The addition of sodium carbonate created water with levels of pH_T (7.93±0.08), carbonate ion (98±14 μM), alkalinity (2360±186 μM), and aragonite saturation (1.51±0.24) that were all significantly greater than the unamended Forge River water ($p<0.05$; Figure 2B; Tables S2), but this treatment did not significantly alter levels of dissolved oxygen (142±57 μM; Figure 2B; Tables S2).

Exposure of larval scallops to estuarine waters that were naturally acidified and low in oxygen caused low rates of growth, survival, and metamorphosis, while raising pH or oxygen levels significantly improved different aspects of their performance. For example, while only 10% of larvae survived a 33 d exposure to water with naturally low pH and low oxygen (Figure 2B), the addition of sodium carbonate (pH increase) and aeration (oxygen and pH increase) each significantly increased larval survival ~ four-fold (to ~40%; Figure 3C; $p<0.005$; Table S8). During the month long experiment, population mortality rates were 3.2, 2.2, and 2.1% d^{-1} in the unamended, carbonate addition, and aeration treatments, respectively. In contrast, aeration alone significantly increased the size and metamorphic rates of scallop larvae, while sodium carbonate alone did not (Figure 3D, 4B; $p<0.001$; Table S9, S10). Specifically, while >95% of surviving individuals in aerated treatments had metamorphosed after 33 d, 0 and 10% had in the unamended and sodium carbonate treatments (Figure 3D). Similarly, while individuals that were aerated had metamorphosed and grown to 835 μm in length after 33 d, those in the unamended and sodium carbonate treatments had not metamorphosed and were <100 μm (Figure 4B).

Hypoxia and acidification had contrasting and age-dependent effects on juvenile hard clams (*M. mercenaria*). In experiments with early stage juveniles (two-month old), levels of dissolved oxygen were significantly lower in the hypoxic and combined low oxygen and acidified treatment (51.5±6.86 μM) relative to the other treatments (262±6.86 μM), while levels of pH were significantly lower in the acidified and combined low oxygen and acidified treatments ($pH_T = 7.51 \pm 0.01$) relative to the other treatments (7.88±0.04; $p<0.05$; Figure 2a). These lower pH levels were accompanied by aragonite saturation states (0.71±0.01) that were significantly lower than the control and low oxygen treatments (1.93±0.07; $p<0.05$; Table S3). Exposure of two-month old juvenile clams to low oxygen significantly reduced their survival (from 87 to 60%; $p<0.05$), while low pH (Figure 2A) did not, and there was no interaction between these treatments (Figure 5A; $p<0.05$, Table S11). In contrast, the low pH water significantly reduced the growth of surviving clams by 60% (from 19 to 7.9 μm d^{-1}; Figure 5B; $p<0.05$; Table S12), while low oxygen did not, and there was no interaction between treatments (Table S12).

In experiments with later stage juvenile clams (four months old), levels of dissolved oxygen were significantly lower in the hypoxic and combined low oxygen and acidified treatment (44.3±4.31 μM) relative to the other treatments (289±4.31 μM) while levels of pH were significantly lower in the acidified and combined low oxygen and acidified treatments ($pH_T = 7.51 \pm 0.01$) relative to the other treatments (7.96±0.03; $p<0.05$; Figure 2a). These lower pH levels were accompanied by aragonite saturation states (0.70±0.01) that were significantly lower than the control and low oxygen treatments (2.08±0.14; $p<0.05$; Table S4). Four month old clams were more resistant to low oxygen and pH conditions, experiencing high survival in all treatments (>95%). Although their growth rates were not altered by either low oxygen or low pH when administered alone, there was a synergistically negative interaction between these factors as the combined exposure to low oxygen and low pH reduced their growth rates by 40% from 12 μm d^{-1} to 7.3 μm d^{-1} (Figure 5C; $p<0.05$; Table S13).

Discussion

Regions of the world's oceans with low levels of dissolved oxygen and pH have become increasingly common and will continue to expand in the near future [16,17], yet the concurrent effects of these two stressors on marine life are largely unknown. Acidification has been shown to inhibit the performance of many calcifying invertebrates [18–20] as well as some vertebrates, including fish [45–48]. Our findings suggest that the true impact of low oxygen zones on marine life likely requires re-evaluation, as most laboratory studies to date have manipulated oxygen with by administering nitrogen gas [e.g. 21–27] resulting in concurrent basification rather than acidification (Figure 1). Given that this approach misrepresents natural marine ecosystems (Figure 1), it may mask the true effects of low oxygen zones on marine animals. For example, while low oxygen severely delayed the metamorphosis and greatly restricted the growth of larval scallops, their survival was only compromised when they were concurrently exposed to the acidified conditions. While juvenile clams responded differently to hypoxia and acidification, the full effects of these processes could not have been predicted by considering only one of these treatments alone. For example, while early stage juvenile clams suffered enhanced mortality under hypoxic conditions, their growth rates were depressed by acidification. More surprising was the synergistically negative effect of concurrent acidification and hypoxia on the growth rates of late stage clams, which was not detected in either individual treatment. While the two species of bivalves studied here have been shown to be differentially susceptible to acidification [19], the present study demonstrates that they also respond differentially to hypoxia and acidification. As such, it is likely that making broad predictions regarding the response of ocean animals to these stressors will prove challenging.

Prior ecosystem studies investigating the distribution of coastal marine organisms within low oxygen zones have often not measured pH or carbonate chemistry and have interpreted findings solely through the lens of oxygen concentrations [e.g. 49–52]. By not concurrently considering acidification within these systems, the processes driving distribution patterns may be misinterpreted. In contrast, a recent open ocean study of oxygen minimum zones (OMZs) found that pteropods migrating into OMZs are more resistant to acidification than those that do not [53], suggesting that acidification may be as important as oxygen in shaping the distribution of these organisms within and around these zones.

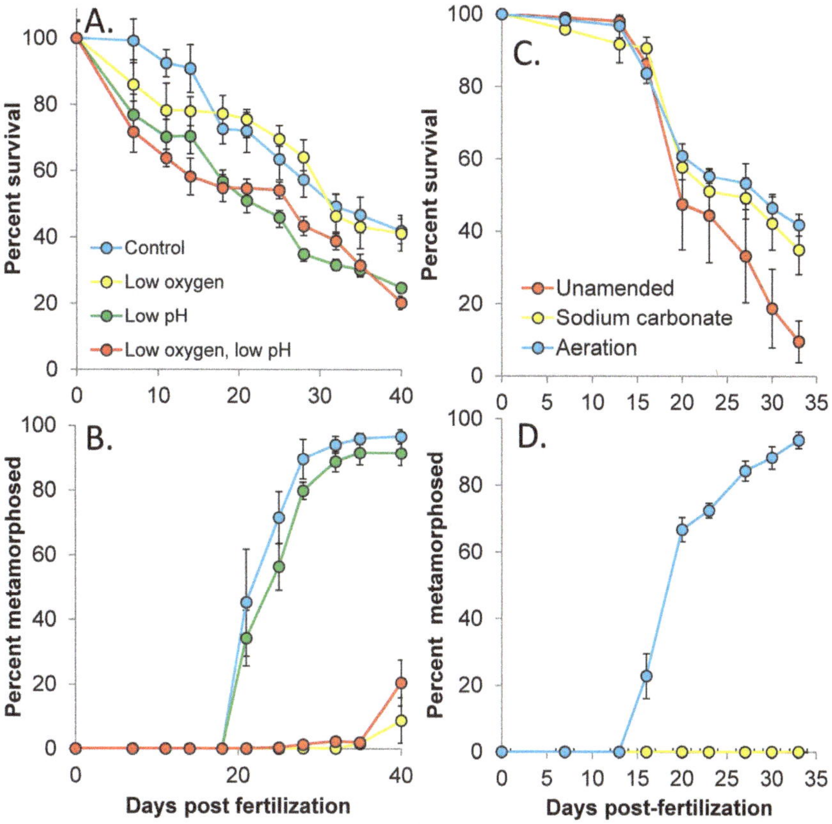

Figure 3. Performance of *A. irradians* larvae exposed to two levels of dissolved oxygen and pH_T. (A) Percent survival and (B) percent metamorphosis of surviving *A. irradians* larvae during exposure to two levels of pH and dissolved oxygen. (C) Percent survival and (D) percent metamorphosis of surviving *A. irradians* larvae in water naturally low in dissolved oxygen and pH (umamended), low oxygen with normal low pH (via the addition of sodium carbonate), or with normal oxygen and pH (via aeration with and without sodium carbonate). There was no metamorphosis in the unamended and sodium carbonate treatments. Points are means ±1 S.D. Dissolved oxygen and pH levels are presented in Figures 2. Complete carbonate chemistry is given in Tables S1 and S2, while statistical analyses appear in Tables S5, S6, S8, and S9.

Recent research has robustly demonstrated the sensitive nature of early life stage bivalves to the levels of ocean acidification projected for the end of this century [19,20]. Concurrently, it has been established that some eutrophic coastal waters are already experiencing seasonal acidification [10–14]. Our novel experiments with water naturally low oxygen and low pH demonstrate that these conditions negatively impact larval shellfish. For example, consistent with our experiments using mixed gases, increasing the pH of acidified coastal waters with sodium carbonate significantly improved larval survival, thereby suggesting that contemporary acidification in eutrophic coastal zones may already be reducing the survival of bivalve larvae and perhaps other organisms. Water used for experiments had levels of pH and pCO_2 (7.36±0.21 and 2620±781 µatm, respectively; Table S2) that are projected to occur in open ocean surface waters under 'business-as-usual' scenarios in hundreds of years [54,55]. As such, eutrophic coastal zones may be relatively less impacted by future atmospheric CO_2 driven acidification than open ocean environments [56] but may also provide early insight into the manner in which future climate change may shape the distribution and abundance of ocean animals [14,57,58]. Interestingly, mean oxygen levels during our experiment with eutrophic coastal water were low (<140 µM) but higher than levels achieved during our mixed gas experiments, yet they still reduced the size and delayed the metamorphosis of larval scallops. Given that larval bivalves with extended metamorphosis times or of smaller sizes are susceptible to higher cumulative predation rates [59,60], mortality rates of larvae exposed to low oxygen conditions in the wild may be even greater than those observed during our experiments. Such elevated mortality among early life stages of bivalves would be expected to influence the distribution and abundance of adult populations [31,44]. While ammonium in the Forge River was not at levels known to be toxic (<25 µM), additional stressors present in hypoxic waters such as sulfides may have further inhibited the performance of scallop larvae. Hence, future research regarding the effects of hypoxic zones on marine life may need to consider stressors beyond oxygen and pH such as elevated sulfides, ammonium, and temperature [3,61,62].

Our findings coupled with prior studies provide insight regarding the differing physiological mechanisms by which hypoxia and acidification may impact early life stage bivalves. Marine invertebrates commonly adapt to environmental stressors by reducing their metabolic rate, and over an extended period of time this can translate in lowered growth rates and smaller sizes [63,64]. Accordingly, acidification and hypoxia both contributed toward smaller individuals in the experiments presented here. In a manner consistent with prior studies [21,22], it was exposure of larval stages to low oxygen, however, that had the largest negative effect on the growth of individuals. Larvae also experienced delayed metamorphosis suggesting these processes are linked and that the lower metabolic state induced by hypoxia does not afford larvae the aerobic scope to initiate metamorphosis. Interestingly,

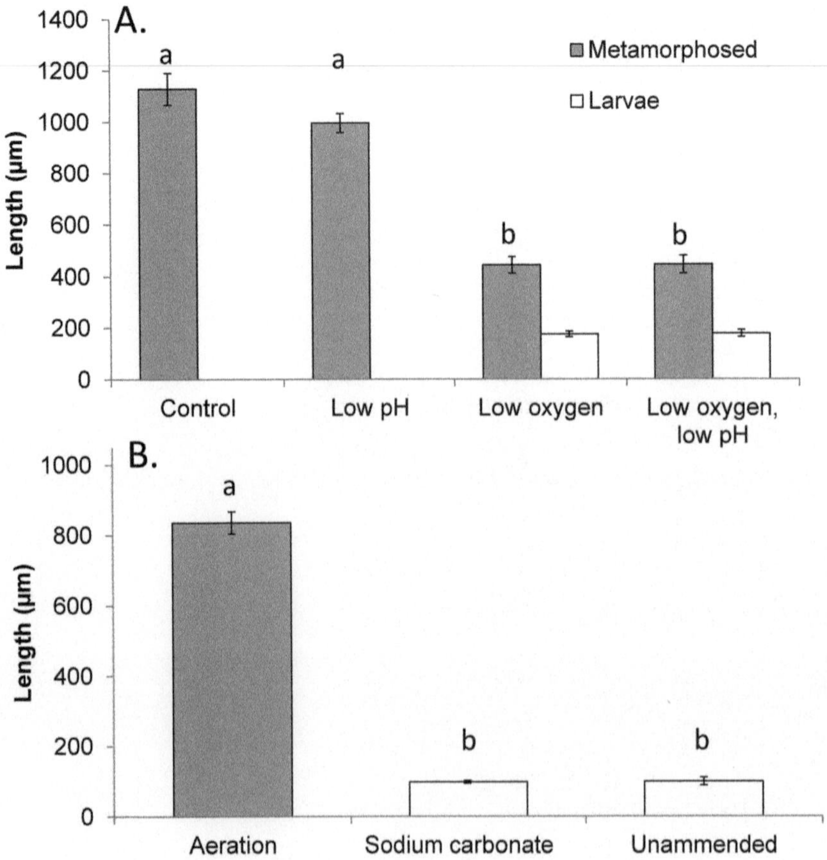

Figure 4. Size of early life stage *A. irradians* larvae, grown under two levels of pH and dissolved oxygen. A. Experiment with mixing gases, and B. Experiment manipulating naturally low oxygen and low pH water with aeration and sodium carbonate Bars are means ±S.D. Complete carbonate chemistry from experiments appear in Tables S1 and S2 while statistical analyses appear in Tables S7 and S10. Shared lower case letters indicate treatments that are not significantly different ($p>0.05$). Statistical comparisons were among metamorphosed individuals only in panel A, but was between individuals of the same cohort that were metamorphosed individuals in the aerated treatment and larvae in other treatments in panel B as no individuals had metamorphosed in those treatments.

larvae exposed to hypoxia (but not anoxia) commonly do not suffer enhanced mortality [21,22, this study], suggesting that the physiological response of reduced metabolism, growth, and metamorphosis are sufficient for sustaining individuals during extended hypoxia, at least during the larval stage.

Regarding acidification, research has demonstrated that lowered pH and the associated reduction in carbonate ion availability can impact multiple physiological pathways in bivalves beginning with calcification [65,66]. Lowered rates of calcification can have cascading effects on larval physiology, inhibiting other biochemical processes such as RNA synthesis and lipid content [19,66] as well as more general indicators of metabolism such as growth and size [20,40]. For some species of bivalve larvae, including *A. irradians* examined here, exposure to undersaturated levels of calcium carbonate during their first week of development can elicit depressed rates of survival [19,66–68]. Consistent with this finding, the largest differences between survival in acidification and normal pH treatments were observed during exposure of larval stage bivalve to acidification. Hence, simply lowering metabolic rates may not be a sufficient physiological response to overcome the compromised calcification associated with acidification in the early stages of some larval bivalves.

There may be a physiological feedback and synergy between low oxygen and low pH in marine invertebrates [63]. As environmental acidification and low oxygen slow metabolic rates, internal gas exchange may slow, intracellular oxygen levels may be reduced, and CO_2 from respiration may accumulate, potentially exacerbating hypercapnia and thus further lowering metabolism. This is consistent with our finding that the combination of low oxygen and low pH, but not either factor alone, inhibited the growth of the later stage hard clams. Finally, our results suggest that, in a manner consistent with the age-dependent effects of ocean acidification on early life stage bivalves [69], the effects of hypoxia and acidification on early life stage bivalves are also age dependent. While larvae exposed to both stressors experienced, up to 75% reduction in survival, early stage juvenile survival was more modestly reduced by the two stressors (20%). The two stressors combined only reduced growth (by 20%) not survival, in later stage juveniles.

Hypoxia and acidification are increasingly co-occurring in the ocean [10,11,14, this study], and understanding the effects of such climate change stressors on marine animals is a significant challenge. In low oxygen regions of the ocean, the tissue demand for oxygen of marine animals is often not met by ambient supply, and thus the ability of these organisms to tolerate secondary stressors such as acidification could be compromised [63,64]. Thus, it is possible that even animals insensitive to acidification alone could be negatively affected by low pH when concurrently exposed to hypoxic conditions. Given that fish are generally more vulnerable to low oxygen than mollusks such as the bivalves

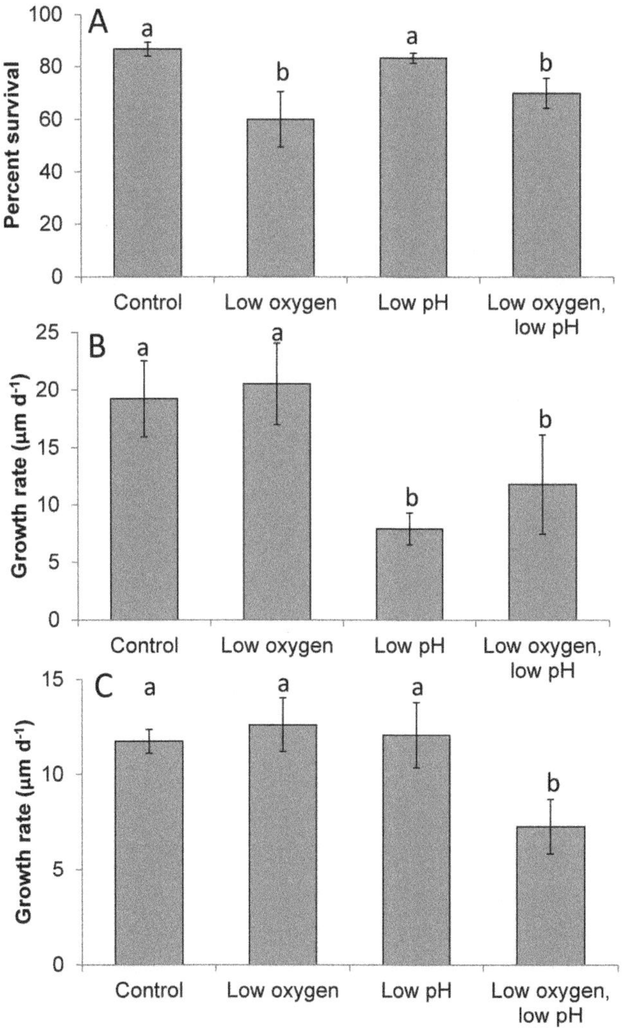

Figure 5. Performance of *M. mercenaria* exposure to two levels of pH and dissolved oxygen. A. Percent survival of two month old individuals, **B.** Growth of two month old individuals, **C.** Growth of four month old individuals. Bars are means ±S.D. Dissolved oxygen and pH levels are presented in Figure 2. Complete carbonate chemistry appears in Tables S3 and S4, while statistical analyses appear in Tables S11, S12, and 13. Shared lower case letters indicate treatments that are not significantly different (p>0.05).

studied here [4] and that some early life stage fish have already been shown to be sensitive to acidification [45–47], the dual effects of hypoxia and acidification on marine teleosts may be profound.

Because hypoxia threatens fisheries and biodiversity in coastal ecosystems [2,4], environmental regulatory agencies often set management goals for dissolved oxygen to protect estuarine animals based on levels known to negatively affect marine life (e.g., 2 mg L^{-1} O_2).[4,70–72]. Given our finding that concurrent acidification and low oxygen significantly and independently depressed growth and survival rates of estuarine bivalves, it is plausible that coastal managerial criteria based strictly on oxygen levels, but not pH, may not adequately protect marine life in some ecosystems. Therefore, future environmental regulations developed to protect estuarine organisms in regions prone to hypoxia should consider the concurrent effects of acidification on these animals, particularly as climate change accelerates the intensity of acidification in coastal zones.

The expansion of hypoxic zones in coastal and open oceans is one of many predicted consequences of intensifying global climate change [9,15,17]. While microbial processes will continue to promote acidification in these regions [11], this will be exacerbated by the loading of anthropogenic CO_2 [16–18]. We suggest that recently discovered high CO_2 sensitivities in many finfish and shellfish [20,45–47], and the compounded effects of high CO_2 and low O_2 in bivalves relative to each individual parameter should prompt a re-alignment of future studies. A comprehensive evaluation of the combined effects of low oxygen and acidification on marine life is critical for understanding how ocean ecosystems respond to these conditions both today and under future climate change scenarios.

Supporting Information

Table S1 Mean temperature, pH, dissolved oxygen, carbonate chemistry, alkalinity, and salinity (±1 SD) during the experiment exposing larval stage *Argopecten irradians* to differing levels of pH and dissolved oxygen achieved via mixing tanked gases.

Table S2 Mean temperature, pH, dissolved oxygen, carbonate chemistry, alkalinity, and salinity (±1 SD) during the experiment exposing larval stage *Argopecten irradians* to differing levels of pH and dissolved oxygen achieved via the addition of aeration and sodium carbonate to estuarine water with naturally low oxygen and low pH.

Table S3 Mean temperature, pH, dissolved oxygen, carbonate chemistry, alkalinity, and salinity (±1 SD) during the experiment exposing two month old *Mercenaria mercenaria* to differing levels of pH and dissolved oxygen achieved via mixing tanked gases.

Table S4 Mean temperature, pH, dissolved oxygen, carbonate chemistry, alkalinity, and salinity (±1 SD) during the experiment exposing four month old *Mercenaria mercenaria* to differing levels of pH and dissolved oxygen achieved via mixing tanked gases.

Table S5 Two-way analysis of variance for *Argopecten irradians* larval survival when exposed to two levels of dissolved oxygen and pH.

Table S6 Two-way analysis of variance for *Argopecten irradians* larval metamorphosis when exposed to two levels of dissolved oxygen and pH.

Table S7 Two-way analysis of variance for *Argopecten irradians* larval size when exposed to two levels of dissolved oxygen and pH.

Table S8 One-way analysis of variance for *Argopecten irradians* larval survival when exposed to hypoxic and acidified Forge River water amended with sodium carbonate or aeration.

Table S9 One-way analysis of variance for *Argopecten irradians* larval metamorphosis when exposed to

hypoxic and acidified Forge River water amended with sodium carbonate or aeration.

Table S10 One-way analysis of variance for *Argopecten irradians* larval size when exposed to hypoxic and acidified Forge River water amended with sodium carbonate or aeration.

Table S11 Two-way analysis of variance for survival of two-month old *Mercenaria mercenaria* exposed to two levels of dissolved oxygen and pH.

Table S12 Two-way analysis of variance for growth rates of two-month old *Mercenaria mercenaria* exposed to two levels of dissolved oxygen and pH.

Table S13 Two-way analysis of variance for growth rates of four-month old *Mercenaria mercenaria* exposed to two levels of dissolved oxygen and pH.

Acknowledgments

We gratefully acknowledge the assistance of H. Mittelsdorf and J. Goleski. We are grateful for the supply of *Mercenaria mercenaria* and *Argopecten irradians* from Cornell Cooperative Extension, Southold, NY and from the East Hampton Shellfish Hatchery, East Hampton, NY.

Author Contributions

Conceived and designed the experiments: CJG ED. Performed the experiments: CJG ELD AWG. Analyzed the data: CJG ELD AWG HB. Contributed reagents/materials/analysis tools: CJG. Wrote the paper: CJG ELD AWG HB.

References

1. Rabalais NN, Turner RE, Wiseman WJ (2002) Gulf of Mexico hypoxia, AKA "The dead zone" Annual Review of Ecology and Systematics 33: 235–263.
2. Diaz RJ, Rosenberg R (2008) Spreading dead zones and consequences for marine ecosystems. Science 321: 926–929.
3. Gray JS, Wu RSS, Or YY (2002) Effects of hypoxia and organic enrichment on the coastal marine environment. Marine Ecology Progress Series 238: 249–279.
4. Vaquer-Sunyer R, Duarte CM (2008) Thresholds of hypoxia for marine biodiversity. Proceedings of the National Academy of Sciences of the USA 105: 15452–15457.
5. Levin LA, Ekau W, Gooday AJ, Jorissen F, Middelburg JJ, et al. (2009) Effects of natural and human-induced hypoxia on coastal benthos. Biogeosciences 6: 2063–2098.
6. Ekau W, Auel H, Portner HO, Gilbert D (2010) Impacts of hypoxia on the structure and processes in pelagic communities (zooplankton, macro-invertebrates and fish. Biogeosciences 7: 1669–1699.
7. Stramma L, Johnson G, Sprintall J, Mohrholz V (2008) Expanding oxygen-minimum zones in the tropical oceans. Science 320: 655–658.
8. Shaffer G, Olsen SM, Pedersen JOP (2009) Long-term ocean oxygen depletion in response to carbon dioxide emissions from fossil fuels. Nature Geoscience 2: 105–109.
9. Keeling RF, Kortzinger A, Gruber N (2010) Ocean deoxygenation in a warming world. Annual Review of Marine Science 2: 199–229.
10. Feely RA, Aline SR, Newton J, Sabine CL, Warner M, et al. (2010) The combined effects of ocean acidification, mixing, and respiration on pH and carbonate saturation in an urbanized estuary. Estuarine, Coastal, and Shelf Science 88: 442–449.
11. Cai WJ, Hu XP, Huang WJ, Murrell MC, Lehrter JC, et al. (2011) Acidification of subsurface coastal waters enhanced by eutrophication. Nature Geoscience 4: 766–770.
12. Paulmier A, Ruiz-Pino D, Garcon V (2011) CO2 maximum in the oxygen minimum zone (OMZ). Biogeosciences 8: 239–252.
13. Waldbusser GG, Voigt EP, Bergschneider H, Green MA, Newell RIE (2011) Biocalcification in the Eastern Oyster (Crassostrea virginica) in relation to long-term trends in Chesapeake Bay pH. Estuaries and Coasts 34: 221–231.
14. Melzner F, Thomsen J, Koeve W, Oschlies A, Gutowska MA, et al. (2012) Future ocean acidification will be amplified by hypoxia in coastal habitats. Marine Biology DOI: 10.1007/s00227-012-1954-1.
15. Justic D, Rabalais NN, Turner RE (1996) Effects of climate change on hypoxia in coastal waters: a doubled CO2 scenario for the northern Gulf of Mexico. Limnology and Oceanography 41: 992–1003.
16. Gruber N (2011). Warming up, turning sour and losing breath: Ocean biogeochemistry under global change. Philosophical Transactions of the Royal Society of London Series A 369: 1980–1996.
17. Doney SC, Ruckelshaus M, Duffy JE, et al. (2012) Climate change impacts on marine ecosystems. Annual Review of Marine Science 4: 11–37.
18. Doney SC, Fabry VJ, Feely RA, Kleypas JA (2009) Ocean acidification: the other CO2 problem. Annual Review of Marine Science 1: 169–192.
19. Talmage SC, Gobler CJ (2010) Effects of past, present, and future ocean carbon dioxide concentrations on the growth and survival of larval shellfish. Proceedings of the National Academy of Sciences of the USA 107: 17246–17251.
20. Gazeau F, Parker LM, Comeau S, Gattuso J-P, O'Connor WA, et al. (2013) Impacts of ocean acidification on marine shelled molluscs. Marine Biology DOI 10.1007/s00227-013-2219-3.
21. Baker SM, Mann R (1994) Description of metamorphic phases in the oyster *Crassostrea virginica* and effects of hypoxia on metamorphosis. Marine Ecology Progress Series 104: 91–99.
22. Wang W, Widdows J (1991) Physiological responses of mussel larvae Mytilus edulis to environmental hypoxia and anoxia. Marine Ecology Progress Series 70: 223–236.
23. Breitburg DL, Loher T, Pacey CA, Gerstein A (1997) Varying effects of low dissolved oxygen on trophic interactions in an estuarine food web. Ecological Monographs 67: 489–507.
24. Gracey AY, Troll JV, Somero GN (2001) Hypoxia-induced gene expression profiling in the euryoxic fish Gillichthys mirabilis. Proceedings of the National Academy of Sciences of the USA 98: 1993–1998.
25. Norkko J, Pilditch CA, Thrush SF, Wells RMG (2005) Effects of food availability and hypoxia on bivalves: the value of using multiple parameters to measure bivalve condition in environmental studies. Marine Ecology Progress Series 298: 205–218.
26. Wang YJ, Hu MH, Cheung SG, Shin PKS, Lu WQ, et al. (2013) Antipredatory responses of Perna viridis (Linnaeus, 1758) under acute hypoxia and low salinity. Journal of Molluscan Studies 79: 42–50.
27. Clark MS, Husmann G, Thorne MAS, Burns G, Truebano M, et al. (2013) Hypoxia impacts large adults first: consequences in a warming world. Global Change Biology 19: 2251–2263.
28. Newell RIE (2004) Ecosystem influences of natural and cultivated populations of suspension-feeding bivalve molluscs: A review. Journal of Shellfish Research 23: 51–61.
29. Cooley SR, Doney SC (2009) Anticipating ocean acidification's economic consequences for commercial fisheries. Environmental Research Letters 4: 024007.
30. Costanza R, Darge R, Degroot R, Farber S, Grasso M, et al. (1997) The value of the world's ecosystem services and natural capital. Nature 387: 253–260.
31. Dupont S, Dorey N, Thorndyke M (2010) What meta-analysis can tell us about vulnerability of marine biodiversity to ocean acidification? Estuarine, Coastal, and Shelf Science 89:182–185.
32. Carriker MR (2001) Embryogenesis and organogenesis of veligers and early juveniles, p. 77–115. In J. N. Kraeuter and M. Castagna [eds.], Biology of the Hard Clam. Elsevier Science.
33. Cragg SM (2006) Development, physiology, behaviour, and ecology of scallop larvae. In: Shumway SE, Parsons GJ Scallops: Biology, ecology, and aquaculture. Elsevier, pp 45–122.
34. Grashoff K, Ehrhardt M, Kremling K 1983) Methods of seawater analysis. Weinheim, Deerfield Beach, Florida, p 419. ISBN:3-527-2599-8 (Weinheim) 0-89573-7 (Deerfield Beach).
35. Roy RN, Roy LN, Vogel KM, Porter-Moore C, Pearson T, et al. (1993) The dissociation constants of carbonic acid in seawater at salinities 5 to 45 and temperatures 0 to 45 C. Marine Chemistry 44: 249–267.
36. Dickson AG (1993) pH buffers for sea-water media based on the total hydrogen-ion concentration scale. Deep-Sea Research I 40: 107–118.
37. Dickson AG, Sabine CL, Christian JR (2007) Guide to best practices for ocean CO2 measurements. PICES Special Publication 3, 191 pp.
38. Swanson RL, Brownawell B, Wilson RE, O'Connell C (2010) What history reveals about Forge River pollution on Long Island, New York's south shore. Marine Pollution Bulletin 60: 804–818.
39. S.C.D.H.S. (2010) Suffolk County Department of Health Services, Annual Report of Surface Water Quality. Suffolk County, NY, USA.
40. Kraeuter JN, Castagna M (2001) Biology of the Hard Clam. Elsevier. Amsterdam, Netherlands.
41. Padilla DK, Doall MH, Gobler CJ, Hartson A, O'Boyle K (2006) Brown tide alga, Aureococcus anophagefferens, can affect growth but not survivorship of Mercenaria mercenaria larvae. Harmful Algae 5: 736–748.
42. Talmage SC, Gobler CJ (2009) The effects of elevated carbon dioxide concentrations on the metamorphosis, size, and survival of larval hard clams

(*Mercenaria mercenaria*), bay scallops (*Argopecten irradians*), and Eastern oysters (*Crassostrea virginica*). Limnology and Oceanography 54: 2072–2080.
43. Arnold WS (2008) Application of larval release for restocking and stock enhancement of coastal marine bivalve populations. Reviews in Fisheries Sciences 16: 65–71.
44. Grosberg RK, Levitan DR (1992) For adults only? Supply-side ecology and the history of larval biology. Trends in Ecology and Evolution 7: 130–133.
45. Munday PL, Dixson DL, Donelson JM, Jones GP, Pratchett MS, et al. (2010) Replenishment of fish populations is threatened by ocean acidification. Proceedings of the National Academy of Sciences of the USA 107: 12930–12934.
46. Baumann H, Talmage SC, Gobler CJ (2012) Reduced early life growth and survival in a fish as a direct response to elevated CO2 levels. Nature Climate Change 2: 38–41.
47. Frommel AY, Maneja R, Lowe D, Malzahn AM, Geffen AJ, et al. (2012) Severe tissue damage in Atlantic cod larvae under increasing ocean acidification. Nature Climate Change 2: 35–37.
48. Bignami S, Enochs IC, Manzello DP, Sponaugle S, Cowen RK (2013) Ocean acidification alters the otoliths of a pantropical fish species with implications for sensory function. Proceedings of the National Academy of Sciences of the USA 110: 7366–7370.
49. Pihl L, Baden SP, Diaz RJ (1991) Effects of periodic hypoxia on the distribution of demersal fish and crustaceans. Marine Biology 108: 349–360.
50. Wannamaker CM, Rice JA (2000) Effects of hypoxia on movements and behavior of selected estuarine organisms from the southeastern United States. Journal of Experimental Marine Biology and Ecology 249: 145–163.
51. Grantham BA, Chan F, Nielsen KJ, et al. (2004) Upwelling-driven nearshore hypoxia signals ecosystem and oceanographic changes in the northeast Pacific. Nature 429: 749–754.
52. Eerkes-Medrano D, Menge BA, Sislak C, Langdon CJ (2013) Contrasting effects of hypoxic conditions on survivorship of planktonic larvae of rocky intertidal invertebrates. Marine Ecology Progress Series 478: 139–151.
53. Maas AE, Wishner KF, Seibel BA (2012) The metabolic response of pteropods to acidification reflects natural CO2-exposure in oxygen minimum zones. Biogeosciences 9: 747–757.
54. IPCC (2007) Intergovernmental Panel on Climate Change. Summary for Policymakers. In The Physical Sciences Basis. Working Group I Contribution to the Fourth Assessment Report of the IPCC. Solomon et al. Cambridge University Press, Cambridge.
55. Le Quere C, Raupach MR, Canadell JG, Marland G, et al. (2009) Trends in the sources and sinks of carbon dioxide. Nature Geoscience 2: 831–836.
56. Duarte CM (2013) Is ocean acidification an open-ocean syndrome? Understanding anthropogenic impacts on seawater pH. Estuaries and Coasts 36: 221–236.
57. Hofmann GE, Smith JE, Johnson KS, Send U, Levin LA, et al. (2011) High-frequency dynamics of ocean pH: A multi-ecosystem comparison. PLoS One 6: e28983.
58. Kelly MW, Padilla-Gamiño JL, Hofmann GE (2013) Natural variation and the capacity to adapt to ocean acidification in the keystone sea urchin *Strongylocentrotus purpuratus*. Global Change Biology 19: 2536–2546.
59. Tamburri MN, Zimmer-Faust RK (1996) Suspension feeding: Basic mechanisms controlling recognition and ingestion of larvae. Limnology and Oceanography 41: 1188–1197.
60. Gosselin LA, Qian PY (1997) Juvenile mortality in benthic marine invertebrates. Marine Ecology Progress Series 146: 265–282.
61. Parker LM, Ross PM, O'Connor WA (2010) Comparing the effect of elevated pCO2 and temperature on the fertilization and early development of two species of oysters. Marine Biology 157: 2435–2452.
62. Talmage SC, Gobler CJ (2011) Effects of elevated temperature and carbon dioxide on the growth and survival of larvae and juveniles of three species of Northwest Atlantic bivalves. PLoS One 6: e26941.
63. Portner HO, Langenbuch M, Michaelidis B (2005) Synergistic effects of temperature extremes, hypoxia, and increases in CO2 on marine animals: From earth history to global change. Journal of Geophysical Research – Oceans 110: C09S10.
64. Sokolova IM (2013) Energy-limited tolerance to stress as a conceptual framework to integrate the effects of multiple stressors. Integrative and Comparative Biology DOI: 10.1093/icb/ict028.
65. Waldbusser GG, Brunner EL, Haley BA, Hales B, Langdon CJ, et al. (2013) A developmental and energetic basis linking larval 1 oyster shell formation to ocean acidification. In press with Journal of Geophysical Research – Oceans.
66. Gobler CJ, Talmage SC (2013) Short and long term consequences of larval stage exposure to constantly and ephemerally elevated carbon dioxide for marine bivalve populations. Biogeosciences 10: 2241–2253.
67. Barton A, Hales B, Waldbusser GG, Langdon C, Feely RA (2012) The Pacific oyster, *Crassostrea gigas*, shows negative correlation to naturally elevated carbon dioxide levels: implications for near-term ocean acidification effects. Limnology Oceanography 57: 698–710.
68. Parker LM, Ross PM, O'Connor WA, Borysko L, Raftos DA, et al. (2012) Adult exposure influences offspring response to ocean acidification in oysters. Global Change Biology (2012) 18, 82–92.
69. Waldbusser GG, Bergschneider H, Green MA (2010) Size-dependent pH effect on calcification in post-larval hard clam Mercenaria spp. Marine Ecology Progress Series 417: 171–182.
70. Renaud ML (1986) Hypoxia in Louisiana coastal waters during 1983: Implications for fisheries. Fisheries Bulletin 84:19–26.
71. USEPA (2000) Ambient Aquatic Life Water Quality Criteria for Dissolved Oxygen (Saltwater): Cape Cod to Cape Hatteras. United States Environmental Protection Agency. Report EPA-822-R-00-012. Washington DC, USA.
72. Llanso RJ, Scott LC, Hyland L, Hyland JL, Russell DE (2002) An estuarine benthic index of biotic integrity for the Mid-Atlantic region of the United States. II. Index Development. Estuaries. 25:1231–1242.

Comparison of Bacterial Communities in Sands and Water at Beaches with Bacterial Water Quality Violations

Elizabeth Halliday[1], Sandra L. McLellan[2], Linda A. Amaral-Zettler[3,4], Mitchell L. Sogin[3], Rebecca J. Gast[1]*

1 Biology Department, Woods Hole Oceanographic Institution, Woods Hole, Massachusetts, United States of America, **2** School of Freshwater Sciences, Great Lakes Water Institute, University of Wisconsin-Milwaukee, Milwaukee, Wisconsin, United States of America, **3** Josephine Bay Paul Center, Marine Biological Laboratory, Woods Hole, Massachusetts, United States of America, **4** Department of Geosciences, Brown University, Providence, Rhode Island, United States of America

Abstract

Recreational water quality, as measured by culturable fecal indicator bacteria (FIB), may be influenced by persistent populations of these bacteria in local sands or wrack, in addition to varied fecal inputs from human and/or animal sources. In this study, pyrosequencing was used to generate short sequence tags of the 16S hypervariable region ribosomal DNA from shallow water samples and from sand samples collected at the high tide line and at the intertidal water line at sites with and without FIB exceedance events. These data were used to examine the sand and water bacterial communities to assess the similarity between samples, and to determine the impact of water quality exceedance events on the community composition. Sequences belonging to a group of bacteria previously identified as alternative fecal indicators were also analyzed in relationship to water quality violation events. We found that sand and water samples hosted distinctly different overall bacterial communities, and there was greater similarity in the community composition between coastal water samples from two distant sites. The dissimilarity between high tide and intertidal sand bacterial communities, although more similar to each other than to water, corresponded to greater tidal range between the samples. Within the group of alternative fecal indicators greater similarity was observed within sand and water from the same site, likely reflecting the anthropogenic contribution at each beach. This study supports the growing evidence that community-based molecular tools can be leveraged to identify the sources and potential impact of fecal pollution in the environment, and furthermore suggests that a more diverse bacterial community in beach sand and water may reflect a less contaminated site and better water quality.

Editor: Zhi Zhou, National University of Singapore, Singapore

Funding: This work was supported by the National Science Foundation grant OCE-0430724, and the National Institute of Environmental Health Sciences grant P0ES012742 to the Woods Hole Center for Ocean and Human Health. E. Halliday was partially supported by WHOI Academic Programs and grants from the WHOI Ocean Ventures Fund and the WHOI Coastal Ocean Institute. The funders had no role in study design, data collection and analysis, decision to publish, or preparation of the manuscript.

Competing Interests: The authors have declared that no competing interests exist.

* E-mail: rgast@whoi.edu

Introduction

Fecal indicator bacteria are increasingly well-documented in sands at a range of freshwater and marine beaches of varied climates [1,2,3,4,5], bringing into question whether their ability to persist in the surfzone environment compromises their utility as proxies for the risk of contact with pathogens in recreational waters. Contaminated sands may impact beachgoers either by negatively contributing to bathing water quality through cycles of deposition and resuspension of bacteria between sand and water [6,7,8], or perhaps be more directly harmful through physical contact with sands or ingestion of sands. Epidemiological studies that examined the health outcomes associated with beach sand activities revealed that increased interaction with sands (e.g. digging in, being buried in) corresponded to increased outcomes of illness [9], most commonly gastrointestinal illness but also including skin, eye, ear and respiratory infections. Interaction with sands that have a higher amount of fecal pollution (as measured by both molecular and culture-based quantification of the marine fecal indicator *Enterococcus*) also corresponded to increased outcomes of illness among beachgoers [10].

To date, a variety of viral, bacterial and eukaryotic human pathogens have been recovered from environmental beach sands [11,12,13,14], but it is difficult to predict which specific pathogens may be present at a beach at any given time due to variations in sources of human and animal fecal pollution, as well as whether the fecal pathogens detected in sands are abundant enough to present a significant risk of illness [15]. In the environment, it is likely that fecal indicators and pathogens respond differently to the complex interactions between environmental variables such as moisture, temperature and sunlight [11,12,16,17,18] as well as to the ecological pressures presented by the indigenous microbial community [19,20]. While intertidal beach sands are frequently in contact with the overlying water, previous studies have indicated that biofilm formation on sand grains effectively maintains separation between bacterial communities in sands, porewater, and overlying water [21] and that sands and sediments have distinctly different communities than those found in overlying water [22,23]. While recent studies have indicated that water quality and sand quality are linked [24,25], we have a limited understanding about how the complex environmental bacterial communities in sands, which may include fecal indicators and pathogens, are related to the putative pollution events that are detected through routine culture-based water quality monitoring. We especially lack information about how episodic water quality violations affect the bacteriological quality of surfzone sands, and whether signatures of human contamination can be identified in the sand during water quality violation events or at other times, which could help us better understand how human contamination persists in the environment and impacts humans [26].

The goal of this study was to leverage pyrosequencing datasets of 16S hypervariable region ribosomal DNA to survey total and constituent components of the bacterial communities at two beaches. Samples of dry sand, intertidal sand, and the overlying waters were collected during two-week periods at locations on the west and east coast of the United States, on days when water quality violations occurred as well as on days representative of baseline fecal indicator bacteria concentrations. At one site (Avalon, CA) waters frequently violate water quality standards and have been identified as being impacted by decaying sewage infrastructure; at the second site (Provincetown, MA) water quality violations are infrequent and of unknown origin, but have a history of occurring during both wet and dry weather.

Short sequence tags were used to examine the total bacterial community composition, the presence of sequence tags belonging to the marine fecal indicator *Enterococcus*, and the presence of a broader group of alternative fecal indicators derived from studies of human sewage [27]. Pyrosequencing was particularly advantageous for this combination of broad and specific queries, since the depth of sampling enables detection of many low-abundance members of the community and can document shifts in the community structure over time [21,28]. Molecular methods (e.g., ARISA, T-RFLP, clone libraries and 454 sequencing) have been used to document bacterial diversity and community structure in similar environments such as submerged marine sediments [29,30], lake sediments [31], and shallow subtidal sands [21,23,32,33], and these previous studies focused on the impact of temporal/spatial, physical, chemical or biological disturbance events on bacterial community structure. The results presented here expand the body of information on marine sand and water microbial communities by utilizing sequence tags to examine the impact of water quality exceedance events due to suspected anthropogenic input.

Materials and Methods

Field Sites and Sample Collection

Water, wet sand (covered in approximately 10 cm of water) and dry sand at the high tide line were sampled at Avalon Bay Beach (Catalina Island, CA, **Figure S1**) from the end of July through the beginning of September, 2007. The site locations and qPCR methods used for enumerating enterococci in sands are described in detail elsewhere [34]. Briefly, sand cores were collected in triplicate by hand in 50 mL sterile Falcon tubes from three sites spanning a 200 m along-shore transect. Water from the sites was filtered onto 47 mm 0.22 µm pore size Durapore® filters (GVWP04700; Millipore) and 100 mL was also filtered to enumerate culturable enterococci using the EPA 1600 method [35]. All samples for DNA analysis were frozen and shipped to Woods Hole, MA, where they were kept at −80°C prior to genomic DNA extraction. *Enterococcus* spp. were quantified via qPCR using primers targeting the 23S rDNA [36]. Wet and dry sand from three days, each a week apart, were chosen for pyrosequencing analysis based on differences in water quality as per official beach closures, and also based on the relative amount of enterococci DNA present in the sands. On August 11th and August 18th the water complied with bacterial health standards, but sands differed in the amount of enterococci as detected by qPCR with August 11th having relatively elevated enterococci and August 18th having a low level of enterococci [34]. The third time point, August 25th, corresponded to a violation of the bacterial water quality standard and elevated enterococci in sands. These samples were designated as Avalon (AV) water, wet sand or dry sand (H/W/D respectively) from day 1, 2, or 3 (detailed in **Table 1**).

In the summer of 2009, the beach at 333 Commercial St. (Provincetown, MA, **Figure S1**) was sampled three days per week from mid June through the end of July. Samples from five days that represented a range of water quality and wet weather conditions were chosen for pyrosequencing analysis. These samples were designated as Provincetown (PTW) water, wet sand or dry sand (H/W/D respectively) from days numbered 1–5 (detailed in **Table 1**). Ancillary environmental data were collected at each sampling event, including the temperature of the sample (water temperature, wet sand temperature, dry sand temperature), the tidal range prior to the sampling event (reflecting variations in spring and neap cycles), the level of the tide during the sampling event, the amount of precipitation within the previous 24 h, and the amount of enterococci cultured from the sample.

Ethics statement. Permits are not required to collect water and beach sand samples from the public beaches in California and Massachusetts. No protected species were sampled.

Genomic DNA Extraction and 454 Pyrosequencing

The UltraClean Mega Prep soil DNA kit (MoBio Laboratories, Inc., Solana Beach, CA) was used to extract genomic DNA from 9.0 g of a wet-weight sand composite of three replicates taken from the sand surface at the beach. DNA was extracted from filtered water samples using a modified combination of hot detergent lysis buffer and mechanical disruption as previously described [37]. Eluted DNA was checked for purity with a NanoDrop spectrophotometer before PCR was used to amplify the V6 hypervariable region. Avalon sand samples were sequenced first, on a Roche Genome Sequencer GS-FLX using standard protocols [38], and at the time were limited toV6 amplicon libraries of tag sequences 60 bp long. By 2009 and the second sequencing run, the sequence tag read length had improved to 250 bp, so combined V4 and V6 amplicon libraries were sequenced. In all samples, sequences of adapters and primers were trimmed and low-quality reads removed as described previously [39]. Taxonomy was assigned through the Global Alignment for Sequence Taxonomy (GAST) using a 16S hypervariable region reference database [39]. GAST assigns taxonomy to a tag based on a two-thirds majority vote of the taxonomy of the nearest full-length relatives using a threshold of >80% sequence similarity. Taxonomical assignments within samples are archived and publicly available for comparison on the Visualization and Analysis of Microbial Population Structures (VAMPS) project website (http://www.vamps.mbl.edu).

Data Analysis

To compare diversity among sample types and sites, 8050 sequence tags were randomly subsampled from each sample to minimize the impact of varied sequencing depth among samples. For all other analyses, sequence tag data was normalized to relative abundance within the sample for analysis and visualization. The statistical software package PRIMER-E [40] was used to analyze the relative abundance data of sequence tags successfully assigned to taxa within our samples, with the one-way ANOSIM testing significance of difference between groups of samples based on differences in site, sample type and water quality violation

Table 1. Environmental samples sequenced in this study.

Archived Sample ID (VAMPS):	#Sequence Tags	Species Richness	Sample	Date:	Abb. in paper:	>104 CFU?	Culturable ENT density
Avalon Bay, CA							
RJG_PTW_Bv6v4_AVB_0001_2007_08_11	35993	686	Water	08-11-2007	AV 1H	No*	55+/−123 CFU/100 mL[1]
RJG_PTW_Bv6v4_AVB_0002_2007_08_18	21647	470	Water	08-18-2007	AV 2H	No*	156+/−176 CFU/100 mL[1]
RJG_PTW_Bv6v4_AVB_0003_2007_08_25	27480	506	Water	08-25-2007	AV 3H	Yes	147+/−42 CFU/100 mL[1]
RJG_BSC_Bv6_AV081107W	9269	363	Wet Sand	08-11-2007	AV 1W	No	256+/−1056 MPN/100 g sand[2]
RJG_BSC_Bv6_AV081807W	8050	342	Wet Sand	08-18-2007	AV 2W	No	124+/−317 MPN/100 g sand[2]
RJG_BSC_Bv6_AV082507W	9437	399	Wet Sand	08-25-2007	AV 3W	Yes	83+/−63 MPN/100 g sand[2]
RJG_BSC_Bv6_AV081107D	9477	391	Dry Sand	08-11-2007	AV 1D	No	N/A
RJG_BSC_Bv6_AV081807D	11006	388	Dry Sand	08-18-2007	AV 2D	No	N/A
RJG_BSC_Bv6_AV082507D	11907	445	Dry Sand	08-25-2007	AV 3D	Yes	N/A
Provincetown, MA							
Archived Sample ID (VAMPS):	**#Sequence Tags**	**Species Richness**	**Sample**	**Date:**	**Abb. in paper:**	**>104 CFU?**	**Culturable ENT density**
RJG_PTW_Bv6v4_PTW_0001_2009_06_29	22757	433	Water	06-29-2009	PTW 1H	Yes	140 CFU/100 mL[3]
RJG_PTW_Bv6v4_PTW_0002_2009_06_30	26481	592	Water	06-30-2009	PTW 2H	No	35 CFU/100 mL[3]
RJG_PTW_Bv6v4_PTW_0003_2009_07_07	27124	624	Water	07-05-2009	PTW 3H	No	30 CFU/100 mL[3]
RJG_PTW_Bv6v4_PTW_0004_2009_07_12	30829	677	Water	07-12-2009	PTW 4H	Yes	185 CFU/100 mL[3]
RJG_PTW_Bv6v4_PTW_0005_2009_07_13	30569	662	Water	07-13-2009	PTW 5H	No	90 CFU/100 mL[3]
RJG_PTW_Bv6v4_PTW_0006_2009_06_29	43267	604	Wet Sand	06-29-2009	PTW 1W	Yes	169 CFU/100 g dry weight sand[3]
RJG_PTW_Bv6v4_PTW_0007_2009_06_30	38079	645	Wet Sand	06-30-2009	PTW 2W	No	95 CFU/100 g dry weight sand[3]
RJG_PTW_Bv6v4_PTW_0008_2009_07_07	38282	648	Wet Sand	07-05-2009	PTW 3W	No	715 CFU/100 g sand[3]
RJG_PTW_Bv6v4_PTW_0009_2009_07_12	41572	668	Wet Sand	07-12-2009	PTW 4W	Yes	858 CFU/100 g sand[3]
RJG_PTW_Bv6v4_PTW_0010_2009_07_13	31806	514	Wet Sand	07-13-2009	PTW 5W	No	217 CFU/100 g sand[3]
RJG_PTW_Bv6v4_PTW_0011_2009_06_29	28326	619	Dry Sand	06-29-2009	PTW 1D	Yes	2927 CFU/100 g sand[3]
RJG_PTW_Bv6v4_PTW_0012_2009_06_30	36023	730	Dry Sand	06-30-2009	PTW 2D	No	2419 CFU/100 g sand[3]
RJG_PTW_Bv6v4_PTW_0013_2009_07_07	28202	600	Dry Sand	07-05-2009	PTW 3D	No	329 CFU/100 g sand[3]
RJG_PTW_Bv6v4_PTW_0014_2009_07_12	29688	605	Dry Sand	07-12-2009	PTW 4D	Yes	10745 CFU/100 g sand[3]
RJG_PTW_Bv6v4_PTW_0015_2009_07_13	33590	660	Dry Sand	07-13-2009	PTW 5D	No	342 CFU/100 g sand[3]

[1]Geometric mean +/− standard deviation of samples collected by SCCWRP at three sites along the beach (Figure 1) at 8am using USEPA1600.
*Based on posted results from health department, which utilized USEPA Method 1600 to detect CFU/100 mL. On the 25th, all water quality samples were >104 CFU/100 mL, whereas the other days were more variable as illustrated in the standard deviation of the SCCWRP results.
[2]MPN/100 g sand, calculated by SCCWRP from a single sample collected on the beach by SCCWRP using the IDEXX Enterolert assay.
[3]Calculated by E. Halliday using USEPA Method 1600 for water and modified for sands as previously described [51], normalizing CFU to dry weight sand.

events. Multidimensional scaling analysis was used to generate graphical representations with Non-metric Multi-dimensional Scaling (NMDS) plots of relative differences in community composition between samples from Avalon and Provincetown. The BIOENV rank-correlation procedure was used with the Provincetown samples to determine which combinations of variables best explain patterns in the sequence tag abundance data. The SIMPER routine was used to identify the specific sequence tags with the greatest contribution to the dissimilarity observed between samples. To assess if fecal organisms present in beach sand were associated with sewage, sequence tags belonging to three orders (Bifidobacteriales, Bacteroidales and Clostridiales) were extracted from the total datasets for comparison between samples. These were directly compared to sewage datasets (SML_SWG_Bv6) [27] archived on the VAMPS website.

Results

Community Diversity and Structure

Thirty-nine phyla were represented among the 630,858 total and 2,349 unique bacterial sequence tags recovered from the twenty-four sand and water samples in this study. Sequencing depth and total species richness within individual samples are documented in **Table 1**, and normalized species richness for each sample is illustrated in the rarefaction curves of **Figure S2**. The Shannon-Weaver diversity index was calculated for each sample, and the average indices for water, wet sand, and dry sand at each of the sites (**Figure 1**) show bacterial communities in sand to be more diverse than those in water, and that greater diversity was present at the Provincetown site than at the Avalon site.

The average distribution of sequence tags among dominant phyla in each sample type is presented in **Figure 2**. Proteobacteria dominate the sequence tags from water samples regardless of site; within this phylum, the orders Alphaproteobacteria and Gammaproteobacteria contain the majority of sequence tags, which are present in a ratio of approximately 2:1 respectively. Sequence tags assigned to SAR11 (*Pelagibacter spp.*) were the most abundant unique tag sequence in water samples accounting for, on average, 15% of the sequence tags in Provincetown water and 25% of the sequence tags in Avalon water samples. Likewise, SAR116 and SAR 86 accounted for 2% and 5% of the total tags in Provincetown water samples, and 4% and 8% of total tags in Avalon water samples. In water samples, the phyla Bacteroidetes followed Proteobacteria in dominance, with strong representation from species in the *Flavobacteriaceae* family. Several species within the *Flavobacteriaceae* family had abundances >1% of the total tags per sample at both sites, and one unique sequence (unidentified to genus) represented 7% and 6% of tags at Avalon and Provincetown as well as 1% of tags in wet sands at both sites and dry sands at Provincetown. The Cyanobacteria were also abundant, including a unique *Synechococcus spp.* sequence that accounted for 2% of total tags in water at both sites. Proteobacteria, Bacteroidetes, Cyanobacteria and the Verrucomicrobia together included approximately 95% of the bacterial sequence tags from waters at Avalon and Provincetown.

In the Avalon beach sands, five phyla collectively contained > 90% of the successfully identified tags (**Figure 2**); these were the Proteobacteria, Bacteroidetes, Actinobacteria, Planctomycetes and Acidobacteria. Three other phyla, the Chloroflexi, Gemmatimonadetes and Verrucomicrobia, were present at appreciable relative abundance (>1% of total phyla tags). Within the Avalon sand Proteobacteria, the Alphaproteobacteria and Gammaproteobacteria were again the most abundantly represented orders, but sequence tags from the Deltaproteobacteria were also relatively abundant, containing approximately 5% of the total sequence tags from sand samples. One unique sequence tag within the Gammaproteobacteria (belonging to the *Sinobacteraceae* family) accounted for 8% and 9% of tags in Avalon wet and dry sand samples, as well as 9% and 3% in Provincetown wet and dry sand samples. Provincetown sands were dominated by the same phyla found at Avalon, but phyla that were minor components at Avalon (Chloroflexi, Cyanobacteria, and Verrucomicrobia) claimed a greater proportion of sequence tags in Provincetown sands, generally at the expense of the Acidobacteria and Actinobacteria.

Among unique sequence tags with abundances accounting for >1% of the total sequence tags in sand types at each of the sites, several species were notably present in both Provincetown and Avalon sands. Within the Acidobacteria, these included a *Chloroacidobacterium* spp. sequence tag which accounted for 2% of tags in wet and dry sand at both sites, and a sequence assigned only to the class Holophage that accounted for 2% of tags in Provincetown wet sand samples and 6% of tags in Avalon wet and dry sand samples.

Within the Actinobacteria, a unique sequence tag belonging to the order Acidimicrobiales accounted for 9% of tags in Avalon wet and dry sand and 3% of tags in Provincetown wet sand. A unique sequence tag identified as *Iamia* spp. accounted for 3% of tags from Avalon wet and dry sand, and 2% and 1% of the sequence tags from Provincetown wet and dry sands respectively. *Nitriliruptor* spp. constituted 1% of sequence tags in Provincetown dry sand as well as in Avalon wet and dry sand.

Within the Bacteroidetes, the Sphingobacteriales were strongly represented in sands including several unique sequence tags that were abundant at all sites; these included a sequence identified as *Haliscomenobacter* that accounted for 3–4% of tags in all sand samples at both sites, a sequence identified to the family *Rhodothermaceae* that included 11% of tags from Provincetown dry sand, 3% of tags in Provincetown wet sand, 2% of Avalon wet tags and 3% of Avalon dry sand tags.

Community Composition based on Shared Sequence Tags

Although the bacterial phyla and some of the species dominating beach sand or water communities at both sites were broadly similar, analysis of the distribution of specific tag sequences among samples yielded a more localized view of community composition. A one-way Analysis of Similarity test (ANOSIM) rejected the null hypothesis that there was no significant difference in total community structure at the level of individual sequences based on the sample type (water, wet sand, dry sand) with a Global R of 0.698 ($p \leq 0.001$). Likewise, a one-way ANOSIM rejected the null hypothesis that there were no significant differences in total community structure at the level of individual sequences between sites (Avalon vs. Provincetown) with a weak but still significant Global R of 0.3 ($p \leq 0.01$). The NMDS ordination (**Figure 3**) illustrates the split between water samples and sand samples (groupings differentiated with >50% sequence tag similarity) and within that, samples are further separated by site (differentiated with >60% sequence tag similarity). Within the Provincetown sand samples, there are further groupings that correspond to wet and dry sand, distinguishing communities that are geographically separated by a tidal range that is greater than Avalon's. Although other temporal influences cannot be precluded, the differences between water samples at Provincetown may in part be attributed to tidal stage, as water samples collected during high (1H and 2H) and low tides (3H, 4H, 5H) are still distinct groups despite being quite similar to each other (<70%).

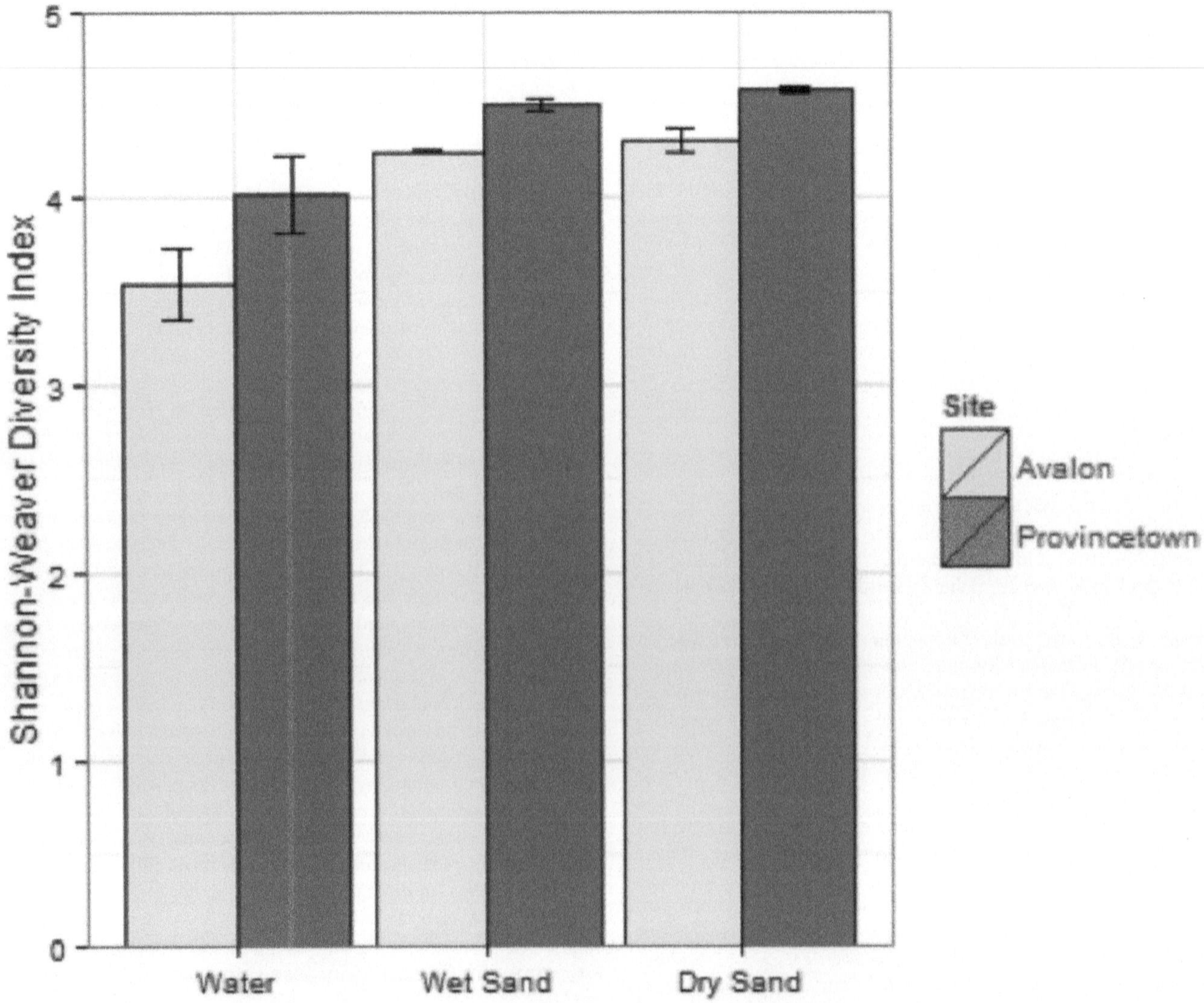

Figure 1. Diversity of bacterial communities in sands and water at Avalon and Provincetown. Error bars indicate two standard errors for the indices calculated for individual samples (n = 3 (days) at Avalon, n = 5 (days) at Provincetown).

Environmental Variables Influencing Total Community Composition at Provincetown

Environmental variables including sample temperature, tidal range, water level, precipitation and amount of enterococci were used in the BIOENV rank-correlation procedure, in which variables or combinations of variables are selected based on how they best explain patterns in the sequence tag abundance data. Among Provincetown water samples, the variables water temperature and tidal range were selected together to produce the best possible $r = 0.758$. The only relationship between the environmental variables and wet sand bacterial samples was the single selection of tidal range with an $r = 0.164$. Dry sand samples had a better relationship to environmental variables with $r = 0.467$ for dry sand temperature. Thus, sample temperature and tidal range emerged as the most important environmental variables shaping total community structure in water, and sand community structure was explained to a lesser extent by these variables.

Community Composition of Alternative Indicator Sequence tag Sample Subsets

A previous study used 454 sequencing technology to examine both human waste and wastewater treatment plant influent, and identified a group of bacteria as potential alternative indicators of fecal pollution belonging to the orders Clostridiales, Bifidobacteriales and Bacteroidales [27]. In our study, sequence tags identified within these orders from the total beach sand and water tag datasets were compared to the sewage tag datasets to determine whether there were trends in any of the sample types that corresponded to water quality violation events. Overall, although all samples contained sequence tags belonging to these orders, there was very little overlap between the specific sequence tags recovered from sewage and environmental samples. The only putative tag signature shared between sewage and environmental samples belonged to the order Clostridiales and was identified as *Roseburia.* This sequence tag was found in Provincetown water and wet sand on the day of and following the dry weather exceedance, and in water and dry sand the day of and following the wet-weather exceedance. This tag was also present in Avalon water

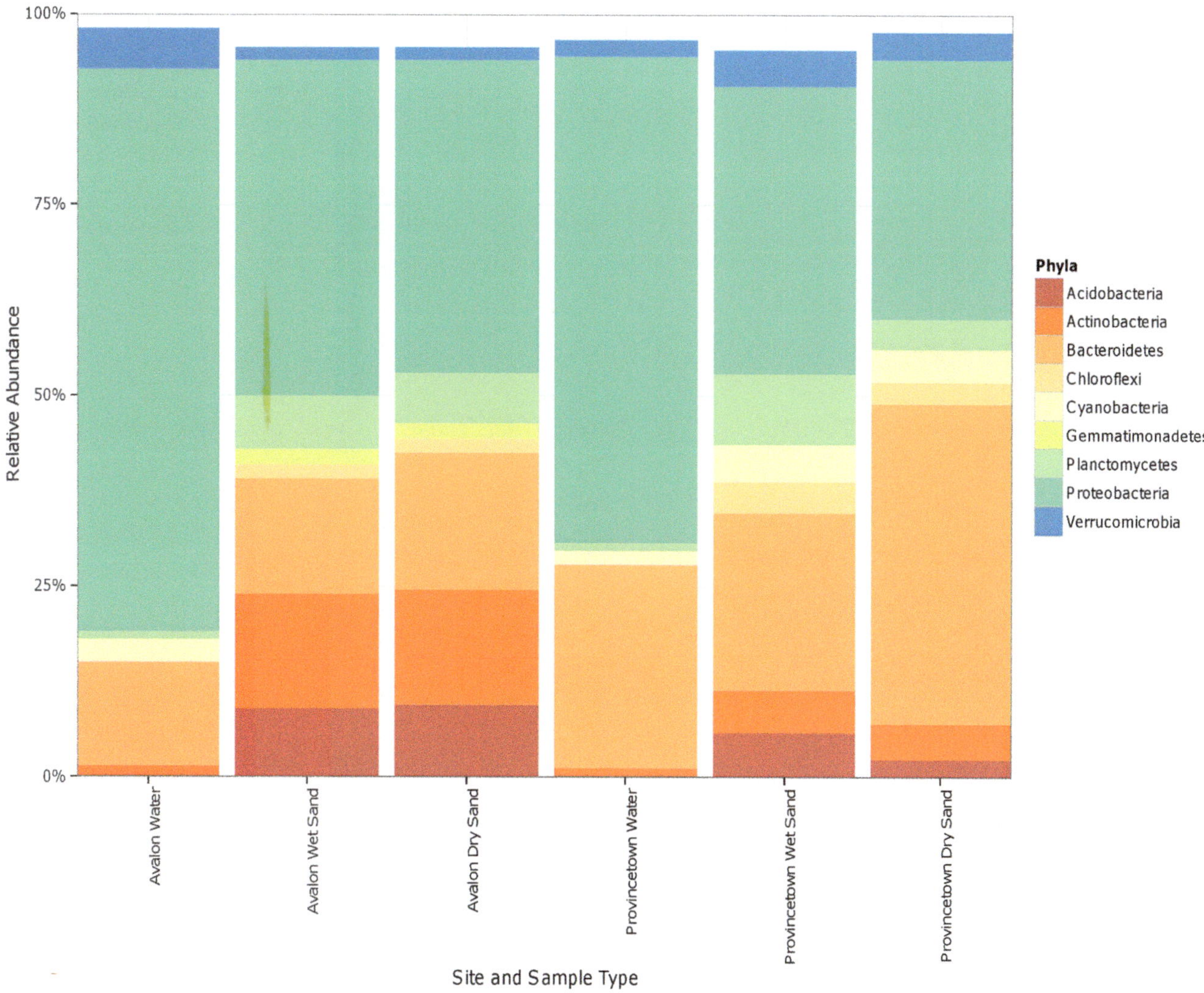

Figure 2. Relative abundance of phyla containing >1% of total sequence tags in water and sand samples.

samples on the days of elevated enterococci and dry weather exceedance events. Consistent with its hypothesized role as a sewage indicator, the relative abundance of the *Roseburia* tag was two to three orders of magnitude lower in environmental samples than would typically be found in sewage.

The one-way ANOSIM showed that there were no significant differences in the distribution of Clostridiales, Bifidobacteriales and Bacteroidales when samples were classified as having acceptable or unacceptable water quality (outlined in **Table 1**). However, ANOSIM showed significant differences among these orders when samples were grouped by site ($R = 0.545$, $p = 0.001$) or sample type ($R = 0.298$, $p = 0.001$). The SIMPER routine revealed that the differences between Clostridiales, Bifidobacteriales and Bacteroidales among sample types were primarily driven by the dominance of specific sewage sequence tags from the Bacteroidales (*Bacteroides*, *Parabacteroides*, *Paludibacter*) and Clostridiales (*Blauthia* and *Fecalibacterium*), all of which were much more abundant in sewage than in sands or waters. Sewage samples from the studies of McLellan et al. [27] were found to be far less heterogeneous than the environmental samples collected in this study, as SIMPER revealed that the within-group similarity of alternative indicators in the sewage community was 77%, compared to 46% similarity within the water samples, and only 32% and 35% similarity for the wet sand and dry sand samples. Alternative fecal indicators differentiating Avalon samples were primarily sequence tags identified to the family *Ruminococcaceae* (genus unresolved), and the genus *Alistipes* within the Bacteroidales. Provincetown samples were differentiated by a relative abundance of tags belonging to Clostridiales (*Robinsonella*, *Fusibacter* and *Acetivibrio*).

Environmental Variables Influence Alternative Fecal Indicators at Provincetown

When the BIOENV procedure was run with the same environmental variables but with the smaller subset of tags in the orders Clostridiales, Bifidobacteriales and Bacteroidales rather than the entire data set, the correlations significantly improved. The variable water temperature was found to best explain the patterns of abundance within the group of potentially sewage-associated orders recovered from water samples ($r = 0.867$), and dry sand temperature best explained the patterns of abundance within the group of potentially sewage-associated orders recovered from dry sands ($r = 0.837$), suggesting that this subset of the total community is more strongly influenced by temperature (or

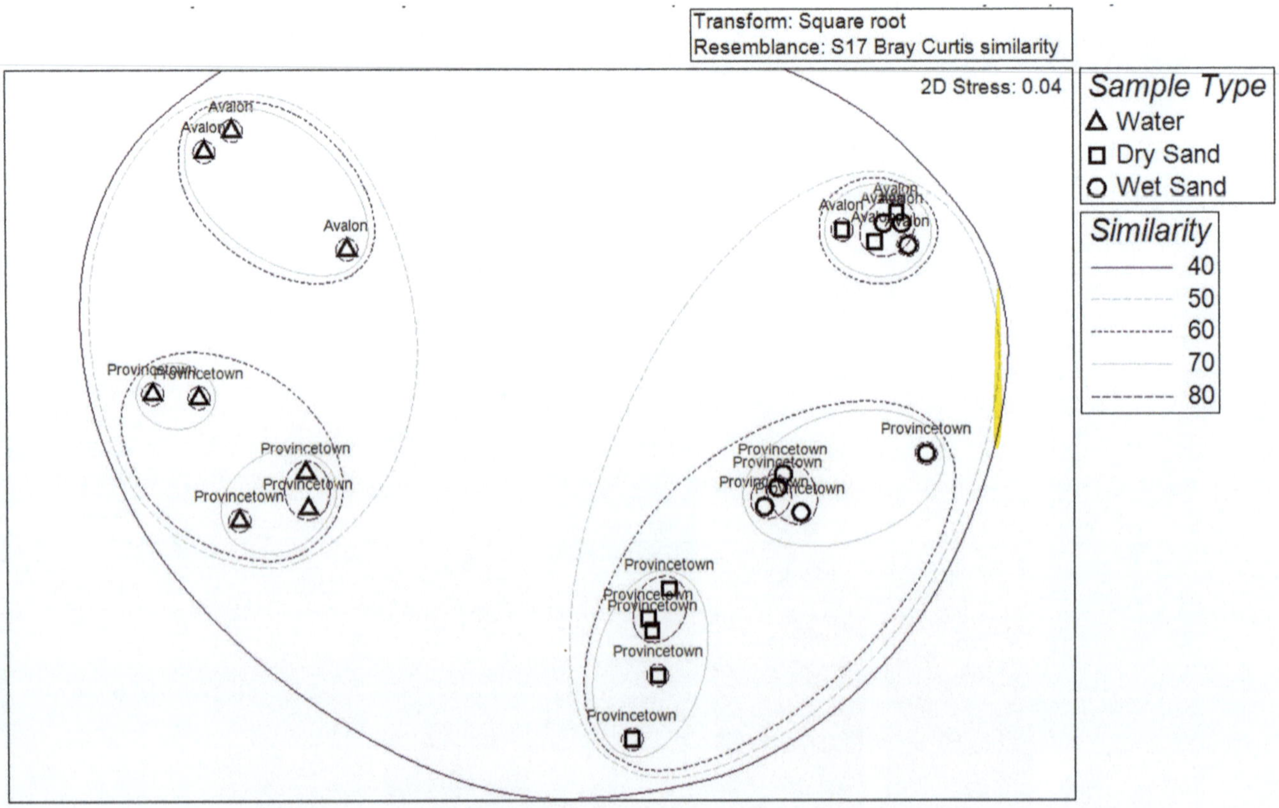

Figure 3. NMDS plot of total bacterial community composition in samples.

perhaps that bacterial loads to the beach covary with temperature) than by traditional fecal indicator abundance.

All of the water samples collected during exceedance events at both sites contained tags assigned to the genus *Enterococcus*. Some of the *Enterococcus* tags from Avalon also had a species level assignation *(E. ratii and E. colombae)* corresponding to GenBank sequence acquisitions from studies of unhealthy rats and pigeons, respectively. Although sands were often enriched in culturable *Enterococcus* compared to water (**Table 1**), only a single sand sample (Avalon dry sand, day of exceedance) contained *Enterococcus* sequence tags.

Discussion

Intertidal sands and the overlying water proved to have distinct bacterial communities, with greater similarity observed between coastal water samples from two distant sites than between the water and sand from the same site. The differentiation between Provincetown wet and dry sand, a phenomenon not observed among Avalon sands, likely reflects the greater tidal range at Provincetown (2–4 m) and thus stronger physical separation and more distinct environmental conditions between the intertidal and upper beach sand bacterial populations. Two recent studies also report observing similar differences in bacterial community structure between dry and wet sand [23,25]. Temperature and daily tidal range appeared to explain some of the variation in community structure in water and dry sand samples at the beach in our study (Provincetown), suggesting that the dry sand community may develop differently from that of the wet sand due to lack of tidal wetting. Furthermore, although the species richness in samples from both sites was similar the differences in community composition between samples collected at Avalon versus those collected at Provincetown may suggest greater ecological health and resiliency to contamination at Provincetown, where bacterial exceedance events occur less frequently and water quality is generally quite good. Additional work comparing beaches with different contamination levels is needed to confirm this observation.

The majority of beach sand sequence tags from both locations belonged to the Acidobacteria, Actinobacteria, Bacteroidetes, Proteobacteria and the Planctomycetes. These are broadly similar to soil communities, as >90% of sequence tags from soils collected around the world have been have been classified within the Actinobacteria, Acidobacteria, Proteobacteria, Bacteroidetes and Firmicutes, with the relative abundance of these groups within samples strongly influenced by soil pH [41]. In particular, beach sand communities are broadly differentiated from both coastal seawater and soil by the relative abundance of the Planctomycetes, which at both sites are relatively more abundant in wet sand than dry sand. The enrichment of the Planctomycetes in wet beach sands may reflect this phyla's frequent affiliation with organic detritus in the marine environment or participation in chitin degradation [42]. Coastal seawater samples at both sites in this study were dominated by tag sequences (such as *Pelagibacter*) that were not abundant in sand samples; this differs from the results of Cui et al. [23] who found subtropical water and intertidal sand shared relatively high abundances of a sequence tag identified as *Pseudoalteromonas*. This tag was present, but not dominant, in our samples; the sum of all unique sequences identified as *Pseudoalteromonas* species ranged from 0.03% to 0.7% of the total sample in both waters and sands. The same study [23] identified four species (*Nitriliruptor*, *Acidobacterium*, *Pseudomonas* and *Paracoccus*) uniquely abundant within subtropical backshore sand samples. These four species were recovered from the sand samples in this study, although they were taken from the upper and lower limits of the intertidal zone and not the "backshore" per se. A unique tag identified as *Nitrilirupter* accounted for 1% of tags in Avalon wet and dry sand as well as 1% of tags in Provincetown dry sand. A variety of *Paracoccus* tags were recovered from water and sand samples at all sites, with the highest relative abundance found in Provincetown dry sands (0.04%). Likewise a variety of *Pseudomonas* tags were recovered from all samples, all at similarly low relative abundances (approximately 0.01%). *Acidobacterium* was found at low relative abundance in Avalon wet sand (0.007%). However, the differentiations between Provincetown wet and dry sand communities was consistent with previous observations of the development of unique bacterial communities in different zones of beach sands [23,25].

In contrast to total community, as a group the alternative fecal indicators were more similar among the sand and water collected at the same site, supporting the theory that while each of these beaches may be anthropogenically impacted, there are possibly regional differences in sewage/source profiles. Furthermore, the complexity of the alternative indicators in environmental samples compared to sewage makes it difficult to interpret the very minor overlap of a few specific sequence tags and may be suggestive of diffuse nonpoint source pollution. Our result is similar to the work by Shanks et al [43] where they determined that fecal associated OTUs were consistent nationwide across sewage samples, but the sewage OTUs identified as infrastructure associated varied among cities with a strong north/south latitudinal separation.

Although massively-parallel pyrosequencing tags of rDNA hypervariable regions provide unprecedented depth of sampling within the bacterial community, the short length of the sequence tags precludes identification of the majority of sequence tags to genera. It simply cannot reliably differentiate between strains, which is often the level of identification required to determine human health risk among a species of bacteria. Sequence tags from the indicator *Enterococcus* were recovered on days that had exceeded water quality standards, but the rarity of *Enterococcus* in this and other pyrosequencing datasets [27,44] illustrates how the concentrations of indicators that cause concern from a monitoring perspective are relatively rare within the total community and not predictive of fecal bacteria in general (at least, based on this group of alternative indicators).

This study adds to the growing evidence that the community-based molecular tools used by microbial ecologists to study spatial and temporal variation and environmental disturbance events can be leveraged to study the sources and potential human health risks of fecal microbial pollution in the environment. In other studies, 16S-based pyrosequencing approaches have been used to broadly survey potential risks within sewage sludge and biosolids [45] and wastewater treatment plant samples [46], and with time these kinds of analyses can be completed with a broader range of potential source material. In terms of sourcing fecal pollution from animals or humans in surface waters, several community-based approaches have been developed that are analogous to current library-based microbial source tracking of single indicators [47]. For example, similarities of T-RFLP profiles of a coastal creek and potential human and animal fecal sources have been used to identify fecal sources and the extent to which contamination upstream impacts sites downstream [48]. In a case involving a limited number of environmental and local source samples, pyrosequencing of 16S rDNA derived from human and animal feces was used to examine the overlap between fecal sources and surface water communities, thereby discriminating which were likely dominant source material to the river [49]. Various approaches have also been used to identify constituent groups of human-specific fecal bacteria in environmental samples, including the pyrosequencing approach that was the basis for the group of alternative fecal indicators we analyzed in this study [27], as well as alternative approaches such as amplification of the V3 region of 16S rDNA combined with capillary-electrophoresis single strand conformation polymorphism (C-ESSCP) to fingerprint human feces and sewage effluents and identify dominant, human-specific bacteria [50]. However, previous studies have not considered the impact of putative sewage contamination events on the microbial community fingerprint of sand, sediment, or wrack. Here we have begun to assess whether such events can be identified in the dynamic marine beach environment by fecal signature tags. Further application of community-based methods to a wide array of environmental samples, sources and reservoirs may ultimately contribute to the diagnoses of bacterial pollution from unknown sources at beaches and in surface waters.

Supporting Information

Figure S1 Sites (indicated by black point) sampled at Avalon Bay, CA (left) and Provincetown Harbor, MA (right). Samples were taken from the left and right of the Green Pleasure Pier in Avalon Bay and composited for this study, as described in the methods. Samples were taken from the left of MacMillan wharf in Provincetown Harbor, corresponding to the street address of 333 Commercial St. in Provincetown, MA. Both maps depict 7.5-minute series from the USGS National Map Viewer (http://viewer.nationalmap.gov/viewer).

Figure S2 Rarefaction Curves. The number of unique species among 8050 randomly subsampled sequence tags from each environmental sample. See Table 1 for sample details.

Acknowledgments

We would like to thank John Griffith, Jennifer Jay and Yiping Cao of the Southern Califronia Coastal Water Research Project (SCCWRP) for beach sand samples from Avalon Bay Beach.

Author Contributions

Conceived and designed the experiments: EH RJG MLS. Performed the experiments: EH LAZ. Analyzed the data: EH SLM LAZ. Contributed reagents/materials/analysis tools: EH RJG LAZ MLS. Wrote the paper: EH SLM RJG LAZ MLS.

References

1. Wheeler Alm E, Burke J, Spain A (2003) Fecal indicator bacteria are abundant in wet sand at freshwater beaches. Water research 37: 3978–3982.
2. Yamahara KM, Layton BA, Santoro AE, Boehm AB (2007) Beach sands along the California coast are diffuse sources of fecal bacteria to coastal waters. Environmental science & technology 41: 4515–4521.
3. Beversdorf LJ, Bornstein-Forst SM, McLellan SL (2007) The potential for beach sand to serve as a reservoir for Escherichia coli and the physical influences on cell die-off. Journal of applied microbiology 102: 1372–1381.
4. Halliday E, Gast RJ (2011) Bacteria in Beach Sands: An Emerging Challenge in Protecting Coastal Water Quality and Bather Health. Environmental science & technology.

5. Bonilla TD, Nowosielski K, Cuvelier M, Hartz A, Green M, et al. (2007) Prevalence and distribution of fecal indicator organisms in South Florida beach sand and preliminary assessment of health effects associated with beach sand exposure. Marine pollution bulletin 54: 1472–1482.
6. Ge Z, Whitman RL, Nevers MB, Phanikumar MS, Byappanahalli MN (2012) Nearshore hydrodynamics as loading and forcing factors for Escherichia coli contamination at an embayed beach. Limnology and Oceanography 57: 362.
7. Zhu X, Wang JD, Solo-Gabriele HM, Fleming LE (2011) A water quality modeling study of non-point sources at recreational marine beaches. Water research.
8. Russell T, Yamahara KM, Boehm AB (2012) Mobilization and transport of naturally occurring enterococci in beach sands subject to transient infiltration of seawater. Environmental science & technology.
9. Heaney CD, Sams E, Wing S, Marshall S, Brenner K, et al. (2009) Contact with beach sand among beachgoers and risk of illness. American journal of epidemiology 170: 164.
10. Heaney CD, Sams E, Dufour AP, Brenner KP, Haugland RA, et al. (2012) Fecal Indicators in Sand, Sand Contact, and Risk of Enteric Illness Among Beachgoers. Epidemiology 23: 95.
11. Abdelzaher AM, Wright ME, Ortega C, Solo-Gabriele HM, Miller G, et al. (2010) Presence of pathogens and indicator microbes at a non-point source subtropical recreational marine beach. Applied and environmental microbiology 76: 724.
12. Yamahara KM, Sassoubre LM, Goodwin KD, Boehm AB (2012) Occurrence and Persistence of Bacterial Pathogens and Indicator Organisms in Beach Sand along the California Coast. Applied and environmental microbiology 78: 1733–1745.
13. Elmanama AA, Fahd MI, Afifi S, Abdallah S, Bahr S (2005) Microbiological beach sand quality in Gaza Strip in comparison to seawater quality. Environmental research 99: 1–10.
14. Goodwin KD, McNay M, Cao Y, Ebentier D, Madison M, et al. (2012) A multi-beach study of *Staphylococcus aureus*, MRSA, and enterococci in seawater and beach sand. Water Research 46: 4195–4207.
15. Shibata T, Solo-Gabriele HM (2012) Quantitative Microbial Risk Assessment of Human Illness from Exposure to Marine Beach Sand. Environmental science & technology.
16. Hartz A, Cuvelier M, Nowosielski K, Bonilla TD, Green M, et al. (2008) Survival potential of Escherichia coli and Enterococci in subtropical beach sand: implications for water quality managers. Journal of environmental quality 37: 898–905.
17. Yamahara KM, Walters SP, Boehm AB (2009) Growth of enterococci in unaltered, unseeded beach sands subjected to tidal wetting. Applied and environmental microbiology 75: 1517.
18. Mika KB, Imamura G, Chang C, Conway V, Fernandez G, et al. (2009) Pilot-and bench-scale testing of faecal indicator bacteria survival in marine beach sand near point sources. Journal of applied microbiology 107: 72–84.
19. Byappanahalli M, Fujioka R (2004) Indigenous soil bacteria and low moisture may limit but allow faecal bacteria to multiply and become a minor population in tropical soils. Water science and technology: a journal of the International Association on Water Pollution Research 50: 27.
20. Feng F, Goto D, Yan T (2010) Effects of autochthonous microbial community on the die-off of fecal indicators in tropical beach sand. FEMS microbiology ecology 74: 214–225.
21. Gobet A, Boer SI, Huse SM, van Beusekom JEE, Quince C, et al. (2012) Diversity and dynamics of rare and of resident bacterial populations in coastal sands. The ISME journal 6: 542–553.
22. Llobet-Brossa E, Rosselló-Mora R, Amann R (1998) Microbial community composition of Wadden Sea sediments as revealed by fluorescence in situ hybridization. Applied and environmental microbiology 64: 2691–2696.
23. Cui H, Yang K, Pagaling E, Yan T (2013) Spatial and temporal variation in enterococcal abundance and its relationship to the microbial community in Hawaii beach sand and water. Applied and Environmental Microbiology 79: 3601–3609.
24. Phillips MC, Solo-Gabriele HM, Piggot AM, Klaus JS, Zhang Y (2011) Relationships between sand and water quality at recreational beaches. Water Research 45: 6763–6769.
25. Piggot AM, Klaus JS, Johnson S, Phillips M, Solo-Gabriele HM (2012) Enterococci Levels are Related to Sediment Biofilms at Recreational Beaches in South Florida. Applied and environmental microbiology.
26. Newton RJ, VandeWalle J, Borchardt MA, Gorelick MH, McLellan SL (2011) Lachnospiraceae and Bacteroidales alternative fecal indicators reveal chronic human sewage contamination in an urban harbor. Applied and environmental microbiology: AEM. 05480-05411v05481.
27. McLellan S, Huse S, Mueller Spitz S, Andreishcheva E, Sogin M (2010) Diversity and population structure of sewage derived microorganisms in wastewater treatment plant influent. Environmental microbiology 12: 378–392.
28. Dethlefsen L, Huse S, Sogin ML, Relman DA (2008) The pervasive effects of an antibiotic on the human gut microbiota, as revealed by deep 16S rRNA sequencing. PLoS biology 6: e280.
29. Hewson I, Vargo G, Fuhrman J (2003) Bacterial diversity in shallow oligotrophic marine benthos and overlying waters: effects of virus infection, containment, and nutrient enrichment. Microbial ecology 46: 322–336.
30. Laverock B, Smith CJ, Tait K, Osborn AM, Widdicombe S, et al. (2010) Bioturbating shrimp alter the structure and diversity of bacterial communities in coastal marine sediments. The ISME journal 4: 1531–1544.
31. Tšertova N, Kisand A, Tammert H, Kisand V (2011) Low seasonal variability in community composition of sediment bacteria in large and shallow lake. Environmental Microbiology Reports 3: 270–277.
32. Böer SI, Hedtkamp SIC, Van Beusekom JEE, Fuhrman JA, Boetius A, et al. (2009) Time-and sediment depth-related variations in bacterial diversity and community structure in subtidal sands. The ISME journal 3: 780–791.
33. Gobet A, Quince C, Ramette A (2010) Multivariate Cutoff Level Analysis (MultiCoLA) of large community data sets. Nucleic acids research 38: e155.
34. Halliday E, Griffith JF, Gast RJ (2010) Use of an exogenous plasmid standard and quantitative PCR to monitor spatial and temporal distribution of Enterococcus spp. in beach sands. Limnol Oceanogr Methods 8: 146–154.
35. U.S.E.P.A. (2002) Method 1600: membrane filter test method for enterococci in water.
36. Ludwig W, Schleifer KH (2000) How quantitative is quantitative PCR with respect to cell counts? Systematic and applied microbiology 23: 556.
37. Gast RJ, Dennett MR, Caron DA (2004) Characterization of protistan assemblages in the Ross Sea, Antarctica, by denaturing gradient gel electrophoresis. Applied and environmental microbiology 70: 2028.
38. Huber JA, Mark Welch DB, Morrison HG, Huse SM, Neal PR, et al. (2007) Microbial population structures in the deep marine biosphere. science 318: 97.
39. Huse SM, Huber JA, Welch DM, Relman DA, Sogin ML (2008) Exploring microbial diversity and taxonomy using SSU rRNA hypervariable tag sequencing. PLoS genetics 4: e1000255.
40. Clarke KR, Gorley RN (2006) Primer v6:user manual/tutorial. Plymouth: Primer-E Ltd.
41. Lauber CL, Hamady M, Knight R, Fierer N (2009) Pyrosequencing-based assessment of soil pH as a predictor of soil bacterial community structure at the continental scale. Applied and environmental microbiology 75: 5111.
42. Fuerst JA (1995) The planctomycetes: emerging models for microbial ecology, evolution and cell biology. Microbiology 141: 1493.
43. Shanks OC, Newton RJ, Kelty CA, Huse SM, Sogin ML, et al. (2013) Comparison of the microbial community structures of untreated wastewaters from different geographic locales. Applied and Environmental Microbiology 79: 2906–2913.
44. Dowd SE, Callaway TR, Wolcott RD, Sun Y, McKeehan T, et al. (2008) Evaluation of the bacterial diversity in the feces of cattle using 16 S rDNA bacterial tag-encoded FLX amplicon pyrosequencing(bTEFAP). BMC microbiology 8: 125.
45. Bibby K, Viau E, Peccia J (2010) Pyrosequencing of the 16S rRNA gene to reveal bacterial pathogen diversity in biosolids. Water research 44: 4252–4260.
46. Ye L, Zhang T (2011) Pathogenic Bacteria in Sewage Treatment Plants as Revealed by 454 Pyrosequencing. Environmental science & technology.
47. Simpson JM, Santo Domingo JW, Reasoner DJ (2002) Microbial Source Tracking: State of the Science. Environmental science & technology 36: 5279–5288.
48. Cao Y, Van De Werfhorst LC, Sercu B, Murray JLS, Holden PA (2011) Application of an Integrated Community Analysis Approach for Microbial Source Tracking in a Coastal Creek. Environmental science & technology 45: 7195–7201.
49. Unno T, Jang J, Han D, Kim JH, Sadowsky MJ, et al. (2010) Use of Barcoded Pyrosequencing and Shared OTUs To Determine Sources of Fecal Bacteria in Watersheds. Environmental science & technology.
50. Wéry N, Monteil C, Pourcher A-M, Godon J-J (2010) Human-specific fecal bacteria in wastewater treatment plant effluents. Water research 44: 1873–1883.
51. Halliday EA (2012) Sands and environmental conditions impact the abundance and persistence of the fecal indicator bacteria Enterococcus at recreational beaches: Massachusetts Institute of Technology.

The Effects of Operational Conditions on the Respiration Rate of Tubificidae

Juqing Lou[1,2], Yongqing Cao[2], Peide Sun[2], Ping Zheng[1]*

1 Department of Environmental Engineering, Zhejiang University, Hangzhou, China, **2** School of Environmental Science and Engineering, Zhejiang Gongshang University, Hangzhou, China

Abstract

Tubificidae is often used in the wastewater treatment systems to minimize the sludge production because it can be fed on the activated sludge. The process conditions have effect on the growth, reproduction, and sludge reduction efficiency of Tubificidae. The effects of the water quality, density of worms, pH, temperature and dissolved oxygen (DO) concentration on the respiration rate of Tubificidae were investigated to determine the optimal conditions for the growth and metabolism of the worms and reveal the mechanisms involving the efficient sludge reduction in terms of these conditions. It was observed that the respiration rate was highest in the water discharged from an ecosystem that included symbiotic Tubificidae and microbes and was lowest in distilled water. Considering density of the worms, the highest rate was 81.72±5.12 mg O_2/g(dry weight)·h·L with 0.25 g (wet weight) of worms in 1 L test flask. The maximum Tubificidae respiration rate was observed at a pH of 8.0±0.05, a rate that was more than twice as high as those observed at other pH values. The respiration rate increased in the temperature range of ~8°C–22°C, whereas the rate declined in the temperature range of ~22°C–30°C. The respiration rate of Tubificidae was very high for DO range of ~3.5–4.5 mg/L, and the rates were relatively low for out of this DO range. The results of this study revealed the process conditions which influenced the growth, and reproduction of Tubificidae and sludge reduction at a microscopic level, which could be a theoretical basis for the cultivation and application of Tubificidae in wastewater treatment plants.

Editor: Hauke Smidt, Wageningen University, Netherlands

Funding: This study was financed by Innovative Research Team in High Educational Institutions of Zhejiang Province, China (No. T200912), Natural Science Foundation of Zhejiang Province, China (No. Y12E080076), and Scientific Research Project of Department of Education of Zhejiang province, China (No. Y201017078). The funders had no role in study design, data collection and analysis, decision to publish, or preparation of the manuscript.

Competing Interests: The authors have declared that no competing interests exist.

* E-mail: pzheng@zju.edu.cn

Introduction

Aerobic wastewater treatment plants (WWTPs) produce large quantities of sludge containing various pollutants like heavy metals, pathogens and micropollutants. The estimated cost of treatment and disposal of waste sludge in WWTPs represents half, or even as much as 60% of the total cost of wastewater treatment [1],[2]. Currently, conventional sludge reduction methods, including mechanical, physical, chemical, electrical and biological methods, are in use [3]; however, these methods are energy and cost intensive [4], [5]. In view of stricter regulations, new technologies should be developed to reduce the production and disposal of waste sludge. Biological wastewater treatment involves several techniques for processing waste sludge, such as lysis-cryptic growth [6], metabolic uncoupling [1], [7] and maintenance metabolism and predation on bacteria [7], [8]. Compared with other strategies, the predation method has attracted recently considerable attention because of several obvious advantages such as energy saving and no secondary pollution production [9], [10]. One of the typical predation methods is the application of aquatic worms for sludge reduction [11], [12], which would decrease the amount and volume of excess sludge by metabolic processes. The volume of sludge is reduced after treatment because the sludge flocs are compacted, and meanwhile, the sludge amount is distinctly decreased [13],[14], because worms eat sludge as food per day several times the their own weight[9]. So, a large amount of sludge is consumed by the worms, and the amount of new worm biomass and worm feces are much lower than the sludge consumed. Indeed, this method may be the most cost-effective and environmentally friendly for sludge reduction in WWTPs [15], [16]. Several worm species present specific characteristics that make them suitable for such applications. Oligochaetes may represent 50–60% and even up to 100% of the benthic biomass [3] and are the largest type of metazoa observed in WWTPs [17], [18]. The principal types of worms found in activated sludge systems are Naididae (such as Tubificidae) [2] and, Aeolosomatidae [8], [9]. The two most common species, *Tubifex tubifex* and *Limnodrilus hoffmeisteri* Clap., are reported to survive in polluted water environment [19], [20] or in highly polluted, oxygen-poor and/or toxic ecosystems [21], [22]. These species being the indicators of organic pollution are capable of growth and stable sludge reduction. Also, they have shown good survival and reproductive rates in activated sludge during wastewater treatment processes [13], [19], [23]. For these reasons, Tubificidae were selected for this study.

The sludge reduction by worms was closely related to the process conditions in applications involving aquatic worms. Some researchers [3], [4], [13] have already reported that the conditions, such as temperature, pH, dissolved oxygen (DO) etc had a clear effect on the worms eating waste sludge. So, high sludge reduction efficiency could only be achieved through stable growth and good reproduction of worms in WWTPs. However, growth, reproduction, and survival of aquatic worms are inevitably influenced by the conditions of aquaculture systems, e.g., temperature and pH [24], [25]. Additionally, the respiration rate is an indicator of metabolism and growth of Tubificidae, which is also affected by these conditions. Therefore, it is necessary to investigate the effects of conditions on the respiration rate, the growth of the worms, and the sludge reduction capability of worms. Although the effects of such pollutants as cadmium and sodium pentachlorophenate on the respiration of tubificid worms [26–29] have been addressed, however, relationship between changes in the respiration rate and natural conditions in the WWTPs environment has never been investigated.

Therefore, this study investigates comprehensively the effects of the water quality, worm density, pH, temperature and DO concentration on the respiration rate of tubificid worms. The results of the study revealed a suitable sludge reduction habitat for Tubificidae and gave insights into the conditions related to sludge reduction efficiency. The results of the study could offer rational explanation for how worms' growth and sludge reduction efficiency were affected by the process conditions, and how application of aquatic worms in sludge reduction can be improved.

Materials and Methods

Materials

Tubificid worms. Tubificidae (primarily *T. tubifex*, with some *L. hoffmeisteri*) specimens were collected from the Shengxin Aquarium, Lanling flower and bird market, Shanghai. The worms to be tested were separated from other organisms using standard graduated sieves and materials. They were kept at approximately 20°C in the laboratory under slow stream of water. Single worm was selected with a medicine dropper after repeated washing with the help of tap water to minimize possible injury. The worms selected were of fairly uniform appearance and size (~6–7 cm), red in color, had no apparent injuries and responded rapidly (the posterior or anterior end of a worm rapidly withdraws and coil their body) to physical stimuli [30], [31]. The animals were not fed during the experiment. The respiration rate tests were performed with discharged water unless the experiment involved differences in the water type.

Test solutions. The experiments were performed in tap water, distilled water and water discharged from system consisting of symbiotic Tubificidae and microbes. The symbiotic system was operated as a sequencing batch reactor (SBR), in which the mixed-liquor suspended solids (MLSS) was about 3500 mg/L, pH ~7–8, temperature was at 25±1°C, the DO concentration was about 3 mg/L, the Chemical Oxygen Demand (COD) was about 50 mg/L and ammonia concentration of discharged water was about 7.2 mg/L.

The discharged water was autoclaved before use to eliminate the effect of microorganisms. The pH of the solutions was adjusted with 1.0 mol/L HCl or 1.0 mol/L NaOH.

Methods

Dry weight and wet weight of worms. The worms were placed on perforated aluminum foil kept over absorbent paper to measure the wet weight (ww) and gentle pressure was applied to the aluminum foil paper to allow the underlying paper to absorb the water from the surface of the worms. The worms were weighed using an analytical balance after drying [14].

Then the worms were heated overnight at ~103–105°C for a constant weight in order to measure their dry weight [4], [32].

The effect of water quality on the respiration rate. Worms samples of 0.25 g (ww) were placed in tap water, distilled water and discharged water from the symbiotic ecosystem. All the tests were performed in triplicate at a temperature of 22±1°C, a pH of 8.0±0.05 and a DO concentration of ~3.5–4.0 mg/L. The original characteristics of distilled water, tap water and discharged water are given in Table 1.

The effect of worm density on respiration rate. Worms samples of 0.25 g, 0.50 g, 0.75 g, 1.00 g and 1.50±0.01 g (ww) were added to the 1 L (standard sample volume) discharged water to determine the effect of worm density on the respiration rate. The pH, temperature and DO were controlled at 8.0±0.05, 22±1°C and ~3.5–4.0 mg/L, respectively.

The effect of pH on respiration rate. The tests were conducted at four different pH values of 6.0, 7.0, 8.0, and 9.0 to investigate the effect of pH on the respiration rate. Each test was performed in triplicate. A Tubificidae sample of 0.25 g (ww) was added to a flask with discharged water. The temperature was controlled at 22±1°C for all experiments, and pH was controlled at the pre-designated set point ±0.05 with the help of NaOH (1.0 mol/L) and HCl (1.0 mol/L).

The effect of temperature on respiration rate. A Worms sample of 0.25 g (ww) was tested individually at six different temperatures 8°C, 15°C, 22°C, 25°C, 27°C and 30°C to investigate the effect of temperature on the respiration rate. All the experiments were performed in discharged water at a pH of 8.0±0.05 and a DO concentration of ~3.5–4.0 mg/L.

Each test was performed in triplicate. The temperature was controlled by the water bath of BI-2000 electrolytic respirometer.

The effect of DO concentration on respiration rate. The DO concentrations were controlled with air and/or nitrogen gas diffusers [4] at 1.5±0.2 mg/L, 2.5±0.2 mg/L, 3.5±0.2 mg/L, 4.5±0.2 mg/L, 5.5±0.2 mg/L, and 6.5±0.2 mg/L. A Tubificidae sample of 0.25 g (ww) was tested in discharged water at a pH of 8.0±0.05 and a temperature of 22±1°C.

Determination of respiration rates. An electrolytic respirometer was often used to monitor oxygen consumption by microorganism [33], [34]. The BI-2000 electrolytic respirometer (Bioscience Inc., Bethlehem, PA) is shown in Figure 1 [35]. The flasks (total volume is 1225 mL, and standard sample volume is 1000 mL) were shaken continuously in a water bath at a constant temperature of 22°C, with the exception of the tests involving temperature effects. Carbon dioxide in the apparatus was absorbed by filter paper strips placed in a KOH trap and saturated with 45% KOH to form a slight vacuum in the reactor (relative to the external atmospheric pressure), and this slight pressure difference was detected by pressure sensor. The electrolysis cell dissociated dilute sulfuric acid to produce oxygen, in an amount proportional to the current, which was added to the reactor until the pressure returned to equilibrium. The hydrogen

Table 1. The original characteristics of distilled water, tap water and discharged water.

	pH	DO (mg/L)	COD (mg/L)	hardness[a] (mg/L)
distilled water	6.07	7.24	0	0
tap water	7.20	6.08	2.11	87.9
discharged water	7.78	2.05	49.2	77.2

[a]The hardness of water was represented as $CaCO_3$, mg/L.

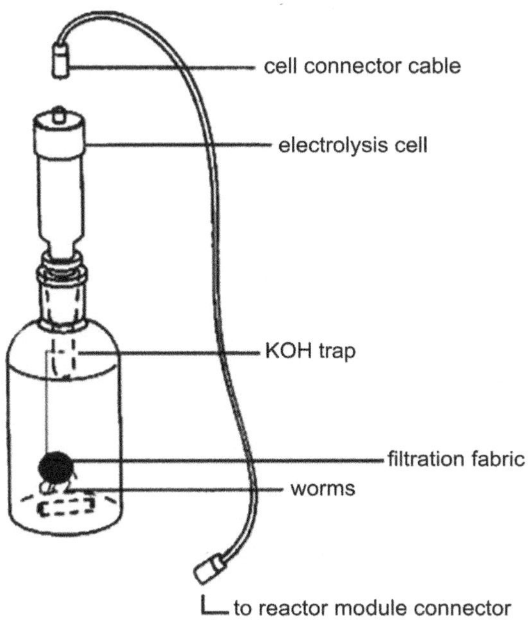

Figure 1. The apparatus used to measure respiration rates.

generated by this reaction was exhausted through a vent. The worms were wrapped in filter cloth and hung in the reactor during the test, which was not initiated until the water bath reached the desired temperature. The respiration rate was expressed as mg/L O_2 consumed per gram (dw) of worms per hour, and it was presented as mean ± SD (standard deviation).

Other item analysis. The dry weight (dw) and wet weight (ww) of Tubificid worms were determined using an EL104 analytical balance. The DO concentration was measured by a DO meter (FE20 Meter Toledo), and pH was measured by Tengine EDO pH meter. The COD and hardness of water were measured according to standard methods (APHA, 1999) [36].

Ethics Statement

The owners of Shengxin Aquarium have permitted to use these worms in this study, and animal procedures were performed according to laboratory animals use guidelines which were approved by the Ethics Committee of Zhejiang Gongshang University, Hangzhou, China.

Results and Discussion

Ratio of dry weight to wet weight of Tubificid worms

The dry weight to wet weight ratio of worms was approximately 16.3% (Figure 2), which was quite close to 0.161±0.015 which has previously been reported for *T. tubifex* [37]. Elissen[13] determined dry weight to wet weight ratio for different types of aquatic worms: 17% for sessile Tubificidae and 13% for *L. variegatus*. The value

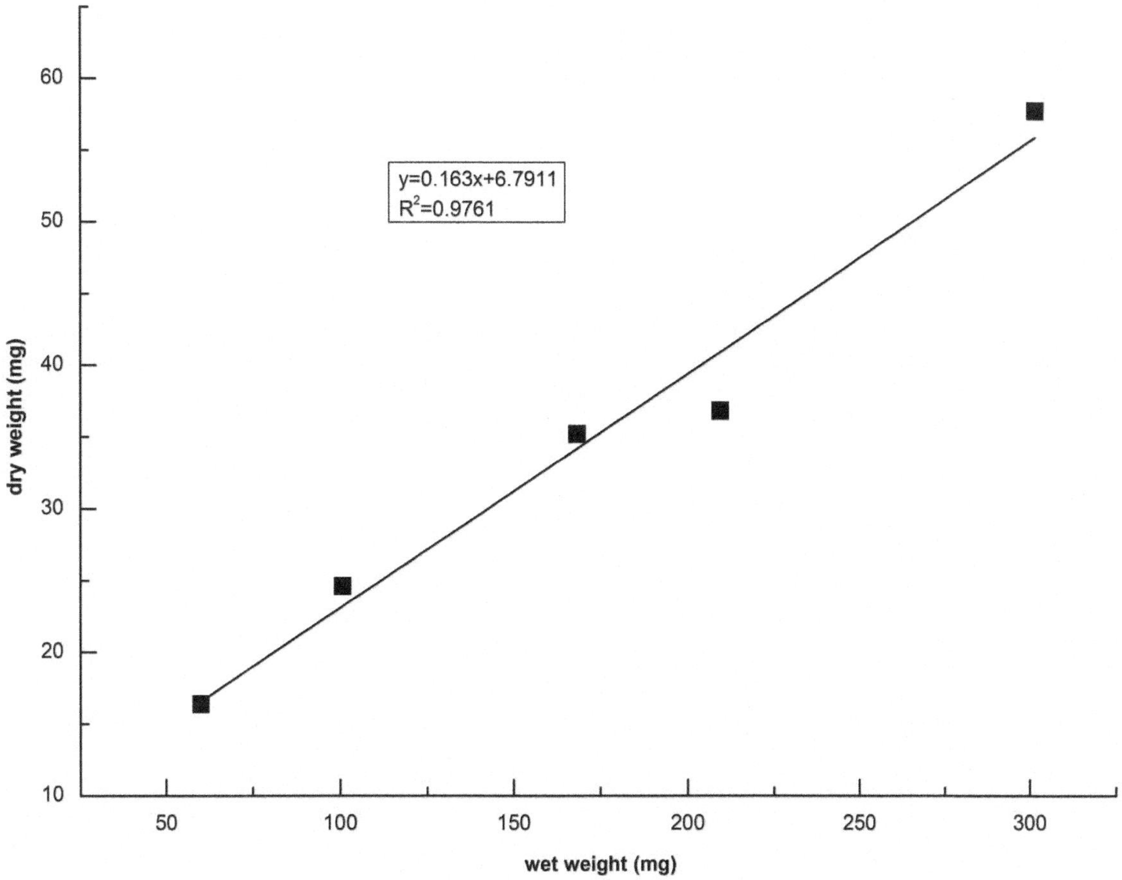

Figure 2. Dry weight vs. wet weight of Tubificidae worms.

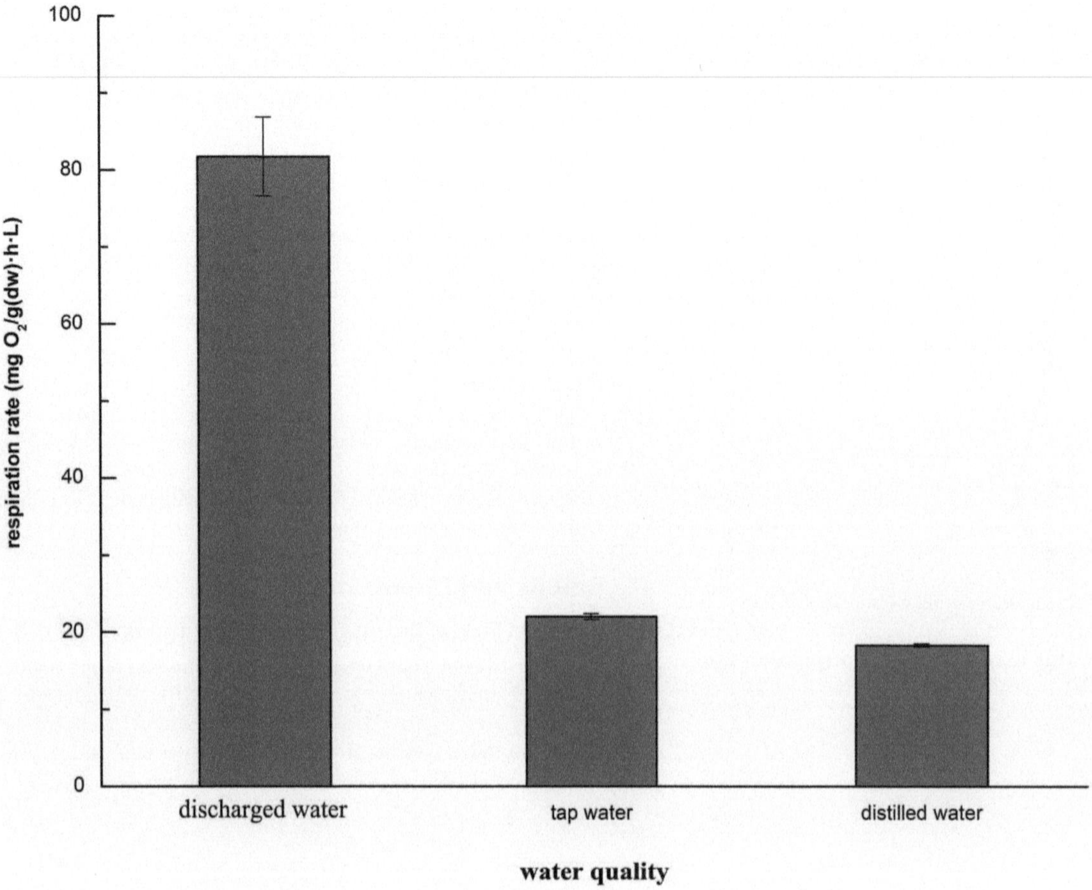

Figure 3. Respiration rates in water with different qualities (n = 3, error bars = SD).

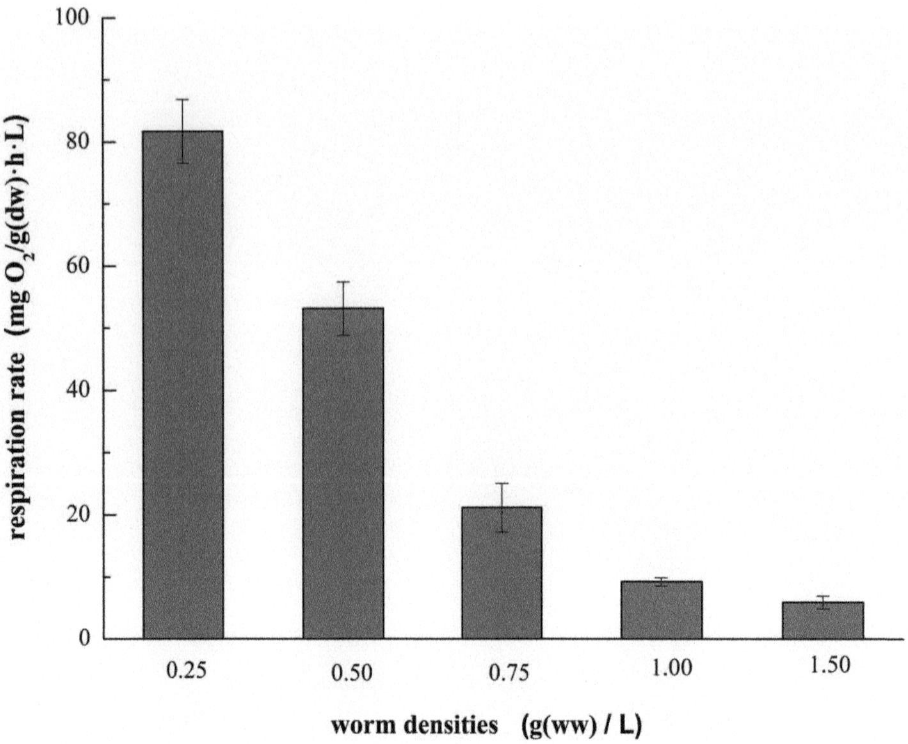

Figure 4. Respiration rates for different worm densities (n = 3, error bars = SD).

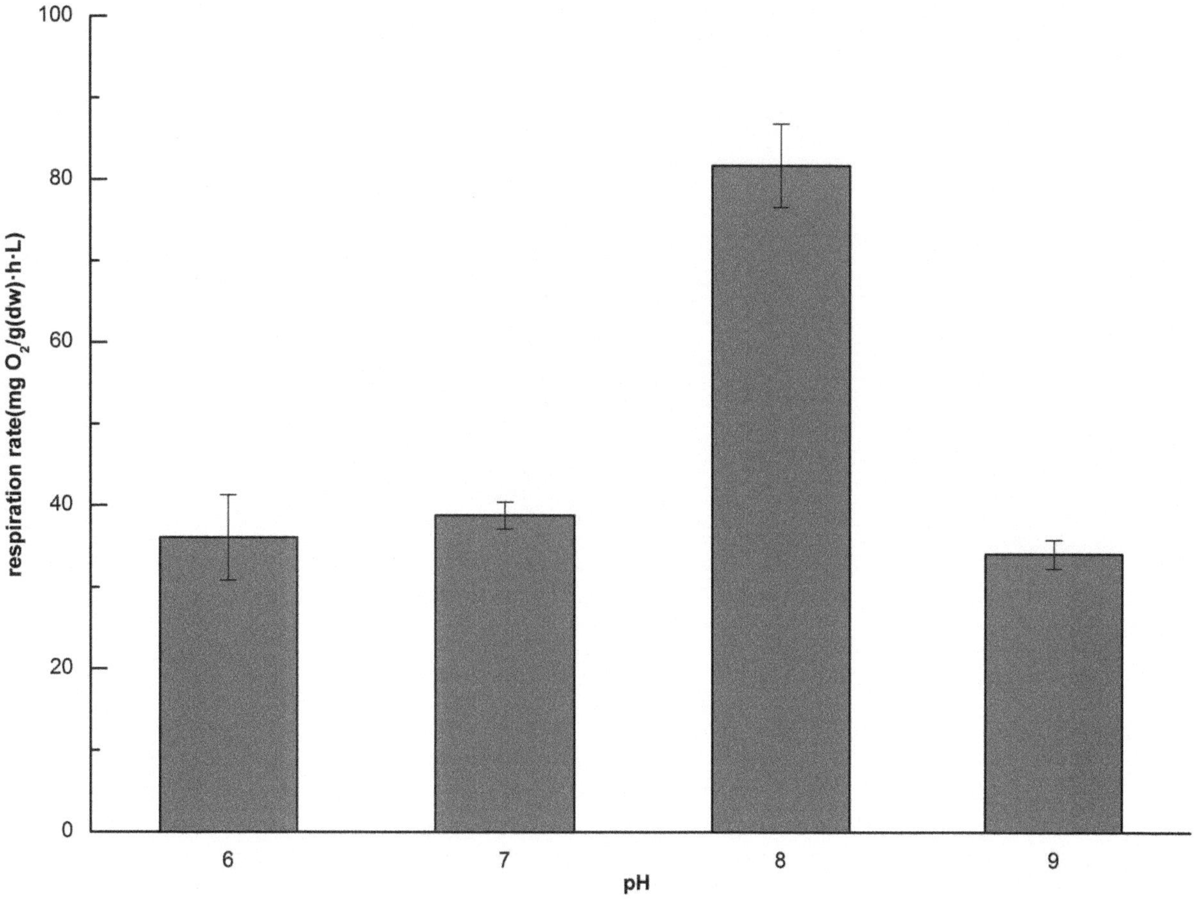

Figure 5. Respiration rates at different pH values (n = 3, error bars = SD).

found by Elissen for Tubificidae differed slightly from that found in the present study, perhaps due to differences in the length, diameter and biovolume of the worms used in two studies [13], [38].

The effect of water quality on respiration rate

The results of the experiment with different water sources are shown in Figure 3. The highest respiration rate was found in the water discharged from the symbiotic ecosystem, which was 81.72±5.12 mg O_2/(g(dw)·h·L), compared to 21.99±0.39 and 18.31±0.22 mg O_2/(g(dw)·h·L) in tap water and distilled water, respectively. These findings showed that the operational sludge reduction system might be favorable for the growth and metabolism of Tubificidae. Distilled water could be considered for Tubificidae growth as a pure substance without any additional components, in the absence of trace elements. Consistently, the respiration rate in tap water was slightly higher than that in distilled water. Moreover, discharged water contained many other micronutrients energizing microbial growth and metabolism, which could be one of the reasons why aquatic worms could grow well in WWTPs, and achieve high sludge reduction efficiency.

The effect of worm density on respiration rate

Figure 4 shows that the worm density had a significant effect on the respiration rate. The respiration rates found with 0.25, 0.50, 0.75, 1.00 and 1.50 g (ww)/L of worms were 81.72±5.12, 53.13±4.34, 21.17±3.90, 9.20±0.69 and 5.89±1.04 mg O_2/(g(dw)·h·L), respectively. It was likely that the markedly decreased respiration rate with an increasing density of worms was a result of the tendency of the worms forming a structure termed "ball cluster", a typical behavior of Tubificidae. Indeed, the increased density of worms resulted in the formation of a larger ball, leading to oxygen transmission and the respiration of the worms was hindered in the middle of the ball. This effect has been reported by previous investigators [39], [40], who found that less oxygen was consumed by the animals if they formed a ball. Wei et al.[41] have even reported that the quality of the effluent could be worse when the worm density was higher. The density of inoculated worms should be maintained within an appropriate range in the practical application.

The effect of pH on respiration rate

Several pH values were investigated experimentally to determine the possible effect of pH on the respiration rate. The respiration rate increased as the pH changed from 6.0 to 8.0 as shown in Figure 5, and reached a maximum rate at pH 8.0, and then decreased when pH increased further. Although the respiration rate in the range of ~6.0–8.0 was slightly higher than that at pH 9.0. A change in pH from 7.0 to 6.0 did not alter the respiration rate significantly, which was in accordance with the research of Fowler [28], there was no difference in the oxygen consumption rate when pH shifted from 7.6 to 6.1. These findings indicated that neutral pH or pH values approaching the alkaline range were beneficial for Tubificidae. The environment under

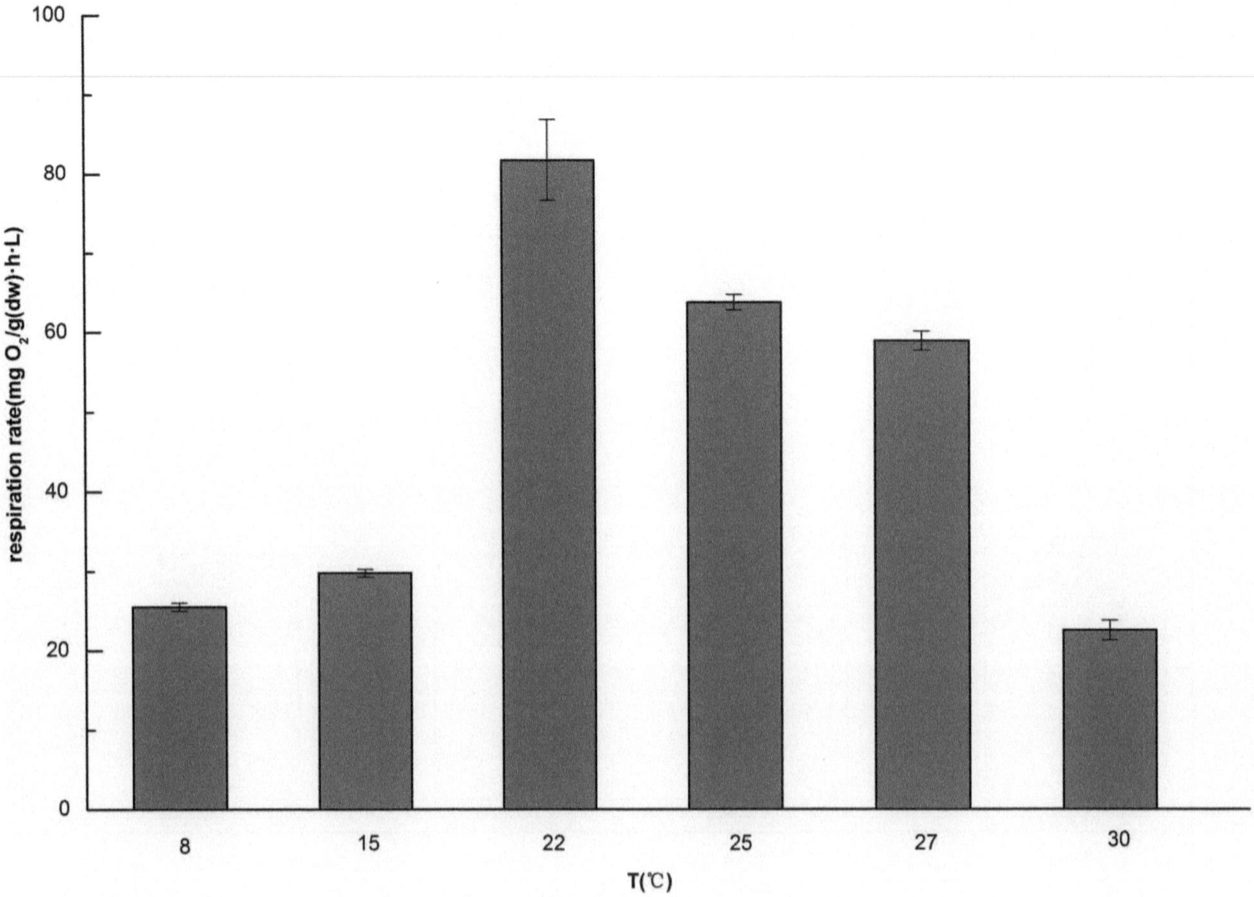

Figure 6. Respiration rates at different temperatures (n = 3, error bars = SD).

these pH values range is suitable for most of microorganisms, such as nitrifying and denitrifying bacteria [42], [43]. The pH in a domestic sewage treatment plant is generally ~6–8 [44], a range that could be benefiting the application of worms and helpful for improving the sludge reduction efficiency.

The effect of temperature on respiration rate

Temperature is an important factor for aquatic organisms. For example, temperature influences the respiration[28], growth and reproduction of Tubificidae [23], [45], [46]. As revealed in this study, the Tubificidae specimens studied were found to be very sensitive to temperature, as shown in Figure 6. The respiration rate increased from 25.46±0.52 mg O_2/(g(dw)·h·L) to 81.72±5.12 mg O_2/(g(dw)·h·L) as the temperature increased from 8°C to 22°C. The respiration rate was maximum at approximately 22°C and then the rate decreased with further increase of temperature. The rate remained relatively high when temperatures varied from ~22–27°C but was considerably low below 22°C. The oxygen consumption rate and activity level decreased significantly when temperature dropped from 25°C to 0°C according to available report [28]. The aggregation of the worms became detached at a temperature of 30°C. The available literature [4] showed that temperature clearly affected both the sludge digestion efficiency and sludge reduction by Lumbriculus variegatus. The optimum consumption rate by the worms occurred at approximately 15°C, whereas the maximum decrease in TSS occurred at 10°C. A gradual decrease in the sludge digestion efficiency occurred when the temperature increased from 10 to 20°C. In contrast, Guo et al.[47] reported that the digestion efficiencies increased as the temperature increased from 15°C to 25°C and that the maximum digestion efficiencies of TSS and VSS occurred at 25°C. Although it is evident from these findings that temperature is an important factor influencing sludge reduction by worms. Because Tubificidae are ectotherms, their metabolism and physiological state [48] are directly dependent upon the water temperature, with a greater metabolic rate at higher temperatures and reduction in the animals' activity and metabolic rate at lower temperatures. But the optimal temperature value remains unclear because a unified conclusion could not be drawn from the available literature. Some researchers think that optimal temperature for growth for Tubificidae usually is in the range of ~20–25°C [3], [46], but others think that the optimum temperature is 25°C[3], [45]. Reynoldson[49] reported that growth of T. tubifex and Limnodrilus hoffmeisteri (Tubifieidae, Oligoehaeta) occurred only over a very similar narrow range at ~10–13°C. So, we could not specify a precise optimum temperature. However, the results of the current study could suggest that the optimal temperature for Tubificidae is at approximately 22°C, a finding that is consistent with the literature [46], [50]. In addition, the temperature varies with the season and climate. However, the temperature of WWTPs is relatively stable, and the range in a typical WWTP is ~20–24°C [51], which is beneficial for sludge reduction by worms. So a temperature greater than 30°C should be avoided and temperature range of ~20–27°C should be maintained in sludge reduction system involving worms.

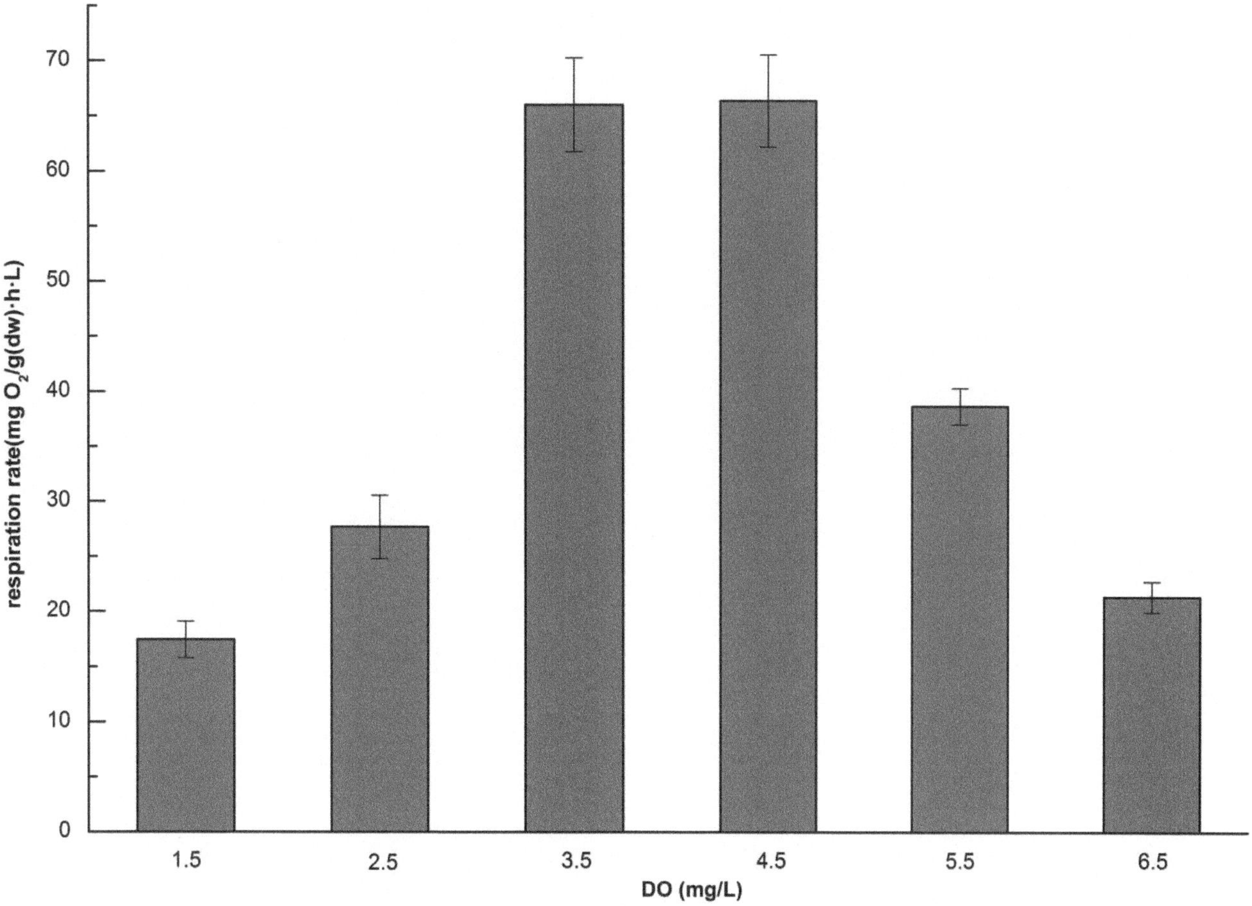

Figure 7. Respiration rates at different DO concentrations (n = 3, error bars = SD).

The effect of DO concentration on respiration rate

DO is important parameter for the growth and reproduction of aquatic worms, for example, it can affect bioturbation by infaunal burrowers [52], and it has also been reported to affect both the sludge reduction efficiency and consumption rate [4]. As shown in Figure 7, changes in the DO concentrations represented an environmental stress for Tubificidae. The respiration rate increased when the DO concentrations increased from 1.5 mg/L to 6.5 mg/L and then the rate decreased gradually. The respiration rate reached its maximum value when DO concentration was at 4.5 mg/L, which was 3.8 and 3.1 times higher than that observed at the lowest and highest concentrations, respectively. In this study, DO concentrations between 3.5 and 4.5 mg/L were found to benefit Tubificidae. However, a high oxygen input represented an increase in energy input, and a strong disturbance in the flow as a result of excessively high agitation might produce a low sludge reduction rate [53]. Fowler [28] has shown that the degree of vigor of undulation and elongation varied directly with the decrease in oxygen pressure. Under low DO condition, worms would extend their bodies and squeeze into each other to compete fiercely for more oxygen, so that they were washed away easily [54]. Also, Tubificidae showed weight loss under anoxic conditions [44], [49], and the Tubificidae reduction efficiencies of TSS and VSS were negative in a DO concentration range of ~0–0.5 mg/L [47]. Clearly, low and high DO concentrations should be avoided. So, an intermediate DO concentration should be identified for practical use which involves both economical benefits and high respiration rate of the worms. In actual installations, the DO level was maintained at approximately 4 mg/L in an aerobic tank, in which worms had a suitable respiration rate and also survive and reproduce well [44].

Conclusions

The respiration rates of Tubificidae in distilled water, tap water and water discharged from the system consisting of symbiotic Tubificidae and microbes were investigated. The highest respiration rate was achieved with the discharged water (i.e., 81.72±5.12 mg O_2/(g(dw)·h·L)), followed by tap water (i.e., 21.99±0.39 mg O_2/(g(dw)·h·L)), and the lowest rate was found with the distilled water (i.e., 18.31±0.22 mg O_2/(g(dw)·h·L)).

The density of worms had a great effect on the respiration rate, which decreased dramatically with the increase in the density of worms. The highest respiration rate was found at a worm density of 0.25 mg (ww)/L, with a maximum value of 81.72±5.12 mg O_2/(g(dw)·h·L). Thus, it is necessary to maintain an appropriate density of worms in the sludge reduction system.

A high respiration rate was observed in the temperature range of ~20–27°C, and the highest rate was three times higher than the rate at 8°C. These results indicated that an optimal temperature of 22°C was beneficial for the survival and metabolism of Tubificidae.

Observations at different pH values showed that the highest respiration rate occurred at a pH of 8.0. Differences among the respiration rates at other pH values were markedly less with the rates at other pH values less than half of the rate observed at pH 8.0.

The respiration rate of the worms was strongly influenced by the DO concentration. The highest rate occurred within an optimum DO range of ~3.5–4.5 mg/L. The rate first increased and then decreased as DO increased from 1.5 to 6.5 mg/L. Excessively high or low DO concentrations were not favorable for the worms.

Author Contributions

Conceived and designed the experiments: JL PZ. Performed the experiments: JL YC. Analyzed the data: PS. Contributed reagents/materials/analysis tools: PZ. Wrote the paper: JL YC.

References

1. Low EW, Chase HA, Milner MG, Curtis TP (2000) Uncoupling of metabolism to reduce biomass production in the activated sludge process. Water Res 34: 3204–3212.
2. Wei YS, Van Houten RT, Borger AR, Eikelboom DH, Fan Y (2003a) Comparison performances of membrane bioreactor and conventional activated sludge processes on sludge reduction induced by Oligochaete. Environ Sci Technol 37: 3171–3180.
3. Ratsak CH, Verkuijlen J (2006) Sludge reduction by predatory activity of aquatic oligochaetes in wastewater treatment plants: science or fiction? A review. Hydrobiologia 564: 197–211.
4. Hendrickx TLG, Temmink H, Elissen HJH, Buisman CJN (2009) The effect of operating conditions on aquatic worms eating waste sludge. Water Res 43: 943–950.
5. Tamis J, Van Schouwenburg G, Kleerebezem R, Van Loosdrecht MC (2011) A full scale worm reactor for efficient sludge reduction by predation in a wastewater treatment plant. Water Res 45: 5916–5924.
6. Häner A, Mason C, Hamer G (1994) Death and lysis during aerobic thermophilic sludge treatment: characterization of recalcitrant products. Water Res 28: 863–869.
7. Mayhew M, Stephenson T (1997) Low biomass yield activated sludge: a review. Environ Technol 18: 883–892.
8. Wei YS, Van Houten RT, Borger AR, Eikelboom DH, Fan Y (2003b) Minimization of excess sludge production for biological wastewater treatment. Water Res 37: 4453–4467.
9. Rensink J, Rulkens W (1997) Using metazoa to reduce sludge production. Water Sci Technol 36: 171–179.
10. Salvado H, Gracia M, Amigo J (1995) Capability of ciliated protozoa as indicators of effluent quality in activated sludge plants. Water Res 29: 1041–1050.
11. Hendrickx TLG, Elissen HJH, Temmink H, Buisman CJN (2011) Operation of an aquatic worm reactor suitable for sludge reduction at large scale. Water Res 45: 4923–4929.
12. Wei YS, Wang YW, Guo XS, Liu JX (2009) Sludge reduction potential of the activated sludge process by integrating an oligochaete reactor. J Hazard Mater 163: 87–91.
13. Elissen HJH (2007) Sludge reduction by aquatic worms in wastewater treatment, with emphasis on the potential application of *Lumbriculus variegatus* [Ph.D. thesis]: Wageningen University 192 p.
14. Elissen HJH, Hendrickx TLG, Temmink H, Buisman CJN (2006) A new reactor concept for sludge reduction using aquatic worms. Water Res 40: 3713–3718.
15. Guo XS, Liu JX, Wei YS, Li L (2007) Sludge reduction with Tubificidae and the impact on the performance of the wastewater treatment process. J Environ Sci (China) 19: 257–263.
16. Lee NM, Welander T (1996) Reducing sludge production in aerobic wastewater treatment through manipulation of the ecosystem. Water Res 30: 1781–1790.
17. Wei YS, Liu JX (2006) Sludge reduction with a novel combined worm-reactor. Hydrobiologia 564: 213–222.
18. Zhao LM, Wen S (2012) Progress and perspectives of microfauna as a powerful method for excess sludge reduction in wastewater treatment. Adv Mat Res 446: 2824–2828.
19. Mosleh YY, Paris-Palacios S, Biagianti-Risbourg S (2006) Metallothioneins induction and antioxidative response in aquatic worms *Tubifex tubifex* (Oligochaeta, Tubificidae) exposed to copper. Chemosphere 64: 121–128.
20. Wiederholm T, Wiederholm A, Milbrink G (1987) Bulk sediment bioassays with five species of fresh-water oligochaetes. Water Air Soil Pollut 36: 131–154.
21. Lucan-Bouché ML, Biagianti-Risbourg S, Arsac F, Vernet G (1999) An original decontamination process developed by the aquatic oligochaete *Tubifex tubifex* exposed to copper and lead. Aquat Toxicol 45: 9–17.
22. Walker JG (1970) Oxygen poisoning in the Annelid *Tubifex tubifex I.* response to oxygen exposure. Biol Bull 138: 235–244.
23. Aston R, Milner A (1981) Conditions required for the culture of *Branchiura sowerbyi* (Oligochaeta: Tubificidae) in activated sludge. Aquaculture 26: 155–160.
24. Schenková J, Helešic J, Jarkovský J (2006) Seasonal dynamics of *Bythonomus lemani* and *Bothrioneurum vejdovskyanum* (Oligochaeta, Annelida) in relation to environmental variables. Biologia, Bratislava 61: 517–523.
25. Lochhead G, Learne MA (1983) The effect of temperature on asexual population growth of three species of Naididae. Hydrobiologia 98: 107–112.
26. Brković-Popović I, Popović M (1977) Effects of heavy metals on survial and respiration rate of Tubificid worms: part II—effects on respiration rate. Environ Pollut 13: 93–98.
27. Chapman PM, Farrell MA, Brinkhurst RO (1982) Effects of species interactions on the survival and respiration of *Limnodrilus Hoffmeisteri* and *Tubifex tubifex* (Oligochaeta, Tubificidae) exposed to various pollutants and environmental factors. Water Res 16: 1605–1608.
28. Fowler DJ, Goodnight CJ (1965) The effect of environmental factors on the respiration of *Tubifex*. Am. Midl. Nat. 74: 418–428.
29. Whitley LS, Sikora RA (1970) The effect of three common pollutants on the respiration rate of tubificid worms. J Water Pollut Control Fed 42: R57–R66.
30. Charles DD (1997) Sublethal effects of environmental toxicants on Oligochaete escape reflexes. Amer. Zool. 37: 346–353.
31. Alsterberg G (1924) Die sinnesphysiologie der Tubificiden. Eine experimentelle Untersuchung mit besonderer Berüchsichtigung der respiratorischen Mechanismen. Lund, Håkan Ohlssons Buchdrückerei (in German).
32. Densem JW (1982) Growth studies of Tubifex tubifex (Oligochaeta: Tubificidae) on sewage sludge. Ph.D. thesis. University of Wales, Institute of Science and Technology. Cardiff, Great Britain.
33. Hawkins S, Robinson K, Layton A, Sayler G (2010) Limited impact of free ammonia on *Nitrobacter* spp. inhibition assessed by chemical and molecular techniques. Bioresour Technol 101: 4513–4519.
34. Villaseñor J, Pérez MA, Fernández FJ, Puchalski CM (2011) Monitoring respiration and biological stability during sludge composting with a modified dynamic respirometer. Bioresour Technol 102: 6562–6568.
35. Ma WG, Sun PD, Song YQ, Guo MX, Shi QY, et al. (2011) Growth kinetics of aquatic worms feeding on activated sludge. Environmental Science 32: 2340–2346.
36. American Public Health Association (APHA) (1999). Standard methods for the examination of water and wastewater (20th ed.). Washington D.C.: APHA, AWWA.
37. Egeler P, Römbke J, Meller M, Knacker T, Franke C, et al. (1997) Bioaccumulation of lindane and hexachlobenzene by Tubificid sludgeworms (Oligochaeta) under standardised laboratory conditions. Chemosphere 35: 835–852.
38. Lafont M (1987) Production of Tubificidae in the littoral zone of Lake Léman near Thonon-les-Bains: A methodological approach. Hydrobiologia 155: 179–187.
39. Collier JG (1947) Relations between metabolism and morphogenesis during regeneration in *Tubifex tubifex*. Biol Bull 92: 167–177.
40. Fox HM (1945) The oxygen affinities of certain invertebrate haemoglobins. J Exp Biol 21: 161–164.
41. Wei YS, Fan YB (2005) Reducing sludge production by predation and its factors analysis in conventional activated sludge process. Environmental Science 26:76–83.
42. Antoniou P, Hamilton J, Koopman B, Jain R, Holloway B, et al. (1990) Effect of temperature and pH on the effective maximum specific growth rate of nitrifying bacteria. Water Res 24: 97–101.
43. Timmermans P, Van Haute A (1983) Denitrification with methanol: Fundamental study of the growth and denitrification capacity of *Hyphomicrobium* sp. Water Res 17: 1249–1255.
44. Lou JQ, Guo MX, Chen MM (2011) Studies on the effect of environmental factors on growth of Tubificidae remote sensing, environment and transportation engineering (RSETE). International Conference on. Nanjing. pp. 4442–4445.
45. Aston R, Sadler K, Milner A (1982) The effects of temperature on the culture of *Branchiura sowerbyi* (Oligochaeta, Tubificidae) on activated sludge. Aquaculture 29: 137–145.
46. Reynoldson TB, Rodriguez P, Madrid MM (1996) A comparison of reproduction, growth and acute toxicity in two populations of *Tubifex tubifex* (Müller, 1774) from the North American Great Lakes and Northern Spain. Hydrobiologia 334: 199–206.
47. Guo MX, Zhu MS, Lou JQ (2011) The effects of environmental factors on sludge reduction by tubificidae. Acta Scientiae Circumstantiae 31: 2657–2662.
48. Heugens EHW, Jager T, Creyghton R, S.Kraak MH, Hendriks AJ, et al. (2003) Temperature-dependent effects of cadmium on Daphnia magna: accumulation versus sensitivity. Environ Sci Technol 37: 2145–2151.
49. Reynoldson TB (1987) The role of environmental factors in the ecology of tubificid oligochaetes - an experimental study. Ecography 10: 241–248.
50. Chapman KK, Benton MJ, Brinkhurst RO, Scheuerman PR (1999) Use of the aquatic oligochaetes Lumbriculus variegatus and *Tubifex tubifex* for assessing the toxicity of copper and cadmium in a spiked-artificial-sediment toxicity test. Environ Toxicol 14: 271–278.

51. Liu YC, Shi HC, Xia L, Shi HM, Shen TG, et al. (2010) Study of operational conditions of simultaneous nitrification and denitrification in a Carrousel oxidation ditch for domestic wastewater treatment. Bioresour Technol 101: 901–906.
52. Sturdivant SK, Robert JD, George RC (2012) Bioturbation in a declining oxygen environment, *in situ* observations from wormcam. PLoS ONE 7: e34539.
53. Zhong X, Sun PD, Song YQ, Wang RY, Fang ZG (2009) The application of Biological-Hydraulic coupled model for Tubificidae-microorganism interaction system; In: AIP Conference Proceedings. Sanya Hainan, China. pp. 133.
54. Tian Y, Lu YB, Chen L, Lin HL (2010) Optimization of process conditions with attention to the sludge reduction and stable immobilization in a novel Tubificidae-reactor. Bioresour Technol 101: 6069–6076.

Designing and Piloting a Program to Provide Water Filters and Improved Cookstoves in Rwanda

Christina K. Barstow[1], Fidele Ngabo[2], Ghislaine Rosa[3], Fiona Majorin[3], Sophie Boisson[3], Thomas Clasen[3,4], Evan A. Thomas[5]*

1 Civil, Environmental and Architectural Engineering, University of Colorado, Boulder, Colorado, United States of America, **2** Maternal and Child Health, Ministry of Health, Kigali, Republic of Rwanda, **3** Department of Infectious and Tropical Diseases, London School of Hygiene and Tropical Medicine, London, United Kingdom, **4** Department of Environmental Health, Rollins School of Public Health, Emory University, Atlanta, Georgia, United States of America, **5** Mechanical and Materials Engineering, Portland State University, Portland, Oregon, United States of America

Abstract

Background: In environmental health interventions addressing water and indoor air quality, multiple determinants contribute to adoption. These may include technology selection, technology distribution and education methods, community engagement with behavior change, and duration and magnitude of implementer engagement. In Rwanda, while the country has the fastest annual reduction in child mortality in the world, the population is still exposed to a disease burden associated with environmental health challenges. Rwanda relies both on direct donor funding and coordination of programs managed by international non-profits and health sector businesses working on these challenges.

Methods and Findings: This paper describes the design, implementation and outcomes of a pilot program in 1,943 households across 15 villages in the western province of Rwanda to distribute and monitor the use of household water filters and improved cookstoves. Three key program design criteria include a.) an investment in behavior change messaging and monitoring through community health workers, b.) free distributions to encourage community-wide engagement, and c.) a private-public partnership incentivized by a business model designed to encourage "pay for performance". Over a 5-month period of rigorous monitoring, reported uptake was maintained at greater than 90% for both technologies, although exclusive use of the stove was reported in only 28.5% of households and reported water volume was 1.27 liters per person per day. On-going qualitative monitoring suggest maintenance of comparable adoption rates through at least 16 months after the intervention.

Conclusion: High uptake and sustained adoption of a water filter and improved cookstove was measured over a five-month period with indications of continued comparable adoption 16 months after the intervention. The design attributes applied by the implementers may be sufficient in a longer term. In particular, sustained and comprehensive engagement by the program implementer is enabled by a pay-for-performance business model that rewards sustained behavior change.

Editor: Rachel A. Nugent, University of Washington, United States of America

Funding: All funding for this study was provided by DelAgua Health Limited www.delaguahealth.com) through a sponsored research grant to Portland State University and the London School of Hygiene and Tropical Medicine. The DelAgua Health Board and CEO had no role in study design, data collection and analysis, decision to publish, or preparation of the manuscript. All such responsibility was devolved to Author Thomas.

Competing Interests: DelAgua Health is the funder as well as implementer of the program described. Authors Thomas and Barstow are compensated by, and responsible to, DelAuga Health for the conceptualization, development, implementation, monitoring and reporting of the program described, in cooperation with the Rwanda Ministry of Health. Author Thomas is also the Principal Investigator at Portland State University, primarily responsible for designing and implementing the sensors described in the paper, and collaborating with other research partners including the London School of Hygiene and Tropical Medicine and Emory University. This does not alter the authors' adherence to all the PLOS ONE policies on sharing data and materials.

* E-mail: evan.thomas@pdx.edu

Introduction

Access to improved drinking water and clean burning stoves could benefit the millions who suffer from diarrheal disease and pneumonia, two of the leading causes of death around the world for children under five. Worldwide, of the 7.6 million deaths in children under 5 in 2010, 64% were associated with infectious diseases including 18% with pneumonia and 11% with diarrhea. Combined, pneumonia and diarrhea kill over 2 million children each year [1].

Some of these deaths may be avoided through interventions to improve indoor air quality and household water quality: pneumonia is often linked to indoor air pollution from biomass fuels [2,3] and diarrhea to deficiencies in water and sanitation, including poor water quality [4]. Many cookstove interventions have shown a reduction in indoor air pollutants such as carbon monoxide and fine particulate matter [5,6]. Similarly, interventions targeted at improving household water quality through the implementation of water treatment strategies such as chemical treatment, boiling, solar disinfection or filtration have been shown to reduce diarrheal disease [7,8].

Even with the fastest annual reduction in child mortality in the world, the Republic of Rwanda still faces challenges related to pneumonia and diarrhea: among deaths of children under 5, pneumonia accounts for 18% and diarrhea for 8% [9]. Cooking practices in a rural Rwandan household may contribute to this pneumonia burden since the predominate fuel and cookstove pairing is wood on a three stone fire [10]. Additionally, while Rwanda has demonstrated significant progress towards the Millennium Development Goals, almost 30% of households do not have access to an improved water source [11], and the improved water sources may become contaminated during collection, transport or storage within the home [12,13]. Once water is in the households, less than half (46.1%) of rural Rwandan families report treating their drinking water, with boiling as the leading treatment method (38.1%) [11], which again can become recontaminated after treatment [14]. The Rwanda Standard for Potable Water states that the microbiological limits for potable water for total CFU/100 ml of total coliforms should be 0 [15]. A baseline water quality assessment of 230 improved water sources, 78 unimproved water sources, and stored water in 468 households across all 30 districts in Rwanda indicated that 27.8% of improved water sources, 80.2% of unimproved water sources, and 58.3% of stored household water supplies exceed this standard [16], falling into the "intermediate", "high" or "very high" risk World Health Organization categories [17] for biological contamination of drinking water supplies. Another study of households within the other 11 districts of the project area prior to the start of the program implementation indicated that 81.1% of households exceed this standard, with 59.1% falling into the "intermediate" or "high" risk categories [18].

DelAgua Health, a for-profit social enterprise, was established to combine household technologies that address environmental health issues with market-based mechanisms. DelAgua Health participates in the United Nations Clean Development Mechanism (CDM) to earn carbon credits associated with the reduced use of, and demand for, fuel wood associated with water treatment and cooking, and then sell those credits to buyers as a way to recover costs and profit [19].

Carbon finance markets facilitate the reduction of greenhouse gas emissions worldwide through economic incentives, while allowing cleaner economic development to take place. Each emission reduction credit represents the non-emission of one tonne of carbon dioxide into the atmosphere. The carbon credits generated under the CDM help Kyoto Protocol Annex I countries to meet their binding targets, and can be traded in the marketplace. However, the carbon markets have yet to be well utilized to finance the distribution of humanitarian technologies in the least developed countries, particularly in Africa. Although the CDM is a multi-billion dollar industry, fewer than 2 percent of projects are registered in African nations [20].

Depending on the project location, structure, methodology and registration mechanism employed, a water treatment and/or cookstove program can earn between approximately 1/2 and 5 carbon credits per household, per year. The carbon credits earned are a function of the approved methodology, referenced to a baseline condition and the current performance of the program, as audited by independent firms. The reported reductions are then issued by the registration authority and are then sold to buyers. Carbon credit buyers may be banks, energy companies, brokers, or sovereign nations who require credits for either regulatory compliance or voluntary social responsibility efforts, or both. Because the carbon credits are issued in proportion to the present adoption and proper use of intervention technologies, this encourages sustained engagement by the program implementer and creates a pay-for-performance model.

In Rwanda, DelAgua Health is partnered with the Ministry of Health since 2012 to distribute free of charge household water treatment and high efficiency cookstoves to approximately 600,000 households (about 3 million people), throughout the country's 30 districts. The project will target *Ubudehe* categories 1 and 2, the government-recognized poorest 30% of the country. *Ubudehe* category is determined by community members based on classifications outlined by the Rwandan Ministry of Local Government [21]. Households categorized as *Ubudehe* 1 and 2 already receive free medical and other assistance through government programs.

A pilot program was initiated in October 2012 to provide input for the full effort, scheduled to start in mid-2014. This pilot was conducted after findings from a preliminary study of 100 households in July 2012. This effort was judged by the implementers to be sufficiently promising for testing at a larger scale. This paper discusses the design, development, implementation, monitoring and periodic modification of the October 2012 pilot program. We summarize the results of surveys collected to evaluate key outputs including intervention uptake and use. Other aspects of the pilot are described elsewhere, including a novel method for assessing intervention use with remotely reporting sensors [22] and a randomized controlled trial to study the impact of the intervention on drinking water quality and household air pollution [23].

Materials and Methods

Design Objectives

The objective of this study was to identify if certain design criteria, integrated together and applied to environmental health technologies, could result in a meaningful proportion of continued use of stoves and water filters. These design attributes are evaluated as a whole, though estimates of relative value are provided in the discussion. The three program design choices considered fundamental were:

1. **Free Distribution.** Free provisioning of high quality stoves and water filters under the authority of the Government of Rwanda and through established community mechanisms including community meetings and community leadership.
2. **Behaviour Change.** A behaviour change messaging and monitoring effort that prioritizes consistent and correct adoption of the stoves and filters through community and household level activities, focusing on both health and non-health benefits.
3. **"Pay for Performance" Public-Private Partnership**. A public-private partnership with the Rwanda Ministry of Health enabled by anticipated carbon credit revenues, which allows sustained, comprehensive community engagement by virtue of future anticipated "pay for performance" carbon credit revenues.

Program Setting and Population

The pilot was conducted in a convenience sample of 15 non-randomly selected villages spread across 11 districts in Western Rwanda (Figure 1). The 15 villages were selected to have at least one village per the 11 districts and the remaining four villages in districts with the largest populations. Additional inclusion criteria included ensuring that no villages were in adjacent sectors (district subdivisions), less than 20% of households in each village served by piped water; less than 60% of households in each village using any water treatment other than boiling; less than 20% of households in each village using cooking fuel sources other than biomass or charcoal; and less than 20% of households in each village using any stove other than a 3-stone fire or two other locally made unimproved stoves (known as *Rondereza* and *Imbabura* stoves). These inclusion criteria were selected by program staff to be representative of typical rural villages in Rwanda, based on rural water service and energy use characteristics identified in the Rwanda 2011 Demographic and Health Survey. Program staff visited each candidate village in advance to confirm with village officials that it met eligibility criteria. All 1,943 households who were registered as members of the 15 villages were eligible to participate in the study. While the full program will consist of distribution to only *Ubudehe* 1 and 2 households, this pilot program consisted of all households, of any *Ubudehe* category, in the 15 villages. The full program originally consisted of distribution to all households in the Western province of Rwanda but was later revised to be a country wide program of *Ubudehe* 1 and 2 households. This program change was directed by the Government of Rwanda Ministry of Health.

Intervention Hardware

The water filter used in this program, the Vestergaard Frandsen LifeStraw Family 2.0 is a point-of-use microbial water treatment system intended for routine use in low-income settings. The system is a table-top unit where the user pours untreated water through a 20 micron pre-filter into a six liter influent water tank. Water is then gravity-filtered through a 0.20 micron hollow-fiber ultrafiltration membrane into a 5.5 liter safe storage container. Water can be dispensed from the safe storage container through a plastic tap, limiting recontamination. The filter is backwashed by squeezing a plastic bulb located on the opposite side of the tap. The membrane can filter up to 18,000 liters of water [24], enough to supply a family of five with microbiologically clean drinking water for three to five years. The system exceeds the 'highly protective' World Health Organization Standard for household water treatment technologies [25,26]. In a recent study, an earlier version of this filter was shown to be highly effective in improving water quality

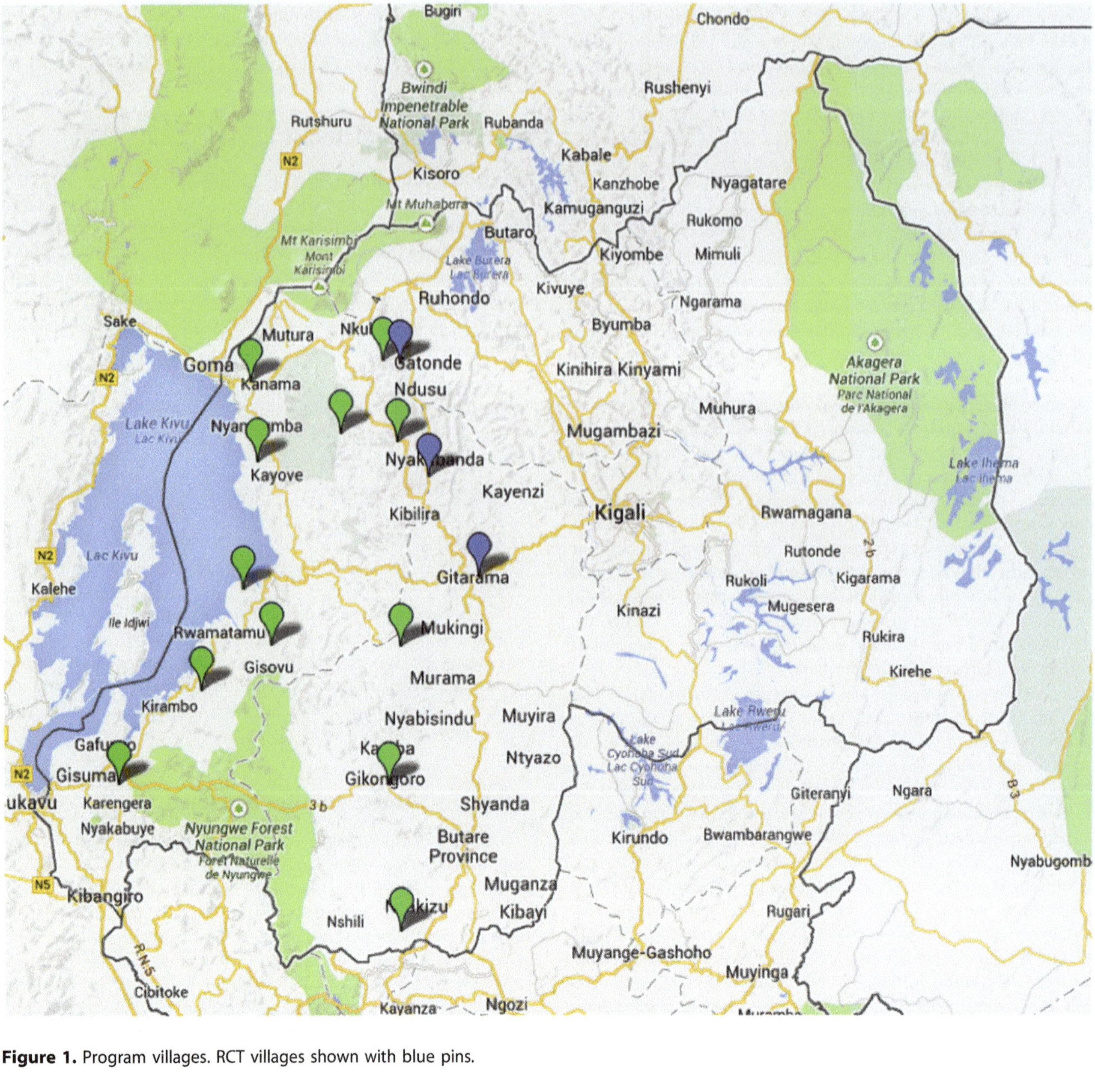

Figure 1. Program villages. RCT villages shown with blue pins.

and was protective against diarrhea among HIV positive individuals, reducing longitudinal prevalence by over 50% [27].

The cookstove used in this program, the EcoZoom Dura, is based on the rocket-stove concept that is designed to concentrate the combustion process while channeling air flow to create a more complete burn. A complete burn of carbon rich material will also result in little to no smoke. Included with the stove are a "stick support" on which fuelwood is placed to promote air flow and a "pot skirt" which increases thermal efficiency. In the field, performance is variable but when properly used, a rocket stove will significantly reduce fuelwood use by at least 50%, although reductions in indoor air pollution vary between designs, fuel types and use [28]. The thermal efficiency of this stove is 38% [29].

Intervention Design

The program is designed leveraging established behavior change theories, including the Diffusion of Innovation theory [30] and the Health Belief Model (HBM) [31]. In particular, the program design assumes that continued, comprehensive engagement is critical in order to effect positive behavior change. The program design takes a hybrid approach, integrating pieces of these theories that apply in a Rwandan context to shape the communication strategy of the program.

Several components of the diffusion of innovation theory are applied to the program. At the initial distribution meeting, community members are informed about the potential health and other benefits of the water filter and cookstove creating 'initial knowledge' around the technologies. The 'persuasion stage' is initiated through demonstrations at both the community meeting and the household. The stove demonstration includes assembling the stove, how to adjust a pot skirt which can be fitted to different sized pots and finally a fire being started in the stove with some demonstrations including boiling a pot of water to show the rapidity of the cooking process. The water filter includes a demonstration of filtering visually dirty water with clear water coming out of the tap and the maintenance procedure, which includes backwashing the filter. Progressing to the 'decision stage' the household is then asked to demonstrate use and maintenance of the technology, allowing them to trial the technology. Households then move into the 'implementation stage' where they can choose to adopt or reject the technology. About a month later, the program implements the 'confirmation stage' where households who have chosen to partially adopt or reject the technology are given additional training and messaging to hopefully reverse their decision.

Village chiefs are promoted as 'early adopters' because of their high degree of influence and respect within their villages. In this program the change agents are Community Health Workers (CHWs). The CHW system in Rwanda includes three CHWs per village who are part-time volunteers of sector health centers and are compensated with a stipend. They provide basic services such as maternal and newborn health monitoring, vaccination advocacy, family planning, treatment of malaria, and sanitation and hygiene education. Through this program CHWs play several important roles including informing households of the need for the devices, encouraging adoption and analyzing potential problems with the technologies. The CHWs play an especially important role with the 'late adopters' and 'laggards' as more effort is needed to change the household's old habit and promote the new behaviors [32].

The Health Belief Model is used to shape messaging. The belief that there is a health threat is compelled by messaging related to clean drinking water and clean indoor air. Households are educated about the reduced risk of diarrheal disease from water borne diseases and the reduced risk of respiratory problems from breathing indoor air pollutants. Additionally an important concept in social cognitive theory is often added to the health belief model, self-efficacy, which states that the user must believe that they can adopt the new behavior [33]. This is facilitated through households gaining confidence in the use of the technologies by having members of the household demonstrate proper use.

While the health belief model provides relevant guidance on behavior change theory it is important that the program also express non-health benefits to users. Previous interventions related to both water quality and improved cookstoves emphasize the need to highlight non-health benefits such as those related to economic and social benefits [32,34]. Thus CHWs educate households on additional benefits such as reduction in medical costs from the water filter and a reduction in cooking time and expenditure on fuel costs for the cookstove.

Adoption and Monitoring Survey

Households were assigned to receive the adoption and monitoring survey in one of six rounds by a random number generator. Approximately 325 households were surveyed each month with the exception of month five where approximately twice as many households were surveyed as rounds 5 and 6 were combined because of time constraints. Environmental Health Officers (EHOs) were responsible for conducting the surveys. Environmental Health Officers (EHOs) are full time employees of the Ministry of Health at sector health centers who are responsible for a range of health interventions, including, food safety, waste management, water, sanitation and hygiene inspection, indoor air pollution and environmental emergency health interventions [21]. Each household was surveyed by EHOs only once during the five month period.

The survey consisted of about 100 questions and was administered using a smartphone in Kinyarwanda by an EHO. Information included household identifying information, demographics, cooking practices, and water treatment and collection practices. The survey included both self-reported use questions to be answered by members of households and observational questions which EHOs answered based on their observations. Observational use of the water filter was measured by checking if water was present in the filter at the time of the visit while observational use of the stove was only confirmed if the stove was actually being used at the time of the visit.

Survey Data Analysis

All survey data was uploaded to the doForms database where it was analyzed using T-SQL and R-Project. Only surveys that fell within 15 to 90 minute survey duration were included in the analysis. All numerical outcomes were additionally analyzed using an outlier analysis where only 1.5 times the upper and lower interquartile range were included in that particular outcome. This outlier exclusion was chosen to be consistent with the program's carbon credit monitoring requirements [35][. Analysis of variance was used to compare group means. Additionally any missing data was excluded from the analysis.

Focus Group Discussions

Three focus group meetings were conducted concurrently with EHOs and CHWs to assess qualitative aspects of the program. A total of 30 participants attended the meetings including one CHW from each of the pilot villages and all EHOs. CHWs were chosen by DelAgua staff as the highest performing CHWs within each village. Topics covered included perceived adoption of technologies within their villages, problems with filter and stove hardware, effectiveness of household messaging and boundaries to exclusive adoption of the filter and stove.

Ethics and Consent

The study was reviewed and approved by the Rwanda National Ethics Committee (IRB #328/RNEC/2012), University of Colorado Institutional Review Board (Protocol #12-0564), and the London School of Hygiene and Tropical Medicine Ethics Committee (Trial registration: Clinicaltrials.gov NCT01882777). Each participating household gave informed, verbal consent after having received complete details regarding the purpose of the survey as well as information regarding privacy of personal information. Enumerators were required to confirm electronically, on their smartphone surveys, if a.) the respondent was over 18, and b.) if they gave verbal consent, before the smartphone allowed the survey to continue. These consent records are kept on a password protected server. Verbal consent was requested and approved by the approving ethics committees, based on the high percentage of illiteracy within the study population. Rwandan residents are often asked about their water and energy habits by community health workers, and the signing of a document adds a level or formality that may mislead participants. Participants were given the opportunity to ask any questions before agreeing to participate. All households were entitled to retain their filters and stoves at the conclusion of the study.Participation in the study was not a prerequisite to receiving the filters and stoves.

Results

Program Delivery

All households that were registered in the 15 villages according to the village chief's list were distributed a stove and filter. The model of distributing at a central location allowed for the implementer to transport the technologies to a location that could be reached by vehicles of which many households could not. It was then members of the household's responsibility to get the stove and filter to their homes.

Household Characteristics

A total of 1943 households participated in the study, from which 1634 (84.1%) valid surveys were included in the analysis. Selected household characteristics are summarized in Table 1. The mean household size was 4.55, consistent with Rwanda's 2011 Demographic and Health Survey household size of 4.5. Approximately one-third (29.3%) of households reported being categorized as *Ubudehe* 1 or 2, 70.5% were classified as 3 or 4 and 0.2% as *Ubudehe* 5 or 6. *Ubudehe* 1 and 2 household size was significantly lower than the entire study population with 3.85 persons per household. However, fuel type and water source were similar with the majority of all households (92.4%) and *Ubudehe* 1 and 2 households (89.1%) using wood as their primary fuel source and all households (41.1%) and *Ubudehe* 1 and 2 households (42.9%) reporting a public tap as their primary drinking water source.

Filter Adoption and Use

Adoption of the LifeStraw filter was measured at approximately 90% or greater by several metrics. Households reported use of the filter had the highest adoption rate with 96.5% of households reporting the water filter as the treatment method for the last water they drank. An observational measure of use through presence of water in the filter showed a slightly lower adoption rate with 9 out of 10 households having water in their filter at the time of household visit (Table 2). Similar adoption rates as measured observationally were seen over the five follow up visits with the first follow up visit having the highest adoption rate of 92.6% and the lowest adoption rate reported as 86.4%. No longitudinal trend in adoption was observed through the five months of the study.

Adoption rates were further analyzed to understand any differences between all households in the study and those who were identified as *Ubudehe* 1 and 2 as well as any differences across the 15 villages. *Ubudehe* 1 and 2 adoption rates were similar to those seen by the full program with an *Ubudehe* 1 and 2 reported adoption rate of 95.6% and an observational adoption rate of 88.5% (Table 2). Observed filter adoption across all 15 villages varied between 74.5% and 98.1% (Table 3) with the lowest adoption rate occurring in one village (Mara) by almost 10% below all other villages. This is likely due to rodents destroying the filter tubes with 21.8% of filter repair of this problem in this village.

Households reported treating an average of 5.06 liters per day in all households and 5.11 liters per day in *Ubudehe* 1 and 2 households with no significant difference between monthly survey rounds. This equates to an average of 1.27 liters per person per day for all households and 1.11 liters per person for *Ubudehe* 1 and 2 households, possibly because of the smaller household sizes (Table 2). Regardless, water quantity consumption is lower than advised by CHWs at 2 liters per person per day. Similar consumption rates of 1 to 1.5 liters per person per day were reported at focus group discussions conducted with community health workers. Primary reasons discussed at the focus group for not consuming more water included not having a container to carry water when leaving the household, an inability to drink 2 liters per day, and a preference for drinking other beverages.

12.8% of households that reported using the filter also reported doing so for purposes other than drinking water. The most common uses were cooking (28.4%), hand washing (28.4%) and washing dishes (27.4%) (Table 2).

Almost two thirds (63.7%) of households reported backwashing their filter every time they treated water as advised in the household visit (Table 2). Not backwashing the filter frequently enough may be the cause of the most common reported problem with the filter, which was that it was clogged and wouldn't filter water (N = 45). The next most common problem reported was damage to rubber tubes because of rodents (N = 33). Overall 11.1% of households over the 5-month period reported any problems with the filter during the household visits. 57 (2.9%) filters were replaced and 366 (18.8%) filters were repaired with the same primary reported problems of tubes being damaged by rodents (N = 119) and filter clogged (N = 124) (Table 4).

The majority of filter problems were addressed through repair and replacement by program staff. Households contacted program staff through phone numbers on informational posters which were provided during the initial household visits. Common repairs included replacing tubes eaten by rodents or power backwashing the filters using a hand pump pressurized canister.

Stove Adoption and Use

As seen with filter adoption, reported primary use of the EcoZoom stove was around 90% for the entire population and *Ubudehe* 1 and 2 households (Table 5). Primary reasons given for stove adoption during focus group meetings included cost savings, time savings and cleanliness of the cook and kitchen when using the EcoZoom. However, 71.5% of these households reported continuing to use their traditional stove as well as their EcoZoom stove. Of households cooking at the time of the follow up visit (20.9%), about two thirds (63.7%) were using the EcoZoom stove, 21.9% a traditional 3-stone fire, and 11.4% cooking on a different

Table 1. Selected demographics and characteristics regarding water and energy practices.

	All households		***Ubudehe* 1 and 2**	
	N	**%**	**N**	**%**
Number of Households	1634		478	
Household size, mean (95%CI)	4.55 (4.46–4.65)		3.85 (3.70–4.00)	
***Ubudehe* Category**				
1 or 2	478	29.3%		
3 or 4	1152	70.5%		
5 or 6	4	0.2%		
Fuel Type				
Wood	1510	92.4%	426	89.1%
Straw/Shrubs/Grass	93	6.3%	42	8.8%
Charcoal	28	1.9%	10	2.1%
LPG/Natural Gas/Biogas	1	0.1%	0	0.0%
Other	2	0.1%	0	0.0%
Drinking water source*	1576		457	
Public Tap	647	41.1%	196	42.9%
Protected Spring	592	37.6%	149	32.6%
Unprotected Spring	184	11.7%	62	13.6%
Surface Water	56	3.6%	26	5.7%
Hand Pump	37	2.3%	6	1.3%
Piped Water in Home	31	2.0%	8	1.8%
Unprotected Well	24	1.5%	8	1.8%
Protected Well	4	0.3%	1	0.2%
Rainwater	1	0.1%	1	0.2%

*Missing 58 (3.6%) answers.

traditional stove. Ten households (2.9%) were also observed using both the EcoZoom stove and a traditional stove.

Using the same metric of observed cooking use, EcoZoom use at the time of the household visit varied from 35.5% – 100.0% across the 15 pilot villages though the sample sizes were low in some villages with only 342 total observed cooking events. The three villages with the lowest observed use were Buhunde, Gisoro and Burorero (Table 3).

The two primary reasons reported during household surveying for not using the EcoZoom stove were inability to use wet wood in the EcoZoom stove (N = 46) and difficulty in using the stove (N = 31). This reported difficulty may refer to the required increased frequency of fire tending while cooking on the EcoZoom stove as 67.1% of households reporting tending the fire more with the EcoZoom stove than with their old stove (Table 4). Additionally the food most frequently reported cooked was beans (53.9%), requiring cooks to tend the fire frequently over a long period of time. Focus group discussions further emphasized these problems with the primary reasons for not using the stove as high frequency of fire tending, difficulty in burning wet wood when dry wood was unavailable and the inability to warm the house. Tending the fire was expressed most frequently as the primary issue since cooks use smaller pieces of wood to keep the fire going.

To assess the degree of EcoZoom stove use compared to other traditional stoves, households were asked the number of times per week they used each stove in their home. The EcoZoom stove was reported being used on average 1.37 times per day for 71.2% of cooking events in a household with a significant difference between monthly survey rounds (Table 5).

Primary reasons for continued use of a traditional stove included needing more than one stove at a time (N = 649) and the inability to cook on the EcoZoom stove when only wet wood was available (N = 330) (Table 4).

Cooking location was assessed because of the program emphasis on cooking outdoors during education and training activities. 342 observations of cooking location at the time of follow up visit were made, where 42.4% were cooking outdoors, 36.0% were cooking indoors and 21.6% in a separate kitchen. Slightly higher rates of cooking indoors were observed in the *Ubudehe* 1 and 2 households with 37.3% cooking outdoors, 44.5% cooking indoors, and 18.2% cooking in a separate kitchen (Table 5).

To quantify wood savings households were asked to report the number of wood bundles they collected or purchased before and after receiving the EcoZoom stove. Of the 1551 valid responses, an average wood reduction of 65.8% was reported across all rounds with no significant difference between the five rounds (Table 5).

A total of 73 stove problems were reported with the two most common problems being the pot skirt screw missing (N = 22) and the pot skirt degrading (N = 8). These were also the two most common reasons for stove repair with 48 pot skirts (2.5%) being replaced due to melting and 11 pot skirt replacements due to missing adjustment screws (Table 4). No stoves were replaced during the five months following distribution.

Table 2. Primary filter adoption and use outcomes.

	All households				*Ubudehe* 1 and 2			
	N	%	Mean (95% CI)	ANOVA p-value	N	%	Mean (95% CI)	ANOVA p-value
Reported Filter Use	1576	96.5%			457	95.6%		
Observed Filter Use	1471	90.0%			423	88.5%		
Other Purposes for Filtered Water	201	12.8%			74			
Cooking	57	28.4%			24	32.4%		
Hand Washing	57	28.4%			20	27.0%		
Washing Dishes	55	27.4%			18	24.3%		
Other	32	15.9%			12	16.2%		
Cleaning Frequency	1576				457			
Everytime	1004	63.7%			265	58.0%		
Daily	234	14.8%			71	15.5%		
Every Other Day	53	3.4%			20	4.4%		
2 Times per Week	236	15.0%			83	18.2%		
Once per Week	48	3.0%			18	3.9%		
Once per Month	1	0.1%			0	0.0%		
Liters per household per day	1210*		5.06 (4.99–5.12)	0.193	304*		5.11 (5.04–5.18)	0.194
Liters per person per day	1436*		1.27 (1.25–1.29)	0.090	304*		1.11 (1.10–1.13)	0.194

*Only records within 1.5 IQR were included.

Table 3. Filter and stove use by village.

	Observed Filter Use			Cooking on EcoZoom only at time of visit		
Village	**Valid observations**	**N**	**%**	**Valid observations**	**N**	**%**
Mara	110	82	74.5%	35	24	68.6%
Kigaga	90	82	91.1%	23	19	82.6%
Buhunde	117	103	88.0%	31	11	35.5%
Rushishi	147	137	93.2%	26	15	57.7%
Nyarubuye	116	102	87.9%	16	10	62.5%
Karambo	177	149	84.2%	34	23	67.6%
Rubona	65	62	95.4%	21	13	61.9%
Burorero	214	191	89.3%	18	7	38.9%
Nyabivumu	61	58	95.1%	22	22	100.0%
Rambura	116	107	92.2%	12	6	50.0%
Kabuga	111	106	95.5%	23	9	39.1%
Rupango	106	102	96.2%	31	27	87.1%
Gisoro	54	53	98.1%	8	3	37.5%
Nyarutovu	81	77	95.1%	31	21	67.7%
Gasumo	69	60	87.0%	11	8	72.7%
All villages	1634	1471	90.0%	342	218	63.7%

Discussion and Conclusions

In the pilot program described here high levels of uptake and continued use of water filters and improved cookstoves were found. The rigorous five-month follow up study was complemented by qualitative assessments by the implementation team periodically over an additional 11 months. 16 months after the intervention, adoption rates of the water filters and cookstoves were assessed to being comparable to those observed during the detailed 5-month study. This outcome may be described through the three design choices outlined previously. These design choices are intended to be applied in the full-scale program scheduled for deployment in 2014 and 2015. A key design difference in the planned full-scale program is that it will reach only the poorest households in a given village.

Behavior Change

The primary purpose of the technologies provided is to realize a health benefit. As a first step, communicating these potential health benefits to a user is often seen as an appropriate prerequisite to adoption. A lack of knowledge of potential health benefits has been shown to result in poor adoption of products like stoves and filters [36]. It has also been demonstrated that knowledge of health benefits alone is not sufficient to result in sustained behavior change in an individual or household [34]. The program studied here uses theories of behavior change such as diffusion of innovation and the health based model with both health based messaging as well as economic and social messaging to promote behavior change within the program.

In the case of the filter, adoption was measured around 90% for the five months of the study. A high adoption outcome has been seen previously with earlier versions of the LifeStraw Family filter with 96% adoption in Zambia [27] after 12 months and 68% adoption in the Democratic Republic of Congo after eight months [37]. Additionally, compared to other point-of-use water methods, filtration often has higher adoption rates [38] possibly because it is seen as easier to use [39] and doesn't result in a change in taste and odor [34]. However, while adoption of the filter was high, the recommended water consumption of 2 liters per person per day was not reached in most cases, suggesting that households may be drinking untreated water at times, a behavior also seen in the study conducted in the Democratic Republic of Congo [37]. Even occasional consumption of untreated water can greatly reduce the potential health benefits from water quality interventions [40,41]. Reasons that households gave for not exclusively drinking clean water included not having clean water when they were away from home and not having any filtered water at the time they were thirsty. Similar reasoning has been found in other studies where water treatment needed to be integrated into the everyday lifestyle of the family [34]. However, guidance from the existing literature is limited to help guide behavior change program development [42] so additional qualitative research is needed to understand people's behavior and preferences for exclusive drinking of treated water to adjust messaging and education activities to be more effective. In collaboration with the manufacturer, the product has been updated based on recipient feedback to protect the soft tubes and the backwash bulb, and to allow for a separable safe water storage container for ease of cleaning.

Adoption of the stove was also around 90%. Non-health benefits such as a cleaner appearance and cooking environment were more highly valued than health or environmental impacts, as observed in other studies. Exclusive use of the EcoZoom stove was only reported in 28.5% of households with most continuing to use their old stove with the EcoZoom stove. The earliest models for stove adoption assumed a "fuel switch" wherein behavior is switched over a short period of time from one stove/fuel combination to another. More recently, continued "stove stacking", where the use of multiple stoves for varying purposes, has been shown to be a more stable behavior, and can result in as high as 90% of stove usage events on the improved stove [43]. Studies have shown that households do not move from older existing methods of cooking such as a 3-stone fire to exclusive use of an improved stove. In

Table 4. Reasons for stove and filter problems, repairs and replacements.

	N	%
Reported reasons for not using EcoZoom	168	
Don't have dry wood	46	27.4%
Difficult to use	31	18.5%
Doesn't warm the house	20	11.9%
Use of a different fuel	13	7.7%
Other (20)	76	45.2%
Reported stove problems	73	
Pot skirt missing screw	22	30.1%
Pot skirt damaged	8	11.0%
Stove too small	6	8.2%
Difficulty in moving pot skirt to another pot	6	8.2%
Ceramic chamber cracked	4	5.5%
Stick support damaged	4	5.5%
Other (12)	23	31.5%
Reported reasons for continued use of old stove	1386	
More than one stove needed	649	46.8%
Don't have dry wood	330	23.8%
Need to warm house	126	9.1%
Pot is too big for stove	73	5.3%
Other (28)	208	15.0%
Reasons for stove repair	67	
Skirt replaced/Skirt damaged	48	71.6%
Skirt replaced/adjustment screws missing	11	16.4%
Stick support replaced/Broken	3	4.5%
Stick support replaced/Missing	1	1.5%
Other	4	6.0%
Reported filter problems	182	
Filter broken or clogged	45	24.7%
Tubes damaged/eaten by rodents	33	18.1%
Tubes are kinked	26	14.3%
Difficulty in backwashing	22	12.1%
Tap is leaking or broken	18	9.9%
Backwash bulb is damaged	17	9.3%
Other (6)	21	11.5%
Reasons for filter repair*	391	
Filter cartridge clogged	124	31.7%
Tubes replaced from rodent damage	119	30.4%
Backwash water not going into container	62	15.9%
Broken tap handle	39	10.0%
Broken tap - leaking	10	2.6%
Backwash leaking	5	1.3%
Backwash bulb replaced	2	0.5%
Other	30	7.7%

*366 total filters repaired - 25 had multiple problems.

order to realize the potential health benefits of improved stoves, exclusive use will need to be further promoted within the program [44,45,46]. Working with the stove manufacturer, the pot skirt has been updated to reduce degradation.

This stove stacking behavior suggests that the true innovation being introduced is not necessarily the stove itself, but the modified cooking practices required to realize the health and other benefits. Previous studies have found that an improved stove must meet the user's traditional cooking needs in order for adoption to occur.

Table 5. Primary stove adoption and use outcomes.

	All households				*Ubudehe* 1 and 2			
	N	**%**	**Mean (95% CI)**	**ANOVA p-value**	**N**	**%**	**Mean (95% CI)**	**ANOVA p-value**
Reported Primary Stove Use	1634				478			
EcoZoom	1479	90.5%			430	90.0%		
Traditional 3-stone Fire	68	4.2%			23	4.8%		
Other Traditional Stove	78	4.8%			22	4.6%		
Other Stove	9	0.6%			3	0.6%		
Stove Using at Time of Visit	342	20.9%			110	23.0%		
EcoZoom Only	218	63.7%			70	63.6%		
Traditional 3-Stone Fire	75	21.9%			28	25.5%		
Other Tradional Stove	39	11.4%			11	10.0%		
EcoZoom and Other Stove	10	2.9%			1	0.9%		
Cooking Location at Time of Visit								
Outdoor	145	42.4%			41	37.3%		
Indoor	123	36.0%			49	44.5%		
Separate Kitchen	74	21.6%			20	18.2%		
Continuing to Use Old Stove	1169	71.5%			316	66.1%		
# of EcoZoom uses per week	1517*		1.37 (1.34–1.40)	<0.000001	438*		1.38 (1.32–1.44)	0.869
% EcoZoom Use	1625*		71.2% (69.8%–72.7%)	0.052	474*		73.7% (71.0%–76.4%)	0.896
% Wood Reduction	1551*		65.8% (65.0%–66.6%)	0.593	443*		65.9% (64.3%–67.4%)	0.082

*Only records within 1.5 IQR were included.

The primary barriers to adoption or exclusive use of the EcoZoom stove in this study center around modifying current practices such as additional fire tending. While the EcoZoom stove is likely to decrease the cooking time of an individual cooking event, additional fire tending compared to a traditional 3-stone fire is often necessary, as reported in the focus group convened for this study. There may be an increased emissions exposure risk associated with greater fire tending that has not yet been characterized. When cooking meals on a 3-stone fire that requires long cooking times, such as beans, the cook will often prepare a large fire and perform other household tasks while the beans are cooking. Use of the EcoZoom stove requires a behavior modification to persuade the cook to stay by the fire while the meal is cooking. Additionally burning of wet wood in an improved stove can be more difficult than a 3-stone fire, so a cook often prefers to use the easier cooking method and therefore it is suggested that a careful examination of cooking practices, and focusing on those practices rather than the intrinsic benefits of the technology may result in higher adoption rates [43].

42.4% of observed cooking events occurred outdoors as instructed through education and training activities. While the EcoZoom stove has the potential to reduce indoor air pollution compared to traditional stoves, the primary reduction may likely come from moving cooking out of the home. As less than half (42.4%) of observed cooking events occurred outdoors as instructed through education and training activities, further messaging targeted at cooking location will need to be performed to increase outdoor usage.

A 65.8% reduction in wood usage by users of the EcoZoom stove is likely to significantly reduce time to collect wood and expenditure on fuelwood. Much of the wood use in Rwanda is with small sticks and branches which burn fast on three stone fires. Wood reduction was calculated through the ratio of wood bundles used before and after receiving the EcoZoom stove. In order to better quantify wood savings, additional methods will need to be employed to better understand wood usage such as the kitchen performance test which can evaluate stove performance in real-world settings [47].

With respect to both the filter and stove, behavior in *Ubudehe* categories 1 and 2 was similar to the overall population. This suggests that similar results could be expected during a large distribution of only *Ubudehe* 1 and 2 households. However, the effect of distributing to only a part of a village while the other households do not receive the technologies is unknown.

Free Distribution

Recent studies have examined cost-sharing for bed-nets, cookstoves and water treatment systems and have found that there is no correlation between free distributions and low adoption rates [48]. Meanwhile, a study examining point-of-use chlorination through marketing campaigns and coupon schemes found these to be ineffective strategies but found free chlorination dispensed at water sources along with community providers as the most effective strategy in potentially preventing diarrheal incidence in areas like rural Kenya [49]. Furthermore, Bensch and Peters determined that a free stove program in Senegal resulted in high uptake of almost 100% of households [50].

These studies suggest that adoption and price are not fundamentally correlated, and that other factors including community engagement, government support and education are worth more careful study. With respect to the private-public partnership described here, the free giveaway nature of the program did not appear to adversely affect technology adoption on a community level, and resulted in a broader population exposure to the interventions than would have been possible via a retail effort over the same time period. The high rate of exclusive use of filters suggest that free distribution did not impact filter use, though it may have impacted intervention stove use.

Public-Private Partnership

The extensive logistical and behavior change messaging components of this program require sustained funding. Rwanda is not yet able to finance all health service activities directly; it relies both on direct donor funding to government programs, as well as careful coordination of programs managed by international non-profits and health sector businesses. By 2002, the government was spending 8.6% of its revenue on health care, which was only a third of the total costs, the remainder covered by donors [47]. Donation based non-profits are not providing services to the target populations serviced by this program. The business model anticipated by the for-profit implementer is designed to recuperate invested costs by the generation and sale of carbon credits associated with the proportion of the intervention that continues to demonstrate successful behavior change. The outcomes observed to-date support the business model in that high adoption rates will correlate to carbon credit generation sufficient to generate sustainable revenue that will allow continued program investment.

Study Limitations

Many of the results described here are from self-reported survey data that may result in over-reporting because of courtesy bias [51]. Over-reporting was measured in this study through the use of remote sensors which revealed over-reporting in frequency of use of both the water filter and cookstove [22]. This contributes to existing evidence of courtesy bias in self-reported outcomes of product distributions. Additionally while the survey directly asked households about their use of other cookstoves, it did not ask about households about supplementing their drinking water from other sources. To fully understand this issue, additional surveying and analysis is necessary. Respondent fatigue may also have been an issue throughout the study as some households were visited several times during a single month. Additionally the short duration of this study (five months) with less rigorous follow up through at least month 16 doesn't allow for complete characterization of the technologies or long-term adoption.

Acknowledgments

The authors thank the important contributions from colleagues and students at the Rwanda Ministry of Health, DelAgua Health, Vestergaard Frandsen, EcoZoom, Portland State University, the University of Colorado at Boulder and the London School of Hygiene and Tropical Medicine.

Author Contributions

Conceived and designed the experiments: CB FN GR SB TC ET. Performed the experiments: CB GR FM ET. Analyzed the data: CB ET. Contributed reagents/materials/analysis tools: CB ET. Wrote the paper: CB TC ET.

References

1. Liu L, Johnson HL, Cousens S, Perin J, Scott S, et al. (2012) Global, Regional, and National Causes of Child Mortality: An Updated Systematic Analysis for 2010 with Time Trends Since 2000. Lancet, 379: 2151–2161.
2. Smith K, Samet J, Romieu I, Bruce N (2000) Indoor Air Pollution in Developing Countries and Acute Lower Respiratory Infections in Children. Thorax 55: 518–532.

3. Ezzati M, Kammen D (2001) Indoor Air Pollution from Biomass Combustion and Acute Respiratory Infections in Kenya: An Exposure-Response Study. Lancet 358: 619–24.
4. Black R, Morris SS, Bryce J (2003) Where and Why are 10 Million Children Dying Each Year. Lancet 361: 2226–2234.
5. Roden C, Bond T, Conway S, Pinel AB, MacCarty N, et al. (2009) Laboratory and Field Investigations of Particulate and Carbon Monoxide Emissions from Traditional and Improved Cookstoves. Atmospheric Environment 43: 1170–1181.
6. Masera O, Edwards R, Arnez CA, Berrueta V, Johnson M, et al. (2007) Impact of Patsari Improved Cookstoves on Indoor Air Pollution in Michoacan, Mexico. Energy for Sustainable Development 11: 45–56.
7. Clasen T, Schmidt WP, Rabie T, Roberts I, Cairncross S (2007) Interventions to improve water quality for preventing diarrhoea: systematic review and meta-analysis. BMJ 334: Epub.
8. Fewtrell L, Kaufmann RB, Kay D, Enanoria W, Haller L, et al. (2005) Water, Sanitation and Hygiene Interventions to Reduce Diarrhoea in Less Developed Countries: A Systematic Review and Meta-Analysis. Lancet Infectious Diseases 5: 42–52.
9. UNICEF (2013). Committing to Child Survival: A Promise Renewed. United Nations Children's Fund.
10. European Union Energy Initiative (2009) Biomass Energy Strategy (BEST), Rwanda - Volume 3 Rural Supply and Demand. Available: http://www.cleancookstoves.org/resources_files/best-volume-3rural-supply.pdf. Accessed 2014 March 1.
11. Republic of Rwanda (2011) Rwanda Demographic and Health Survey 2010. National Institute of Statistics of Rwanda. Kigali.
12. Gundry S, Wright J, Conroy R (2004) A Systematic Review of the Health Outcomes Related to Household Water Quality in Developing Countries. Journal of Water and Health 1: 1–13.
13. Onda K, LoBuglio J, Bartram J (2012) Global Access to Safe Water: Accounting for Water Quality and the Resulting Impact on MDG Progress. Int. J. Enviorn. Res. Public Health 9: 880–894.
14. Wright J, Gundry S, Conroy R (2004) Household drinking water in developing countries: a systematic review of microbiological contamination between source and point-of-use. Tropical Medicine and International Health 9: 106–117.
15. Rwanda Bureau of Standards (2011) Potable Water Specification,RS 435:2009, Rev 1. Kigali.
16. Kirby M (2013) Water Quality Testing in 30 Districts of Rwanda. Unpublished.
17. WHO (1997) Guidelines for drinking water quality, 2nd edition. Volume 3: Surveillance and control of community supplies. Geneva. Available: http://www.who.int/water_sanitation_health/dwq/gdwqvol32ed.pdf. Accessed 2014 March 1.
18. Rosa G (2012) Water Quality Testing in 11 Districts in Rwanda. Unpublished.
19. Thomas E (2012) Leveraging carbon financing to enable accountable water treatment programs. Global Water Forum. Available: http://www.globalwaterforum.org/2012/09/23/leveraging-carbon-financing-to-enable-accountable-water-treatment-programs. Accessed 2014 March 1.
20. Williams M, Murthy S (2013) Reconciling the Carbon Market and the Human Right to Water: The Role of Suppressed Demand Under Clean Development Mechanism and the Gold Standard. Environmental Law 43: 517–562.
21. Rwanda Ministry of Health (2012) Third Health Sector Strategic Plan July 2012- June 2018. Ministry of Health. Kigali.
22. Thomas E, Barstow C, Rosa G, Majorin F, Clasen T (2013) Use of remotely reporting electronic sensors for assessing use of water filters and cookstoves in Rwanda. Environmental Science & Technology 47: 13602–13610.
23. Rosa G, Majorin F, Boisson S, Barstow C, Johnson M, et al. (2014) Assessing the impact of water filters and improved cookstoves on drinking water quality and household air pollution: A randomised controlled trial in Rwanda. PLOS One In Press.
24. Clasen T, Naranjo J, Frauchiger D, Gerba C (2009) Laboratory assessment of a gravity-fed ultrafiltration water treatment device designed for household use in low-income settings. American Journal of Tropical Medicine and Hygiene 80: 819–823.
25. WHO (2011) Evaluating household water treatment options: Health-based targets and microbiological performance specificiations.
26. Narajo J, Gerba C (2011) Assessment of the LifeStraw Family Unit using the World Health Organization Guidelines for Evaluating Household Water Treatment Options: Health-based Targets and Performance Specifications. University of Arizona.
27. Peletz R, Simunyama M, Sarenje K, Baisley K, Filteau S, et al. (2012) Assessing water filtration and safe water storage in households with young children of HIV-positive mothers: A randomized, controlled trial in Zambia. PLOS One doi:10.1371/journal.pone.0046548.
28. Johnson M, Pilco V, Torres R, Joshi S, Rajeev M, et al. (2013) Impacts on household fuel consumption from biomass stove programs in India, Nepal, and Peru. Energy for Sustainable Development 17: 403–411.
29. Aprovecho Research Center (2012) Results of Water Boiling Test 41 Testing of the EcoZoom Dura Stove.
30. Rogers E (2003) Diffusion of Innovations 5th Edition. New York: Free Press.
31. Becker M (1974) The Health Belief Model and Personal Health Behavior. Health Education Monographs 2: 324–473.
32. Aboud FE, Singla DR (2012) Challenges to Changing Health Behaviors in Developing Countries: A Critical Overview. Social Science and Medicine 75: 589–594.
33. Rosenstock I, Stretcher V (1988) Social Learning Theory and the Health Belief Model. Health Education Behavior 15(2): 175–183.
34. Figueroa M, Kincaid D (2010) Social, cultural and behavioral correlates of household water treatment and storage. USAID.
35. United Nations Clean Development Mechanism (2013) DelAgua Public Health Program in Eastern Africa, 2012. Available: http://cdm.unfccc.int/ProgrammeOfActivities/Validation/DB/32Z3TT7LC70PMDWOEW1706HJ0ZJLN7/view.html. Accessed 2014 March 1.
36. Lewis J, Pattanayak S (2012) Who adopts improved fuels and cookstoves? A systematic review. Environmental Health Perspectives 120(5): 637–645.
37. Boisson S, Kiyombo M, Sthreshley L, Tumba S, Makambo J (2010) Field Assessment of a Novel Household-Based Water Filtration Device: A Randomized, Placebo-Controlled Trial in the Democratic Republic of Congo. PLoS ONE doi:10.1371/journal.pone.0012613.
38. Hunter P (2009) Household Water Treatment in Developing Communities: Comparing Different Intervention Types Using Meta-Regression. Environmental Science and Technology 43(23) 8991–8997.
39. Albert J, Luoto J, Levine D (2010) End-User Preferences of Competing POU Water Treatment Technologies Among the Rural Poor of Kenya. Enviornmental Science and Technology 44(12) 4426–4432.
40. Brown J, Clasen T (2012) High Adherence is Necessary to Realize Health Gains from Water Quality Interventions. PLoS One doi:10.1371/journal.pone.0036735.
41. Enger KS, Nelson KL, Rose JB, Eisenberg JN (2013) The Joint Effects of Efficacy and Compliance: A Study of Household Water Treatment Effectiveness Against Childhood Diarrhea. Water Research doi:10.1016/j.watres.2012.11.034.
42. Fiebelkorn P, Person B, Quick RE, Vindigni SM, Jhung M, et al. (2012) Systematic Review of Behavior Change Research on Point-of-Use Water Treatment Interventions in Countries Categorized as Low-to Medium-Development on the Human Development Index. Social Science and Medicine doi:10.1016/j.socscimed.2012.02.011.
43. Ruiz-Mercado I, Masera O, Zamora H, Smith KR (2011) Adoption and sustained use of improved cookstoves. Energy Policy 39: 7557–7566.
44. Geary CW, Prabawanti C, Aristani C (2012) A Field Assessment of Adoption of Improved Cookstove Practices in Yogyakarta, Indonesia: Focus on Structural Drivers. Fhi360 Available: http://www.fhi360.org/sites/default/files/media/documents/CookstovePracticeFieldAssessment101512.pdf. Accessed 2014 March 1.
45. Masera OR (2000) From Linear Fuel Switching to Multiple Cooking Strategies: A Critique and Alternative to the Energy Ladder Model. World Development 28(12) 2083–2103.
46. Redman A (2011) Stove Stacking: The Integration of the Justa Cookstove in One Rural El Salvadorian Community. EHOS Conference 2011. Kirkland.
47. Shell Foundation (2007) Kitchen Performance Test.
48. Jameel Poverty Action Lab (2011) The price is wrong: Charging small feels dramatically reduces access to important products for the poor. Jameel Poverty Action Lab. Availible: http://www.povertyactionlab.org/publication/the-price-is-wrong. Accessed 2014 March 1.
49. Kremer M, Miguel E, Mullainathan S (2008) Source dispensers and home delivery of chlorine in Kenya. Innovations for Poverty Action. Availible: http://www.povertyactionlab.org/evaluation/source-dispensers-and-home-delivery-chlorine-kenya. Accessed 2014 March 1.
50. Bensch G, Peters J (2012) A recipe for success? Randomized free distribution of improved cooking stoves in Senegal. Ruhr-Universität Bochum (RUB). Availible: http://repec.rwi-essen.de/files/REP_12_325.pdf. Accessed 2014 March 1.
51. Wood L, Egger M, Gluud LL, Schulz K, Juni P, et al. (2008) Empirical Evidence of Bias in Treatment Effect Estimates in Controlled Trials with Different Interventions and Outcomes: Meta-Epidemiological Study. BMJ. http://dx.doi.org/10.1136/bmj.39465.451748.AD. doi:10.1136/bmj.39465.451748.AD.

23

How Much Is Enough? Minimal Responses of Water Quality and Stream Biota to Partial Retrofit Stormwater Management in a Suburban Neighborhood

Allison H. Roy[1]*, Lee K. Rhea[2], Audrey L. Mayer[3], William D. Shuster[2], Jake J. Beaulieu[2], Matthew E. Hopton[2], Matthew A. Morrison[4], Ann St. Amand[5]

1 U. S. Geological Survey, Massachusetts Cooperative Fish and Wildlife Research Unit, University of Massachusetts, Amherst, Massachusetts, United States of America, **2** U. S. Environmental Protection Agency, Office of Research and Development, Cincinnati, Ohio, United States of America, **3** School of Forest Resources and Environmental Science and Department of Social Sciences, Michigan Technological University, Houghton, Michigan, United States of America, **4** SABIC, Mt. Vernon, Indiana, United States of America, **5** PhycoTech, Inc., St. Joseph, Michigan, United States of America

Abstract

Decentralized stormwater management approaches (e.g., biofiltration swales, pervious pavement, green roofs, rain gardens) that capture, detain, infiltrate, and filter runoff are now commonly used to minimize the impacts of stormwater runoff from impervious surfaces on aquatic ecosystems. However, there is little research on the effectiveness of retrofit, parcel-scale stormwater management practices for improving downstream aquatic ecosystem health. A reverse auction was used to encourage homeowners to mitigate stormwater on their property within the suburban, 1.8 km^2 Shepherd Creek catchment in Cincinnati, Ohio (USA). In 2007–2008, 165 rain barrels and 81 rain gardens were installed on 30% of the properties in four experimental (treatment) subcatchments, and two additional subcatchments were maintained as controls. At the base of the subcatchments, we sampled monthly baseflow water quality, and seasonal (5×/year) physical habitat, periphyton assemblages, and macroinvertebrate assemblages in the streams for the three years before and after treatment implementation. Given the minor reductions in directly connected impervious area from the rain barrel installations (11.6% to 10.4% in the most impaired subcatchment) and high total impervious levels (13.1% to 19.9% in experimental subcatchments), we expected minor or no responses of water quality and biota to stormwater management. There were trends of increased conductivity, iron, and sulfate for control sites, but no such contemporaneous trends for experimental sites. The minor effects of treatment on streamflow volume and water quality did not translate into changes in biotic health, and the few periphyton and macroinvertebrate responses could be explained by factors not associated with the treatment (e.g., vegetation clearing, drought conditions). Improvement of overall stream health is unlikely without additional treatment of major impervious surfaces (including roads, apartment buildings, and parking lots). Further research is needed to define the minimum effect threshold and restoration trajectories for retrofitting catchments to improve the health of stream ecosystems.

Editor: Maura (Gee) Geraldine Chapman, University of Sydney, Australia

Funding: The project was largely funded by the U. S. Environmental Protection Agency (USEPA) through the Office of Research and Development (ORD). Field sampling was conducted by USEPA employees and volunteers and supported in part by Pegasus Technical Services, Inc. under EPA Contract # EP-C-05-056. The auction, installation, and maintenance of rain gardens and rain barrels was funded by the USEPA's ORD under the Scientific, Technical, Research, Engineering and Modeling Support (STREAMS) contract EP-C-05-061, Task Order 8 to TetraTech, Inc. Biotic identification was funded by the USEPA's ORD through simplified acquisitions to PhycoTech, Inc. in 2003–2004 (3C-R312-TTSA) and 2007–2010 (EP10C000111), to EnviroScience, Inc. in 2005–06 (EP05C00134), and Rhithron Associates, Inc. in 2007–2010 (EP07C000170 and EP10C000110). As such, the funders had a role in study design, data collection and analysis, decision to publish, and preparation of the manuscript.

Competing Interests: The work was undertaken by USEPA employees. One of the authors, Ann St. Amand, conducted lab analysis of periphyton for this study under a contract with the USEPA. The paper has undergone peer review through USEPA and the U. S. Geological Survey to ensure that there is no misrepresentation of findings, and the paper makes no recommendations about policy or management. The use of trade names or products does not constitute endorsement by the U.S. Government. The views expressed in this paper are those of the authors and do not necessarily reflect the views or policies of the USEPA. This does not alter the authors' adherence to all the PLOS ONE policies on sharing data and materials.

* E-mail: aroy@eco.umass.edu

Introduction

Rapid urbanization and the ongoing conversion of landscapes from natural habitats to industrial, commercial, and residential land uses to support a growing human population remain the most salient threats to natural ecosystems [1–3]. Aquatic ecosystems that drain urban areas are particularly vulnerable due to their low position in the landscape [4]. In most urban and suburban areas, untreated stormwater runoff from impervious surfaces is typically routed directly into rivers, lakes, and oceans. This conventional design of urban drainage systems reflects concerns about human health and safety, but largely ignores threats to aquatic ecosystem health that stem from stormwater runoff [5,6].

The urban stream syndrome describes changes in stream ecosystems associated with urbanization, a subject that has been increasingly studied in the last few decades (see reviews by [7–9]). These changes primarily arise from stormwater runoff from impervious cover—particularly impervious cover that is directly

Table 1. Catchment area and land cover based on cumulative piped catchments draining to each site.

Site	Total Area (ha)	Forest Cover[1] (%)	TIA[2] (%)	DCIA[3] (%)			2007 Installations[4]			2007 & 2008 Installations		
				Before (<2007)	During (2007–08)	After (2008)	Barrels	Gardens	Density (#/ha)	Barrels	Gardens	Density (#/ha)
Sub1	28.0	43.8	19.9	11.6	11.2	10.4	50	20	2.5	84	33	4.2
Sub2	57.9	46.3	16.3	9.0	8.7	8.1	68	27	1.6	123	55	3.1
Sub3	68.9	68.0	13.1	7.3	7.2	7.1	32	23	0.80	42	26	1.0
Catch	183	62.6	13.1	7.4	7.2	7.0	100	50	0.82	165	81	1.3
Sub4	24.9	66.4	11.2	5.4	5.4	5.4	1	1	0.08	1	1	0.08
Sub5	34.8	64.2	12.1	7.3	7.3	7.3	0	0	0.00	0	0	0.00

[1]Forest cover is based on topographic catchment.
[2]Total impervious area (%) includes all impervious surfaces identified by air photos and site visits [38].
[3]Directly connected impervious area (DCIA) was calculated by subtracting the rooftop area draining to rain barrels installed in 2007 and 2008.
[4]Stormwater management installations includes number of rain barrels and rain gardens installed in 2007 and overall (2007 & 2008), and the total density of barrels and gardens within each subcatchment.

connected to streams by stormwater pipes [6]— which alters stream hydrology, water chemistry, and biotic communities. High magnitude, flashy flows in urban streams can scour stream beds and erode stream banks, thus reducing habitat quality. Furthermore, the extreme high flows can wash out aquatic biota and low lying riparian vegetation, whereas reduced base flows can reduce in-stream habitat and alter stream ecosystem function [10]. Runoff that enters urban and suburban streams often contains increased toxicants, ions, and nutrients, along with higher temperatures, reduced oxygen saturation, and organic material that can alter biotic structure and ecosystem function (production, nutrient uptake, leaf breakdown, etc.) relative to streams in natural landscapes [7]. While there are other catchment stressors (e.g., point sources, septic systems, riparian degradation) and in-stream stressors (e.g., impoundments, water withdrawals, stream burial) associated with urbanization [9], stormwater runoff is a dominant source of impairment of ecological structure and function in most urban catchments. As such, comprehensively managing stormwater runoff with the goals of mimicking pre-disturbance flow regimes, improving water quality, and ultimately improving ecosystem health is a leading approach to urban stream restoration [5,6,11,12].

There is an increasing movement throughout the world to address runoff through decentralized stormwater management [6,13–15]. This management approach can include small-scale tools that capture and detain (e.g., detention and retention basins), infiltrate (e.g., pervious pavements, rain gardens), and filter (e.g., biofiltration swales, wetlands) runoff on individual parcels throughout a landscape [16–18]. These tools are implemented through new development (often referred to as low impact development, LID) [15] or retrofitting of already-developed areas.

To date, there have been no reports of the effectiveness of decentralized stormwater management for improving stream water quality and biota in suburban catchments. Most assessments are limited to measurements of hydrology and water quality within individual treatment practices [18,19] (see also www.bmpdatabase.org) and modelled effects of installations throughout catchments [20–22]. There have been some catchment-scale studies comparing LID and conventional practices in new developments [23,24], but no catchment-scale retrofits of existing developments. This is due in part to the distribution of impervious surfaces within catchments, and the legal, economic, and logistical difficulty of implementing stormwater management practices at a scale appropriate for improvement [25,26]. The research presented here and another study currently underway in Australia [27] are, to our knowledge, the first attempts at assessing stream responses to retrofit stormwater management at the catchment scale.

Like many large, older US cities, the metropolitan area of Cincinnati, OH has an aging stormwater infrastructure that uses common combined sewer overflows, which lead to both an environmental and legal (e.g., consent decree) need to address stormwater management [28]. Thurston et al. [29] used Cincinnati to demonstrate that decentralized stormwater abatement was likely to be less expensive than a centralized (e.g., deep tunnel) solution to the water quality and quantity problem. Thus, a multidisciplinary study was designed that 1) assessed the legal, economic, and scientific challenges associated with decentralized stormwater management [25], and 2) developed and tested a stormwater management strategy within a small, suburban catchment. Rain barrels (up to four per property) and rain gardens (one per property) were offered to eligible residents (i.e., owner-occupied and within the experimental area) through a voluntary, reverse auction where bids consisted of the stormwater management practice(s) and a financial subsidy (if desired). The lowest cost bids at locations with the highest potential environmental benefits were prioritized. Winning homeowners received the bid amount, free stormwater management practices, installation, and maintenance for three years [30]. The voluntary nature of the auction avoided private property rights issues while providing financial incentives to property owners for installation and increasing stakeholder ownership [25]. The project placed 81 rain gardens and 165 rain barrels on ca. 30% of the eligible properties within the headwaters of the Shepherd Creek catchment in 2007 and 2008 [30]. The rain barrels resulted in an overall decrease in directly connected impervious area (DCIA) from 7.4% to 7.0% in the catchment, and 11.6% to 10.4% (Sub1), 9.0% to 8.1% (Sub2), and 7.3% to 7.1% (Sub3) in the Experimental subcatchments (Table 1). Rain gardens did not change the DCIA, but offered additional capacity to capture overland runoff.

The objective of our study was to determine if the retrofit stormwater management imposed as a result of the economic auction would result in measurable shifts in the ecological condition of streams in the Shepherd Creek catchment. We assessed habitat conditions, baseflow water quality, and the biological assemblage (periphyton and macroinvertebrates) at various sites along tributary streams within the catchment. For the hydrometric portion of the Shepherd Creek study, Shuster & Rhea [31] demonstrated a small but significant effect of decentralized stormwater management on stream hydrology, reflected by a decrease in runoff volume in treatment subcatchments. Given the small change in hydrology and small reduction in % DCIA, combined with levels of impervious cover above published thresholds of impairment [6,32], we expected minimal or no responses of water quality and biota to stormwater management.

Methods

Study design

We used a modified before–after–control–intervention (BACI) study design, where the intervention was the installation of treatments (rain barrels and rain gardens) on select parcels within the catchment. The modified paired-catchment BACI design relied on comparison of the difference between responses of Control and Experimental subcatchments for three periods to determine if significant treatment effects were present [33–35]. Typically only a pre-treatment and experimental period are used in a BACI design, but because implementation of the treatments spanned 16 months, we also included a transition period. Six study sites were sampled within the Shepherd Creek catchment, including four Experimental sites (Sub1, Sub2, Sub3, Catch) and two Control sites (Sub4, Sub5; Figure 1). Access to field sites Sub1, Sub2, Sub3, and Catch was granted by private landowners. Sites Sub5 and Sub5a were in the Mt. Airy Forest, and permission was granted by the Cincinnati Park Board. Sub4 was in the road right-of way. Multiple Experimental and Control sites were used to minimize potential confounding of location-specific differences with treatment effects [36,37]. Total impervious area (TIA) in the Shepherd Creek catchment was 13.1%, with just over half (7.4%) of the TIA directly connected to stormwater or combined sewer pipes (Table 1) [38]. The five subcatchments (24.9–68.9 ha) ranged from 11.2–19.9% TIA, with a corresponding range of 43.8–68.0% forest cover (Table 1).

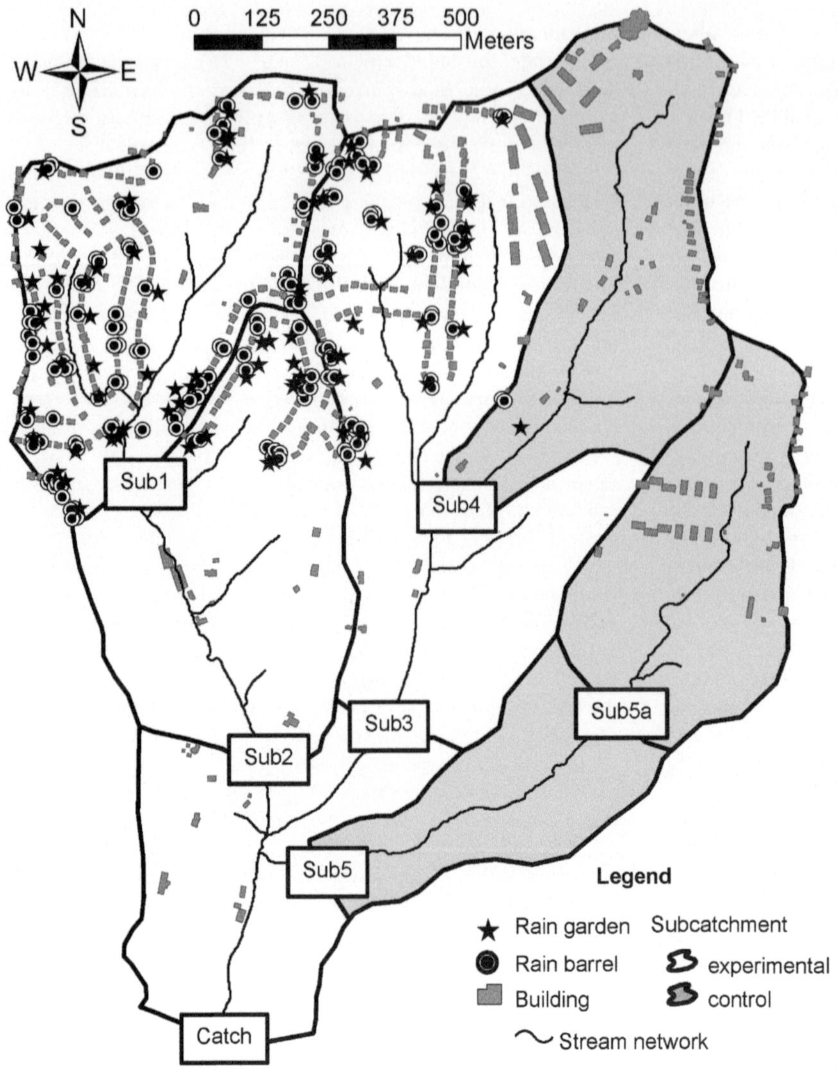

Figure 1. Shepherd Creek catchment and subcatchments in Cincinnati, Ohio (USA). Sub1 was nested within Sub2, Sub4 was nested within Sub3, and all of the subcatchments drained to the catchment outlet (Catch). Sub 5a was discontinued in 2006. Rain barrels and rain gardens were installed throughout the catchment except in Sub4 and Sub5.

Table 2. Variables used in statistical models.

Variable	Type	Description
Period	fixed, numeric	study period (Before treatment = 0, During treatment = 0.5, After treatment = 1)
Site	fixed, class	study sites (Sub1-5, Catch); nested in Group
Group	fixed, numeric	Control (Sub4, Sub5 = 0) and Experimental (Sub1, Sub2, Sub3, Catch = 1) sites
Round	random, class	sample collection dates grouped in 7-day windows
Sample	random, class	sample event (year and month) (ordination analyses only)
Group*Period	fixed, class	interaction of study period and group; indicates significant effect of stormwater treatment
Axis	fixed, class	NMS axis number (ordination combined axes analysis)
Axis*Period	fixed, class	interaction of NMS axis and study period (ordination combined axes analysis)
Axis*Group	fixed, class	interaction of NMS axis and study sites (ordination combined axes analysis)

Sites were sampled five times per year from April through October from 2003 through 2010, with a few exceptions. Sub5 was moved downstream in 2005 due to lack of sufficient flow at Sub5a. Sub4 was added in April 2005 to provide an additional Control, and Sub1 was discontinued in October 2009 following sewer repair that drastically changed stream baseflow hydrology. Stormwater treatment installation occurred in June through September 2007 (Phase 1) and July through September 2008 (Phase 2). Sample periods were therefore designated as Before (prior to June 2007), During (June 2007 through September 2008), and After (October 2008 and later), in respect to treatment installation. There was no new road or building construction in the study area during the study period, but there were some potentially significant events, including: road and sewer repair in Sub1 in 2006 and 2008, roadside vegetation clearing along Sub4 in summer 2007, and invasive riparian plant removal in Sub5 in summer 2007.

Physical and chemical characteristics

Basic morphometric, geomorphic, and water quality parameters were measured five times/year within the 61-m sample reach during biotic sampling. We calculated approximate values for average width, average depth, wetted area, and surface velocity based on field measurements. Additional physical attributes included estimates of: % riffle, % pool, % run, large wood density, % small wood, % large wood, % detritus, bed texture (% bedrock, cobble, gravel, sand, silt), and % canopy cover following the EPA Rapid Bioassessment Protocols (RBP) Physical Characterization data sheet [39]. Water quality measurements were taken with a YSI 6600 data sonde (YSI, Inc., Yellow Springs, OH, USA), and included: water temperature, conductivity, dissolved oxygen, pH, oxidation-reduction potential, and turbidity. YSI data sondes were calibrated within 24 hours of use and tested immediately following return from the field. Measurements from water quality probes that did not pass post-deployment calibration checks were not used. Finally, we calculated two visual assessment habitat evaluation scores: EPA's Rapid Bioassessment Protocols Quantitative Habitat Assessment (QHEI) [39], and the Primary Headwater Habitat Evaluation Form (HHEI) [40] that is specifically designed for streams with water depths <40 cm.

Water quality sampling was also conducted monthly during baseflow conditions. In addition to measuring the water quality parameters described above, grab water samples were filtered with a 0.45-µm filter for total dissolved phosphorus (TDP), dissolved organic carbon (DOC), and dissolved metals (Al, Fe, Mn, Cu, Zn). TDP and DOC were preserved with sulfuric acid, and dissolved metals were preserved in nitric acid. Unfiltered grab samples were collected for analysis of nutrients (nitrate/nitrite nitrogen, ammonium nitrogen, dissolved inorganic nitrogen, total Kjeldahl nitrogen, total phosphorus, ortho-phosphate, TDP), total organic carbon (TOC), total recoverable metals (Al, Fe, Mn, Cu, and Zn), and base cations (Na, Mg, K, Ca). Nutrients and TOC were preserved with sulfuric acid, and metals and base cations were preserved with nitric acid. Suspended sediment concentration (SSC), anions (Cl, Br, F, SO_4, NO_3, and ortho-PO_4), and alkalinity samples were unpreserved. For SSC, the 250-mL sample was filtered through a pre-washed and pre-weighed 1.5-µm glass-fiber filter, dried to a constant weight, and re-weighed following ASTM Method D3977-97 [41]. Analyses for nutrients, metals, anions, and cations were performed by EPA Region 5 Central Regional Laboratory (Chicago, IL) using standard EPA protocols as follows: EPA Method 353.2 for nitrate/nitrite nitrogen, EPA Method 350.1 for ammonia nitrogen, EPA Method 351.2 for total Kjeldahl nitrogen, EPA Method 365.4 for total and dissolved phosphorus, EPA Method 415.1 for total organic carbon, EPA Methods 200.2 (digestion) and 200.7 (analysis) for metals and base cations, and CRL Method AIG045 for anions, based closely on EPA Method 300.1.

Periphyton

Periphyton samples were collected from submerged rocks throughout the 61-m study reach. Cobbles were removed from the stream and a 13.2-cm^2 area on each rock, designated with a PVC ring, was brushed with a toothbrush for 2 min. Rocks and brushes were then rinsed with stream water into a 500-mL bottle. Algae from all rocks within a reach were composited into a single bottle and placed in the dark on ice.

In the laboratory, 20–30 mL of the periphyton slurry was filtered onto each of two glass fiber filters and frozen for subsequent analysis of chlorophyll *a* using a multi-wavelength spectrophotometer following EPA's Method 446.0 [42]. An additional 50 mL of sample was preserved in 1% gluteraldehyde for biomass analysis. The sample was later filtered onto a pre-ashed glass fiber filter (47 mm, PALL Type A/E, 1-µm pore size). Filters were dried at 105°C to a constant weight, weighed for dry weight, ashed in a muffle furnace for 1.5 hours at 500°C, wetted, re-dried at 105°C, and re-weighed to obtain ash-free dry mass (AFDM). The remaining algal sample was preserved in 1% gluteraldehyde for identification. All algae (diatoms and soft algae) were identified and enumerated to the genus level by PhycoTech,

Table 3. Physical and chemical variable summary statistics and ANOVA results for Group*Period interaction.

Variable	Lambda[1]	N	Min	Max	Mean	St Dev	P^2	
Habitat[3]								
DO (mg/L)	1.45	206	0.350	13.15	7.11	2.98	0.087	
HHEI score[4]	1.80	204	51.0	99.0	81.3	8.55	0.722	
Oxidation reduction potential (mV)	1.35	190	1.80	392	195	77.3	0.501	
Conductivity (mS/cm)	−1.55	210	0.280	2.53	1.00	0.394	<0.001	***
Temperature (°C)	1.90	210	5.68	25.3	15.8	4.68	0.218	
Turbidity (NTU)	−0.250	154	0.100	19.9	8.71	5.40	0.042	*
Closed canopy (%)	2.50	199	5.00	90.0	60.9	22.4	<0.001	***
Mean depth (m)	−6.25	201	0.019	0.600	0.074	0.066	0.067	
Filamentous algae (score)	−1.25	88	1.00	4.00	1.76	0.830	0.017	*
Wetted reach area (m^2)	−1.30	203	4.30	366	100.07	61.75	0.744	
Riffle habitat (%)	1.90	179	5.00	90.0	50.9	20.2	0.331	
QHEI score[5]	1.50	204	55	175	113.97	23.53	0.702	
Water Quality[6]								
Alkalinity (mg $CaCO_3$/L)	−7.00	293	100	350	251	40.7	0.240	
Chloride (mg/L)	−1.80	273	0.100	856	147	125	0.228	
Calcium (mg/L)	3.10	279	1.40	182	99.4	42.4	0.010	*
Dissolved organic carbon (mg/L)	0.55	282	1.90	35.2	9.95	7.59	0.216	
Iron (mg/L)	−5.70	285	0.002	3.42	0.287	0.342	<0.001	***
Magnesium (mg/L)	−1.00	255	0.100	35.4	19.4	6.22	0.104	
Nitrate (mg/L)	−6.55	271	0.020	10.7	1.40	1.58	0.858	
pH	0.500	278	6.07	10.2	7.79	0.507	0.623	
Sulfate (mg/L)	−5.20	242	0.100	222	83.3	37.6	0.015	*
Suspended sediment concentration (mg/L)	−1.75	286	−8.22	45.9	7.84	8.75	0.842	
Total dissolved phosphorus (mg/L)	−2.20	277	0.040	0.620	0.185	0.099	0.539	
Total organic carbon (mg/L)	−2.05	281	1.80	37.3	11.6	8.74	0.526	
Temperature (°C)	2.40	272	0.090	25.2	12.9	7.02	0.831	
Zinc (dissolved) (μg/L)	−2.10	232	0.010	127	21.0	18.4	0.482	

[1]Lambda is value for the exponential transformation.
[2]***$P<0.001$, *$P<0.05$.
[3]Habitat variables (including some water quality variables) were sampled five times per year during biotic sampling events.
[4]HHEI score from Ohio Environmental Protection Agency Primary Headwater Habitat Evaluation Index [40].
[5]QHEI score from Rapid Bioassessment Protocols Quantitative Habitat Assessment for high gradient streams, and filamentous algae score (range 0–4) is from RBP benthic macroinvertebrate field sheet [39].
[6]Water quality variables were sampled monthly during baseflow conditions.

Inc., consistent with Standard Methods 10200 and 10300 [43]. Three permanent slide mounts were made with 2-hydroxypropyl methacrylate (HPMA), and all slides were examined using a stratified counting procedure (200× and 400× for soft algae, 1000× for diatoms and picoalgae) to a minimum of 400 natural units [44].

Algal indices were calculated based on densities of cells per cm^2 and included total density, density of major orders (Bacillariophyta, Chlorophyta, and Cyanophyta), relative proportions of the major orders, taxa richness, Shannon diversity, and percent of sample in the dominant taxon. A periphyton index of biotic integrity (PIBI) developed for the mid-Atlantic region of the United States was calculated that incorporated nine metrics (phosphatase activity metric excluded) [45].

Macroinvertebrates

Macroinvertebrates were collected using two methods: 1) a triangular dip net (500-μm mesh) used to collect a composited, multi-habitat sample in the entire 61-m reach, and 2) a bucket sampler (0.053-m^2) within three replicate depositional/riffle habitats. The net samples were considered ideal for capturing macroinvertebrate diversity and relative abundance [39], whereas the bucket samples were used to determine macroinvertebrate densities in riffle habitats that are most sensitive to disturbance [46]. Net samples were collected five times/year in conjunction with periphyton and habitat sampling, and bucket samples were collected during three sampling events per year (spring, summer, autumn). Bucket samples were taken by pushing an open bucket into the bed sediment, surrounding the bucket-sediment interface with a custom-made canvas skirt to enclose the area, and scrubbing each large rock into the water contained in the bucket. The bed sediment was then disturbed for 10 sec. using a trowel,

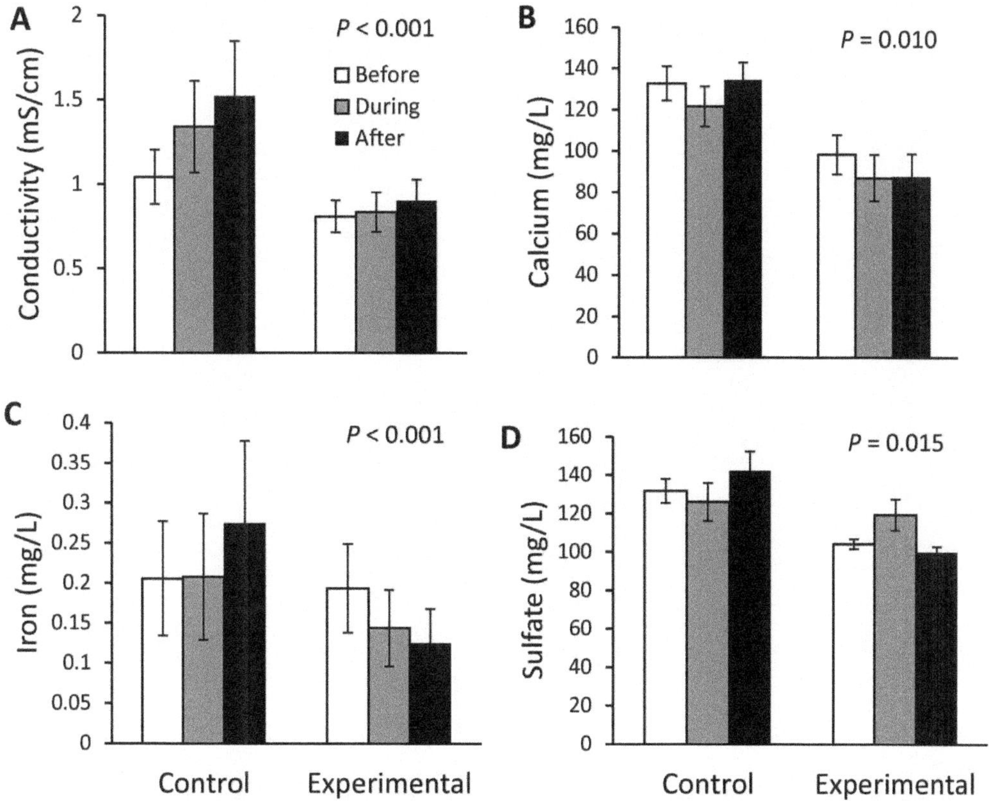

Figure 2. Water chemistry before, during, and after treatment for control and experimental sites. Mean (± SE) back-transformed values are reported. Conductivity (A) was sampled during seasonal biotic monitoring, and calcium (B), iron (C), and sulfate (D) were sampled during monthly baseflow water quality monitoring. *P*-values reflect results of ANOVA for Group*Period interaction for the strongest models (Table 3).

followed by 10 sec. of sweeping with a small dip net (500-µm mesh), and repeated for a total of three times. All samples were emptied into wash basin, elutriated, poured through a 500-µm sieve, and preserved in 70% ethanol.

In the lab, macroinvertebrates were subsampled to a minimum of 300 individuals and identified to lowest possible taxonomic unit (typically genus or species), enumerated, and measured (body length or shell width). All midges (Diptera: Chironomidae) and oligochaetes were slide-mounted for identification. To address differences in taxonomic resolution among three contractors, we lumped taxa to a common taxonomic level (e.g., genus or family) or assigned lower classification levels as appropriate (e.g., where there was only one genus in a family found at our sites). Macroinvertebrate biomass was calculated using published length-mass relationships (e.g., [47]) to generate AFDM for each taxon. Several macroinvertebrate indices were calculated for analysis. Abundance, relative abundance, richness, and biomass were calculated for Chironomidae, EPT (Ephemeroptera, Plecoptera, and Trichoptera), insects only, and all taxa. We calculated the abundance, relative abundance, and biomass of the isopod Asellidae (typically the most abundant taxon in each sample), the proportional abundance and biomass of the dominant taxon in each sample, and Shannon diversity.

Statistical analysis

Data analyses were performed using SAS version 9.2 (SAS Institute, Cary, NC, USA). Prior to any analysis, each analyte (e.g., water quality parameters, habitat measures, periphyton and macroinvertebrate indices) was screened for outliers using scatter plots and histograms; Box-Cox transformed for normality using SAS proc Transreg; normalized using SAS proc Standard; and analyzed using SAS proc Mixed and SAS proc HPmixed. Less than 10 percent of data were excluded as outliers at this stage. Each analyte was then evaluated separately using a simplified "screening" model including study period (Period as Before, During, or After treatment implementation), sample site (Site), and sample group (Group as Control or Experimental) with the "influence" option selected to identify suspicious data points. For the most part, we could not differentiate outliers due to human error from system noise. We believe that the outliers do more to obscure real signals than reflect actual conditions, and therefore have we have omitted them from further analysis.

After data cleaning was completed, separate analyses were performed for each analyte to assess the responses of individual variables to treatment using SAS proc Mixed. Model parameters included Period, Site, Group, Group*Period interaction, and sampling round (Round as sampling dates grouped in 7-day windows) (Table 2). Period and Group were coded in the model as a numerical, fixed main effects, Site was coded as a fixed effect nested within Group, and Round was coded as a random effect. The Group*Period interaction was used to assess for significance of treatment effects.

We used non-metric multidimensional scaling (NMS) to ordinate taxonomic data and express differences in assemblage structure across samples. Species-specific periphyton abundances and macroinvertebrate abundances and biomass (for bucket samples only) were log (x+1) transformed and extreme biomass outliers (3 samples) were removed prior to ordination. The NMS

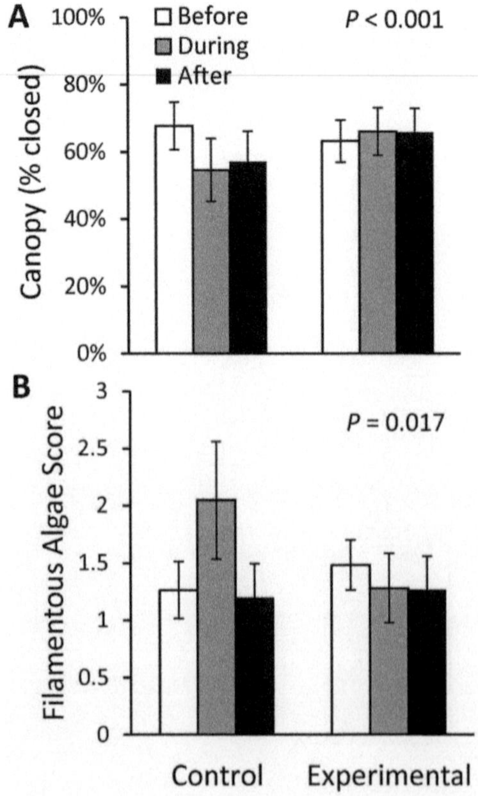

Figure 3. Canopy cover (A) and filamentous algae score (B) before, during, and after treatment for control and experimental sites. Mean (± SE) back-transformed values are reported based on qualitative data collected during biotic sampling. *P* values reflect results of ANOVA for Group*Period interaction (Table 3).

ordination was configured with the Sorensen distance measure and step down in dimensionality, and run in PC-ORD™ (Version 6, MjM Software Design, Gleneden Beach, OR, USA). The ordination axes were tested for fixed effects of Period, Group, Site (nested in Group), and Group*Period interactions, and random effect of Sample (month, year) using SAS proc Mixed. The combined axes for each analyte type (periphyton, macroinvertebrate abundance, and macroinvertebrate biomass) ANOVA included fixed effects for Period, Group, Group*Period, Axis, Axis*Period, and Axis*Group. Due to the ordination step, the statistical model was slightly altered, but equivalent to that used for non-ordinated data. In all cases, a significant Group*Period interaction indicated a significant effect of stormwater treatment installation (Table 2).

For all results, we used a $P<0.05$ cut-off for designating significance. Following the recommendations of Moran [48] for diverse ecological data and in the interest of maintaining detailed analyses, we did not correct for multiple comparisons (e.g., using sequential Bonferroni). Therefore, we caution the reader to tend toward a more conservative interpretation of tests that may be affected by Type I error, even for P-values<0.05. All raw data and SAS codes are available in EPA's STORET database (http://www.epa.gov/storet/).

Results

Landscape conditions, habitat, and water quality

Most of the headwater streams had a mix of gravel, cobble, and boulder substrate with high amounts of fine sediment deposition in the pools. The QHEI habitat scores reflected the mix of substrates and high proportion of riffle habitats, although most sites had suboptimal conditions due to high sediment deposition, poor vegetative protection, and moderately unstable banks (Table 3). Baseflow water quality varied considerably across sites and seasons, but on average streams had elevated nutrients (nitrate = 1.40 mg/L; total dissolved phosphorus = 0.620 mg/L), and high conductivity. Both natural (e.g., limey parent material as calcium = 99.4 mg/L) and anthropogenic (e.g., road salting and domestic wastewater as chloride = 147 mg/L) sources of ions likely contributed to high conductivity (Table 3).

There was a significant effect of treatment (rain garden and rain barrel installations) on several habitat and water quality parameters (Table 3). Conductivity (Fig. 2A), iron (Fig. 2C), and sulfate (Fig. 2D) increased in the Control sites through time, with no apparent change or a decrease (iron) over the corresponding periods in Experimental sites. Calcium concentrations were similar in Control sites Before and After installations, but decreased in Experimental sites through time (Fig. 2B). Canopy cover was lower in the During and After periods (vs. Before), corresponding to the vegetation removal that occurred at the Control sites in 2007 (Fig. 3A). The qualitative filamentous algae score was much higher in the During period than Before and After treatment at the Control sites, whereas the Experimental sites did not experience a similar fluctuation through time (Fig. 3B).

Periphyton

Over seven years, 107 periphyton genera were identified. Average periphyton densities were highest in April prior to leaf-out (15.4 M cells cm^{-2}) and lowest May/June (8.0 M cells cm^{-2}). Chlorophyll *a* was also highest in April (52.6 mg m^{-2}) compared to other seasons (average range: 5.31–9.29 mg m^{-2}). In contrast, algal AFDM was similar across seasons (range 43.2–49.2 g m^{-2}), but varied across sites (range 38.3–62.2 g m^{-2}, Sub3–Sub5). The periphyton were numerically dominated by Cyanophyta (cyanobacteria) in the family Chroococcaceae, encompassing 83% of the total cell density. This dominance was consistent across sites (site averages 78–85%), but increased throughout the growing season from 63% in April (before leaf-out) to 92–94% in June through October. Numerically, the cyanobacteria were dominated by small pico-periphyton <2 µm. Within Bacillariophyta (diatoms), *Navicula* (51.6%), *Nitzchia* (14.0%), *Amphora* (9.3%), *Gomphonema* (9.1%), and *Achnathes* (5.7%) were the most common genera. Average Shannon diversity (1.43–1.54) and PIBI scores (21.4–24.8) were similar across sites, and generally reflected degraded stream conditions.

Analysis of the five individual periphyton metrics that met the statistical criteria revealed no significant treatment effects (Table 4). NMS ordination of cell densities of periphyton taxa revealed a visible shift between samples Before, During, and After treatment installations in the 3-dimensional solution (Fig. 4). Further analysis of the individual axes showed significant Period effects for all three axes separately, and a significant effect of Group (Control vs. Experimental) for axis 2 (Table 5). However, there were no significant effects of Group*Period (treatment) for any of the NMS axes separately or combined (Table 5).

Table 4. Biotic variable summary statistics and ANOVA results for Group*Period interaction.

Variable	Lambda[1]	N	Min	Max	Mean	St Dev	P^2	
Periphyton								
% Dominant diatom	−0.400	218	20.2	100	55.0	17.9	0.375	
% Motile diatom	1.00	212	3.90	100	52.2	25.8	0.602	
PIBI[3]	1.75	216	0.100	37.8	21.3	6.79	0.415	
Shannon diversity	0.050	218	0.740	2.78	1.48	0.376	0.218	
% Eutraphentic diatom	3.00	210	4.30	100	63.2	26.0	0.402	
Macroinvertebrates[4]								
Abundance (bucket)								
% Chironomidae	−2.75	102	1.04	89.9	29.6	24.8	0.088	
Chironomidae richness	0.550	102	1.00	19.0	8.87	4.38	0.053	
% EPT[5]	−7.00	77	0.101	35.4	5.17	6.69	0.324	
EPT richness	−3.70	77	0.500	8.00	2.79	2.13	0.786	
Shannon diversity	2.35	104	0.220	3.08	2.14	0.561	0.267	
Insect richness	1.20	104	3.00	33.0	16.1	6.22	0.002	**
% Asellidae	−3.15	103	0.179	83.9	26.8	24.5	0.400	
Total richness	0.800	104	7.00	45.0	24.0	6.70	0.009	**
Abundance (net)								
% Chironomidae	−4.20	215	0.276	96.2	21.1	21.1	0.254	
Chironomidae richness	−1.05	215	1.00	23.0	7.18	4.16	0.705	
EPT richness	−2.25	144	0.500	7.00	2.42	1.51	0.420	
Shannon diversity	1.70	220	0.176	3.14	1.72	0.702	0.143	
Insect richness	−0.45	220	1.00	32.0	13.9	6.56	0.932	
% Asellidae	0.050	210	0.293	97.4	44.5	32.3	0.503	
Total richness	−0.15	220	5.00	42.0	21.3	7.53	0.924	
Biomass (bucket)								
Chironomidae richness	0.700	102	1.00	20.0	9.21	4.52	0.058	
% EPT	−7.00	76	0.007	70.4	7.72	12.1	0.478	
Shannon diversity	1.05	104	0.188	2.63	1.46	0.512	0.019	*
% Insecta	−1.15	104	0.561	96.5	41.0	30.9	0.318	
% Dominant taxon	−0.550	104	19.4	97.3	55.0	19.3	0.029	*
% Asellidae	−2.70	103	0.015	85.4	28.4	26.3	0.243	

[1]Lambda is value for the exponential transformation.
[2]**$P<0.01$, *$P<0.05$.
[3]PIBI is the Periphyton Index of Biotic Integrity [45].
[4]Macroinvertebrate variables were calculated separately for multi-habitat net samples (based on abundance data) and bucket samples in riffle habitats (represented as abundance and biomass).
[5]EPT represents taxa in the orders Ephemeroptera, Plecoptera, and Trichoptera (considered sensitive to disturbance).

Macroinvertebrates

We collected 189 unique macroinvertebrate taxa across all samples and sites. Assemblages were dominated by the Asselid isopod *Lirceus* that composed 60% of the abundance of all net samples and 39% of bucket samples. Oligocheata worms (8.7% and 28.8%, respectively) and the chironomid Tanytarsus (5.1% and 12.6%, respectively were the second and third most abundant taxa, respectively). Other common taxa (composing >2% of the abundance) included the chironomids *Diamesa*, *Paratendipes*, and *Orthocladius*, and Ostracoda crustaceans. On average (± SD), the bucket samples were composed of 29.6±24.8% Chironomidae, 26.8±24.5% Asellidae, and 5.2±6.7% EPT taxa (Table 4). Across the six sites, there was a range in average total richness (20–27), EPT richness (0.5–5), and Shannon diversity (1.61–2.29) in bucket samples, with Sub1 consistently reflecting the most degraded conditions, and Catch having the highest diversity and richness of sensitive EPT taxa. Although net samples captured more individuals, richness and diversity metrics were slightly higher in the quantitative, bucket samples. Density of macroinvertebrates within bucket samples ranged from 94 to 48,648 individuals m^{-2} (average density 2,892 individuals m^{-2}) and biomass ranged from 5.5 to 81,951 mg AFDM m^{-2} (average biomass 1,986 mg AFDM m^{-2}) across samples.

There were few significant effects of treatment (Group*Period) on the individual macroinvertebrate abundance and biomass variables (Table 4). Insect richness (Fig. 5A) and total richness (Fig. 5B) from bucket samples tended to increase through time in the Control sites, whereas the Experimental sites had lower richness in the During period compared to Before and After. Shannon diversity (based on biomass from bucket samples) was

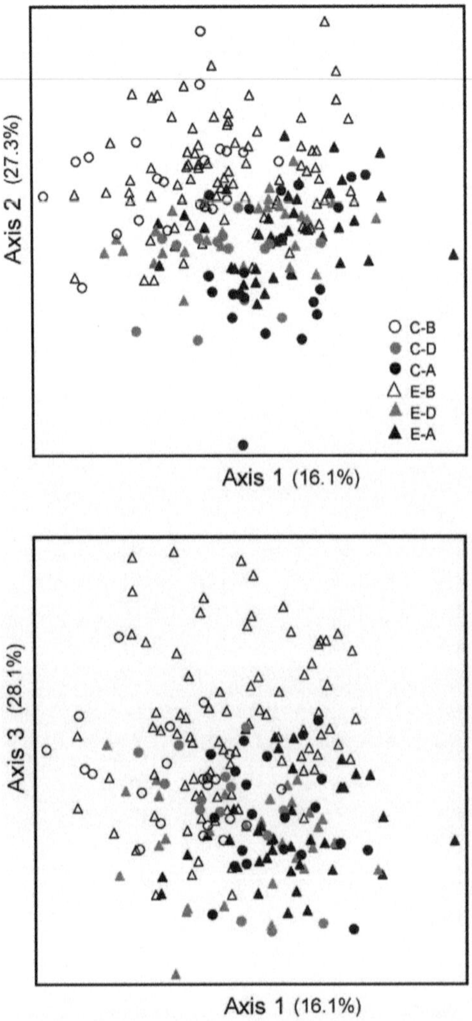

Figure 4. Non-metric multidimensional scaling ordination for periphyton abundances. Assemblages were different for Control (C) and Experimental (E) sites (Axis 2) and comparing Before (B), During (D), and After (A) installation of stormwater management (Axes 1, 2, and 3; Table 5). The 3D solution explained 71.5% of the variation, and the final stress was 19.3.

Table 5. ANOVA results for ordination axes.

Effect[1]	Axis	*P*[2]	Effect	Axis	*P*
Periphyton					
Group	1	0.151	Site(Group)	1	0.495
Group	2	0.006 **	Site(Group)	2	0.232
Group	3	0.510	Site(Group)	3	0.108
Group	All	0.370	Site(Group)	All	–
Period	1	0.014 *	Group*Period	1	0.913
Period	2	<0.001 ***	Group*Period	2	0.146
Period	3	<0.001 ***	Group*Period	3	0.311
Period	All	0.755	Group*Period	All	0.983
Macroinvertebrate Abundance[3]					
Group	1	0.085	Site(Group)	1	0.434
Group	2	<0.001 ***	Site(Group)	2	0.009 **
Group	3	<0.001 ***	Site(Group)	3	<0.001 ***
Group	All	0.032 *	Site(Group)	All	–
Period	1	0.295	Group*Period	1	0.497
Period	2	0.532	Group*Period	2	0.977
Period	3	0.163	Group*Period	3	0.002 **
Period	All	0.024 *	Group*Period	All	0.815
Macroinvertebrate Biomass					
Group	1	0.291	Site(Group)	1	0.006 **
Group	2	<0.001 ***	Site(Group)	2	<0.001 ***
Group	3	0.463	Site(Group)	3	0.311
Group	All	0.237	Site(Group)	All	–
Period	1	0.008 **	Group*Period	1	0.624
Period	2	0.231	Group*Period	2	0.097
Period	3	0.369	Group*Period	3	0.934
Period	All	<0.001 ***	Group*Period	All	0.989

[1]See Table 2 for variable descriptions.
[2]***$P<0.001$, **$P<0.01$, * $P<0.05$. – indicates effect not tested in combined axis model.
[3]Macroinvertebrate abundance and biomass were based on bucket samples in riffle habitats.

highest in the During period for the Control sites and Before treatment for the Experimental sites (Fig. 5C). Percent dominant taxon based on biomass from bucket samples (a variable that should increase with disturbance) was lowest in the During period for the Control sites and highest During treatment for the Experimental sites (Fig. 5D).

Macroinvertebrate assemblages collected from riffle habitats were distinct based on Group (Control vs. Experimental) and Site (Table 5) based on the ordination of taxa abundance (Fig. 6) and biomass (Fig. 7). There was also a significant Period effect for abundance (combined axes) and biomass (axis 1 and combined axes; Table 5). Only macroinvertebrate abundance axis 3 revealed a significant effect of stormwater treatment (Group*Period, $P=0.002$, Table 5).

Discussion

Stream responses to rain garden and barrel installations

As expected, the installation of rain barrels and rain gardens on 30% of the properties in the Experimental catchments resulted in very few responses in stream water quality, periphyton, and macroinvertebrate metrics relative to Control sites. The few significant results that were detected should be interpreted with caution, given the high number of comparisons and potential risk of Type I error. Despite the high number of samples, with only four experimental and two Control sites there was low statistical power, a challenge of BACI designs [35]. Nonetheless, the few detected responses are notable given the study design and relatively small amount of stormwater runoff mitigated in the catchment.

There was a statistically significant effect of stormwater treatment on a few baseflow water quality variables, generally reflecting reduced water quality through time at Control sites. The small reduction in runoff volume from Before to After treatment [31] may have stabilized the water quality in the Experimental sites over a period of time when conditions in the Control sites were deteriorating. Rain gardens and barrels captured runoff, thereby reducing the likelihood of pollutant mobilization and transport, and potentially decreasing the total mass of pollutants delivered to streams during higher-flow storm conditions [49,50].

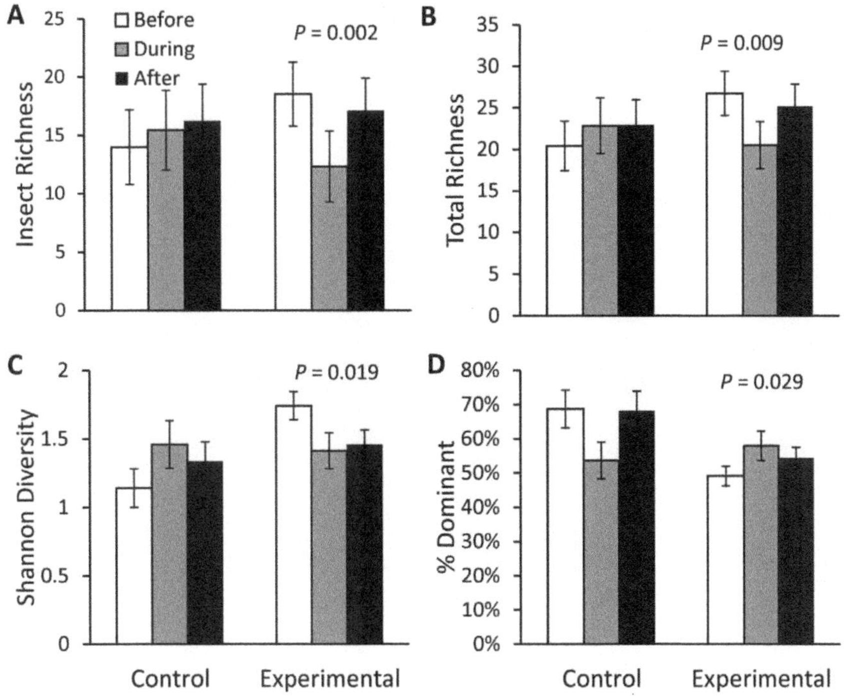

Figure 5. Macroinvertebrate variables before, during, and after treatment for control and experimental sites. Mean (± SE) back-transformed values are reported for insect richness (A) and total richness (B) based on riffle bucket samples, and Shannon diversity (C) and % dominant (D) based on biomass values for riffle bucket samples. *P*-values reflect results of ANOVA for Group*Period interaction; only significant biotic models ($P<0.05$) are included (Table 4).

In addition to capturing and detaining stormwater, rain gardens can play a role in filtration of pollutants [51,52], and may contribute to overall reduced pollutant loading to streams, with some potential for improved baseflow water quality [49], although this mechanism was untested in our study.

There was no significant effect of treatment on the algal community, as examined through individual periphyton metrics and ordination of cell densities by taxon. This is not surprising given the lack of a response of nutrients (nitrogen and phosphorus) to treatment, and that these systems are not significantly nutrient limited [53]. The qualitative filamentous algal score was much higher in the Control sites in the During period (June 2007 through September 2008), which corresponded to the removal of trees and shrubs within the riparian zones at both Control sites. Loss of riparian cover increased the duration and intensity of light reaching the stream, potentially also increasing stream temperature locally, and could have triggered the increase in the relative proportion of filamentous green algae in the Control sites [53,54]. The lower scores for the After period may be a reflection of algal sloughing during the After period, which had higher precipitation and flows compared to the During period.

The few significant treatment effects on the macroinvertebrate assemblage were not intuitive, and may be explained by multiple factors independent of the stormwater management. In the Control sites, there was an increase in richness and diversity through time, and lower percentage of dominant taxa in the During period. These patterns may reflect the periphyton responses, especially if increases in filamentous algae provided habitat, food, or increased nutrient uptake to support new taxa and higher diversity in the Control sites. In contrast, the Experimental sites demonstrated lower richness and diversity, and higher percentage of dominant taxa in the During period, compared to Before and After. It is possible that the reductions can be attributed to differences in low flow hydrology over the course of the study. Whereas high storm flows can directly alter macroinvertebrate communities through physical washout [55], small streams such as those in our study are more likely structured by seasonal variation in stream flow. The two Control streams (Sub4 and Sub5) had two of the three smallest catchment areas, and dried to pools nearly every summer, whereas the other sites remained perennial, which may explain why the Control sites had the lowest overall richness and diversity. However, in 2007 and 2008 (when the stormwater management devices were being installed), all six streams dried to pools in the summer. The lack of permanent flow and associated fluctuations in temperature likely resulted in loss of taxa in the Experimental sites that require flowing water, and these taxa may have already been missing from the Control sites [56,57].

Why were there so few responses to stormwater management?

Although the installation of rain gardens and rain barrels represented a widespread retrofit management effort, it is likely the number and capacity of installations were simply insufficient to elicit any response from the water quality or biotic measures. The management effort targeted runoff from rooftops and driveways on private properties, which comprised a majority (53.2%) of the total impervious area in the Shepherd Creek catchment. However, even if the 30% of properties that received treatments captured all of the runoff from rooftops and driveways on those properties (which we know was not the case), it would not reduce the effective impervious area (EIA) or DCIA in the subcatchments to below the threshold (1–14% EIA [6], 2% TIA [32]) of expected biotic impairment (See Fig. 2.2.2 in [58]). The range in impervious

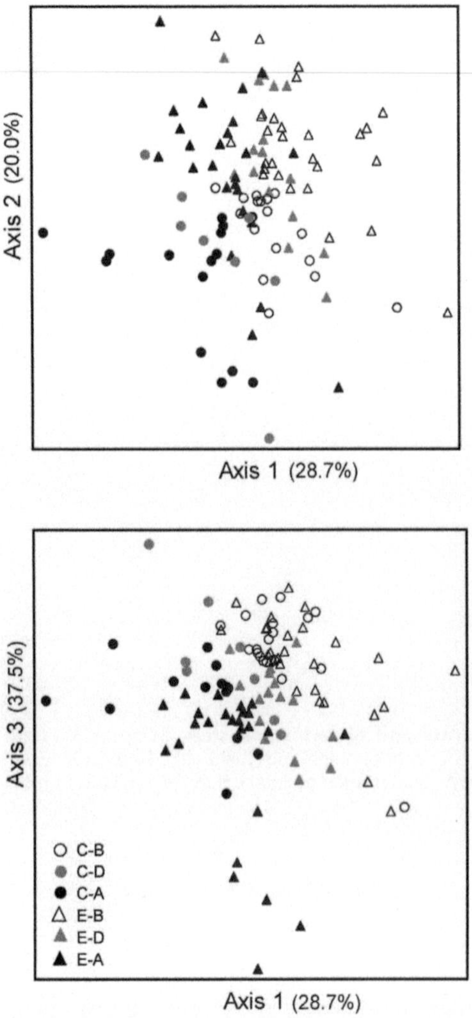

Figure 6. Non-metric multidimensional scaling ordination for macroinvertebrate abundances from bucket samples. Assemblages were different for Control (C) and Experimental (E) sites (Axes 2 and 3) and comparing Before (B), During (D), and After (A) installation of stormwater management (all axes combined; Table 5). The 3D solution explained 86.2% of the variation, and the final stress was 16.1.

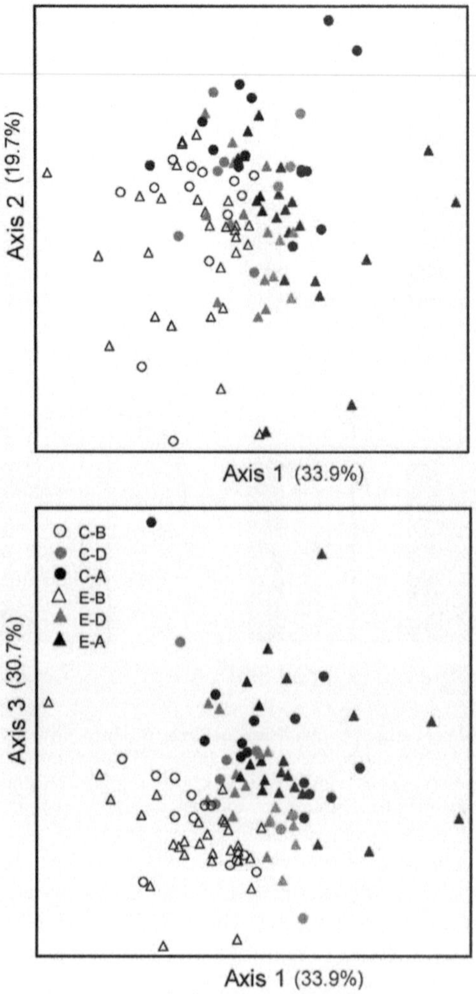

Figure 7. Non-metric multidimensional scaling ordination for macroinvertebrate biomass from bucket samples. Assemblages were different for Control (C) vs. Experimental (E) sites (Axis 2) and comparing Before (B), During (D), and After installation of stormwater management (Axes 1, 3, and all; Table 5). The 3D solution explained 84.3% of the variation, and the final stress was 15.8.

(11.2–19.9%) and forest (43.8–68.0%) cover across sites may have also masked detection of responses to treatments. Furthermore, our management approach did not address runoff from streets, which comprised 22.7% of the total impervious area, but had a proportionally higher amount of impervious cover directly connected to storm sewers [38]. Streets were therefore likely to have a disproportional impact on streams, and their lack of treatment may have masked the benefits provided by the rain gardens and rain barrels. Walsh et al. [59] demonstrated that it is possible for streams with ~10% imperviousness to have good ecological condition if stormwater is infiltrated throughout the catchment. Thus, it is conceivable to achieve in-stream improvements in the Shepherd Creek catchment if we (1) increase the number of properties with management practices, (2) ensure all impervious surfaces on the property are routed to rain gardens, (3) increase the capacity of management devices, and (4) mitigate runoff from streets with high proportions of connected impervious cover.

In addition to the lack of hydrologic capacity of installed stormwater treatment devices, there are other possible reasons for the lack of water quality and biotic responses. First, there could have been an overwhelming influence of other stressors, despite the reduced stormwater volume. Although in-stream hydrology is tightly linked to water quality and biotic health in suburban and urban streams, water quality and other stressors (e.g., dispersal barriers, riparian forest loss, channelization) can shape biotic communities independently of stormwater runoff [9]. In the Shepherd Creek catchment, some of the properties have private septic tanks, and poorly maintained or malfunctioning systems can increase nutrients and bacteria, especially during low-flow conditions [60]. Road salt inputs were extremely high in the catchment, and although this can be partially mitigated by capturing runoff, salt concentrations and conductivity remained high despite restoration efforts. It is possible that these and other aquatic and terrestrial stressors were not mitigated by restoration efforts, thus preventing improvement in periphyton and macroinvertebrate assemblage integrity.

Over the seven-year study, there were many changes in the catchment unrelated to the project that may have masked responses associated with stormwater management. The Shepherd Creek project involved county and city organizations (e.g., Hamilton County Soil and Water Conservation District, Hamilton County Engineers Office, Cincinnati Metropolitan Sewer District, and Cincinnati Parks) in effort to maximize potential effectiveness of the project. Despite this, a few road maintenance, sewer maintenance, and tree removal projects occurred during the study period. These changes can increase variability in response variables and reduce the potential to detect improvements in the catchment. Because tree removal (to improve road visibility in Sub4 and for management of invasive species in Sub5) occurred at the same time as installation of stormwater management devices, it was difficult to separate the causes of any biotic responses. In addition, landowners likely made changes in their landscaping, watering, and other practices independently of the project. Given that it is unrealistic to prevent these non-target changes in suburban catchments, future studies will likely have to do additional improvements, matching the scope of management to the type and extent of disturbance, in order to detect a response.

The lack of detectable responses may also be explained by the high spatial and temporal variability of the biotic variables, which is typical in small, hydrologically-complex urban catchments [61]. In the spatial dimension, there are large differences in macroinvertebrate [46] and periphyton [62] taxa found across habitats. By targeting sample collection to riffle habitats (macroinvertebrate bucket samples) and hard substrates (periphyton samples), our study minimized this within-reach variation, although habitat was still likely an important factor influencing differences across sites. We observed high intra- and inter-annual variability in both the periphyton and macroinvertebrate assemblages. Although these were accounted for in the statistical model (Round), the additional variable in the model can minimize the power to detect a response, a disadvantage of this hybrid designed-observational study with numerous unavoidable nuisance effects. Moreover, research suggests that weather variability (on a small scale) and climate variability (on a larger scale) may drive assemblages [56,57,63,64] and override responses to localized hydrologic management in some years. As mentioned before, we experienced drought conditions during the installation phase (late 2007 & early 2008), which may explain biotic responses in the During period. It is likely the low flows, combined with other stressors (water quality, temperature, sedimentation) in the catchment, interacted in complex ways to control biotic assemblages and ultimately prevent any detectable response to stormwater management.

Whereas improvements in hydrology were expected almost immediately following restoration, subsequent improvements in water quality and biotic integrity may take much longer. This study included three years of post-restoration monitoring after the initial installations (Phase 1), and only two years of monitoring after all of the installations were complete (i.e., following Phase 2). Existing, sediment-bound pollutants may take several years to process before streams experience improved water quality from reduced loading [65] and contaminated groundwater reservoirs can maintain high levels of contaminates in streams decades after the pollutant source has been eliminated [66]. Although periphyton have relatively short life cycles and are more sensitive to short-term shifts in water quality than macroinvertebrates [39,62], their assemblages are structured by habitat and substrate, neither of which changed during our study. Similarly, macroinvertebrates may display a delayed response to increased detention of stormwater runoff because it takes time for critical resources (e.g., food, habitat) to improve [67]. Furthermore, the multi-year life cycles of many macroinvertebrates and their modes of dispersal suggests that recovery may take several years [39]. Even in studies where in-stream habitat enhancement has restored habitat diversity, some researchers have shown limited recovery of macroinvertebrates that can be explained by other, persistent stressors [67–69]. Even if instream conditions are suitable, aquatic invertebrates that have terrestrial adults need good riparian habitat and dispersal corridors across the landscape to persist [70].

Conclusions

Although this study represents a sizeable effort to control stormwater runoff on private properties throughout a catchment, the stream responses to the retrofit management were limited to localized responses of a few variables. These results are not surprising given the number of rain gardens and rain barrels and the capacity of these stormwater devices relative to the impervious surfaces in the catchment. There is an obvious need for additional controlled studies where stormwater management practices are installed at higher densities to capture a greater volume of stormwater to determine the extent of stormwater management necessary to improve ecosystem health. A large-scale stormwater restoration is currently underway in the Little Stringybark Creek catchment in Melbourne, Australia [27], and more studies of this scope are needed despite the logistical and financial challenges of implementing and monitoring catchment-scale restoration [26].

The focus of this paper was on the effects of stormwater management on stream water quality and biota. However, the success of stream restoration should not only be measured in terms of improved ecosystem health [71], but in other benefits that can be derived. This study took a multidisciplinary approach to designing and implementing catchment-scale management, providing economic and social benefits that extend beyond the ecosystem responses [28]. The auction revealed substantial landowner interest and the potential for mitigating stormwater at much lower costs than centralized options. There were additional contributions to ecosystem services such as flood protection and water supply that extend beyond stream ecosystem benefits. For example, the 165 rain barrels installed resulted in water savings in cases where residents used the outside water source for watering that would otherwise come from potable sources. The 81 rain gardens included only native plant taxa, so the installations contributed to increases in native flora and wildlife habitat in the neighborhood. Additional benefits included generating public awareness of stormwater issues and the connection between human activities and environmental quality. Overall, it is clear that management efforts designed to mimic natural ecosystems will provide a variety ecosystem and other benefits, yet the extent of retrofit stormwater management necessary to restore healthy streams remains to be determined.

Acknowledgments

The design, planning, implementation, and contract management for the Shepherd Creek project took place over nearly 10 years with the help of numerous people in the USEPA: H. Cabezas, M. Clagett, T. Hoagland, P. Parikh, M. Taylor, J. Templeton, and H. Thurston, and from county and city offices: M. Alam, B. Bohl, M. Flanders, T. Gilday, J. Godby, L. Parker, H. Utrata-Halcomb, and W. Wimmer. Thanks to the following individuals for assisting with field sampling, laboratory processing, and data entry: B. Morris, R. Burrows, M. Clagett, A. Dybas, N. Fudge, M. Goss, K. Head, D. Kowalski, H. Lubbers, J. Milanovich, N. Patel, D. Pennington, J. Rebholz, J. Shaffer, K. Taulbee, Y. Zhang.

Author Contributions

Conceived and designed the experiments: WDS ALM AHR MAM. Performed the experiments: AHR ALM MAM JJB AS WDS MEH. Analyzed the data: LKR AHR. Wrote the paper: AHR LKR ALM WDS JJB MEH MAM AS.

References

1. Alig RJ, Kline JD, Lichtenstein M (2004) Urbanization and the US landscape: looking ahead in the 21st century. Landsc Urban Plan 69: 219–234.
2. Grimm NB, Faeth SH, Golubiewski NE, Redman CL, Wu J, et al. (2008) Global change and the ecology of cities. Science 319: 756–760.
3. United Nations (2011) World Urbanization Prospects: The 2010 Revision. New York: UN Department of Economic and Social Affairs, Population Division. Available: http://esa.un.org/unpd/wpp/index.htm. Accessed 3 August 2012.
4. Allan JD (2004) Landscapes and riverscapes: the influence of land use on stream ecosystems. Annu Rev Ecol Evol Syst 35: 257–284.
5. Walsh CJ (2000) Urban impacts on the ecology of receiving waters: a framework for assessment, conservation and restoration. Hydrobiologia 431: 107–114.
6. Walsh CJ, Fletcher TD, Ladson AR (2005) Stream restoration in urban catchments through redesigning stormwater systems: looking to the catchment to save the stream. J North Am Benthol Soc 24: 690–705.
7. Paul MJ, Meyer JL (2001) Streams in the urban landscape. Annu Rev Ecol Evol Syst 32: 333–365.
8. Walsh CJ, Roy AH, Feminella JW, Cottingham PD, Groffman PM, et al. (2005) The urban stream syndrome: current knowledge and the search for a cure. J North Am Benthol Soc 24: 706–723.
9. Wenger SJ, Roy AH, Jackson CR, Bernhardt ES, Carter TL, et al. (2009) Twenty-six key research questions in urban stream ecology—An assessment of the state of the science. J North Am Benthol Soc 28: 1080–1098.
10. Konrad CP, Booth DB (2005) Hydrologic changes in urban streams and their ecological significance. American Fisheries Society Symposium 47: 157–177.
11. Booth DB, Jackson CR (1997) Urbanization of aquatic systems: degradation thresholds, stormwater detection, and the limits of mitigation. J Am Water Resour Assoc 33: 1077–1090.
12. Burns MJ, Fletcher TD, Walsh CJ, Ladson AR, Hatt BE (2012) Hydrologic shortcomings of conventional urban stormwater management and opportunities for reform. Landsc Urban Plan 105: 230–240.
13. Sieker F (1998) On-site stormwater management as an alternative to conventional sewer systems: a new concept spreading in Germany. Water Sci Technol 38: 65–71.
14. Ando AW, Freitas LPC (2011) Consumer demand for green stormwater management technology in an urban setting: the case of Chicago rain barrels. Water Resour Res 47: W12501. doi:10.1029/2011WR011070.
15. Ahiablame LM, Engel BA, Chaubey I (2012) Effectiveness of low impact development practices: literature review and suggestions for future research. Water Air Soil Pollut 223: 4253–4273.
16. United States Environmental Protection Agency (1991) Construction site stormwater discharge control: an inventory of current practices. EPA 883-R-91-100. Washington, DC: US EPA Office of Water. Available: http://www.epa.gov/npdes/pubs/owm017.pdf. Accessed 17 April 2013.
17. Urbonas B, Stahre P (1993) Stormwater best management practices and detention for water quality, drainage, and CSO management. New Jersey: PTR Prentice Hall. 449 p.
18. Dietz ME (2007) Low impact development practices: a review of current research and recommendations for future directions. Water Air Soil Pollut 186: 351–363.
19. Hunt WF, Davis AP, Traver RG (2012) Meeting hydrologic and water quality goals through targeted bioretention design. J Environ Eng (New York) 138: 698–707.
20. Perez-Pedini C, Limbrunner JF, Vogel RM (2005) Optimal location of infiltration-based best management practices for storm water management. J Water Resour Plan Mgmt 131: 441–448.
21. Gilroy KL, McCuen RH (2009) Spatio-temporal effects of low impact development practices. J Hydrol 367: 228–236.
22. Lee JG, Selvakumar A, Alvi K, Riverson J, Zhen JX, et al. (2012) A watershed-scale design optimization model for stormwater best management practices. Environ Model Softw 37: 6–18.
23. Bedan ES, Clausen JC (2009) Stormwater runoff quality and quantity from traditional and low impact development watersheds. J Am Water Resour Assoc 45: 998–1008.
24. Line DE, Brown RA, Hunt WF, Lord WG (2012) Effectiveness of LID for commercial development in North Carolina. J Environ Eng (New York) 138: 680–688.
25. Parikh P, Taylor M, Hoagland T, Thurston H, Shuster W (2005) At the intersection of hydrology, economics, and law: application of market mechanisms and incentives to reduce stormwater runoff. Environ Sci Policy 8: 133–144.
26. Roy AH, Wenger SJ, Fletcher TD, Walsh CJ, Ladson AR, et al. (2008) Impediments and solutions to sustainable, watershed-scale urban stormwater management: lessons from Australia and the United States. Environ Manage 42: 344–359.
27. Fletcher TD, Walsh CJ, Bos D, Nemes V, RossRakesh S, et al. (2011) Restoration of stormwater retention capacity at the allotment-scale through a novel economic instrument. Water Sci Technol 64: 494–502.
28. Mayer AL, Shuster WD, Beaulieu JJ, Hopton ME, Rhea LK, et al. (2012) Building green infrastructure via citizen participation: a six-year study in the Shepherd Creek (Ohio). Environ Pract 14: 57–67.
29. Thurston HW, Goddard HC, Szlag D, Lemberg B (2003) Controlling stormwater runoff with tradable allowances for impervious surfaces. J Water Resour Plan Mgmt 129: 409–418.
30. Thurston HW, Taylor MA, Shuster WD, Roy AH, Morrison MA (2010) Using a reverse auction to promote household level stormwater control. Environ Sci Policy 13: 405–414.
31. Shuster WD, Rhea LK (2013) Catchment-scale hydrologic implications of parcel-level stormwater management (Ohio USA). J Hydrol 485: 177–187.
32. King RS, Baker ME (2010) Considerations for analyzing ecological community thresholds in response to anthropogenic environmental gradients. J North Am Benthol Soc 29: 998–1008.
33. Green RH (1979) Sampling design and statistical methods for environmental biologists. Chichester: Wiley. 257 p.
34. Stewart-Oaten A, Murdoch WW, Parker KR (1986) Environmental impact assessment: Pseudoreplication in time? Ecology 67: 929–940.
35. Faith DP, Humphrey CL, Dostine PL (1991) Statistical power and BACI designs in biological monitoring: comparative evaluation of measures of community dissimilarity based on benthic macroinvertebrate communities in Rockhole Mine Creek, Northern Territory, Australia. Aust J Mar Freshwater Resour 42: 589–602.
36. Underwood AJ (1991) Experimental designs for detecting human environmental impacts on temporal variations in natural populations. Aust J Mar Freshwater Resour, 42, 569–587.
37. Underwood AJ (1992) Beyond BACI - the detection of environmental impacts on populations in the real, but variable, world. J Exp Mar Bio Ecol 161: 145–178.
38. Roy AH, Shuster WD (2009) Assessing impervious surface connectivity and applications for watershed management. J Amer Water Resour Assoc 45: 198–209.
39. Barbour MT, Gerritsen J, Snyder BD, Stribling JB (1999) Rapid bioassessment protocols for use in streams and wadeable rivers: periphyton, benthic macroinvertebrates, and fish. Second Edition. EPA 841-B-99-002. Washington, DC: Environmental Protection Agency. Available: http://water.epa.gov/scitech/monitoring/rsl/bioassessment/index.cfm. Accessed 18 April 2013.
40. Ohio Environmental Protection Agency (2002) Field Evaluation Manual for Ohio's Primary Headwater Habitat Streams. Version 1.0. Ohio EPA Division of Surface Water, Columbus, Ohio. Available: http://www.in.gov/idem/files/headwater_oh_phwh_man.pdf. Accessed 18 April 2013.
41. ASTM (1997) Standard test methods for determining sediment concentration in water samples. Philadelphia: American Society for Testing and Materials.
42. Arar EJ (1997) Method 446.0, In vitro determination of chlorophylls a, b, c1, +c2 and pheopigments in marine and freshwater algae by visible spectrophotometry. Revision 1.2. Washington, DC: U. S. Environmental Protection Agency.
43. Rice EW, Baird RB, Eaton AD, Clesceri LS (2005) Standard methods for the examination of water and wastewater. 21st Edition. American Public Health Association, American Water Works Association, and Water Environment Federation. Available: http://www.standardmethods.org/. Accessed 18 April 2013.
44. PhycoTech (2011) Technical approach to algal identification. PhycoTech, Inc., St. Joseph, Michigan. Available: http://www.phycotech.com/technical.html. Accessed 3 August 2012.
45. Hill BH, Herlihy AT, Kaufmann PR, Stevenson JR, McCormick FH, et al. (2000) Use of periphyton assemblage data as an index of biotic integrity. J North Am Benthol Soc 19: 50–67.
46. Roy AH, Rosemond AD, Leigh DS, Paul MJ, Wallace JB (2003) Habitat-specific responses of stream insects to land cover disturbance: biological consequences and monitoring implications. J North Am Benthol Soc 22: 292–307.
47. Benke AC, Huryn AD, Smock LA, Wallace JB (1999) Length-mass relationships for freshwater macroinvertebrates in North America with particular reference to the southeastern United States. J North Am Benthol Soc 18: 308–343.
48. Moran MD (2003) Arguments for rejecting the sequential Bonferroni in ecological studies. Oikos 100: 403–405.
49. Davis AP, Hunt WF, Traver RG, Clar M (2009) Bioretention technology: overview of current practice and future needs. J Environ Eng (New York) 135: 109–117.
50. Li H, Sharkey LJ, Hunt WF, Davis AP (2009) Mitigation of impervious surface hydrology using bioretention in North Carolina and Maryland. J Hydrol Eng 14: 407–415.

51. Hatt BE, Fletcher TD, Deletic A (2009) Hydrologic and pollutant removal performance of stormwater biofiltration systems at the field scale. J Hydrol 365: 310–321.
52. Trowsdale SA, Simcock R (2011) Urban stormwater treatment using bioretention. J Hydrol 397: 167–174.
53. Beaulieu JJ, Arango CP, Balz AD, Shuster WD (2013) Continuous monitoring reveals multiple controls on ecosystem metabolism in a suburban stream. Freshwater Biol. doi:10.1111/fwb12097.
54. DeNicola DM, Hoagland KD, Roemer SC (1992) Influences of canopy cover on spectral irradiance and periphyton assemblages in a prairie stream. J North Am Benthol Soc 11: 391–404.
55. Poff NL, Ward JV (1989) Implications of streamflow variability and predictability for lotic community structure: a regional analysis of streamflow patterns. Can J Fish Aquat Sci 46: 1805–1818.
56. Williams DD (1996) Environmental constraints in temporary freshwaters and their consequences for the insect fauna. J North Am Benthol Soc 15: 634–650.
57. Clarke A, MacNally R, Bond N, Lake PS (2010) Flow permanence affects aquatic macroinvertebrate diversity and community structure in three headwater streams in a forested catchment. Can J Fish Aquat Sci 67: 1649–1657.
58. Thurston HW, Roy AH, Shuster WD, Cabezas H, Morrison MA, et al. (2008) Using economic incentives to manage stormwater runoff in the Shepherd Creek Watershed, Part I. EPA/600/R-08/129. Washington, DC: U.S. Environmental Protection Agency.
59. Walsh CJ, Fletcher TD, Burns MJ (2012) Urban stormwater runoff: a new class of environmental flow problem. PLOS ONE 7: e45814. doi: 10.1371/journal.pone.0045814.
60. Arnscheidt J, Jordan P, Li S, McCormick S, McFaul R, et al. (2007) Defining sources of low-flow phosphorus transfers in complex catchments. Sci Total Environ 382: 1–13.
61. Kaushal SS, Belt KT (2012) The urban watershed continuum: evolving spatial and temporal dimensions. Urban Ecosyst 15: 409–435.
62. Rosen BH (1995) Use of periphyton in the development of biocriteria. In: Davis WS, Simon TP, editors. Biological assessment and criteria: Tools for water resource planning and decision making. Boca Raton: Lewis Publishers. pp. 209–215.
63. Bradley DC, Ormerod SJ (2001) Community persistence among stream invertebrates tracks the North Atlantic Oscillation. J Anim Ecol 70: 987–996.
64. Jackson JK, Fureder L (2006) Long-term studies of freshwater macroinvertebrates: a review of the frequency, duration, and ecological significance. Freshwater Biol 51: 591–603.
65. Meals DW, Dressing SA, Davenport TE (2010) Lag time in water quality response to best management practices: a review. J Environ Qual 39: 85–96.
66. Hamilton SK (2012) Biogeochemical time lags may delay responses of streams to ecological restoration. Freshwater Biol 57: 43–57.
67. Louhi P, Mykra H, Paavola R, Huusko A, Vehanen T, et al. (2011) Twenty years of stream restoration in Finland: little response by benthic macroinvertebrate communities. Ecol Appl 21: 1950–1961.
68. Lepori F, Palm D, Brannas E, Malmqvist BB (2005) Does restoration of structural heterogeneity in streams enhance fish and macroinvertebrate diversity? Ecol Appl 15: 2060–2071.
69. Palmer MA, Menninger H, Bernhardt E (2010) River restoration, habitat heterogeneity and biodiversity: a failure of theory or practice? Freshwater Biol 55 (Supplement 1): 205–222.
70. Smith RF, Alexander LC, Lamp WO (2009) Dispersal by terrestrial stages of stream insects in urban watersheds: a synthesis of current knowledge. J North Am Benthol Assoc 28: 1022–1037.
71. Bernhardt ES, Palmer MA (2007) Restoring streams in and urbanizing world. Freshwater Biol 52: 738–751.

PERMISSIONS

All chapters in this book were first published in PLOS ONE, by The Public Library of Science; hereby published with permission under the Creative Commons Attribution License or equivalent. Every chapter published in this book has been scrutinized by our experts. Their significance has been extensively debated. The topics covered herein carry significant findings which will fuel the growth of the discipline. They may even be implemented as practical applications or may be referred to as a beginning point for another development.

The contributors of this book come from diverse backgrounds, making this book a truly international effort. This book will bring forth new frontiers with its revolutionizing research information and detailed analysis of the nascent developments around the world.

We would like to thank all the contributing authors for lending their expertise to make the book truly unique. They have played a crucial role in the development of this book. Without their invaluable contributions this book wouldn't have been possible. They have made vital efforts to compile up to date information on the varied aspects of this subject to make this book a valuable addition to the collection of many professionals and students.

This book was conceptualized with the vision of imparting up-to-date information and advanced data in this field. To ensure the same, a matchless editorial board was set up. Every individual on the board went through rigorous rounds of assessment to prove their worth. After which they invested a large part of their time researching and compiling the most relevant data for our readers.

The editorial board has been involved in producing this book since its inception. They have spent rigorous hours researching and exploring the diverse topics which have resulted in the successful publishing of this book. They have passed on their knowledge of decades through this book. To expedite this challenging task, the publisher supported the team at every step. A small team of assistant editors was also appointed to further simplify the editing procedure and attain best results for the readers.

Apart from the editorial board, the designing team has also invested a significant amount of their time in understanding the subject and creating the most relevant covers. They scrutinized every image to scout for the most suitable representation of the subject and create an appropriate cover for the book.

The publishing team has been an ardent support to the editorial, designing and production team. Their endless efforts to recruit the best for this project, has resulted in the accomplishment of this book. They are a veteran in the field of academics and their pool of knowledge is as vast as their experience in printing. Their expertise and guidance has proved useful at every step. Their uncompromising quality standards have made this book an exceptional effort. Their encouragement from time to time has been an inspiration for everyone.

The publisher and the editorial board hope that this book will prove to be a valuable piece of knowledge for researchers, students, practitioners and scholars across the globe.

LIST OF CONTRIBUTORS

Simon F. Thomas, Fabian Rudin and Paul T. Mason
Plymouth Martine Laboratory Applications Ltd, Plymouth, United Kingdom

Paul Rooks, Sov Atkinson, Rachel Bransgrove and Michael J. Allen
Plymouth Marine Laboratory, Plymouth, United Kingdom

Paul Goddard
Protein Technologies Ltd, Manchester, United Kingdom

Òscar Guadayol, Megan J. Donahue and Florence I. M. Thomas
Hawai'i Institute of Marine Biology, School of Ocean and Earth Science and Technology, University of Hawai'i at Mānoa, Kāne'ohe, Hawai'i, United States of America

Nyssa J. Silbiger
Hawai'i Institute of Marine Biology, School of Ocean and Earth Science and Technology, University of Hawai'i at Mānoa, Kāne'ohe, Hawai'i, United States of America
Department of Biology, University of Hawai'i at Mānoa, Honolulu, Hawai'i, United States of America

Bojan Tamburic, Supriya Guruprasad, Dale T. Radford, Milán Szabó, Ross McC Lilley, Anthony W. D. Larkum, Jim B. Franklin, Martin Schliep and Peter J. Ralph
Plant Functional Biology and Climate Change Cluster, University of Technology Sydney, Sydney, New South Wales, Australia

David M. Kramer
Department of Biochemistry and Molecular Biology, Michigan State University, East Lansing, Michigan, United States of America

Susan I. Blackburn
Marine and Atmospheric Research, Commonwealth Scientific and Industrial Research Organisation, Hobart, Tasmania, Australia,

John A. Raven
Plant Functional Biology and Climate Change Cluster, University of Technology Sydney, Sydney, New South Wales, Australia
Division of Plant Sciences, University of Dundee at the James Hutton Institute, Invergowrie, Scotland, United Kingdom

Mauricio M. Petrucio
Graduate Program in Ecology of Department of Ecology and Zoology, Institute of Biological Sciences, Federal University of Santa Catarina, Florianó polis, Santa Catarina, Brazil

Renan S. Rezende
Graduate Program in Ecology of Department of Ecology and Zoology, Institute of Biological Sciences, Federal University of Santa Catarina, Florianó polis, Santa Catarina, Brazil
Department of Ecology, Institute of Biology, University of Brasília, Brasília, Federal District, Brazil

José F. Gonc¸alves, Jr.
Department of Ecology, Institute of Biology, University of Brasília, Brasília, Federal District, Brazil

Meha Jain, Yili Lim and María Uriarte
Department of Ecology, Evolution and Environmental Biology, Columbia University, New York, New York, United States of America

Javier A. Arce-Nazario
Department of Biology, University of Puerto Rico in Cayey, Cayey, Puerto Rico, United States of America

J. Marcus Drymon, Matthew J. Ajemian and Sean P. Powers
Department of Marine Sciences, University of South Alabama, Mobile, Alabama, United States of America
Center for Ecosystem Based Fishery Management, Dauphin Island Sea Lab, Dauphin Island, Alabama, United States of America

Bin Zhu and Kelly L. Fancher
Department of Biology, University of Hartford, West Hartford, Connecticut, United States of America

Michael S. Ellis
Department of Biology, Hobart and William Smith Colleges, Geneva, New York, United States of America

Lars G. Rudstam
Department of Natural Resources, Cornell University, Ithaca, New York, United States of America

Alexander Golberg
Centre for Engineering in Medicine, Massachusetts General Hospital, Harvard Medical School, Shriners Burns Institute, Boston, Massachusetts, United States of America

Nina Stawski
Fuels Synthesis Division, Joint BioEnergy Institute, Emeryville, California, United States of America
Physical BioSciences Division, Lawrence Berkeley National Labs, Berkeley, California, United States of America

Gregory Linshiz and Nathan J. Hillson
Fuels Synthesis Division, Joint BioEnergy Institute, Emeryville, California, United States of America
Physical BioSciences Division, Lawrence Berkeley National Labs, Berkeley, California, United States of America
DOE Joint Genome Institute, Walnut Creek, California, United States of America

Ilia Kravets
Department of Computer Science, Technion Institute of Technology, Haifa, Israel

Martin L. Yarmush
Centre for Engineering in Medicine, Massachusetts General Hospital, Harvard Medical School, Shriners Burns Institute, Boston, Massachusetts, United States of America
Department of Biomedical Engineering, Rutgers University, New Jersey, United States of America

Robert S. Marks
Department of Biotechnology Engineering, The National Institute of Biotechnology in Negev, Ben Gurion University, Beer-Sheva, Israel
School of Materials Science and Engineering, Nanyang Technological University, Singapore
NRF CREATE program for Nanomaterials in Energy and Water Management, Singapore

Tania Konry
Department of Pharmaceutical Sciences, School of Pharmacy Bouvé College of Health Sciences, Northeastern University, Boston, Massachusetts, United States of America

Andreas Tornevi and Bertil Forsberg
Occupational and Environmental Medicine, Department of Public Health and Clinical Medicine, Umea° University, Sweden

Olof Bergstedt
Department of Civil and Environmental Engineering, Water Environment Technology, Chalmers University of Technology, Gothenburg, Sweden

Ghislaine Rosa, Fiona Majorin, Sophie Boisson and Miles Kirby
Department of Disease Control, London School of Hygiene and Tropical Medicine, London, United Kingdom

Christina Barstow
Department of Civil, Environmental and Architectural Engineering, University of Colorado at Boulder, Boulder, Colorado, United States of America

Michael Johnson
Berkeley Air Monitoring Group, Berkeley, California, United States of America

Fidele Ngabo
Ministry of Health, Government of Rwanda, Kigali, Rwanda

Evan Thomas
Department of Mechanical and Materials Engineering, Portland State University, Portland, Oregon, United States of America

Thomas Clasen
Department of Disease Control, London School of Hygiene and Tropical Medicine, London, United Kingdom
Department of Environmental Health, Emory University, Atlanta, Georgia, United States of America

Emily G. I. Payne, Ana Deletic and Belinda E. Hatt
Monash Water for Liveability, Department of Civil Engineering, Monash University, Victoria, Australia

Tim D. Fletcher
Department of Resource Management and Geography, Melbourne School of Land and Environment, The University of Melbourne, Victoria, Australia

Douglas G. Russell, Michael R. Grace, Perran L. M. Cook and Victor Evrard
Water Studies Centre, School of Chemistry, Monash University, Victoria, Australia

Timothy R. Cavagnaro
School of Biological Sciences, Monash University, Victoria, Australia

Georgina G. Gurney
Australian Research Council Centre of Excellence for Coral Reef Studies, James Cook University, Townsville, Queensland, Australia
School of Zoology, University of Tasmania, Hobart, Tasmania, Australia

Jessica Melbourne-Thomas
Australian Antarctic Division, Department of Sustainability, Environment, Water, Population and Communities, Kingston, Tasmania, Australia
Antarctic Climate & Ecosystems Cooperative Research Centre, University of Tasmania, Hobart, Tasmania, Australia

Rollan C. Geronimo
Conservation International Philippines, Quezon City, Philippines

Perry M. Aliño
Marine Science Institute, University of the Philippines, Quezon City, Philippines

Craig R. Johnson
Institute for Marine and Antarctic Studies, University of Tasmania, Hobart, Tasmania, Australia

Jeonghwan Jang, Doris Y. W. Di and Anna Lee
School of Environmental Science and Engineering, Gwangju Institute of Science and Technology, Gwangju, Republic of Korea

Hor-Gil Hur
School of Environmental Science and Engineering, Gwangju Institute of Science and Technology, Gwangju, Republic of Korea
International Environmental Analysis and Education Center, Gwangju Institute of Science and Technology, Gwangju, Republic of Korea

Tatsuya Unno
College of Molecular Life Sciences, Jeju National University, Jeju,Republic of Korea

Michael J. Sadowsky
Department of Soil, Water and Climate, University of Minnesota, St. Paul, Minnesota, United States of America
BioTechnology Institute, University of Minnesota, St. Paul, Minnesota, United States of America

Matthew S. Bird
DST/NRF Research Chair in Shallow Water Ecosystems, Nelson Mandela Metropolitan University, Port Elizabeth, South Africa

Jenny A. Day
Freshwater Research, Department of Biological Sciences, University of Cape Town, Private Bag, Rondebosch, South Africa

Yoichi Hatamoto and Hiroyuki Sagayama
Graduate School of Sports and Health Science, Fukuoka University, Fukuoka, Japan

Yosuke Yamada
The Fukuoka University Institute for Physical Activity, Fukuoka, Japan
Laboratory of Applied Health Science, Kyoto Prefectural University of Medicine, Kyoto, Japan
Research Fellow (SPD), Japan Society for the Promotion of Science, Tokyo, Japan

Yasuki Higaki, Akira Kiyonaga and Hiroaki Tanaka
The Fukuoka University Institute for Physical Activity, Fukuoka, Japan
Faculty of Sports and Health Science, Fukuoka University, Fukuoka, Japan

Jinliang Huang
Coastal and Ocean Management Institute, Xiamen University, Xiamen, People's Republic of China
Fujian Provincial Key Laboratory for Coastal Ecology and Environmental Studies, Xiamen University, Xiamen, People's Republic of China

Yaling Huang and Zhenyu Zhang
Fujian Provincial Key Laboratory for Coastal Ecology and Environmental Studies, Xiamen University, Xiamen, People's Republic of China

Matthew P. Hammond and Jurek Kolasa
Department of Biology, McMaster University, Hamilton, Ontario, Canada

Joleah B. Lamb and Bette L. Willis
ARC Centre of Excellence for Coral Reef Studies, School of Marine and Tropical Biology, James Cook University, Townsville, Queensland, Australia
AIMS@JCU, Australian Institute of Marine Science and James Cook University, Townsville, Queensland, Australia

F. Joseph Pollock
ARC Centre of Excellence for Coral Reef Studies, School of Marine and Tropical Biology, James Cook University, Townsville, Queensland, Australia
AIMS@JCU, Australian Institute of Marine Science and James Cook University, Townsville, Queensland, Australia
Australian Institute of Marine Science, PMB 3, Townsville, Queensland, Australia

Britta Schaffelke and David G. Bourne
Australian Institute of Marine Science, PMB 3, Townsville, Queensland, Australia

George Shedrawi
Department of Parks and Wildlife, Marine Science Program, Kensington, Western Australia, Australia

Stuart N. Field
Department of Parks and Wildlife, Marine Science Program, Kensington, Western Australia, Australia
Oceans Institute, University of Western Australia, Crawley, Western Australia, Australia

Scott F. Heron
NOAA Coral Reef Watch, Townsville, Queensland, Australia,

Marine Geophysical Laboratory, School of Engineering and Physical Sciences, James Cook University, Townsville, Queensland, Australia

Christopher J. Gobler, Elizabeth L. DePasquale, Andrew W. Griffith and Hannes Baumann
School of Marine and Atmospheric Sciences, Stony Brook University, Southampton, New York, United States of America

Elizabeth Halliday and Rebecca J. Gast
Biology Department, Woods Hole Oceanographic Institution, Woods Hole, Massachusetts, United States of America

Sandra L. McLellan
School of Freshwater Sciences, Great Lakes Water Institute, University of Wisconsin-Milwaukee, Milwaukee, Wisconsin, United States of America

Mitchell L. Sogin
Josephine Bay Paul Center, Marine Biological Laboratory, Woods Hole, Massachusetts, United States of America

Linda A. Amaral-Zettler
Josephine Bay Paul Center, Marine Biological Laboratory, Woods Hole, Massachusetts, United States of America
Department of Geosciences, Brown University, Providence, Rhode Island, United States of America

Ping Zheng
Department of Environmental Engineering, Zhejiang University, Hangzhou, China

Juqing Lou
Department of Environmental Engineering, Zhejiang University, Hangzhou, China
School of Environmental Science and Engineering, Zhejiang Gongshang University, Hangzhou, China

Yongqing Cao and Peide Sun
School of Environmental Science and Engineering, Zhejiang Gongshang University, Hangzhou, China

Christina K. Barstow
Civil, Environmental and Architectural Engineering, University of Colorado, Boulder, Colorado, United States of America

Fidele Ngabo
Maternal and Child Health, Ministry of Health, Kigali, Republic of Rwanda

Ghislaine Rosa, Fiona Majorin and Sophie Boisson
Department of Infectious and Tropical Diseases, London School of Hygiene and Tropical Medicine, London, United Kingdom

Thomas Clasen
Department of Infectious and Tropical Diseases, London School of Hygiene and Tropical Medicine, London, United Kingdom

Department of Environmental Health, Rollins School of Public Health, Emory University, Atlanta, Georgia, United States of America

Evan A. Thomas
Mechanical and Materials Engineering, Portland State University, Portland, Oregon, United States of America

Allison H. Roy
U. S. Geological Survey, Massachusetts Cooperative Fish and Wildlife Research Unit, University of Massachusetts, Amherst, Massachusetts, United States of America

Lee K. Rhea, William D. Shuster, Jake J. Beaulieu and Matthew E. Hopton
U. S. Environmental Protection Agency, Office of Research and Development, Cincinnati, Ohio, United States of America

Audrey L. Mayer
School of Forest Resources and Environmental Science and Department of Social Sciences, Michigan Technological University, Houghton, Michigan, United States of America

Matthew A. Morrison
SABIC, Mt. Vernon, Indiana, United States of America

Ann St. Amand
PhycoTech, Inc., St. Joseph, Michigan, United States of America

Index

O

P

R

S

T

V

W

Y